Signals And Communications Technology

For other titles published in this series, go to
http://www.springer.com/series/4748

Ramjee Prasad (Ed.)

My Personal Adaptive Global NET (MAGNET)

Springer

Editor
Ramjee Prasad
Aalborg University
CTIF
Niels Jernes Vej 12
9220 Aalborg
Denmark
prasad@es.aau.dk

ISSN 1860-4862
ISBN 978-94-007-3133-2 e-ISBN 978-90-481-3437-3
DOI 10.1007/978-90-481-3437-3
Springer Dordrecht Heidelberg London New York

Cover design: eStudio Calamar S.L.

Printed on acid-free paper

Springer is part of Springer Science+Business Media (www.springer.com)

To the Technical Managers of MAGNET and MAGNET Beyond

Juha Saarnio
Mikael Latvala
Karsten Vandrup
Liljana Gavriloska
Albena Mihovska (Deputy)

Preface

सहजं कर्म कौन्तेय सदोषमपि न त्यजेत् ।
सर्वारम्भा हि दोषेण धूमेनाग्निरिवावृताः ॥

saha-jaṁ karma kaunteya
sa-doṣam api na tyajet
sarvārambhā hi doṣeṇa
dhūmenāgnir ivāvṛtāḥ

Every endeavour is covered by some fault, just as fire is covered by smoke.
Therefore one should not give up the work born of his nature, even if such work is full of fault.

– The Bhagvad-Gita (18.48)

This book is the outcome of the research and development contributions of partners from three different continents, Asia, Europe, America, coming from universities, research centers, industrial partners and SMEs (Small and Medium Enterprise), all of them collaborating in MAGNET (My Adaptive Personal Global Net) and MAGNET Beyond project supported by European Commission within the Sixth Framework Programme (FP6). The project was focusing on a secure user-centric approach developing secure Personal Networks in multi-network, multi-device, and multi-user environments.

The innovative concept of Personal Network (PN), which was introduced and developed in MAGNET, finds in this book the first confirmation of the success that the future of wireless communications is bound to achieve. The importance of this book is not only related to being the first work on PNs, it also gives an overview of operation of a big project, like MAGNET, and in fact the organisation of the book reflects how then project itself has been structured.

The book summarize all the steps taken from the introduction of a user-centric perspective until the implementation of PN-Fs, outlining the applications and commercialisations of the new concepts carried out of the project. The starting point has been the concept of Personal Network coming out like an extension of the local scope of Wireless Personal Area Networks (WPAN) by addressing virtual personal environments that span a variety of infrastructures. The new element was that the composition, organisation, and topology of a PN have determined by its context and the geographical location, the time, the environment and the explicit wishes to use particular services determined which device and network element have been incorporated in a PN. The PN can be defined as a dynamics collection of personal nodes and device not only centered around a person, but also personal devices on remote locations. A PN is composed of multiple clusters, where the communication is between remote clusters that have a common trust relationship. To extend the PN solutions to enable interactions between multiple PNs, it have been introduced the concept of PN Federation (PN-F). A PN Federation can be defined as a secure cooperation between different PNs, making selected service(s) and resource(s) available to selected receiver(s) for the purpose of achieving a common goal.

The project started in January 2004, and was divided in two phases, in the first, named MAGNET (January 2004–December 2005), the objectives were to design, develop, demonstrate and validate the concept of a flexible Personal Network that supports resource-efficient, robust, ubiquitous service provisioning in a secure, heterogeneous networking environment for nomadic users. There were 37 partners, 13 industrial, 7 research centres, 14 universities, and 3 SMEs coming from 16 different countries around three different continents: Austria, Belgium, China, Denmark, Finland, France, Germany, Greece, India, Italy, Netherlands, Spain, Sweden, Switzerland, United States, and UK. In the second phase, MAGNET Beyond (January 2006–June 2008) the interest was concentrated on the interactions between multiple PN users with common interests for various services. MAGNET Beyond had 30 partners from 15 countries, the same involved in MAGNET except United States:

- Twelve Universities
- Seven Research Centres
- Nine Industrial Partners
- Two SMEs

The cooperation from several partners from all over world and from different organization was a hard task but, at the same time, the level of the discussions was always very high, and very interesting results were obtained. MAGNET/MAGNET Beyond had a significant influence in specifying the PN and PN-F, offering to the community patents, demo-platform, pilots and test bed useful for next industrial commercialization. This was possible because of the collaboration among all the partners, which coming from different organization highlighted different points of view and achieving results that led directly to the future wireless technologies known as 4G.

The intent of this book is to disseminate the concept of PN and PN-F among with the activities and achievements carried out in MAGNET/MAGNET Beyond to encourage new projects and academic initiatives toward personalized, ubiquitous communications. We tried to make our best to write each chapter as self-contained as possible in order to allow the reader to read them independently.

Any remarks to improve the text and correct any errors or typos would be highly appreciated.

Acknowledgements

The material in this book originates from the EU project MAGNET/MAGNET beyond. Therefore, the editor would like to thank all the colleagues involved in the project for their collaboration and dedication that made the success of the project and also helped to finalize this book. The editor also hopes that the personal relations established during these years will remain and make possible future collaborations.

In the first place, the editor would like to thank the Project Officer, Rémy Bayou, for his remarkable support to our work.

The editor would like to acknowledge the contributions from Aalborg University, Advanced Communications Research and Development S.A, ALCATEL Italia, Brunel University, Centre Suisse d'Electronique et de Microtechnique – Recherche et Development SA, Commissariat à l'Energie Atomique, Danmarks Tekniske Universitet, Delft University of Technology, France Telecom R&D, Fraunhofer Institut FOKUS, Forschungszentrum Telekommunikation Wien Betriebs GmbH, Groupe des Ecoles des Télécommunications – Institut National des Télécommunications, Institute of Communication and Computer Systems (ICCS) of the National Technical University of Athens, Interuniversitair Micro-Elektronica Centrum vzw, INTRACOM S.A. Hellenic Telecommunications and Electronics Industry, Lund University, National Institute of Informational and Communication Technology, NEC Europe Ltd., Nokia Corporation OYJ, NXP Semiconductors Netherlands B.V, Shanghai Institute of Microsystems and Information Technology/CAS, Tata Consultancy Service, TeliaSonera, Telefónica Investigación y Desarrollo Sociedad Anónima Unipersonal, Universidad de Cantabria, The University of Surrey, University of Rome "Tor Vergata", Technical Research Centre of Finland, Twente Institute of Wireless and Mobile Communications, University of Kassel.

Finally, the editor likes to express his special thanks to Antonietta Stango and Juan J. Sanchez for their patience and cooperation in freeing from the enormous editorial burden.

Acknowledgements

Contents

List of Partners in MAGNET and MAGNET Beyond

List of partners	Country	Magnet	Magnet beyond
Aalborg University	Denmark	*	*
Advanced Communication Research and Development, S.A.	Spain	*	*
ALCATEL Italia	Italy		*
Alcatel Sel Ag	Germany	*	
Beijing University of Posts and Telecommunications	China	*	
Brunel University	UK	*	*
Centre Suisse D'electronique Et De Microtechnique Sa – Recherche Et Development	Switzerland	*	*
Commissariat A L'energie Atomique	France	*	*
Danmarks Tekniske Universitet	Denmark	*	*
Eidgenoessische Technische Hochschule Zuerich	Switzerland	*	
Fraunhofer Institut FOKUS	Germany	*	*
Forschungszentrum Telekommunikation Wien Betriebs-Gmbh	Austria	*	*
France Telecom	France	*	*
Groupe Des Ecoles Des Telecommunications	France	*	*
Institute of Communication and Computer Systems – National Technical University of Athens	Greece	*	*
Interuniversitair Micro-Electronica Centrum Vzw	Belgium	*	*
Intracom S.A. Hellenic Telecommunications and Electronics Industry	Greece	*	*
Lucent Technologies Inc.	United States	*	
Lucent Technologies Nederland Bv	The Netherlands	*	

(continued)

List of partners	Country	Magnet	Magnet beyond
Lund University	Sweden	*	*
National Institute of Information and Communications Technology	Japan	*	*
NEC Europe Ltd.	Germany		*
Nokia Corporation	Finland	*	*
Nokia Gmbh	Germany	*	
NXP Semiconductors Netherlands B.V	The Netherlands		*
Pcom: I^3 Aps	Denmark	*	
Rheinisch-Westfaelische Technische Hochschule Aachen	Germany	*	
Samsung Electronics (UK) Limited	UK	*	
Shanghai Institute of Microsystem and Information Technology, Chinese Academy of Sciences	China	*	*
Tata Consultancy Service	India		*
Tata Sons Limited	India	*	
Tata Sons Limited, Europe		*	
Technical Research Centre of Finland	Finland	*	*
Technische Universiteit Delft	The Netherlands	*	*
Telefonica Investigacion Y Desarrollo Sa Unipersonal	Spain	*	*
Teliasonera Sverige Aktiebolag	Sweden	*	*
The University of Surrey	UK	*	*
Twente Institute of Wireless and Mobile Communications	The Netherlands		*
Universidad De Cantabria	Spain	*	*
Universita Degli Studi Di Roma "Tor Vergara"	Italy	*	*
University of Kassel	UK		*

About the Editor

Ramjee Prasad is currently a Professor and Director of Center for Teleinfrastruktur (CTIF), and holds the chair of wireless information and multimedia communications. He was coordinator of European Commission Sixth Framework Integrated Project MAGNET (My personal Adaptive Global NET) and MAGNET Beyond. He was involved in the European ACTS project FRAMES (Future Radio Wideband Multiple Access Systems) as a project leader in Delft University. He was also project leader of several international, industrially funded projects of Technology. He has published over 700 technical papers, contributed to several books, and has authored, co-authored, and edited over twenty five books. His latest book is "*Introduction to Ultra Wideband for Wireless Communications*".

Professor Prasad has served as a member of the advisory and program committees of several IEEE international conferences. He has also presented keynote speeches, and delivered papers and tutorials on WPMC at various universities, technical institutions, and IEEE conferences. He was also a member of the European cooperation in the scientific and technical research (COST-231) project dealing with the evolution of land mobile radio (including personal) communications as an expert for The Netherlands, and he was a member of the COST-259 project. He was the founder and chairman of the IEEE Vehicular Technology/Communications Society Joint Chapter, Benelux Section, and is now the honorary chairman. In addition, Professor Prasad is the founder of the IEEE Symposium on Communications and Vehicular Technology (SCVT) in the Benelux, and he was the symposium chairman of SCVT'93. Presently, he is the Chairman of IEEE Vehicular Technology/Communications/Information Theory/Aerospace and Electronics Systems/Society Joint Chapter, Denmark Section.

In addition, Professor Prasad is the coordinating editor and editor-in-chief of the *Springer International Journal on Wireless Personal Communications.* He was the technical program chairman of the PIMRC'94 International Symposium held in The Hague, The Netherlands, from September 19–23, 1994 and also of the Third Communication Theory Mini-Conference in Conjunction with GLOBECOM'94, held in San Francisco, California, from November 27–30, 1994. He was the conference chairman of the fiftieth IEEE Vehicular Technology Conference and the steering committee chairman of the second International Symposium WPMC, both held in Amsterdam, The Netherlands, from September 19–23, 1999. He was the general chairman of WPMC'01 which was held in Aalborg, Denmark, from September 9–12, 2001, and of the first International Wireless Summit (IWS 2005) held also in Aalborg, Denmark on September 17–22, 2005. He was the General Chair of the First International Conference on Wireless Communication, Vehicular Technology, Information Theory and Aerospace and Electronic Systems Technology (Wireless VITAE) held on May 17–20, 2009 in Aalborg.

Professor Prasad was also the founding chairman of the European Center of Excellence in Telecommunications, known as HERMES and now he is the honorary chairman. He is a fellow of IEEE, a fellow of IETE, a fellow of IET, a member of The Netherlands Electronics and Radio Society (NERG), and a member of IDA (Engineering Society in Denmark). Professor Prasad is advisor to several multinational companies. He has received several international awards; one of this is the "Telenor Nordic 2005 Research Prize" (website: http://www.telenor.no/om/).

Abbreviations

3GPP	Third Generation Partnership Project
AAF	Anti-Aliasing Filter
ACL	Access Control List
ActCom	Activity Based Communication Concept
AES	Advanced Encryption Standard
AGC	Automatic Gain Control
AI	Air interface
AIPN	All-IP networks
AMC	Adaptive Modulation and Coding
AN	Ambient Networks
APF	All Pass Filter
API	Application Programming Interface
ARPU	Average Revenue Per User
ARQ	Automatic Repeat Request
AWA	Alternating Wireless Activity
AWGN	Additive White Gaussian Noise
BAN	Body Area Networks
BC	Business Card
BER	Bit Error Rate
BI	Beacon Interval
BiCMOS	Bipolar Complementary Metal Oxide Semiconductor
BMA	Berlekamp-Massey
BO	Beacon Order
BP	Beacon Period
CA	Certificate Authority
CA	Context Agent
CAC	Context Agent Controller
CAC	Context Aware Component
CALA	Context Access Language
CAM	Context Access Manager
CAN	Community Area Network
CAP	Contention Access Period
CASD	Context Aware Service Discovery

CASM	Context Aware Security Manager
CC/PP	Composite Capabilities/Preferences Profile
CCIB	Computational Complexity per Information Bit
CDMA	Code Division Multiple Access
CFP	Contention Free Period
CID	Cluster Identifier
CLH	Cluster Head
CMN	Context Management Node
CMOS	Complementary metal oxide semiconductor
CP	Control Point
CPFP	Certified PN Formation Protocol
CPNS	Converged Personal Network Service
CRC	Cyclic Redundancy Check
CSI	Channel State Information
CSMA/CA	Carrier Sense Multiple Access/Collision Avoidance
CTAP	Channel Time Allocation Period
DAA	Detect and Avoid
DAC	Digital Analog Converter
DDS	Direct Digital Synthesiser
DEV	Device
DEVID	Device ID
DH	Diffie-Hellman
DHCP	Dynamic Host Configuration Protocol
DHT	Distributed Hash Table
DME	Device Management Entity
DoS	Denial of Service
DQPSK	Differential Quadrature Phase Shift Keying
DSA	Data Source Abstraction
DSAL	Data Source Abstraction Layer
DSAM	DSA Manager
DSN	Data Sequence Number
EAP	Extensible Authentication Protocol
EC	European Commission
ECC	Elliptic Curve Cryptography
ECDH	Elliptic Curve Diffie-Hellman
ECDSA	Elliptic Curve Digital Signature Algorithm
ECMA	European Computer Manufacturers Association
ECMQV	Elliptic Curve Menezes-Qu-Vanstone
EEA	Extended Euclidean Algorithm
ESD	Electrostatic Discharge
ETSI	European Telecommunications Standards Institute
FCS	Frame Check Sequence
FCSL	Frame Convergence Sub Layer
FDMA	Frequency Division Multiple Access
FEC	Forward Error Correction

FER	Frame Error Rate
FFD	Full Function Device
FFT	Fast Fourier Transform
FIFO	First In First Out
FM	Federation Manager
FMC	Fixed Mobile Convergence
FM-UWB	Frequency Modulation Ultra Wide Band
FSB	Frequency-Spreading Blocks
FSK	Frequency Shift Keying
FSMC	Finite-State Markov Channel
FTD	Fixed Time Delay
GENA	Generic Event Notification Architecture
GF	Galois Field
GSM	Global System for Mobile communications
GSMA	GSM Association
GTS	Guaranteed Time Slots
GUI	Graphical User Interface
GUP	Generic User Profile
HCS	Header Check Sequence
HDR	High Data Rate
HTTP	Hyper Text Transfer Protocol
IAWA	Improved AWA
ICMP	Internet Control Message Protocol
IDFT	Inverse Discrete Fourier Transform
IdP	Identity Provider
IEEE	Institute of Electrical and Electronic Engineers
IETF	Internet Engineering Task Force
IF	Intermediate Frequency
IFS	Inter Frame Space
IMS	IP Multimedia Subsystem
IMT-A	International Mobile Communication-Advanced
INR	Intentional Name Resolver
INS	Intentional Naming System
IP	Internet Protocol
IPsec	IP security
ISM	Industrial, Scientific and Medical
ISO	International Organization for Standardization
ISO/IEC	International Organization for Standardization/ International Electrotechnical Commission
IST	Society Technology
ITU	International Telecommunication Union
KDF	Key Derivation Function
LAN	Local Area Network
LDC	Low Duty Cycle
LDR	Low Data Rate

LIFS	LongIFS
LLC	Logical Link Control
LNA	Low Noise Amplifier
LOS	Line of Sight
LPF	Low Pass Filters
M C	Modulation and Coding
MAC	Message Authentication Code
MAC	Medium Access Control
MAGNET	My personal Adaptive Global NET
MAS	Medium Access Slots
MC-CDMA	Multi-carrier CDMA
MCDU	MAC Command Data Unit
MC-SS	Multi Carrier Spread Spectrum
MCTA	Management Channel Time Allocation
MFR	MAC Footer
MIC	Message Integrity Code
MIFS	Minimum Inter Frame Space
MIMO	Multiple-Input and Multiple-Output
MITM	Man-in-the-Middle
MLME	MAC (sub)Layer Management Entity
MMC	Multi Media Card
MMS	Multimedia Messaging Service
MNO	Mobile Network Operator
MOD	Modality environment
MOPED	Mobile Grouped Device
MOS	Metal Oxide Semiconductor
MOSFET	Metal Oxide Semiconductor Field Effect Transistor
MPDU	MAC Protocol Data Unit
MPEG	Moving Picture Experts Group
MSDP	MAGNET Service Discovery Platform
MSDU	MAC Service Data Unit
MSMP	MAGNET Service Management Platform
MUP	MAGNET User Profile
NAT	Network Address Translation
NF	Noise Figure
NGN	Next Generation Networks
NGWS	Next-Generation Wireless Systems
NIC	Network Interface Card
NoC	Network on Chip
OA	Output Amplifier
OFDM	Orthogonal Frequency Division Multiplexing
OMA	Open Mobile Alliance
OSAL	Operating System Abstraction Layer
OSGi	Open Service Gateway initiative
OSI	Open Systems Interconnection

OSS	Operation Support System
OWL-DL	Ontology Web Language – Description Logics
P S	Processing and Storage
P2P	Peer to Peer
PAC	Authenticated Channel
PACWOMAN	Power Aware Communications for Wireless Optimised Personal Area Network
PAN	Personal Area Network
PDA	Personal Digital Assistant
PDE	Personal Distributed Environment
PE	Policy Engine
PeP	Personalization Provider
PER	Packet Error Rate
PFP	PN Formation Protocol
PGZ	Peterson-Gorenstein-Zierler
PHY	Physical Layer
PIP	Personal Identity Provider
PKI	Public Key Infrastructure
PLL	Phase Lock Loop
PMH	Personal Mobile Hub
PN	Personal Network
PNC	Piconet Coordinator
PNCA	PN Certificate Authority
PNDS	Personal Network Directory Service
PN-F	Personal Network Federation
PNID	Piconet Identifier
PNM	Personal Network Management
POS	Personal Operating Space
P-PAN	Private Personal Area Network
PTAT	Proportional to Absolute Temperature
PU	Processing Unit
PUCC	The P2P Universal Computing Consortium
QoS	Quality of Service
RAF	Repository Access Function
RD	Radio Domain
RDF	Resource Description Framework
RF VCO	Radio Frequency Voltage-Controlled Oscillator
RFC	Request for Comments
RFD	Reduced Function Devices
RFID	Radio Frequency Identification
RI	Radio Interfaces
RPC	Remote Procedure Call
RRM	Radio Resource Manager
RS	Reed Solomon
RTP	Real Time Protocol

SAM	Slot Allocation Matrix
SAN	Service Assistance Node
SAP	Service Access Point
SB	Stuff Bits
SCE	Service Creation Environment
SCIM	Service Capability Interaction Manager
SCM	Service Control Module
SCMF	Secure Context Management Framework
SCP	Sub Carrier Processing
S-CSCF	Serving-Call Session Control Function
SD	Superframe Duration
SD	Service Discovery
SDAL	Service Discovery Adaptation Sub-layer
SDM	Service Discovery Module
SGN	Service Gateway Node
SGSN	Serving GPRS Support Node
SHA	Secure Hash Algorithm
SHAMAN	Security for Heterogeneous Access in Mobile Applications and Networks
SIFS	ShortIFS
SIG	Signature
SiGe:C	Silicon Germanium:Carbon
SIM	Subscriber Identity Module
SIP	Session Initiation Protocol
SK	Secret Key
SLA	Service Level Agreement
SLEE	Service-Logic Execution Environment
SLP	Service Location Protocol
SME	Small and Medium Enterprise
SMMM	Service Mobility Management Module
SMN	Naming System Service
SMN	Service Management Node
SMS	Short Message Service
SNR	Signal to Noise Ratio
SO	Superframe Order
SOA	Service Oriented Architecture
SOAP	Simple Object Access Protocol
SOCM	Service Orchestration and Composition Module
SORM	Service Ontology and Reasoner Module
SP	Service Proxy
SPN	Service Provider Network
SR	Service Ranker
SRC	Source
SSCS	Service Specific Convergence Sublayer
SSDP	Simple Service Discovery Protocol

SSID	Service Set Identifier
SSL	Secure Sockets Layer
SSMM	Service Session Management Module
STF	Special Task Force
TCP	Transmission Control Protocol
TDMA	Time Division Multiple Access
TISPAN	Telecommunications and Internet converged Services and Protocols for Advanced Networking
TLS	Transport Layer Security
TTP	Trusted Third Party
UAProf	User Agent Profile
UCL	Universal Convergence Layer
UDN	Unique Device Name
UDP	User Datagram Protocol
UI	User Interface
UMA	Unlicensed Mobile Access
UML	Unified Modelling Language
UMTS	Universal Mobile Telecommunication System
UPnP	Universal Plug and Play
USIM	Universal Subscriber Identity Module
UWB	Ultra Wide Band
VB	Virtual Badge
VBR	Variable Bit Rate
VID	Virtual Identity
VoIP	Voice over Internet Protocol
VPN	Virtual Private Network
W3C	The World Wide Web Consortium
WAN	Wide Area Network
WCDMA	Wideband Code Division Multiple Access
WHERE	Wireless Hybrid Enhanced Mobile Radio Estimators
WLAN	Wireless Local Area Network
WP	Work Package
WPAN	Wireless Personal Area Network
WWAN	Wireless Wide Area Network
WWRF	Wireless World Research Forum
XCAP	XML Configuration Access Protocol
XDM	XML Document Management
XML	Extensible Mark-up Language

List of Figures

List of Tables

Chapter 1
Introduction

Ramjee Prasad

This book builds on the achievements of the EU-funded projects MAGNET and MAGNET Beyond in the area of personal area networks and related technologies. Wireless connectivity has already enabled computer users to profit from a new convenient mobile lifestyle. Consumers are now demanding the same simplicity throughout their homes, connecting personal computers (PCs), personal digital recorders, MP3 recorders and players, and every kind of digital and electronic devices to each other in versatile domestic wireless personal area networks (WPAN) and also the possibility to be connected with any body area networks (WBAN) if needed. However, current wireless local area network (WLAN) and WPAN technologies cannot yet meet the needs of tomorrow's connectivity for the host of emerging consumer electronic devices that offer full mobility while requiring low power, quality of service (QoS) and security. So, as computing, communications and consumer applications converge to provide domestic consumers with extensive new services in an intelligent ambient environment, there is an urgent need to develop short-range user-centered wireless networks. This challenge was undertaken by the EU-funded IST projects MAGNET and MAGNET Beyond.

1.1 The Concept of Personal Networks

The concept of PAN (Personal Area Network) refers to a space of small coverage around the person where ad hoc communication occurs. To extend the local scope of PANs a new kind of network has been developed: Personal Network. The concept of the Personal Network (PN) goes beyond the concept of a PAN by addressing virtual personal environments that span a variety of infrastructures (as well as ad hoc networks) [1].

Personal Networks is a concept that supports the professional and private activities of users without being obtrusive and while safeguarding their privacy and

---

R. Prasad (✉)
Aalborg University - CTIF, Niels J. Vey 12, Aalborg 9220, Denmark
e-mail: prasad@es.aau.dk

R. Prasad (ed.), *My Personal Adaptive Global NET (MAGNET)*, Signals and Communication Technology, DOI 10.1007/978-90-481-3437-3_1,

security. A PN may operate in both ad hoc and infrastructure-based networks and is dynamic and diverse in composition, configuration and connectivity depending on the time, place and circumstances as well as the required resources [2–17].

PNs comprise potentially "all of a person's devices capable of network connection in the real or virtual vicinity". In PNs, users interact with various companion-, embedded-, or invisible computers not only in their close vicinity but potentially anywhere. They also need to interact with other persons having their own PNs, leading to group communication and federation of PNs to achieve particular tasks. PNs constitute a category of distributed systems with very specific characteristics. This requires major extensions of the present Personal Area Networking.

The PN concept has been researched by various groups and from different perspectives. Examples are found in "Scenarios for Ambient Intelligence in 2010" [12], "The Book of visions – Visions of the Wireless World" [13], "Telecom Scenarios in 2010" [14], and the vision of the Association of Computing Machinery (ACM) in "The Next 1000 Years" [15]. EU-funded IST projects such as the projects PACWOMAN [16] and MOBILIFE [17] addressed users and the wireless vision in different ways. The projects MAGNET and MAGNET Beyond [2] exercised a different approach in order to identify and represent user requirements in the PN-development process, which would provide a better design and identify the path towards novel business models speeding up their adoption and successful deployment. In MAGNET the methodology to describe and develop understanding for the implementation of an efficient PN-solution in a heterogeneous multimodal environment has been carried out involving 'technology', 'user needs' and 'economics' requirements. A key element of 'user needs' was the perceived QoS associated with given private or business activities and its relation to the technical solutions. Furthermore, the user requirements were derived from real user involvement in the process in all stages.

The actual introduction, implementation, and commercialisation of PN services derived a unique and enhanced understanding of the combination between user requirement and technology developments, business models, market strategies and socio-economic aspects that are necessary to give a holistic picture of the PN concept and its possibilities to secure the European communication needs in the future.

PNs are configured in an ad hoc fashion, as the opportunity and the demand arise to support a person's private and professional applications. These applications may run on a user's personal device, but also on foreign devices. PNs consist of communicating clusters of personal digital devices, possibly shared with others, and connected through various suitable communications means. This is shown in Fig. 1.1. Unlike PANs, with a limited geographically coverage, PNs have an unrestricted geographical span, and may incorporate devices into the personal environment regardless of their geographic location.

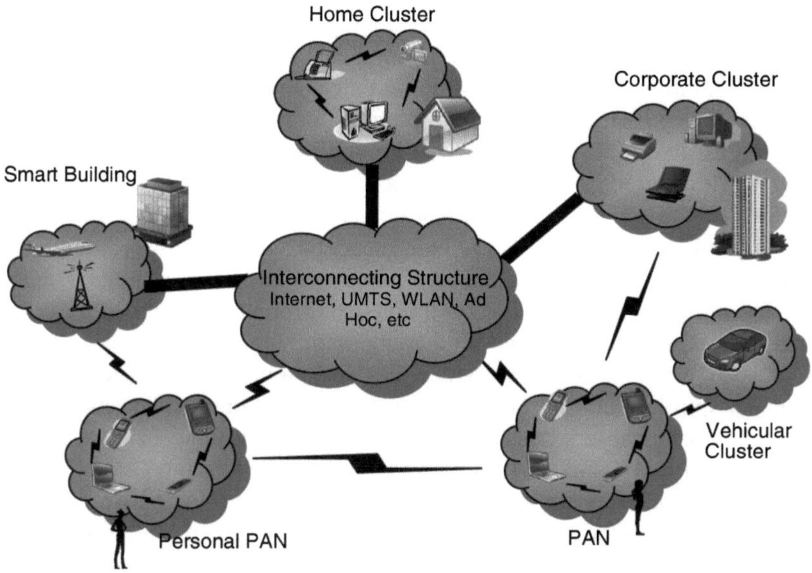

Fig. 1.1 The PN concept

1.1.1 PN Networking

Current radio technologies offer, up to a certain extent, self-organisational capabilities at the link layer:

- IEEE 802.11 provides link-level self-organisation
- Bluetooth networks organise themselves by forming pico-nets or even scatternets

Self-organisation at the network layer is also receiving a lot of attention in the context of mobile networks (e.g., ad hoc, MANETs, cooperative communications), in which nodes need to cooperate to organise themselves and to provide network functionality, due to the absence of any fixed infrastructure or simply to provide for autonomic resources. However, in the context of PNs, the problem has a completely different dimension, as self-organisation spans over multiple network technologies and strongly builds on the concept of trust between the personal nodes and devices.

The field of mobile ad hoc networks has seen a rapid expansion due to the proliferation of wireless devices, witnessed by the efforts in the IETF MANET working group [18]. A lot of attention has been given to the development of routing protocols, with the MANET group working on the standardization of a general reactive and proactive routing protocol, and, in a lesser extent, to address Internet connectivity [19].

When analysing the characteristics of a PN and its communication patterns, a number of similarities with mobile ad hoc networks can be observed. A PN should be self organising and self maintaining, handling mobility and, thereby, providing its own addressing and routing mechanisms for its internal communication. So, similar to ad hoc networks, PNs require self organizing and self maintaining networking capabilities that can deal with their dynamic behaviour. Therefore, developing PN networking solutions can be considered an extension of ad hoc networking techniques and concepts. However, existing solutions for mobile ad hoc networks cannot be adopted as is, due to the specific nature and the context of PNs. A PN has a specific wireless/wired geographically dispersed network topology, which, to a certain extent, can rely on the fixed infrastructure (e.g., edge routers), for providing networking solutions. Also, PNs are built around a specific trust relation concept, on which higher layer protocols can rely, which is absent in traditional ad hoc networks. The architecture developed for the PN concept and described further in this book is a novel one and a step further than the traditional concepts.

As PNs support mobility of individual devices, mobility of complete *clusters* of devices and splitting and merging of these clusters, efficient solutions are needed when dealing with these types of mobility. Worth mentioning in this context are the activities on mobile networks within the Mobile IP Working Group [20] of the IETF, the work on extensions of mobile IP for mobile ad-hoc networks interconnection [21] and the work within the NEMO working group that is concerned with the mobility of an entire network [22]. Mobility solutions for PNs can borrow from this work, but should be adapted to fit the proposed PN architecture and addressing schemes.

1.1.2 Service and Context Discovery

Routing is one of the main processes on the networking abstraction level, which is responsible for the finding and establishment of the routes among the communicating nodes. Current ad hoc routing protocols inherently trust all participants. Most ad hoc routing protocols are cooperative by nature and depend on neighbouring nodes to route packets. This simple trust model allows malicious nodes to paralyze an ad hoc network by inserting erroneous routing updates, replaying old messages, changing routing updates or advertising incorrect routing information. None of the protocols such as AODV, DSR, Ariadne, ARAN, SAR, SRP, etc. provide a solution to the requirements of certain discovery, isolation or Byzantine robustness.

The routing process must be shielded by solutions that grant the integrity and the availability of the networking procedures.

The capability to provide secure context transfer is essential in achieving fast performance in a wireless environment. Secure fast context transfer in handovers between heterogeneous access technologies/network types is needed. Furthermore, providing context-aware, adaptive and personalised services to the user, poses many opportunities, challenges and risks. Perhaps the greatest challenge is the ability to offer secure, intuitive and easy to use solutions for accessing contextual services that have to be location-aware and time-sensitive; personal preference and network bandwidth aware, and finally, device-capability aware.

Self organisation and routing aspects are fundamental aspects in the PN point of view, requiring investigation in order to provide schemes for devices and services discovery.

In a PN world, trust, identity management and privacy need considerable effort if we want to talk about an end-to-end security. Thus, a mechanism of enabling extension of the trust between personal nodes needs to be defined. Also, protection of user location, identity and privacy need to be considered. The user's location, identity and privacy requirements must be taken into account by the mobility procedures. The precise nature of these requirements may have a considerable impact on the mobility procedures. The PN world should bring concepts of anonymity and pseudonymity. Also privacy, resistance to denial of service and performance requirements is a crucial issue that needs to be considered. The project MAGNET starts with this considerations developing new concepts for service and context discovery.

1.1.3 Advances in the State of the Art of PNs

Commercially viable PNs were enabled by the joint efforts of a number of key academic and industrial players organized in the frames of the EU-funded project MAGNET and MAGNET Beyond [2]. The developed concept enabled attractive, affordable and beneficial for end users PN services in their everyday life. The MAGNET Beyond project constituted a system approach to what is expected to be one of the most important telecom related growth markets of the future, the Personal Area Network style networking. The main achievement of MAGNET Beyond was that it produced concepts and technologies that did not treat the PAN networking in isolation: the concept was extended into that of a PN by interconnecting PANs with other networks and, in particular, with wireless wide area networks to access the rich services available on and through these networks, including the interconnection to other PANs.

The following advances were made in relation to the PN:

- Research-based, comprehensive, short-term and long-term solutions for the technologies and protocols needed to build Personal Networks that meet the user requirements, in particular in terms of the quality, security, and trust requirements
- Technology roadmaps for the evolution of PNs
- System specification for first generation PNs
- Effective platforms that optimally and cost-effectively meet the short- and long-term communication requirements for personal devices
- A pilot PN system and pilot services
- An assessment of the market potential of the PN based on PN services usage, usability and acceptation tests

The project MAGNET Beyond introduced pilot services, obtained real-market and user feedback and provided the basis for the business of personal services over PNs. This had helped promote the PNs and related technologies and provided input and recommendations to standardisation and regulatory bodies and fora.

Wireless personal and body area networks are set to play an increasing role in applications such as health, personal safety, secure wireless data exchange or home entertainment. The PN concept addresses the challenge to deliver the next generation of ubiquitous and converged network and service infrastructures for communication, computing and media. It provides a new type infrastructure that can overcome the scalability, flexibility, dependability and security bottlenecks of current ones and permits the emergence of dynamic and, pervasive and robust new communication technologies. This is achieved by the extension of the PN to the concept of the PN-Federation (PN-F).

1.2 The Concept of the PN-Federation

In order to extend their reach, PNs need the support of infrastructure-based, and also ad-hoc networks. The cooperation between PNs belonging to different people in a federation is shown in Fig. 1.2.

In PN-Fs, PNs of different users cooperate for a certain purpose by sharing information and services. The daily life of persons does not involve their personal network only, but persons also need to communicate and collaborate with groups of people. Figure 1.2 shows how constituents from various PNs are federated in overlays to establish trusted groups and communities.

In such a scenario of networking of people, the needs in collaborative working, resource sharing or common interest groups such as family members, friends,

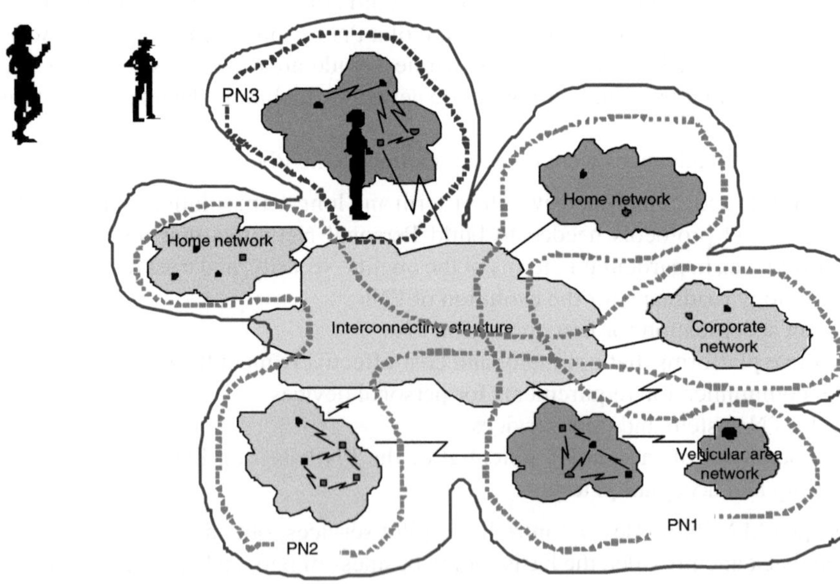

Fig. 1.2 The concept of the PN-F

kids at school, colleagues or public servants are all addressed. In such contexts, networking and security are confronted with far greater challenges. Designing enablers for user-centric personal networking and for creating a secure architectural framework suitable and viable for PN services become essentials.

To this end the concept of the long term or permanent trust relation between personal devices belonging to a single user should be extended to group trust between personal services shared by a group of users. In contrast to the single-user PN-concept, where secure communication exists between all personal devices constituting the PN, secure communication needs in the PN-F need to be established between a subset of personal devices belonging to different PNs, hereby creating a multi-user virtual private network overlay in a federation of multiple co-operating PNs.

A PN Federation as introduced by MAGNET Beyond is meant for a well defined goal and sets certain rules and policies for participation in the federation, defined by the creator of the Federation. Key management issues at PN Federation level for different scenarios can be supported by means of the PN-F Formation Protocol (PNFFP) [23].

1.3 Optimised Air Interfaces for PAN, PN and PN-F Communications

The PAN as a basic component of the PN relies on suitable air interfaces to ensure the communication process. Even though wireless has exploded in the last decade, wireless standards are dominated by a few protocol types. For example, most cellular networks use fixed-capacity channels, while data networking standards (e.g., IEEE802.11, IEEE802.15) are often contention-based so they can exploit statistical multiplexing of traffic. The use of simple, traffic-specific protocols has helped the rapid growth of mobile networks, but it stifles innovation and has lead to inefficient spectrum use. Today, basically, three wireless technologies, besides satellite communications, have made an impact: WLANs, WPANs, and wireless wide area networks (WWANs).Work in that direction is on-going in the various standardisation activities supported by the European Telecommunication Standardisation Union (ETSI) and the Institute of Electrical and Electronic Engineers (IEEE). Currently, the standardized WPAN technologies are BLUETOOTH, HIPERPAN and IEEE 802.15. These technologies are used for short distance ($\sim$10 m) with low data rates for different QoS requirements. It is envisaged that the WPANs will exist in all mobile terminals in the near future. The WPAN standards, IEEE 802.15.3 and 3a have developed and work is ongoing for paving the way towards broadband WPANs with envisioned data rates up to about 1 Gbps. IEEE 802.15.4 is focusing on very low data rate solutions, which will work at a few or a few hundred Kbps, which is the first step towards body area networks (BANs). Ultra wideband (UWB) schemes are considered for both IEEE 802.15.3 and IEEE 802.15.4. The working group IEEE 802.15.3a proposed direct-sequence (DS) UWB for low and medium data rates and multiband orthogonal frequency-division multiplexing (OFDM) for high data rates.

The latter is based on a transmission over 14 overlapping OFDM channels each having a bandwidth of 528 MHz for 128 subcarrier signals.

The specifics of the PAN radio environment (i.e., user proximity, user dynamics, radio co-existence with legacy and emerging communication systems, terminal/device sizes and their use cases), affect the choice of a proper channel model and consequently the air interface configuration. Appropriate and accurate radio channel and radio interference models, based on previous results and from new investigations, were investigated in the context of PNs with the objective to approximate the real time varying PAN radio environment. The proposed MAGNET PAN radio access solutions were taken as a basis for the optimisation of the air interfaces for typical PAN scenarios to ensure a favourable trade-off between user satisfaction (QoS) and system complexity.

MAGNET Beyond proposed air interfaces for high data rate (HDR) and low data rate (LDR) applications. The HDR applications are enabled by a multi-carrier spread spectrum (MC-SS) air interface solution. The only other available solution with similar capabilities at the moment is WiMedia, a radio platform standard for high-speed UWB wireless connectivity. For LDR applications, a low-power, low-complexity frequency modulation based UWB (FM-UWB) air-interface solution was proposed compatible to standards such as BLUETOOTH, ZigBee, and WiBree. The medium access control (MAC) of these two is based on the IEEE 802.15.3 and IEEE 802.15.4 standards. The FM-UWB approach was adopted after being studied and compared with other solutions like ZigBee and Bluetooth. Accordingly, the MC-SS scheme was compared to the orthogonal frequency-division multiplexing (OFDM) based UWB PHY scheme in a WiMedia system. Results are reported in details in and show that the developed air interfaces fulfil the requirements for next generation technologies.

Broadband wireless access is the third wireless revolution, after cell phones and Wi-Fi. The broadcast nature of wireless transmission offers ubiquity and immediate access for both fixed and mobile users, clearly a vital element of next generation quadruple play (i.e., voice, video, data, and mobility) services. Unlike wired access (copper, coax, fiber), a large portion of the deployment costs is incurred only when a subscriber signs up for service. An increasing number of municipal governments around the world are financing the deployment of multi-hop wireless networks with the overall aim of providing ubiquitous Internet access and enhanced public services. Broadband wireless access is an inherent feature of next generation communication systems. Therefore, PAN and PN solutions as proposed by the projects MAGNET and MAGNET Beyond will be the additional component together with IMT-Advanced (International Mobile Communication-Advanced) candidate systems that would complete the equation for the realisation of the next generation communication systems. In Fig. 1.3 is shown the overall structure of the wireless telecommunications, including the past and the future.

Efficient implementation of the transceivers for PANs is a key driver for enabling low cost, low power portable hand-held devices. The efficiency of the implementation relies on architectural choices. For example, most of the power in a transceiver, especially for LDR, are consumed in the RF part. An intensive research activity is

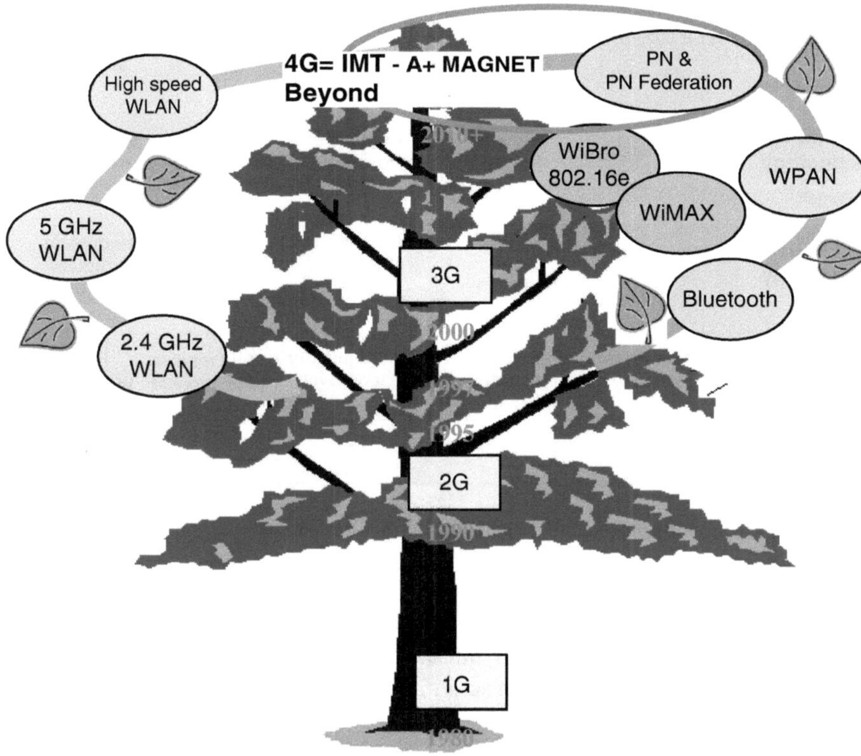

Fig. 1.3 Tree of communication standards evolution towards next generation systems

required in order to optimise the power figures. This is particularly true for UWB solutions, on which designers have less background than on the classical narrow-band systems. New architectures using high data rate digitiser have been introduced recently. They open the door to a software defined radio (SDR) approach where the RF section is reduced to a low noise amplifier (LNA) and fast sampler. Since all processing is then performed in the digital domain, reconfigurability can be introduced more easily. On the other hand more analogue solutions can bring some interesting features in terms of complexity and power consumption figures for LDR air interfaces [24–28]. For HDR, new architectures such as *networks on chip (NoC)* have been applied to MC-CDMA techniques [28–30]. This kind of architecture can be promising for the PAN HDR air interfaces that need more computational power than LDR solutions. Such schemes were evaluated and compared to more classical *system on chip (SoC)* approaches to propose the optimal compromise between flexibility and power consumption figures. Besides, the use of deep submicron technology may enable the design of monolithic approaches for the mass market target transceiver using the resulting advanced architectures.

Figure 1.4 proposes a roadmap for the realisation of the PAN-optimised air interfaces. Currently, as a result of the research and development effort put forward

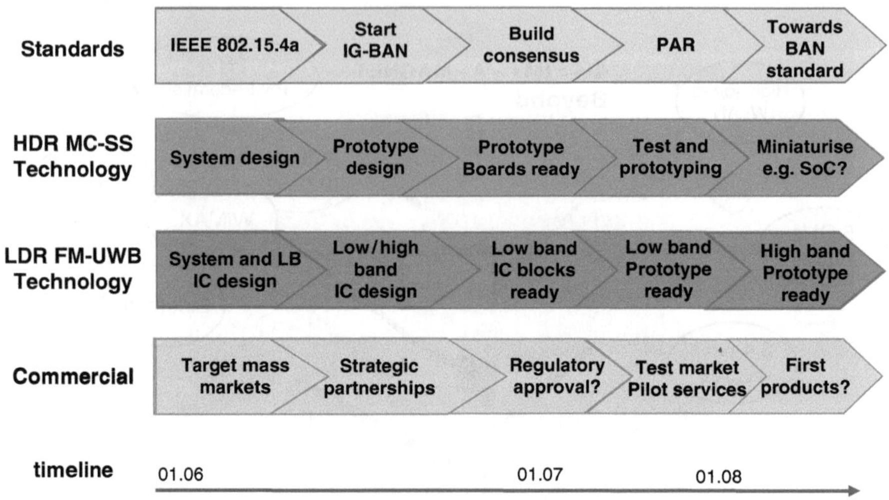

Fig. 1.4 Proposed roadmap for commercialization of the PN concept

by the consortium members of the projects MAGNET and MAGNET Beyond, the integrated prototypes for the two air interface solutions are a reality. These have been also fed into the standardisation activities of the ETSI and IEEE802.15 bodies.

1.4 Security, Privacy and Trust

Security, availability, and reliability are three key requirements for the successful deployment of the MAGNET Beyond concept. With a multitude of wireless standards in use, it is very important to ensure the dependability of the connections established by means of PNs and PN-Fs. One of the reasons why PNs can support a large variety of applications is that in PNs different types of access technologies can work hand in hand to deliver services to the users. The PN in Fig. 1.5 is configured in ad ad-hoc fashion, as the opportunity and the demand arise to support personal applications.

It consists of communicating clusters of personal and foreign devices, possibly shared with others, and connected through various suitable communication ways.

In order to access a device or service, the user needs to provide an identity that can be authenticated and authorised by the PN components. The provision of such an identity needs to be user friendly. In addition it should be possible to exchange the identity between service providers without affecting the privacy of the user. Concepts of anonymity and pseudonymity must be adapted to the PN and PN-F architecture to develop a coherent identity management solution, which is interoperable with the existing addressing, naming and identity management systems. Scalable and efficient methods for protection of user identity must be defined.

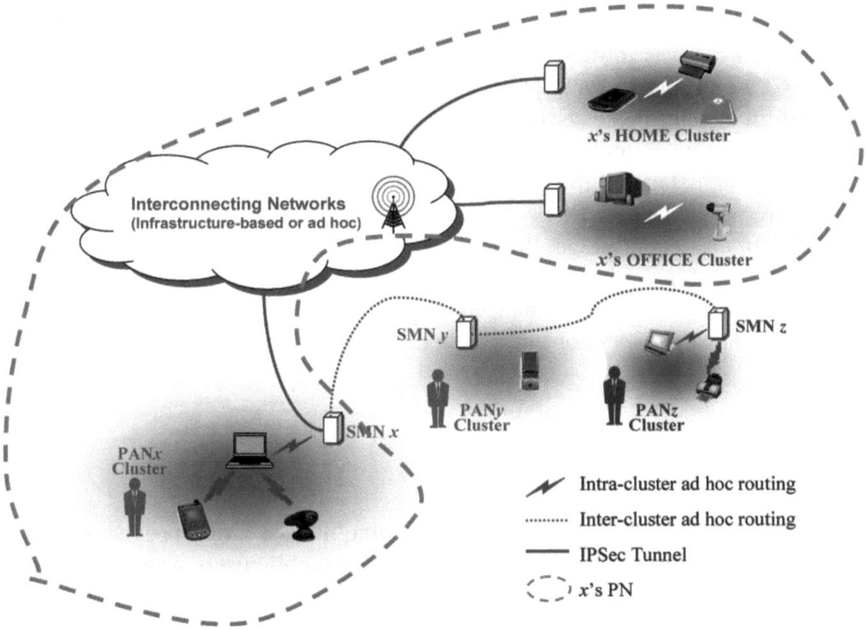

Fig. 1.5 Secure communications in a PN [31]

The vision of MAGNET Beyond of PNs combines two types of trust relationships: a priori *trust* inside the PN, which is managed by the user, which is ensured through proper authentication based on credentials; and the hand trust between PNs, which is an a posteriori *evolutionary* trust, as authentication (and identities) schemes in such a scenario are meaningless.

Methods to protect user privacy, including investigation of use of virtual identities protection of location of user and devices must also be developed. Protection of disclosing mobility behaviour, would, for example, require solutions for identity management, trust and privacy in PNs.

Communication with low-weight devices like sensors will obviously play a major role in the upcoming important market of PNs and on the background of the vision defined for the Future of Internet by the European Commission. For example, one such area is the application of mobile health in body area networks in which people will be equipped with several biosensors to continuously monitor their medical data such as glucose level, blood pressure and temperature. In these scenarios, these external devices are rather resource scarce in terms of processing and communication capabilities and it is necessary to support them with light-weight key exchange mechanisms.

MAGNET Beyond proposed novel solutions for physical encryption applicable to the PN-F security architecture. The solutions included an efficient hybrid protocol that secures the federation. Further, a physical layer encryption mechanism for both LDR and HDR was designed.

In the PN level a new key agreement protocol (i.e., the Certified PN Formation Protocol (CPFP)) was introduced. CPFP is based on the Elliptic Curve Cryptography (ECC) and the personal public key infrastructure (Personal PKI) in which instead of global certificates issued by a trusted third party, the local certificates issued by PN certificate authority (PNCA) can be applied. CPFP has two different stages, in the first stage all the PN devices get imprinted with PNCA, i.e., equip to its signature public key as the PN root key and get a certificate on their long term public key. In the second stage, PN nodes use their certificates to authenticate each other and establish pairwise keys based on the ECMQV protocol. CPFP is scalable to larger PNs and provides an enhanced level of authentication and non-repudiation with ease of the key revocation and key update.

1.5 PN Platforms

The concept of a flexible PN that supports ubiquitous service provisioning in a secure heterogeneous networking environment for mobile users was a challenging objective for MAGNET. PNs, apart from link level platforms, involve several heterogeneous networking and security components that must cooperate in order to make a reality such a concept.

The validation of such a concept cannot be provided only by simulations and it was necessary to implement a real testbed where the validity of this concept could be tested by users and industry. This testbed was the support for the real pilot services developed and specified within the frames of the project MAGNET Beyond. Testing as well as the identification of future optimisations that could be achieved by enhancing the collaboration between the different components comprising the whole system were another development activity in this context.

Well deployed operating system embedded platforms are key for supporting the PN networking components functionalities as introduced in the previous sections.

1.6 Preview of This Book

Figure 1.6 shows the collaboration of the various PN technologies described above in the scope of the IST project MAGNET Beyond that are also the basis for the organization of this book.

The organization of the book depends also on the division of the tasks among the Work Packages (WPs) involved in the projects. Every chapter is the summary of the achievements earned from the WPs, highlighting the efforts and the collaborations necessary to reach the excellent result obtained.

This book is organized as follows.

Chapter 2 discusses in details the concept, challenges and solutions for the provision user-centric personalised communications. In particular it describes

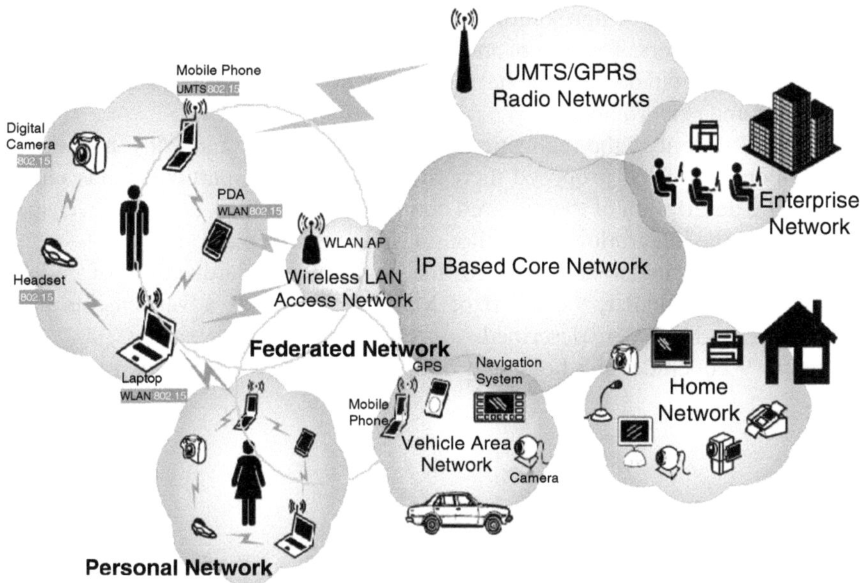

Fig. 1.6 Collaboration of MAGNET Beyond Technologies for realising a number of personalised applications

the user requirements to be considered, including requirements related to the user-friendliness of the personal device, the management of user profiles and the required business models for the successful deployment of personalised communications. Further, it proposes evaluation scenarios for the validation of the proposed requirements and business models.

Chapter 3 discusses in details the concept and advances in the area of PNs and PN-Fs. In particular, it proposes solutions for the realisation of self organisation at the network level (e.g, the network overlay approach), solutions for PN-aware service discovery and life cycle management, and it discusses the topic of user collaboration. Here, the focus is on the establishment of networking and services when joining of PN-Fs.

Chapter 4 proposes connectivity solutions for PNs and PN-Fs. In particular, it proposes advanced air interfaces for low and high-data rates (LDR and HDR, respectively), optimized for user-centric communications and provides benchmarking results as a proof-of-performance. Further, novel concepts related to interference mitigation and spectrum efficiency are proposed in support of the communication process. Issues such as multi-mode operation and PAN-to-PAN communications are also discussed.

Chapter 5 proposes solutions related to security, privacy and trust challenges in PNs and PN-Fs. In particular, the proposed solutions relate to the secure communication between personal nodes, the encryption and encoding for PAN air interfaces, and the architecture for management and enforcement of security policies.

Chapter 6 proposes design solutions for the PN connectivity concepts proposed in the preceding chapters. The design and prototyping of the LDR and HDR interfaces are described in detail down to the basic components. Results are represented related to the measured performance.

Chapter 7 describes the realization of the complete PN and PN-F testbed as a proof-of-concept of the proposed theoretical solutions. In particular, this chapter provides the description of the required components with high-and low-level specifications, and the integration of the pilot services onto the platform.

Chapter 8 discusses advances in the area of standardization of WPANs and BANs. In particular, the effort of MAGNET towards advancements in the IEEE.802.15 and ETSI are described.

Chapter 9 concludes the book and outlines the future challenges for PNs and PN-Fs.

References

1. R. Prasad, Personal network, Guest Editorial Telektronikk (Jan 2007)
2. IST Project MAGNET and MAGNET Beyond (2004–2008), www.ist-magnet.org
3. J. Saarnio, N.R. Prasad, *Foolproof Security Mechanisms and Challenges Within*, Int. J. Wireless Pers. Commun. (Kluwer, the Netherlands, 2004)
4. N.R. Prasad, A novel secure multi hop routing protocol for personal networks. WPMC 2004, Abano Therme, Italy, 12–15 Sept 2004
5. J. Lilleberg, R. Prasad, Research challenges for 3G and paving the way for emerging new generalisation. Wireless Pers. Commun. **17**, 355–362 (2001)
6. R. Prasad, M. Ruggieri, *Technology Trends in Wireless Communications* (Artech House Publishers, Boston, MA, 2003), ISBN 1-58053-352-3
7. S. Hara, R. Prasad, *Multicarrier Techniques for 4G Mobile Communications* (Artech House Publishers, Boston/London, 2003), ISBN 1-58053-482-1
8. R. Prasad, L. Munoz, *WLANs and WPANs Towards 4G Wireless* (Artech House Publishers, London), ISBN 1-58053-090-7
9. I.G. Niemegeers, S.M.H. de Groot, *From Personal Area Networks to Personal Networks: A User Oriented Approach*, Special issue J. Wireless Pers. Commun. (Kluwer, Hingham, MA, May 2002)
10. I.G. Niemegeers, S.M.H. de Groot, Research issues in ad-hoc distributed personal networking. Special issue Wireless Pers. Commun. **26**(2–3), 149–167 (2003)
11. I.G. Niemegeers, S.M.H. de Groot, FedNets: Some ideas for applying concepts of cognitive networks. Dagstuhl Seminar on Cognitive Networks and Radios, Dagstuhl, Germany, 18–21 Oct 2004, http://www.dagstuhl.de/04431/Materials/
12. K. Ducatel et al., Scenarios for Ambient Intelligence in 2010. IST Advisory Group (ISTAG), European Commission, Brussels, www.cordis.lu/ist/istag.htm, 2001
13. W. Mohr et al. (eds.), The book of Visions 2000 – Visions of the wireless world. Version 1.0, Wireless Strategic Initiative (Nov 2000), www.wireless-world-research.org
14. J. Zander et al., Telecom Scenario's in 2010. PCC, KTH, Sweden, 1999
15. ACM, The next 1000 years. Special issue Commun. ACM **44**(3), 50–52 (Mar 2001)
16. IST PACKWOMAN, http://www.imec.be/pacwoman
17. IST MOBILIFE, http://www.ist-mobilife.org
18. IETF MANET Working Group, http://www.ietf.org/html.charters/manet-charter.html
19. J. Hoebeke, I. Moerman, B. Dhoedt, P. Demeester, An overview of mobile ad hoc networks: Applications and challenges. J. Commun. Netw. **Part 3**, 60–55 (July to Sept 2004)

20. IP Routing for Wireless/Mobile Hosts, http://www.ietf.org/html.charters/mobileip-charter.html
21. U. Jönsson, F. Alriksson, T. Larsson, P. Johansson, G.Q. Maguire Jr., MIPMANET – Mobile IP for mobile ad hoc networks, in *Proceedings of the IEEE/ACM Workshop on Mobile and Ad Hoc Networking and Computing*, Boston, MA, Aug 2000
22. Network Mobility (NEMO), http://www.ietf.org/html.charters/nemo-charter.html
23. IST-027396 MAGNET/WP4/D4.2.1, First solutions for implementation of key management and crypto techniques (Dec 2006)
24. K. Marsden et al., Low power CMOS re-programmable pulse generator for UWB systems. IEEE Conference on UWB Systems and Technologies, Reston, VA, Nov 2003, pp. 443–447
25. S. Bagga et al., A PPM Gaussian monocycle transmitter for ultra-wideband communications. By IEEE Joint International Workshop of UWBST and IWUWBS, May 2004, pp. 130–134
26. T. Tong, T. Larson, Concept and architecture of integral receiver for Low Data Rate Ultra-Wide-Band System, in *Proceedings of Magnet Workshop*, Shanghai, China, 11/12 Nov 2004
27. J.F.M. Gerrits, J.R. Farserotu, J.R. Long, UWBFM: A low and medium data rate constant envelope UWB communications system with localisation potential, in *Proceedings of Magnet Workshop*, Shanghai, China, 11/12 Nov 2004
28. U. Hanke, A. Bøifot, J. Gamag, F. Bekkadal, Integrated reconfigurable radio front-end technology, URSI/COST 284 (2004)
29. S.B. Slimane, A low complexity antenna diversity receiver for OFDM based systems. IEEE ICC2001 **4**, 1147–1151 (2001)
30. K. Strohmenger, M. Laugeois, D. Noguet, B. Oelkrug, K. Seo, Architectures for digital physical layer implementation in multi-mode 3G terminals, IST-SUMMIT'04
31. A. Mihovska, N. Prasad, Adaptive security architecture based on EC-MQV algorithm in a personal network (PN), in *Proceedings of PERNETS'07*, Philadelphia, PA, Aug 2007

Chapter 2
Users, Pilot Services and Market

Knud Erik Skouby, Lene Sørensen, Henning Olesen, Allan Hammershøj, Anders Henten, and Iwona Windekilde

2.1 Introduction

Working within the overall purpose of MAGNET/MAGNET Beyond one of the specific challenges that is elaborated on in this chapter has been to represent and include a direct and clear user centred focus. The user centricity was firmly agreed to be ever present both in the development process in the focus areas of the project and as direct involvement of users at different stages in the systems development process. The basic idea has been to identify and build up relevant user requirements as the basis for formulation of systems requirements.

The MAGNET system focuses in particular on the user concept in five categories: user requirements, user case studies, user scenarios and use cases, evaluation and business models. The links between the five categories and the rest of MAGNET are illustrated in Fig. 2.1.

1. *User Requirements*. The user requirements elicitation process is part of the overall project synthesis process running from identifying preliminary user themes over selected themes or cases to establishment of user workshops, user scenarios and expert workshops all contributing to the identification of user requirements.
2. *User Case Studies*. Through idea generation based on work with selected themes or cases, initial user scenarios has been created as input to user workshops and expert workshops. These resulted in identification of a number of user cases relevant for demonstration of the MAGNET idea.
3. *User Scenarios and Use Cases*. Out of the user cases two user cases were selected to clearly demonstrate the MAGNET concepts and elements: MAGNET.Care and Nomadic@work. Idea creation as basis for scenario writing took place differently in the two cases. For the MAGNET.Care case, workshops were carried out in a lab while in the Nomadic@work case, a cultural probe was used to capture the nomadic perspectives of the users. In both cases, however, the result was elaborate story board-based scenarios outlining potential use situations challenging the MAGNET system to deliver relevant services to the users. A new approach

K.E. Skouby (✉), L. Sørensen, H. Olesen, A. Hammershøj, A. Henten, and I. Windekilde
CMI/Aalborg University, Lautrupvang 15, Ballerup 2750, Denmark
e-mail: skouby@cmi.aau.dk

R. Prasad (ed.), *My Personal Adaptive Global NET (MAGNET)*, Signals and Communication Technology, DOI 10.1007/978-90-481-3437-3_2,

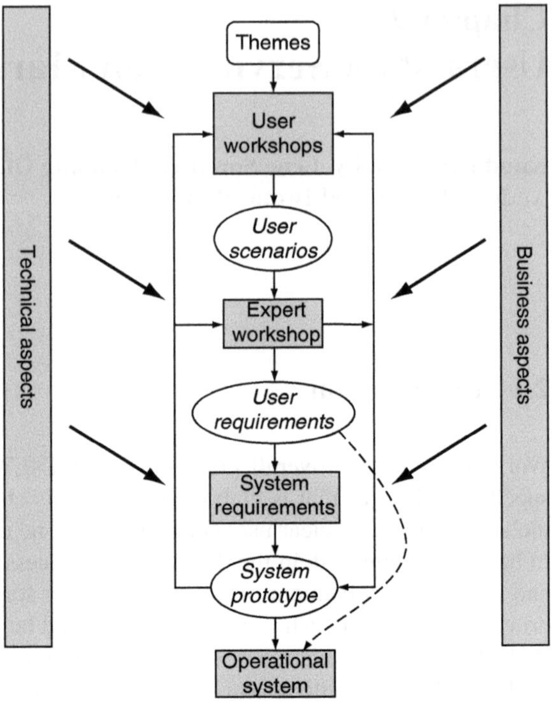

Fig. 2.1 Overall synthesis process

for user interaction on communication devices was developed, an "activity based communication concept" (ActCom). A key element in the concept is user profiles which again are as an essential part of the general Personal Network (PN) framework. User profiles connect the user's preferences, the context of the user and any other relevant information to optimize services for the user in any given situation. This makes management of the user profile a central issue including several aspects, e.g. updating or adding data content in the already defined user profile, and the supporting technology needed to get the user profile to work in a system. Policies play an important role and a profile management system must ensure that only as much information as needed is revealed (e.g. to a service provider) in order to have a value-added and personalized service delivered to the user. To actually enable the Nomadic@Work and MAGNET.Care different aspects of the two associated scenarios were technically described in details and implemented as two sub-scenarios or pilot services: Icebreaker and LifeStyle Companion respectively.

4. *Evaluation*. A focal point in MAGNET has been to define the usability and user experience of the technologies when in play. Evaluations have taken place at two levels: low fidelity prototype evaluations and high fidelity prototype evaluation. In both tests, the pilot services applications were used as specific examples and as basis for development of a GUI (Graphical User Interface) structure. It turned out, that in general the MAGNET Beyond concepts and the pilot services were accepted by the test persons.

5. *Business Models.* In order to analyse different aspects of the relations between user centricity and business models, a business model concept with a conceptual differentiation between the use value of a product (service and/or good) and the commercial value that it may have to the supplier of PNs is introduced. Another important differentiation is made between the intrinsic and extrinsic value of a product. The intrinsic value concept denotes the 'inherent' core value offered – meaning, for instance, that the intrinsic value of a piece of software is the immediate use value that it has to a user. The extrinsic value is the 'additional' value offered – in the case of software, the value that users derive from the fact that many other users have implemented the same software and that they, therefore, easily can exchange files, etc.

2.2 User Requirements

One central focus point in MAGNET Beyond has been development of user requirements in relation to PN services. User requirements have been identified at different stages throughout the project period as a consequence of project interests and specifications of project goals. The focus on the user and user requirements have within MAGNET Beyond been perceived as a focus on the user need and acceptance of the Personal Services concepts but also as a direct involvement of users to elicit requirements and later to test the results of the project (this last perspective is discussed more in the following Section 2.4).

In relation to elicitation of user requirements, the goal of obtaining the user's perspectives on MAGNET concepts and technologies has been done through several different ways:

1. Formation of User scenarios that describe users and their actions as well as their requirements for PN services
2. Development of specific Use Cases to clearly demonstrate MAGNET concepts
3. Development of storyboards to visualize how MAGNET Beyond technologies can be helpful and useful in daily life situations and as illustrations of user scenarios and user requirements
4. Introduction of Participatory Design [1] as concept for user involvement
5. Different types of user involvement to elucidate user requirements and the abovementioned scenarios, use cases, and storyboards. Involvement of users took place through creative workshops, interviews, and development of a mobile probing tool kit as well as through low-fidelity tests of first drafts of GUI's displaying different types of PN services identified as part of the use cases and scenario developments.

Overall, the user requirements elicitation process can be seen as a part of the overall project synthesis process as displayed in Fig. 2.1, which shows the process from identifying preliminary user themes (areas within which users carry out daily activities and that all constitute a special case; such as transport, health, shopping, etc.), from selected themes or cases to establishment of user workshops, user scenarios and expert workshops (workshop in which persons from the MAGNET project have been participating in order to work with user requirements) all contributing to the identification of user requirements. Throughout the process, technical potentials and constraints and business aspects provide input to shape, check and complement the user requirements as illustrated in the figure. The technical aspects outline the sphere of possible PN services whereas the business aspects characterize the economically viable services. The final results of the process are prototypes and pilot services. Some user requirements cannot be directly transformed to system requirements, but must be fed directly into the operational system (illustrated by the dotted line in Fig. 2.1).

In the following the work around user requirements will be presented in more detail.

2.2.1 Participatory Design

User-centred design does not necessarily mean that users actually participate in the design process. However, from the beginning of the MAGNET project, it was decided to involve users in different stages in the project. Participatory design [1] was therefore applied as an overall frame for the inclusion of users. Applying participatory design as a design method, a number of different techniques and stages can be identified in capturing and exploring user needs and requirements. These are

- Idea generation and initial requirements
- User scenarios and use cases
- Low fidelity prototyping and simple mock-ups

The overall idea with Participatory Design is to ensure that the finally developed applications and services will be adaptive to the users and not the other way around. Users are not brought into the design process individually but in teams to provide variations in feedback and to build on the learning that may take place in such teams. Therefore, it is often necessary to establish a basis for a common communication language. That is in particular if the team members do not share the same background and education. In MAGNET Beyond, these perspectives were considered in the workshops that were carried out. These were based on establishment of a shared design language by use of external cognitive aids such as pictures and different kinds of elements for prototyping. These perspectives were inspired by the PICTIVE approach that guides users with help of predefined elements [2], and from [3], where the content of cultural probes is obtained by getting users to create their own personal stories. More on the Participatory Design process that was followed in MAGNET can be found in [4].

2.2.2 *The Elicitation Process*

The requirements elicitation process took place through the above-mentioned steps. Two specific focus areas were selected within the project as cases. These were the so-called MAGNET.Care and Nomadic@work cases. The MAGNET.Care case refers to the situation where users have an interest in managing their own health. This may be in relation to normal health such as managing food intake and weight for example. However, it can also include management of an illness such as diabetes, where a user needs to monitor and manage the illness on a daily basis. The Nomadic@work case focuses on a group of users who have high demands on availability of high quality of data and communication links. In this particular case, it may be a travelling journalist who for example at all times would like to be able to produce high quality broadcasts – perhaps also with reference to old material already existing in the home database.

For each of the cases, the user requirement elicitation process has been different as a result of the users and the focus of the project. In the following examples on elements of the elicitation process are given.

2.2.2.1 Idea Creation

Idea creation took place differently in the two cases. For the MAGNET.Care case, workshops were carried out in a lab while in the Nomadic@work case, a cultural probe was used to capture the nomadic perspectives of the users.

Idea Creation in MAGNET.Care

For the MAGNET.Care case, the so-called creative user workshops were applied as an essential part of user centricity. The overall focus of the creative workshops was for the users to develop a conceptual text-based scenario landscape relating to their situation and their needs and requirements. A scenario landscape shall be understood as a conceptual, physical paper landscape showing different situations and pictures of how users think about their situation and about future technology solutions to an improvement of their situation. During a workshop, the users were given additional external cognitive prototyping aids in the form of so-called image elements consisting of pictures, words or short sentences. These were produced on the basis of the case study's conceptual scenario landscape, the case study context, human activities and important high-level user requirements identified beforehand. Image elements would typically represent the related context and human activities.

Figure 2.2 shows an overview of some scenario landscapes produced at a creative workshop with participants having diabetes.

Image elements and questions used at the workshop represented predefined contexts and user situations as well as user requirements. The predefined contexts and user situations covered: shopping, education, travel, community, collaborative work,

Fig. 2.2 Overview of scenario landscape and image elements (text in Danish)

surveillance, emergency, health care, society in general, transportation and home. Each of these contexts was represented by a number of pictures intended to stimulate the participants in remembering and discussing their needs and requirements in these situations. They covered areas such as; usability perspectives, personalization, user experience, user interface, economy, ethical issues, security and legal issues.

More details on this workshop as well as the outcome of it can be found in [4].

Idea Creation in Nomadic@work

In order to address nomadic users, a mobile probing kit was developed in order to capture ideas, user requirements and needs for the nomadic workers as they would encounter needs or ideas in their everyday. That is, to facilitate the idea-generation in everyday situations and capture the ideas and requirements in the situations they would occur. Applied in the MAGNET Beyond project the probing kit was referred to as the so-called IDEA-MAGNET. The probing kit was a notebook (inspired by the Hawkins lump-of-wood [5]). The size of the notebook was 5 cm times 7 cm with a metal cover and an integrated pen (see Fig. 2.3). Additionally, a few stickers were added to the inside of the metal cover which could be used when taking notes.

Fig. 2.3 The pocket size (5 × 7 cm) probing kit notebook with integrated pen

The participants were instructed to carry the probing kit as much as possible during the day for a total of three weeks. They were asked to use the probing kit to write down situations, problems or future activities where they could envision the use (benefit from) from technologies, services or applications which are not available today, e.g. situations where the use of technologies could help or assist them in improving their everyday working life and secondly also their family and leisure activities.

Taking into consideration the use of the probing kit and the importance of clarifying and making sure that participants understood (were guided in the right direction), small notes or 'bumper stickers' representing different usages scenarios were included in the probing kit as reminders to the participants to think about as they made use of the probing kit. The bumper stickers were meant as a way of reminding the participants of certain aspects of their daily life involving the use of technology, services and applications that they should record in their probing kit.

More details on the approach and results of applying the Idea MAGNET can be found in [6].

2.2.2.2 User Scenarios

For both cases, the idea generation resulted in a large number of different ideas and requirements. For example, the use of the Idea MAGNET resulted in more than 175 ideas (from 10 users), which after a critical selection came to around 65 ideas and requirements, which were relevant for the MAGNET project.

Different user scenarios were developed for MAGNET.Care and Nomadic@Work to show how the technologies described in the project could be used. The scenarios are, in general, scenes that describe users and their interactions with the different technologies and services of MAGNET Beyond. The scenarios present ideas of how a typical user would make use of the services and represent the diversity of users' needs and requirements and try to represent this in the target group as closely as possible. The scenarios were developed using a scenario framework, which can be seen in [4]. Examples on scenarios developed can be found in [7] and [6].

2.2.2.3 Use Cases in Summary

From the scenarios, a number of use cases were derived with the aim of illustrating and showcasing the MAGNET Beyond technologies such as PN formation, PN Federation, Security and Collaborative work. These original use cases were then put through a through screening process with overlapping issues of technology, users and business prospects in mind and the final cases that have been selected for the pilot services are the outcome of this. One example of a use case, directly derived from the user identified ideas and requirements can be seen in the following. This example was derived from the Nomadic@work case and was, later on, implemented as part of the pilot services [6].

Use Case: Exchanging Business Cards

User A and B agree to exchange business cards. A short lived, low-trust, lightweight, reactive PN-fed is set up between the two of them in order to pass the business cards.

1. User A and user B both select 'Digital Business Card'.
2. The menu show: [Store name of user xx in your contacts] (user xx represents a list of the PN-Identities with which the user has been in contact with recently), [Send your business card] (active search for recipient of business card et al.)
3. User A and B both select [Send your business card]
4. The menu show: [standard], [context based], [my own cards], [attach file], [new card], > [OK]
   (a) [standard] sends the default business card
   (b) [context based] sends a generated context aware business card that is customized with relevant, non-confidential information about the users, the situational context for their meeting (the topic of the trade fair), and eventual a commercial

(c) [new card] lets the user select which information based on the default business card that should be added and which should be removed. It is stored in [my own cards] for later reuse or modification.
(d) [attach file] lets the user attach a document that is not restricted from public distribution

5. When the user presses [OK]

   (a) Selection of recipient: [User xx], [Find person]
      - [User xx]: If User A and User B have a common recent history of interaction (in this case from the Ice Breaker service) they are by default chosen as recipients.
      - [Find person]: If User A and User B haven't had a common interaction history, they would need to scan for each other among the nearby people. A list of PN-identities in the physical vicinity with visible PN-Identities [make me visible] is shown, preferably with as many identifying details as possible: Full name, Business, Photo, Phone number, Mail addresses, etc. Search criteria can be used to limit the list. This [Find Person] has generic similarities with the Icebreaker function: Icebreaker and digital business card may be part of the same application.
      - As the Digital Business Card application was installed by the users, allowance for establishing low trust, temporary PN-feds with very restricted mutual access rights was default granted.

6. A short lived, low-trust PN-fed is established between User A and B.
7. When receiving the business card, the other user is prompted to confirm [receive business card from user xx?] [yes], [no]

   (a) If [yes] is selected a temporary restricted trust relation is build between the two PNs and the digital business card file is transferred.
   (b) When the transaction is successfully completed the PN-Fed is torn down.

More use cases can be found in [6].

2.2.3 *The Activity-Based Concept*

As a direct consequence of the user involvement in the project, a new approach for user interaction on communication devices was developed [8]. The approach is referred to as an "activity based communication concept" (ActCom). The purpose of this approach is to organize the interaction around the notion of activities that is the activities that the user is carrying out on a daily basis. From a user point of view people carry out activities, rather than use devices and services as such. The devices and services are just part of these activities. Each task takes place in a certain context and has its special requirements, which includes the devices and services needed to accomplish these tasks. So ActCom aims to make available to the user all the information, services and devices that are needed in order to carry out an activity in a context. This means the necessary devices; services and information should be easily available to the user for him/her to focus on the activity at hand rather than worrying about the details of configurations of the devices, services or network to accomplish the task. The user should be able to access and switch between several different activities, as occurrences or interruptions from the surrounding may impose such changes in user focus, and thus changes in current activity. Taking this approach is inspired by work in activity theory and especially the concept of ABC (Activity Based Computing) [9].

The ActCom approach was used to develop the GUI's of the MAGNET pilot services (see Section 1.4), and was later evaluated as a special usability area in the MAGNET final user evaluations (see Section 1.5).

Furthermore, it should be mentioned that a low-fi test and simple mock-ups were used during the user requirements elicitation process. However, this part is mentioned in more detail in the Evaluation section (Section 1.5).

2.3 User Profiles and Profile Management

2.3.1 Personalization and Service Adaptation

The work on user profiles is seen as an essential part of the general Personal Network (PN) framework. Being equipped with a PN, the users are empowered and assisted in carrying out their tasks under varying conditions in their everyday life. The objective is to take advantage of knowing the user's preferences, the context of the user and any other relevant information to optimize services for the user in any given situation. This must be done while safeguarding the user's privacy and keeping the user in full control of his or her resources and personal information.

The following definitions have been used in the project [10, 11]:

- *User Profile*. the total set of user-related information, preferences, rules and settings, which affects the way in which a user experiences terminals, devices and services [12].
- *Context*. any information that can be used to characterize the situation of an entity. An entity is a person, place, or object that is considered relevant to the interaction between a user and an application, including the user and application themselves [13].

To be able to make better use of services, especially in situations where the user is on the move, has his focus on other activities, or has a device with limited input/output capabilities, the services need to adapt to the situation and how the user typically uses the service, most likely as a combination of both, i.e., how the user uses the service in a given situation. The information to adapt the services in this way can be found in the user profile and the context information, which can be seen as two sides of the same coin, as both are needed to adapt the service for providing a better user experience. Therefore, it also makes sense to use the same middleware for making them accessible, which is why we decided in MAGNET Beyond to store user profile in the Secure Context Management Framework (SCMF) and use the same mechanisms to access the information [10, 11]. The SCMF is the key element within the PN, which acquires and stores the user profile and context information and controls the access to this information and the sharing of personal resources (cf. Sections 3.6 and 7.2.11).

Services can be adapted in different ways to user profile and context information. In the following, we list some examples:

- The information presented to the user can be adapted based on profile and context information, e.g., relevant information may be different when the user is at work or at home, and again different when the user is on a trip, e.g., present the local weather, possibly in addition to the weather at home.
- How the information is presented may differ according to the situation, preferences, and the device available. For example, navigation information could be displayed differently on a PC than on a mobile phone, and also differently depending on the situation of the user, whether she is standing or running, in which case the output could be reduced to easy-to-grasp arrows.
- Available services may be pre-configured with parameters used in the past or which are relevant in the current situation, e.g., the wake-up call in a hotel could be preconfigured to the room of the user.
- Services may be executed automatically depending on the user profile and the situation, e.g., calling a doctor in an emergency situation.

In the general case, it is a complex undertaking to decide which part of the entire available user profile and context information that is relevant and useful for performing service adaptation. Organizing the information in an ontology (cf. Section 2.3.3) is a step on the way, as it supports reasoning and decision-making, but there is a lot more research to be done on how to combine this with intelligent application logic and policies that provide a proper protection of user privacy.

Two main scenarios are considered throughout the project:

- PN Federations (PN-Fs), which can be seen as an advanced, well-controlled peer-to-peer interaction between two or more users within the same or different domains
- Access to foreign or third party services, where improved personalization and service adaptation is facilitated by the PN

In the PN-F scenario, Fig. 2.4, the user has full control over the resources that he or she wants to share with the federation in order to achieve the common goal of the federation, in other words avoiding exposing or revealing more personal information and content than needed. More details about PN-Fs are given in Section 3.5.

In case of accessing foreign services (push or pull type), a MAGNET-enabled user also has much better control of the personal information and can decide where the right balance lies between protection of privacy and revealing of personal information. Today, a large number of web sites offer users or subscribers a basic level

Fig. 2.4 Basic PN-F scenario

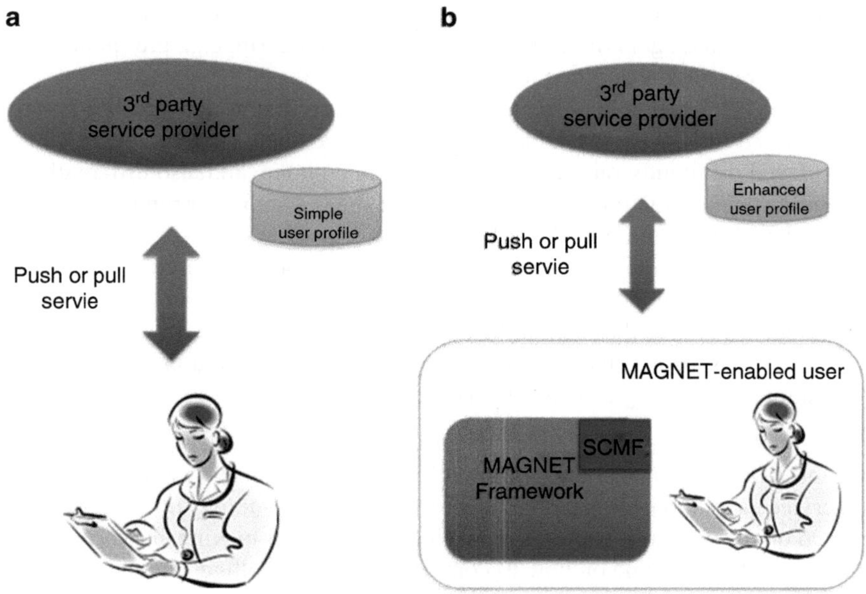

Fig. 2.5 Access to third party services. (**a**) Basic personalization targeting a standard user. (**b**) Enhanced personalization targeting a MAGNET-enabled user

of personalization, cf. Fig. 2.5a). This can be initiated, when the user signs up for the first time, where typically a set of personal data such as name, address, e-mail address, phone number(s) etc. may be requested, and the user chooses a user-ID and password to access the personalized services later on. Furthermore, the user is often given the option of ticking various preferences or areas of interest.

More sophisticated services will collect data about the usage history and based on this perform some "intelligent" processing in order to provide relevant information or offers to the user.

For a MAGNET-enabled user, Fig. 2.5b, we may envision that the service provider will be informed when dealing with a "more sophisticated" user, which in turn will enable a better personalization. The template of the user profile might be publicly available, so a service provider would know, what kind of personal information is *potentially* available, and hence be able to query the user for a certain part of this. However, the information may not have been filled in, or it may not be accessible because of the policies attached to the user profile. But if the information can be accessed, the service provider can use it to customize or personalize the service to this particular user. The MAGNET framework can assist the user in filtering and navigating huge amounts of contents, services and offerings. This provides better value for the user and better revenue options for the service provider. To fully take advantage of the PN, a service provider would need to adapt services significantly, but some benefits of PNs in terms of personalization and service adaptation are readily available as discussed above.

As the name implies, Personal Networks are personal, i.e. they belong to a user, and there is only one user in a PN. However, users often deal with "non-personal" networks or collections of resources, e.g. facilities in an office environment or in a conference centre. Instead of being personal these resources may belong to the premises, and they are typically managed by a system administrator. In order to extend the management framework of MAGNET Beyond to cover such cases as well, we have introduced to concept of a **Service Provider Network (SPN)** [11]. We can then apply similar procedures to govern access to and sharing of resources between a user's PN and an SPN as in a PN-F between two or more users.

2.3.2 Modelling of User Profiles

In accordance with the definition given above, the user profile is a record of preferences, rules, settings and other relevant user information that are saved and changed dynamically so as to provide the appropriate personalized behaviour to the device, the services and the whole PN. Dynamics of the PN composition is very important when devices and services come and go.

Previous work done on user profiles was probably missing an important requirement, namely the formation of federations of user communities, which care about all personal clusters and devices linked with them. The user centricity in MAGNET Beyond implies that the user becomes an entire communication cluster made by the user himself with his personal resources (devices, personal clusters and personal federations). The user profile should therefore be able to accommodate:

- Heterogeneity of access, communication infrastructures and domains
- Multi-device scenarios
- Personal Networking
- Federations of PN user communities
- User centricity
- Personalisation
- Preferences
- Third party services and access policies

Most of the above requirements, apart from the PN-related ones, are to some extent already discussed and proposed in 3GPP [14], ETSI [12], Liberty Alliance [15], W3C and the DAIDALOS project[1]. Instead of defining a new user profile concept, the approach has been to extend existing architectures defined in other projects or standardization bodies and adapt them to match the PN scenarios. The proposed structure can thus be seen as an evolution of existing scientific or industrial approaches in defining user profiles towards a global profile including personalization and federation concepts.

The conceptual user profile structure from MAGNET Beyond is shown in Fig. 2.6. The user profile is composed of several parts, each corresponding to different parts of MAGNET Beyond. It is organized in a tree structure and consists of several subcomponents, which are placed throughout the Personal Network (PN). Most of the user profile subcomponents are placed locally in the user's devices (with an online backup repository), whereas the extended user profile is placed only in the repository and thus accessible only when connectivity is available.

The user information is accessed through the "User profile" subcomponent, which contains references to the other subcomponents. Policies are retrieved and used, when the user browses through content, either on the Internet as web pages or in third party services. The user profile consists of the following parts:

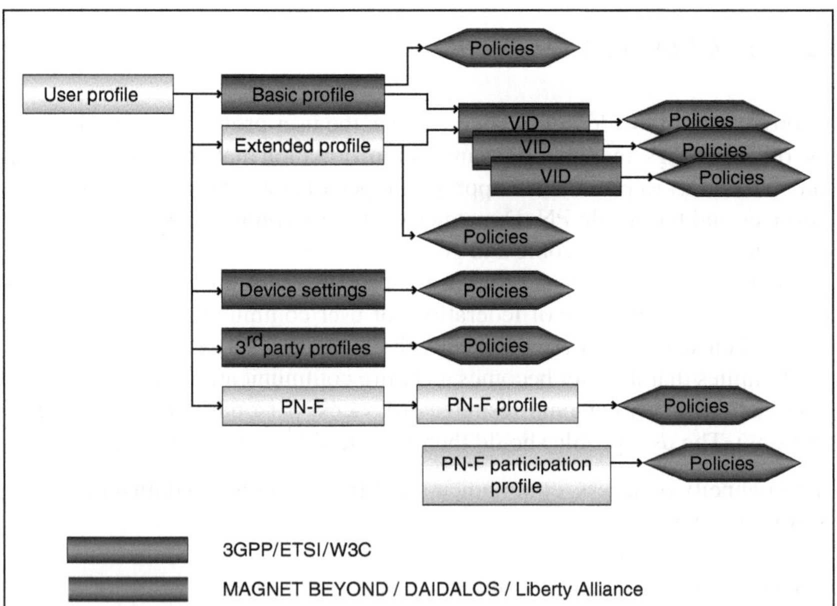

Fig. 2.6 MAGNET user profile in a conceptual representation displaying the different categories and dependencies compared to state-of-the-art (Adapted from [11])

[1] http://www.ist-daidalos.org

- The top-level **user profile**. The top level of the user profile contains the user profile ID, obtained security clearances etc.
- The **basic profile**. This part of the user profile contains the basic information, such as e-mail address(es), phone number(s) etc. Several instances of this information constitute Virtual Identities (VIDs), which they user may take on and use for specific purposes [11].
- The **extended profile**. Contains generic user settings and preferences that are based on the individuality of a user, but are not permanent and can change according to the user's will and needs, along with a reference to the user history log. This is where usage patterns can be used to help adapting the user profile.
- **Device settings** (information and settings). Device profiles may be generally available, but the user will often want to apply personal preferences and settings for each of his or her devices.
- Service-specific or **third party profiles**. Preferences or settings for third party services.
- The **PN-F profile** contains all the information about the user's PN-Fs. The PN-F profile is a data structure that is created, stored and maintained by the federation creator and describes the entire PN-F. The **PN-F participation profile** includes security settings, administrative rights, other preferences etc. The creator or administrator of a PN-F has a copy of all participation profiles and administer each other participant's rights. Strictly speaking, only the participation profile is part of the user profile.

The user profile has a unique identifier, which is used for security clearance, similar to Single Sign-On systems. The profile is a container for all the information about the user and represents the user as an individual in the system. The user profile does not directly convey login to MAGNET Beyond; rather, the user must do so with a specific identity as defined in the basic profile within the user profile. The reason for this is that a user must assume a (virtual) identity, VID,[2] upon login, and this identity is registered as the user's identity, when logged in on MAGNET Beyond. A VID consists of an identifier that the user selects (a sort of nickname) and a set of policies, which determine what information or services may be disclosed during the usage of a VID. Having a certain identity in MAGNET Beyond implies having a certain level of clearance in different systems, with which one interacts. Thus, the VID related to a basic profile data set derives one or more clearances from the user profile. Upon VID creation, the user selects which of the already obtained levels of clearance should be active when using the relevant identity.

The basic profile component of the user profile contains the basic information about the user, such as e-mail address(es) name, address, gender etc. This information does not necessarily have to be provided; rather it is up to the user to fill in, even with false information, e.g. in VIDs, should he/she wish to remain (partly)

[2] The VID concept [16], [17] was introduced by DAIDALOS to meet the privacy purposes of protecting the user identity in personalized and mobile environments.

anonymous. The basic profile is a rigid set of information, which is provided to the user upon creation of the profile. The basic profile will be available throughout the user's devices, thus independent of connectivity.

The basic profile data collection is identical to the user's identity at a given time. That is, the basic profile may contain several sets of basic profile data, each with a more or less distorted and elaborate version of the user's real identity. When the user decides to switch identity, this is technically done by swapping the credentials with those of the desired basic profile data set.

The extended user profile contains information that is generated over time; that is, the entries are not present upon profile creation. Thus, the extended profile is dynamic and highly generic, allowing for the introduction of new entries later on. The possible entries in the extended profile are managed by MAGNET Beyond alone and are publicly available through online schemas. Third party service providers may then access the information in the individual user profiles, provided that the user grants them this access.

2.3.3 Common Ontology for User Profiles and Context Information

Although user profiles and context information are differentiated in MAGNET Beyond, the two types of information do share some common attributes which makes it possible to treat them as similar when considering the distribution of the information. Building a common ontology and a common management framework for user profiles and context information has been an important objective of the project. This has been successfully accomplished and applied in the pilot services.

Figure 2.7 shows the core concepts of the Integrated SCMF Ontology for context and user profile information. It is used as a basis for storing context and user profile information in the SCMF.

The underlying idea of this ontology is to define a hierarchy of entity types, facilitating a type-based access to context and user profile information. Its top-level concept is the *MagnetEntity*. The *MagnetEntity* concept introduces the property *hasIdentifier*. Any entity that can be uniquely identified using an identifier can thus be modelled as a *MagnetEntity*. Based on the unique identifier an index can be built to provide the basis for efficiently accessing context information in all cases in which the specific entity is known.

The *MagnetEntity* concept has two subconcepts, the *SpatialEntity* and the *VirtualEntity*. The *SpatialEntity* concept introduces the *hasLocation* property. The *VirtualEntity* concept comprises all types of entities that are not associated with a geographical location. *VirtualEntity* has a subsconcept *Profile*, which in turn has a subconcept *UserProfile*.

The attributes of MAGNET Beyond entities are modelled as properties in the ontology. Properties can either have simple types supported as base types in the ontology such as Strings or Integers, or they can be complex types, in which case

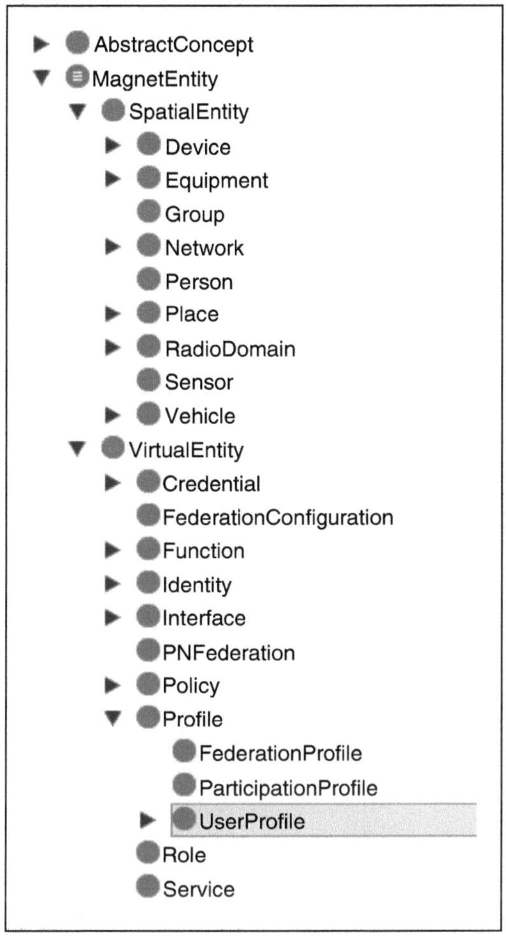

Fig. 2.7 Overview of the Integrated SCMF Ontology

they are modelled as an AbstractConcept. For the user profiles we have made heavy use of these AbstractConcepts as they determine the units that can be retrieved by the SCMF. For example, if the user profile should contain a property "home address", there must be a complex structure for the whole address. It is not sufficient to model street, post code, city, etc. separately. Especially if there could be multiple instances of home address in the same profile, it needs to be clear which information belongs to which address. On the other hand, modelling an address as a separate entity would have the effect that two subsequent requests to the SCMF would be needed for retrieving the information. Figure 2.8 shows a simple example with details of the user profile part of the ontology (BasicUserProfile and ExtendedUserProfile). This has been used for the pilot services (see Section 2.4.2.2).

The *FitnessCenterProfile* has the properties shown in Fig. 2.9.

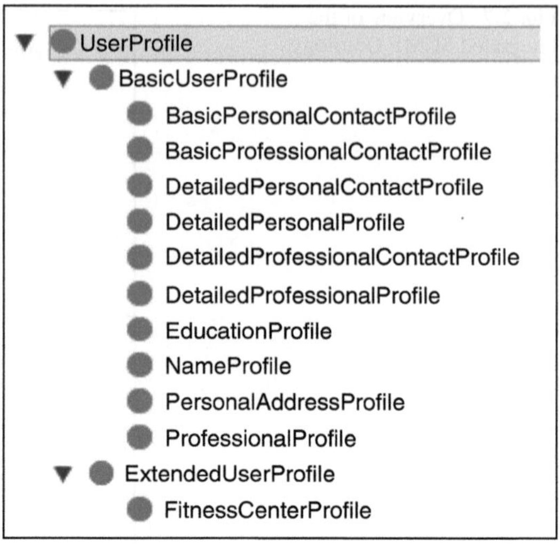

Fig. 2.8 User profile part of the Integrated SCMF Ontology

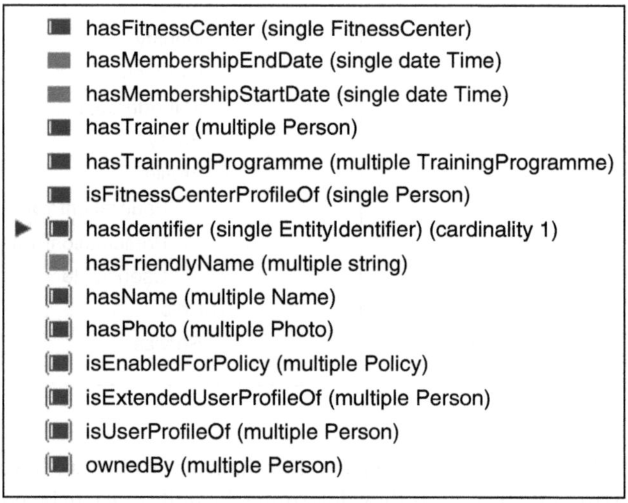

Fig. 2.9 Properties of the FitnessCenterProfile

2.3.4 Profile Management

Management of the user profile includes several aspects, e.g. updating or adding data content in the already defined user profile, and the supporting technology needed to get the user profile to work in a system. Many of these supporting technologies and concepts are described and handled by different task forces or forums on the Internet working towards standardisation.

Internally (within the PN) user profile and context information are jointly managed by the Secure Context Management Framework. User profile data are stored in the Processing & Storage module of the SCMF, and insertion, updates and queries are handled by the Context Access Language (CALA) on the user's Context Agent [11].

The key functionality related to user profiles is its capability of storing the user profiles and making them available to all nodes in the SCMF, and when further coupled with the interaction of the MAGNET User Profile (MUP) system described in Section 2.3.4.1, the framework provides a powerful and efficient access to user profile data distributed in the PN.

As already discussed, MAGNET Beyond has dealt with two main type of user interactions:

- Interaction with other users (PN-Federations)
- Interaction with an external service provider offering services to the user

In either case the user profile (or selected parts of it) serve to optimize the interaction and make it user-friendlier. It is therefore important that the user profile is well structured and managed. Policies play an important role and a profile management system must ensure that only as much information as needed is revealed (e.g. to a service provider) in order to have a value-added and personalized service delivered to the user.

Considering the trade-off between utility and privacy and how to keep the user in control, it is obvious that:

- On the one hand the user must always have access to his or her profile data in order to manage and update them as desired, but
- On the other hand user profile data must be revealed to others in order to be useful. An isolated user profile kept on the user's own device(s) would only facilitate interaction, where no other persons are involved, e.g. between the user and a system

These considerations imply that we need to operate both a local and (partly) federated user profile.

Figure 2.10 illustrates the concept of the "Digital Butler", which has been proposed by MAGNET Beyond. It displays the different security layers of the federated user profile, relating to the fact that a user could have different levels of trust towards different service providers (the "Onion Model"). The layers of "onion" are meant to illustrate different levels of importance or sensitivity of the personal information contained in the federated profile. The outer layers are least sensitive, meaning limited loss of privacy, whereas getting closer to the core means more sensitive data and stronger policy enforcement.

The local profile on the user's Context Agent (CA) is synchronized with the online federated profile, which is managed by the "Digital Butler". As stated earlier the "butler" would require a federated user profile making it a trusted service. If an entry in the user profile has been updated that also applies to the federated user profile, it would require a strong secure synchronization between the user profile in

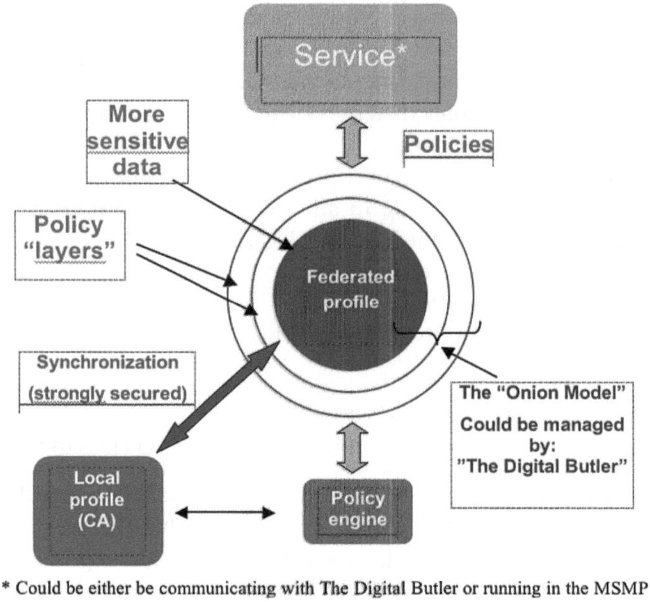

Fig. 2.10 Conceptual view of a federated user profile from a security point of view. The *grey arrows* represent exchange of policies [11]

the SCMF and the distributed part of the profile. The concept of trust is the main issue as the "butler" is actually keeping parts of a user's profile, and it is important only to provide the information to a third party service provider that is in the interest of the user. It is defined in the policy parts of the user profile and enforced in the policy engine.

2.3.4.1 Subscriber Data Management

Profile management is closely related to identity management and – from the operators' perspective – to subscriber data management. Many of the ideas and concepts already developed can be extended to cover user profiles in general rather than just identities or subscriber data.

3GPP has released a series of technical specifications [14], which define a Generic User Profile (GUP) for 3G mobile systems. The ulterior goal of those specifications is to enable harmonized usage of user-related information originating from different domains. They aim at facilitating user preference management, user service customization, user information sharing, and terminal capability management as well as profile key access. The GUP is the collection of data which is stored and managed by different entities such as the User Environment, the Home Environment, the Visited Network and Value Added Service Provider, which affects the

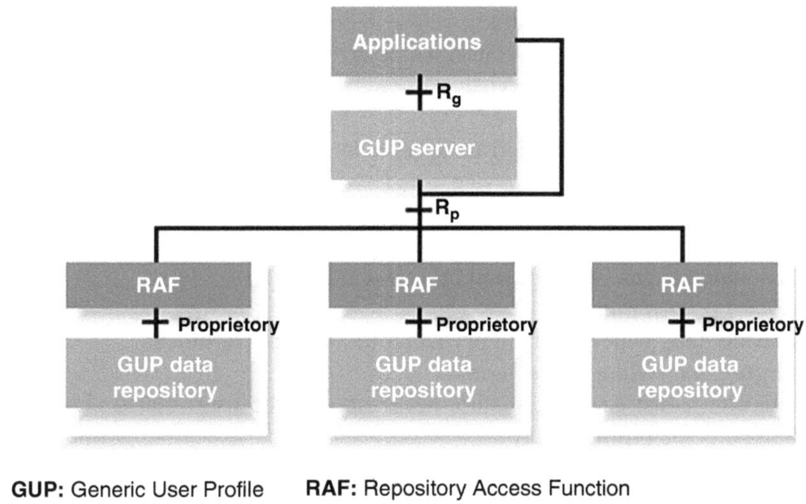

Fig. 2.11 The basic GUP architecture [18]

way in which an individual user experiences services. An individual service may make use of a number of **User Profile Components** from the GUP.

The distributed nature of the GUP system architecture is displayed in Fig. 2.11. Different applications (like third party services or others) query information about the user through the GUP server. The GUP server does not contain the actual user profile data, but knows where the newest information is available from. It also acts as a gatekeeper by authorizing or denying access to profile data. The GUP server either operates in **proxy mode** (collects the requested data and provides it to the requestor), or in **redirect mode** (provides the addresses of the data repositories to the requestor). It acts therefore as a data federator and offers a single point of entry to the Operation Support System (OSS). The GUP server can then (based on the implementation) get the data from the repository using the Repository Access Function (RAF) of the different repositories. The interface to the repository itself can be proprietary, but the communication with the RAF is standardized. This distributed concept has been adapted in MAGNET Beyond, and a security layer with policy enforcement has been added making all user profile queries secured to prevent leakage of unwanted user profile information.

The following information is typically stored in the GUP:

- *Authorised and subscribed services information*
  These kinds of data are generally owned by the home operator and allow management and interrogation of subscription information.
- *General user information*
  Data owned by the user, which are not specific to individual services, but may be useful for any service. These would be data like: settings (e.g. name, postal address), preferences (e.g. language), Registered Service Profiles of the user, indicating the currently active Service Profile of the user.

- *PLMN specific user information*
  Data owned by the home operator, which are not specific to individual services, but may be useful for any service.
- *Privacy control data of the user*
  Data owned by the user, which are specific to individual services and which control privacy settings of that service.
- *Service specific information of the user*
  Data owned by the user or value added service provider, which are specific to individual services (standardized or non-standardized).
- *Terminal related data*
  These are data, which relate in particular to the user's terminals.
- *Charging and billing related data*
  These data consist of information necessary for the user related charging and billing, e.g. the billing policy.

Building on the idea of a GUP server handling subscription and notification and access to relevant data repositories leads to the proposal of the more decentralized MUP architecture. The MUP function does not store any user profile data itself, but uses its metadata to find the mapping into the various requests to the concerned data repositories, as shown in Fig. 2.12. So the MUP server only knows where to get (store) the data, it does not actually keep the data. An application requiring some data asks the MUP server and the MUP server will query the appropriate repositories, assembles the results and provide the application with the response.

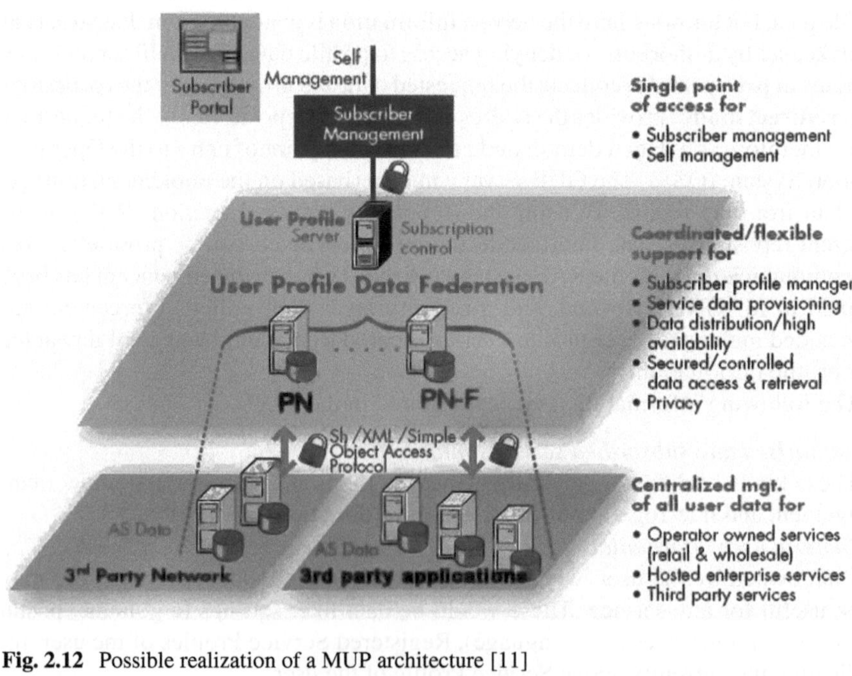

Fig. 2.12 Possible realization of a MUP architecture [11]

As already discussed, all information that the users might need even without his connection should be placed within the PN. This mainly includes basic information about the PN-F and users (only keys to MUP and some replicas). Participation profiles can be either included in PN-F or GUP. Some PN-Fs shall also be associated with specific services. Roles of users in a PN-F should be stored in the PN-F.

However, some "administration & professional" roles should only be assignable by the administrator of the PN-F (if it exists). Other social and secondary roles could be edited directly by the user himself/herself. Assuming a certain role or presence status is also solely interesting, when connectivity is available, since it relates to the user's presence and preferences in MAGNET Beyond, and therefore only available online.

The architecture shown in Fig. 2.13 uses the functionality offered by the PN-F SCMF gateway to interact with the external MUP server. This requires that the MUP server uses CALA in both directions, i.e., can get user profile information from the SCMF as well as provide access to its user profile information for the SCMF.

The MUP realized in the project represents the main access point for retrieving user profile data, synchronization between the local and the remote instances of the basic user profile and an interface to query the OWL-DL ontology based on the standard SPARQL language,[3] and an external interface (CALA client) to manage specific user data based on the CALA language. This client can be installed on other nodes in the PN for queries and updates.

The middleware and the databases are strongly based on the use of ontologies in a seamless way to access all different data repositories. In fact, it does not hold any data, but gives the impression of holding all the data by being able to answer queries. An important database included into this architecture is the GUP. Based on this architectural approach, MAGNET Beyond has specified and designed IP

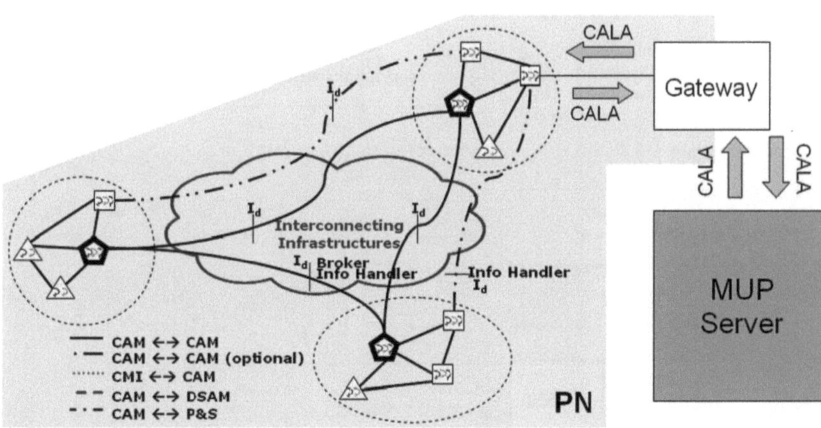

Fig. 2.13 PN agents forming the SCMF and communicating with the MUP server through a gateway using CALA [19]

[3] http://www.w3.org/TR/rdf-sparql-query/

Multimedia Subsystem (IMS) pilot applications for health care and professional sectors utilizing external profile server (MUP) storing profile information of PN users. MAGNET Beyond offers a service platform that leverages new ways for IMS technology to deliver improved, context aware and personal services for end-users increasing revenue opportunities for service providers and operators who wish to turn their commodity-priced service bundling into a highly competitive one. The opportunity for an operator to provide quadruple play (triple plus mobile) enhanced with context-aware PN capabilities may become the key success point in a Web 2.0 Internet world.

2.3.4.2 Identity Management and the "Digital Butler"

The concept of "single sign-on" and federated identities has been studied intensively, e.g. by Liberty Alliance, and is already widely used. It relies on having a trusted identity provider to manage the user's federated identity. Combining this with the "Digital Butler" idea leads to the next step of having not only an Identity Provider (IdP), but a Personal Identity Provider (PIP) that manages the user profile and assists the user in receiving personalized services. Furthermore, it would be natural to take advantage of the well-established GUP framework and extend it to manage the entire set of profile data, not just the subscriber data needed by the operators. This is illustrated by the high-level architecture model in Fig. 2.14. With relevant user profile information the Digital Butler surfs different third party

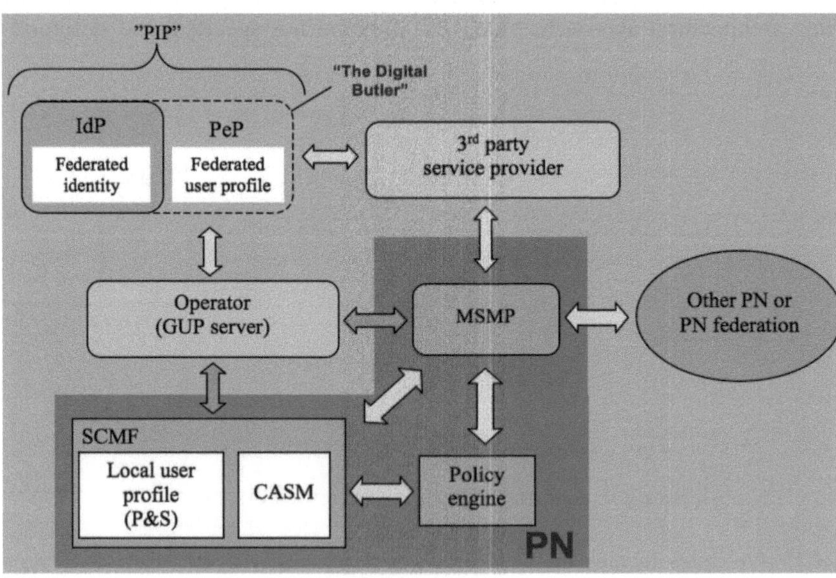

Fig. 2.14 Overview of a MAGNET-enabled user with an optional "Digital Butler" communicating with a third party service provider. The *orange arrows* are only meant as the components having connectivity [11]

services and reveals only disclosed user information to the service provider with the intention to personalize or value-add the service before presenting it to the user.

Other projects have also combined the concept of personalization and identity providers. These are also referred to as Personal Identity Provider (PIP). An example of a PIP is VeriSign. However in many cases the PIP is acting passively depending solely on the user interaction and not proactively predicting the user's needs.

Having the user profile available online would also help on two other aspects. One is the aspect of power consumption on handheld devices, as this entity would require a lot of processing. The other aspect is that keeping the entity online would make it a more 24/7 value-adding service discoverer adapting relevant services to suit the user.

2.3.5 Business Opportunities

The "butler" could be a part of a user's PN but it is not a requirement. It could also be a third party service provider acting as a personalization provider (PeP) working in collaboration with the relevant IdPs. It could actually be one of the IdPs making it more like an autonomous PIP. This is a potential business opportunity of MAGNET Beyond. Anyway, if one looks with security glasses on the "Digital Butler", it is not a pure third party provider. As stated earlier the "butler" would require a federated user profile making it a trusted service. If an entry in the user profile has been updated that also apply to the federated user profile, it would require a strong secure synchronisation between the user profile in the SCMF and the distributed part of the profile.

2.3.5.1 Stakeholders

Looking at a full-blown MAGNET Beyond scenario a lot of stakeholders will be involved. Some of the main actors are:

- User
- Operator
- Service providers (could be the operators themselves)
- Third party service providers
- Identity/personalization provider

The users themselves are prime stakeholders in the personal networking concept. The user profile is an important part of the way towards simplicity for users in the IT and telecom world – with the growing complexity for most appropriate services and applications in each situation and context. A specific set of user related information included in a user profile would help and make the complicated selection of services, applications and devices for each network, access and context situation almost autonomous.

The operator is the stakeholder taking care of the access, networking and management roles as the provider of connectivity. The operator can be a traditional mobile network operator, a specific local access provider or an actor combining these roles for PN connectivity and infrastructure interconnection. This means that the operator is an important stakeholder for realisation of PN connectivity through MAGNET Beyond specific networking solutions.

The operator is also a possible stakeholder for management and storage of user profiles, which in that case is communicated through the MAGNET Service Management Platform (MSMP).

The service provider is among the stakeholders that are most dependent on the content of the user profile as many services can be adjusted towards it. The service provider could be the network operator itself or an independent actor. In any case, the user profile data is communicated through the MSMP.

The third party service provider is also a stakeholder of the user profile, although not really focused on in this work.

One aspect of the role of the stakeholder identity/personalization provider is single-sign-on and that is discussed below.

2.3.5.2 Single Sign-on and Personalization Aspects

A major issue that is addressed in MAGNET Beyond is the opportunity to access relevant information from a single point of entry with a single sign-on function. This data might be found in the PN, through a PN-F or a service provider, etc. As long as there are no security/law violations and no hidden billings, the routing to the file should be transparent to the user. In MAGNET Beyond the service provisioning is a key issue. If a user has to create separate profiles at each service provider the entire concept of service discovery based on personalized user data would fall apart. Many projects address the problem of a single sign-on function and different solutions have been presented with various security aspects.

As described in [10] the Liberty Alliance project has presented a solution that solves the single sign-on aspect but goes beyond this [15]. A user logs on to authenticate himself to an identity provider. An IdP is defined as a computer system that issues credentials to a user and verifies that the issued credentials are valid. An IdP may operate one or more credential services, each of which issues end user credentials based on standards for identity verification and operations defined by the National Institute of Standards and Technology (NIST). A user can hold credentials from multiple IdPs and a "Federation" of IdPs is also possible. In short, one could say that a user logs on to the IdP and can then be automatically authenticated to all service providers or other IdPs that have been trusted by this IdP (a "circle of trust"). The different service providers, however, are not allowed to communicate any information about the user between each other. They can exchange information relating to this user only with the IdP that can access the MUP for relevant data, if it is available. Overall, LA standardizes functions for authentication, authorization, security/privacy control and service discovery. In other words LA can grant access for a service provider to offer services to an identified user or a representative of the user – say a sort of a "Digital Butler" – if this service provider is accepted by either the IdP or the user. To make the service personal, however, specific content from the user profile is needed to adapt the service. This not treated by the concept of an IdP and is not specified in LA.

The concept of personalizing services making them value-added is not new. It has been described thoroughly in many projects and one project worth mentioning is TV-Anytime [20]. In 2004 they joined forces with LA bringing the concept of IdPs into the project of TV-Anytime and by using metadata to make a standard for digital video recording and thereby open the opportunities for video-on-demand services. The TV-anytime project introduces the concept of a personalization provider that helps the user find and present his or her wanted media.

2.4 Implementation of GUIs, General Services and Pilot Services

The activity-based concept (see Section 2.2.3) developed during MAGNET Beyond [8] is supported by the MAGNET user profile and the conceptual description originally presented in [11]. As already described, everything the user wants to do with a device is called activities, and how these are accessed and navigated through by the user is shown in Fig. 2.15.

Here the activity concept once and for all make up with the concept of everything being application dependent for the user. In the project, all applications are instead called **tools**, and services can enable the tools, as they are needed. The user gives or edits a presentation, in contrast to a traditional operating system like Windows, where the user opens a Microsoft PowerPoint file or so. Depending on what activity you are in, the amount and types of tools can vary. This does not mean that the tools are only available in a given activity, but they are rated individually, depending on their relevance for the given activity, and are per default hidden for the user. The user can select, whether the tools should be available or not. For example, if a user is currently in the activity "At work", tools relevant to the work are visible, but tools concerning private issues are not. The content, which is available using the tools, is also dependent on the activity. If the user wants to write an e-mail in the activity "At work", the e-mail will be sent using the business signature and card. Also mails relating to work are displayed. If searching for pictures only pictures relevant to the activity are found and not private pictures, if the activity is still "At work".

The tools can also be shared with other PN users if they are in a PN-F. This concept is shown in Fig. 2.15b, where a tool called "Calendar" has been added at some time, either by a service or manually by the user, when this was needed. The tool is shared with three MAGNET-enabled friends and the individual calendars can be read depending on the security settings of the PN-F. The tool is only visible in one activity, but can however at any time in any given activity be accessed with a few extra clicks. The different tool settings and the different activities with specific

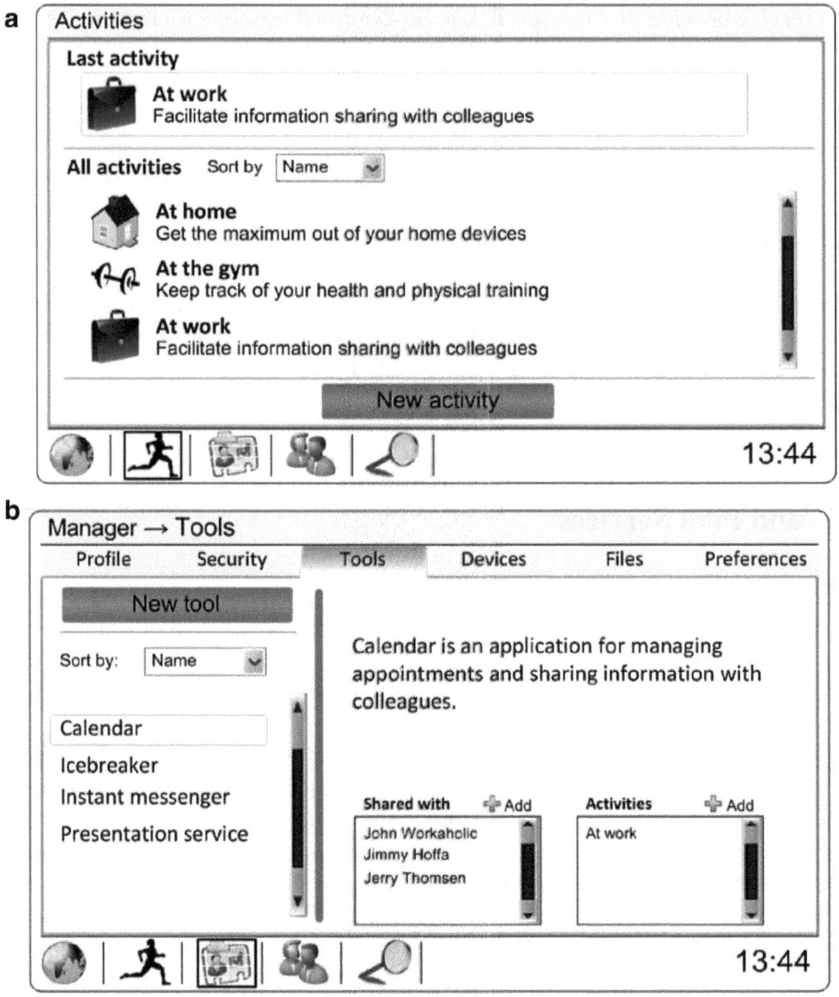

Fig. 2.15 (**a**) The activity menu on the user's device. Last activity was "At work". (**b**) The manager screen. A tool called "Calendar" is selected. This tool is shared with three people and only visible in the activity "At work"

attributes all go into the basic user profile. However, if some of the services have extra data (apart from those defined that need to be stored), it will go into the third party profiles of the user profile. This could e.g. be something like the user's history with the specific service.

The overall concept also goes for having different contacts and devices that relate to different activities. However, they all contain a lot of specific extra data for each entry.

An example of a MAGNET user available to another and how this user is handled is displayed in Fig. 2.16. All MAGNET-enabled contacts are stored in the "Basic

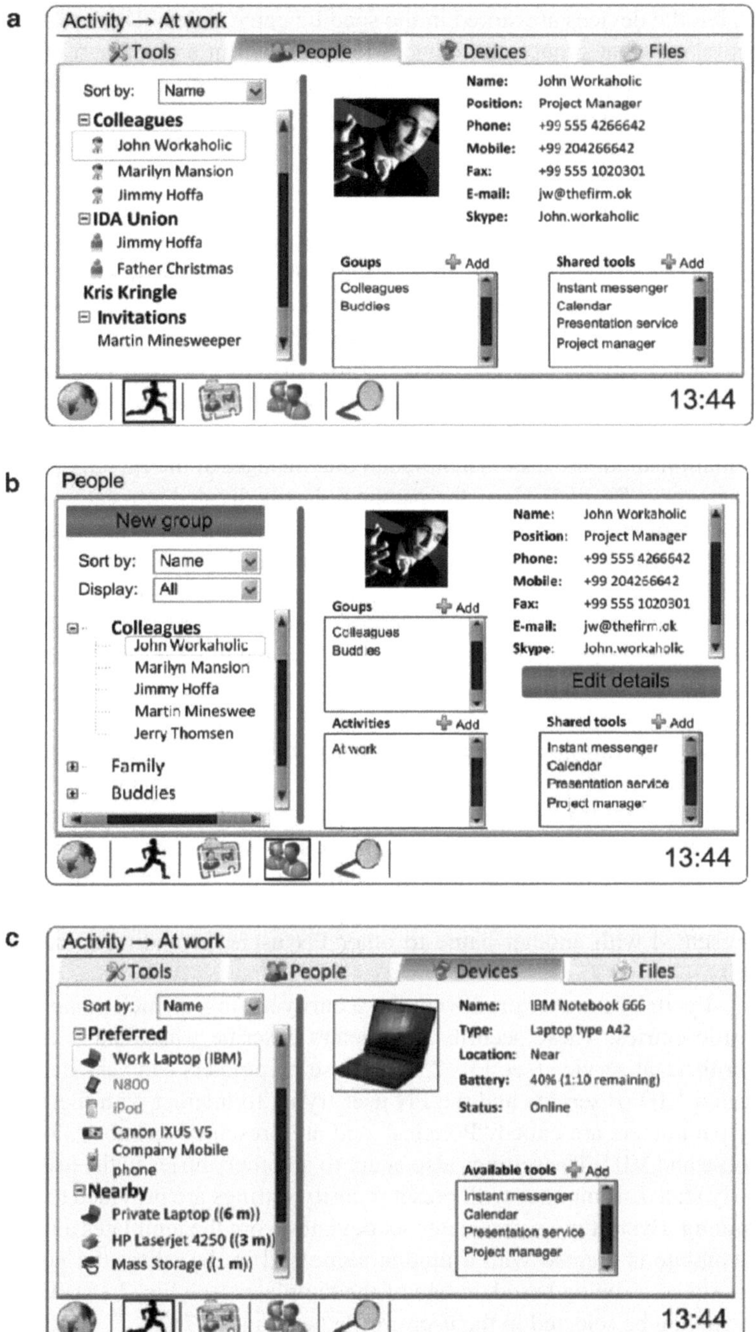

Fig. 2.16 Screen displays. (**a**) The different MAGNET users available in different groups. The user selected is available in two groups and has a lot of shared tools. (**b**) The manager of the same person where specific information can be edited. (**c**) An example of a MAGNET-enabled device with attributes and tools available

profile", but the devices are stored in the specific entry called "Devices". However, information on what groups and devices to be shown in a given activity is stored in the activity entry in the "Basic profile". The term "groups" refer to the titles "Colleagues" and "IDA Union" in Fig. 2.16a. They are called groups to the users, but they are technically speaking camouflaged PN-Fs, meaning that the users displayed in the different groups are members of a given PN-F with all necessary attributes stored in the PN-F profile and PN-F participation profile (PN-F part. prof.) as shown in Fig. 2.16b. When a new group is created by the user, the user can choose members and specific security settings, which all go into the PN-F related entries of the user profile. The device screen in the same figure shows an example of a laptop that is available in the given activity. It is called a preferred device, and this information is stored in the activity profile. The device profile displayed here is just user-friendly information. A lot more metadata on screen resolutions and other hardware profiling is stored in the "Device profiles" entry of the user profile (see Fig. 2.16c).

Information about the user is handled in the manager of the MAGNET GUIs under the category "Profile". Here, the personal data is divided into categories, which fit the user profile in Fig. 2.6, as they have the same names. These categories or entries are called "Basic", "Extended" and "Virtual Identity". However, as an exception, the editable entry of third party services goes into the entry of the same name in the user profile. The virtual identities are subsets of the basic profile with some data from the extended profile also, such as payment information etc. However, they are still stored in the same entry called VID with a unique entry per virtual identity.

The basic user profile information shown in the editor consists of personal information such as name, phone number and general contact information. In the extended profile information of payment methods and attributes relating to specific services are stored. The VIDs can be based on information from the basic and extended profile but can be fully customized if the user wants them to be (see Fig. 2.17). They even have an attribute called "Display name", if the user wants to be presented with another name to other PN users, providing some degree of anonymity.

The last parts of the user profile are the security settings, which relate to all other user profile entries. These security parameters describe what data is available to whom or to what service (see Fig. 2.18). These parameters can vary depending on the selected VID or service and the PN user trying to interact with the user. These security parameters are called "Policies" and are presented as subsets of the basic user profile and VID. The settings also adapt to all other entries in the user profile as previously stated. Templates with preset security settings are provided to the user to select among. Every time a parameter set deviates from the templates a new version of the template is created with a unique name and in the editor the new security profile is stated as being based on one of the templates (see Fig. 2.18). The security profile can then be selected in the user profile (see Fig. 2.17).

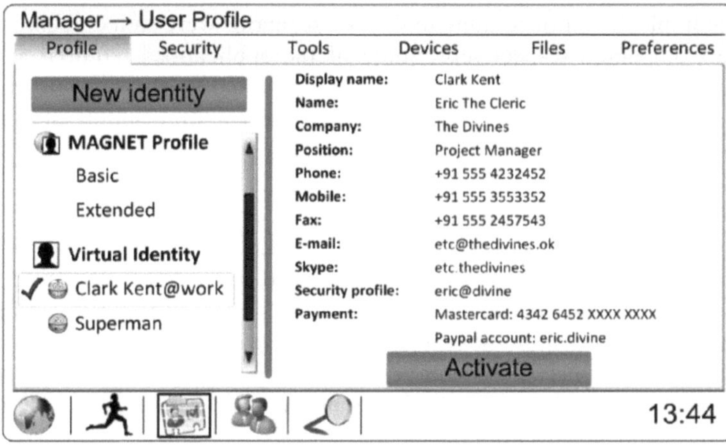

Fig. 2.17 User profile editor for personal information about the user. The screen shows an example of metadata in the "Virtual Identity" entries. This is partly composed of information from the MAGNET user profile and specific data to the VID

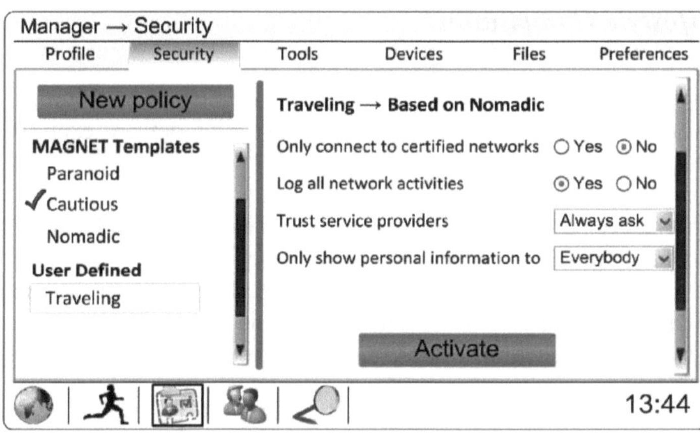

Fig. 2.18 Security editor for setting policies in the user profile. This is a small example of the concept of having security templates to help the user find the right settings

2.4.1 General Service Architecture

As earlier mentioned the MAGNET Beyond project was among many other things also to implement and demonstrate the concepts. To actually enable the Nomadic@Work and MAGNET.Care (see Section 2.1) the different use cases of the two scenarios were thoroughly examined and software in the shape of small support applications was identified. These applications called tools from the activity concept point of view were named after functionality and technically described in details.

The actual implementation relating to the two scenarios were called: Icebreaker and LifeStyle Companion respectively. Some of the applications required additional functionality from other applications to make them work. These where called pilot core applications and will not be described further. An example of one of the applications is a file browser to open and store data in your PN.

All of the applications were programmed to be discovered and launched from the service discovery GUI in the tools menu. The applications were all implemented as client and server components communicating via the MSMP for service discovery and session control. The services are invoked by direct service calls from the service client. The pilot services all communicated using a PN federation with either another user or a service provider in a so-called Service Provider Network (SPN) (see Section 2.3.1) which is the non-personal version of a PN used in a company, exhibition hall or so. Technically the solutions are identical and the functionality is basically the same, however not personal.

In the following sections the different applications implemented and supporting the Lifestyle companion and the Icebreaker is described in more details.

2.4.2 Lifestyle Companion

The LifeStyle Companion pilot service is basically an exercise-guiding system for use in a workout centre. It needs the user to have a predefined exercise program made on his mobile device.

Upon entering the gym, the user's device forms a PN-F with the gym, which holds his/her exercise program. It then guides the user through the program by telling, which equipment is needed, which exercises are to be carried out, and registering the user's performance for later evaluation. To enable the functionality and provide a real-life demonstration as stated earlier the system has been split into corresponding service applications. These are the following:

- Check-in
- Exercise guiding
- Weight measuring

The service offers a "personal trainer" functionality by which the service acts as a fitness trainer guiding the user through fitness programs in fitness centre keeping track of repetitions, load settings, etc. This service comprises the following core MAGNET functionalities:

- Proximity-based PN formation (enabling the user to easily interconnect an amount of MAGNET-enabled nodes into a PN).
- Location/context-aware service-discovery (providing the user with service-related information based on the current physical location of the user). The position is estimated with the help of localization retriever of SCMF, which uses T-Motes with IEEE 802.15.4 stack.

- Activation of MAGNET air interfaces using wireless transmission of low-rate data between MAGNET-enabled nodes.
- Automated proximity-based PN federation establishment.

The service applications, which together make up the LifeSyle Companion pilot service, include also core MAGNET components, which are needed by all MAGNET applications. The ones, which are specific to the LifeStyle Companion, are described in detail in the following subsections.

2.4.2.1 Check-In

This support application handles electronic check-in based on proximity to some place or event, both the device-device communication and any GUI-based user-interaction.

From the users' point of view, the Check-In service is similar in both pilot services: it simply grants them access to a place and informs them graphically. From a back-end point of view, the two systems are quite different, as there are specific hardware and interfaces for each case.

An example of the GUI for the *check-in* application is displayed in Fig. 2.19.

2.4.2.2 Exercise Guiding

This application is primarily the implementation of a GUI, which guides the user through the exercises (such as warming up, exercises - with or without machines, stretching) in his/her exercise program.

The third party service knows the user's workout program. When the user enters the gym (check-in or manually starting the application) the user's device activates a third party gym application that is provided with the workout program for the user

Fig. 2.19 Example of GUI for Check-In to a fitness centre

by the gym place. Specific data for this third party application has been stored in the users third party instance of the user profile when subscribing to the gym place.

This program is provided to the user through the gym's SPN in a PN-F with the user and the physiotherapist (represented by the fitness center). When the user is ready to work out, the application starts from the first exercise in the program, telling the user which type of equipment/machine is needed (if any). The first and last exercises to be carried out are weight measurement using a MAGNET-enabled personal scale.

2.4.2.3 Weight Measuring

The aim of this application is to measure the weight of the user, and store this data into the user's MAGNET user profile. It is the first 'exercise' in the user's training program. When the user enters the fitness centre, a list of all available gym equipment in the centre is displayed to the user. In order to perform the Weight Measuring, the user is asked to find a MAGNET-enabled scale.

The exercise guiding and the weight measuring pilot support applications where adapted to work with the MAGNET air interfaces using LDR as one of many possible communication technologies. As the MAGNET enabled fitness device a bicycle compliant with the CSAFE protocol[4] was chosen and adapted to work with the applications. For the scale, a version using serial communication and a proprietary protocol was chosen. This was adapted to communicate with a MAGNET service server, which made the devices available in the LifeStyle companion GUIs.

2.4.3 Icebreaker

The idea behind the Icebreaker in general was to bring a common title for different applications created to do automatic matchmaking and interaction between different PN users with MAGNET technologies. However another application demonstrating other technical aspects of MAGNET was also put under this title even though it was specific for giving digital presentations like PowerPoint. To explain what the demonstrator was all about a story about a journalist was invented. This journalist has signed up to an event in advance and upon arrival to the event, the mobile device works as an access card. When meeting potential new business contacts at the event, he/she can exchange digital business cards with these. The information on the business is policy enforced. The journalist can also subscribe to an additional matching service, where the journalist sets up some criteria based on public information on the user profile. The service will then notify whenever there are some interesting people nearby, who match the user's criteria. In the story the journalist

[4] Available from Internet: http://www.fitlinxx.com/csafe/ [cited 8. December 2008; 15:30]

needs to make a presentation in a showroom, where the MAGNET enabled terminal discovers MAGNET enabled presentation equipment and this is used to show a presentation directly from the journalist's terminal or a presentation from the journalist's PN. The personal device controls the slides. Everyone in the audience with a MAGNET enabled device can also join and store the different presentations directly in their PNs or simply view the presentation remotely. It is also possible to make electronic booking of the equipment, and to set up a list of presenters in advance.

For the implementation of the pilot service again the entire scenario was broken into small supporting applications needed in the different use cases of the pilot. These were called:

- Check-In
- Matching Service
- Community Building
- Presentation Service

2.4.3.1 Check-In

The Check-in service application is used for participating in the event. The event organiser will create a PN-F corresponding to the event. Subscription to the event means joining the event PN-F. It is only after joining the event that the users may proceed to browse virtual badges of the nearby people at the event. The virtual badges are provided by the matching service according to matching criteria given by the user. By selecting a virtual badge, the user may further engage in business cards exchange via the community building service. Let us next explain in detail how check-in is used to join the event.

Subscription to an event is expected to be made in advance. This way authentication at the entrance can be made, based on the MAGNET id of the participant, and make the user's device aware of the event, to for example receive announcements before, under and after the event. It can also make it possible for a participant to search the list of participants in advance for people, he wants to find at the event, and set up explicit search criteria for them for the Matching service.

The proximity detection can be carried out using any technology capable of detecting identity and close-range proximity, for example LDR, RFID or WLAN.

As the event organiser creates the PN-F, his/her computer starts advertising the PN-F within the wireless neighbourhood at the venue. The users notice this via GUI, see Fig. 2.20.

2.4.3.2 Matching Service

The part of the matching service on the user's device is provided as a third party application. It provides the user interface to setting up some matching criteria on the public available information about other MAGNET users, additional information

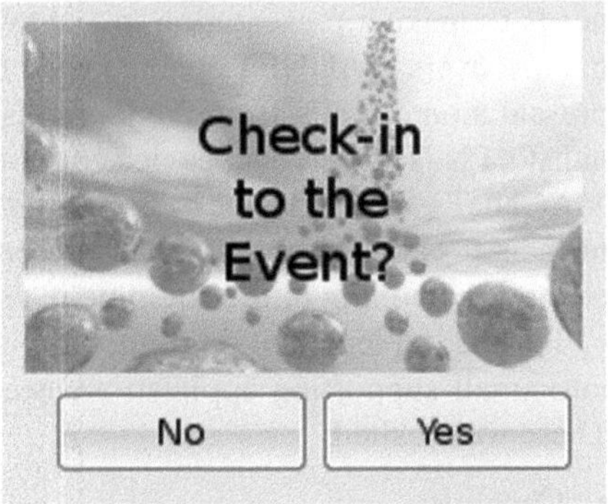

Fig. 2.20 Example GUI for Check-In Application

about the user which the specific matching service needs (third-party part of MAGNET user profile in Fig. 2.6), and some kind of notification setting (one-time or notify-on-match).

As a third-party service provider, the matching service at a given event utilizes the MAGNET user profile to match user-information against the matching criteria (such as physical distance, line of business, etc.). The matching application is then notified, and providing the user the opportunity to add the matching profile to for example a contact list, or initiates real-world contact.

2.4.3.3 Community Building (CB)

The *community building (CB)* is about management and exchange of contact information corresponding to an extended business card in digital form, or a virtual badge. The Virtual Badge consists of the user's name and picture and is provided by the matching server through the integration of the CB and the matching service, while the business-card includes fields of information such as: *Name, Job title, Company, Education, Address, Telephone number, Date of exchange of VB, Place of exchange of VB, Actual matching criteria.*

2.4.3.4 Presentation Service

The audio/video equipment in a showroom is MAGNET-enabled through a computer, which also includes the application to show presentations. This could include a combination of slide show, audio and video. It also contains the software to communicate with a user's control software, such that gaining session-wise read access to files in the user's PN, and the remote-controlling from the user's device can be established.

The software on the user's device provides the user the possibility to:

1. Book a conference room with equipment in advance (in the pilot however only available through a web browser application on the device)

2. Initiate a presentation from his PN to be shown on the showroom equipment and remote-control the presentation with a mobile device
3. Spectators can discover the presentation service and watch it remotely on their respective mobile devices

The final implementation of the Presentation Service ended up working with OpenOffice's Impress, which was more or less compatible with Microsoft PowerPoint 97–03 presentations.

2.5 Evaluation

Throughout the MAGNET Beyond project period a focal point has been to define the usability and user experience when MAGNET Beyond technologies come into play. Evaluations have taken place at two levels: low fidelity prototype evaluations and high fidelity prototype evaluation [21]. The low fidelity evaluation can be seen to be a part of the user requirement elicitation process while the high fidelity prototype evaluation in MAGNET has been part of the usability testing of the final pilot services. In both tests, the pilot services applications (mentioned in Section 2.4) were used as specific examples and as basis for development of a GUI structure. This section presents each of the two evaluations as well as the results of the user involvement.

2.5.1 Low Fidelity Evaluation

Central MAGNET concepts were identified as the basis for the low fidelity test. These were [22]:

- Personal Networks (PNs)
- PN Service discovery
- PN Federation
- User profile management
- PN management
- Privacy and security.

As part of the conceptual evaluation, the ActCom concept was developed and implemented as the underlying design for navigating on the MAGNET device.

An important part of the low fidelity prototyping was the identification and development of the navigation design and structure that would secure that the users could test the above-mentioned MAGNET concepts. The overall frame for the GUI design can be seen in Fig. 2.16.

As overall menu structure, "My Activities", "Me", "Devices" and "Search" were identified. "My Activities" emphasized the ActCom concept developed, and the tap would allow the users to organize different functions/activities such as get overview of configuration or status of devices, PN-F memberships, services and files and to provide easy access to these. An overview of all activities could be seen in an Activity list. "Me" represents all management entities of the user's device. Any item that is related to the managing of the user's communication, information gathering and personal choices would be included in the "Me" menu. The menu "Me" also comprehends the user profile manager, managing general privacy and sharing settings, cost/quality settings for network connections, and setting politics needed generally by the SCMF. The "devices" menu item accesses the PN manager. The devices can be ranked according to how far they are situated in relation to the current physical position of the user. Here can also be given information about the owner of the device, the present status of the devices etc. "Search" is to enable the user to search for everything like PNs, devices, services and files.

All screens for navigation and for carrying out the pilot services scenarios were made in a paper form, as shown in Fig. 2.21. For details on how the different screens were organized, see [8]. The screens were bundled and tiered to a (non working) Nokia N770/800 to give the user a conceptual feeling of they were navigating on a mobile device. Since the bundle of paper screens was rather big, little flyers were placed on the right side for the test persons and the facilitator to find different places in the screen structure.

The actual testing of the MAGNET concepts and the overall GUI menu structure took place through two different types of tests; a simulated and a situated environment. The purpose of the simulated environment setup was to include visual context as parameters while maintaining the advantages of a controlled laboratory environment. The setup was established by placing test participants along with the low-fi paper prototype in a closed environment (one half of a large tent shutting out exterior

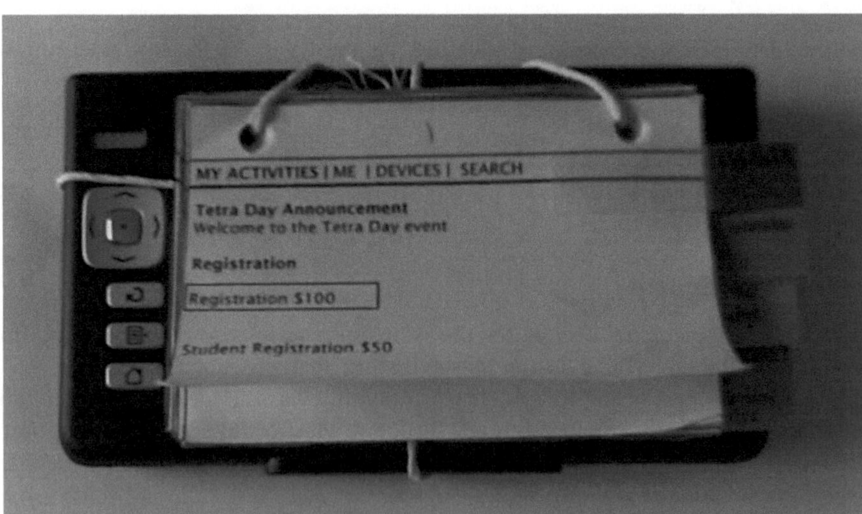

Fig. 2.21 Low-Fi prototype

light). Video, recorded in a first-person-view, was then projected onto a canvas in front of the participant establishing a sensation of being "present" in the projected environment. In the other half of the tent, a test conductor was situated controlling the setup including behavioral of the video stream (according to the participant's interaction with the prototype) as well as taking notes during the evaluation. Another facilitator was placed with the user, to change the papers according to the clicks on the buttons done during the user interaction. In the simulated test, the MAGNET concepts were discussed, as well as the "Lifestyle Companion" pilot service (focusing on the "Weight Measuring" scenario, [6]) was played out. Details on questions and the setup can be found in [8].

The situated environment setup was carried out in situ in relation to a real event; a conference with focus on a specific technology, TETRA. Also in this test, a mix between a dialogical and a scenario approach was used. Here the dialogical approach was interpreted in the way so that a facilitator would follow a participant (with the agreement of the participant) during one session of the conference and in breaks, ask the participant to carry out different tasks using the low-fi prototype. In the scenario setup, participants were asked (randomly selected) to engage in the test and to envision themselves playing out a specific scenario made for the day. The scenario would ask test persons to perform elements related to the "Icebreaker" pilot service scenario. Details on the tasks and the setup of this event can be seen in [8].

A total of 18 persons went through the tests with an even distribution of savvy and non-savvy IT users. The overall results of the test can be summarized here:

- The predominant majority of test users understood the six MAGNET concepts
- A majority of the test users consider privacy and security to be of utmost importance
- The activities concept was, in the beginning, unclear to many test participants
- The menu structure and the naming used in the menu structure were unclear to most participants. It was for example unclear both what the "My Activities" and "Me" menus would mean

More results of the test can be found in [8].

As a direct result of the low fidelity prototype the conceptual design of the GUI menu structure was redesigned. The redesign was tested in the final, high fidelity tests. Menu structure, tabs and different functionalities of the final redesign can be seen in Section 2.4.

2.5.2 Final Usability Test (High Fidelity Test)

When planning and doing user evaluations it has been beneficial to distinguish between usability and user experience, and how they are interrelated. They can in short be understood as the more objective (usability) versus subjective (user experience) measures based on users' interactions with a given product in a given context and setting. For instance a usability measure may be how long time it takes to complete a given task and a user experience measure may be whether the user finds the product exciting to use. Therefore when dealing with the final usability testing of the pilot services it must be noted that both usability and user experience goals was tested.

As already described (in Section 2.4), the pilot services applications were implemented on the Nokia N770/800 tablet, which was then used for the user evaluations.

As with the low fidelity evaluation, MAGNET core concepts were the overall aim for the evaluation. For the final evaluation, the following MAGNET core concepts were evaluated:

- Service discovery/pull-push (Service/Network Discovery)
- PN/PN-F (Personal Networks Federations)
- User profile management/Virtual Id
- PN management
- Privacy and security/Ethical issues
- Activity based communication approach
- Context awareness

Full description and how they are linked to the different pilot services scenarios can be found in [23].

Since the two MAGNET cases, Nomadic@work and MAGNET.Care, focus on different user situations, they were tested separately in different situated environments but following the same set of questions and tests.

2.5.2.1 Common Test for Icebreaker and Lifestyle Companion

The first part of the evaluations was aimed at testing the conceptual understanding of the MAGNET core concepts. Four workshops were carried out with a total number of 35 users present. All users were students (with average age of 23 years) from two universities in Denmark. The users were found by advertising for test persons and they received a small fee for their participation.

Each workshop presented a MAGNET Beyond flash movie (http://www.ist-magnet.org/pr) describing the overall pilot services scenarios, the MAGNET core concepts were then explained to the users, and finally, the users carried out an exercise to conceptually show how they understood the concepts of PN/PN-F.

After this followed a test where the users were to go through the scenarios developed for the pilot services (see Section 2.4). More specifically, the users were to (details in [23]):

- Set up a PN, managing devices (to test PN, PN, management)
- Prepare for the event (activating MAGNET and select tools)

Following this exercise, individual tests were carried out for "Icebreaker" and "Lifestyle Companion" individually.

2.5.2.2 Test with Icebreaker and Presentation Service

The users (a total of 24 test persons) were asked to envision themselves to be at a job fair at the university. Such a job fair was known to the students and there had been one just a few weeks before the test. Physically, the students were situated in a room at the university to carry out the whole test. The students were asked to carry out a number of tasks using the N800 device and navigating the devise. They were asked a set of questions and were left (in the first round) to find out how it was to be carried out in practice. Tests included:

- Log in – using the device menu/MAGNET login. Logged in, the user could see a list of predefined activities. They were asked to navigate through the activity 'At work' which gathers resources relevant for professional use.
- Business card/profile
- Matching
- Exchange card
- Look for nearby users
- Free match

Tasks with the Presentation service

- Registration
- Presentation

During tests, observations were made of the users while they carried out the activities. This was followed by several questionnaires to test their understanding of the concepts. Details on how the screens look and the scenario has already been described in Section 2.4.

2.5.2.3 Test with Lifestyle Companion

For this test, test persons (11 students) were placed in a real life environment (a fitness centre), where the MAGNET Beyond concepts were illustrated through a scenario simulating the use of the MAGNET Beyond technologies. The users were here again asked to perform some tasks which both illustrated the MAGNET Beyond concepts and the technologies supporting them.

When the users were done with the practical test in the fitness center, the users were asked to fill in a questionnaire, spilt in two parts. The first set of questions assessed their understanding of the MAGNET Beyond concepts discussed during the conceptual discussion, while the second set of questions dealt with issues the user faced during the practical evaluation.

Primary concepts involved:

- Service discovery
- User profile

- Context awareness
- Security and privacy

Again, the test persons were asked to play out a scenario. Details on these can be found in Section 2.4 or in [23].

2.5.3 Final Test Results

In general the users had a positive attitude towards the MAGNET Beyond concepts and technologies. They found the concepts innovative and demonstrated a clear interest in using the technologies in their everyday life. However, some concepts were new to most users, which required some explanations before they could understand the presented ideas. Since tests took place over just a few hours (6 h in total) the learning period for how to for example navigate on the Nokia N770/800 tables was short. However, because of the test persons' average age (23 years old on average) the users could understand and use most concepts after a little while. Additionally, the terminology and metaphors used to describe the MAGNET concepts were not all intuitive to the users and required further explanations.

Most users understood and liked the concept of Personal Network, granting them easier access and management of their devices. The PN-Federation (Groups on their device) concept was also considered as a good way to structure the connections between people and to be able to share information between them, especially with regard to security issues. However, the users related this concept to existing applications (on their mobile phone or on their laptops) and their functionalities and therefore did not fully understand some of the main characteristics of the PN-F.

An important aspect of the communication concept in MAGNET Beyond is the activity-based approach. The general meaning of "activities" was understood by the users, but only a part of users preferred this approach instead of the traditional approach for organizing the different resources separately. Nevertheless, although the test persons were challenged in identifying the difference between the two approaches, the Activity concept itself was well accepted. The difficulties experienced by some users are most probably related to the way people think of and organize their lives: some think of activities and some think of devices, applications, services and files.

During the evaluations the users expressed concerns about privacy and security with regards to sharing their profile and other personal information. Only few of the test persons admitted trusting third parties including service providers in keeping their data safe. The users wanted to be able to control what information is to be shared, with whom and how they interact with service providers. However, it must be noted that the users felt more comfortable with sharing personal information when experiencing a real-life service they can benefit from. For instance, the fact that personal information relevant to the used service is accessible when the application starts, pleases most users, as long as they can decide which service (and therefore which service provider) can access such information. On the contrary, when facing an unsolicited (yet relevant) offer from a third party, some users strongly reacted against the service's intrusive behaviour. This reaction emphasizes the users' need for controlling the way services interact with service providers.

Finally, the pilot services evaluation gathered the users' opinion about the examples of MAGNET Beyond services, which demonstrated the MAGNET Beyond concepts in practice. The users liked the social aspects of the Icebreaker service and the possibilities to control presentation from the Nokia N800 device, even though they criticized some of the features and the GUIs. Additionally, the users reacted positively to the Lifestyle Companion pilot services (Exercise Guiding including Weight Measuring). They referred to them as helpful and easy to use.

In general the MAGNET Beyond concepts and the pilot services were accepted by the test persons. However, due to the relatively short testing period (typically between 4–6 h), some users did not intuitively understand the concepts of PN-F and ActCom. Both concepts are profoundly differently from the interaction mechanisms, menus and functionalities on current mobile phones and pc's and should most likely be tested over a longer time period so that test persons could get used to the concepts and gain a more long term understanding.

More details on the final test can be found in [23].

2.6 PN Business Models

In order to analyse different aspects of business models regarding Personal Network solutions, a business model concept including service design, technology design, organisation design, and the finance design is used. Such a concept of business models has evolved during the past few years developed, e.g., by Osterwalder et al. [24] and Faber et al. [25].

According to [25], there are four interrelated design domains, which are shown in Fig. 2.22. Each of these will have to be looked at separately and in relation to one another in order to design the best business model for each of the companies in the value network.

Briefly, the four domains are described here:

- Service Design: Description of the service (value proposition), which this network of companies will offer to a target group of users.
- Organisation Design: Description of the network of different actors that is required to deliver the services to the end users. Also the roles played by each actor in the network.
- Technology Design: Description of the fundamental organisation the technical system and technical architecture needed to deliver the services.
- Finance Design: Description of revenue that is intended to be obtained or earned from the services - includes risks, investments and revenue division amongst the different actors.

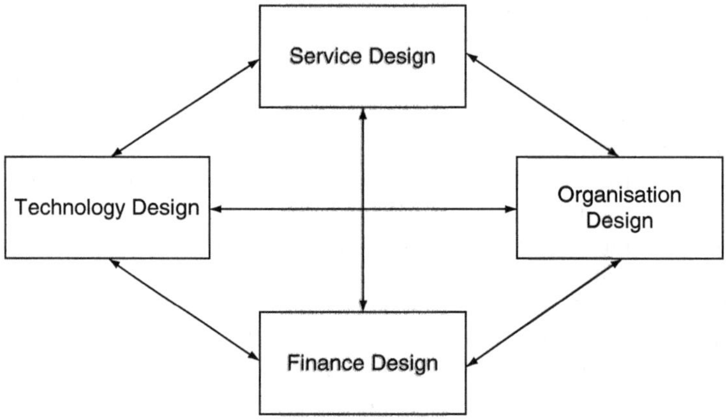

Fig. 2.22 The four inter-related design domains [25]

2.6.1 Conceptual Framework

The basic conceptual differentiation made in the section is between the use value of a product (service and/or good) and the commercial value that it may have to the supplier of PNs. The two aspects are, obviously, connected, as it will not be possible to appropriate the commercial value of a product if it does not have any use value to the user. The focus in this section is, however, only on the use value and how it is adopted by users.

Another important differentiation is made between the intrinsic and extrinsic value of a product. The intrinsic value concept denotes the 'inherent' core value offered – meaning, for instance, that the intrinsic value of a piece of software is the immediate use value that it has to a user. The extrinsic value is the 'additional' value offered – in the case of software, the value that users derive from the fact that many other users have implemented the same software and that they, therefore, easily can exchange files, etc.

However, it should be noted that it is difficult to draw a sharp line between intrinsic and extrinsic values in connection with communication services. The basic intrinsic value of communication is the communication with others, but this value increases when more users are connected to the network, as it will then be possible to contact or be contacted by more users (which is traditionally conceived as an extrinsic value). There are, however, other intrinsic values of mobile/wireless communication. The most important one is the mobility of communication, and another one is the personalisation of the terminal and, therefore, also the communication [26].

Furthermore, an additional differentiation has to be made regarding extrinsic value. In the case of many information and communication services, one can distinguish between direct and indirect network effects.[5] A direct network effect is found, for instance, in communication networks where users will benefit from additional users joining the network. Indirect network effects relate to situations where there are effects on goods or services which are complementary to the network effects of other goods or services, for example if more mobile services are offered to users because of the growth of mobile communication systems.

In the context of new mobile and wireless communication, the differentiation between direct and indirect network effects is important. The reason is that some of the services that mobile/wireless users get access to are information services – which can be either one-way or have some degree of interactivity. In relation to voice services, the network effects are direct. However, in relation to, for instance, broadcast services, the network effect are indirect – but could also be classified as virtual. The users of broadcast services do not directly benefit from other users having access to the same broadcast services. But there can be indirect effects related to the fact that there is a social value in having watched/listened to the same transmission. Furthermore, the more users, the more money producers will be able to make and use on the productions to increase the quality.

Finally, a differentiation should be made between the intended, delivered, expected, and perceived value of a product. The idea is that a producer may have an understanding of what s(he) intends to deliver. In fact, however, what is actually delivered is different from what was intended. It will not necessarily be 'less' than intended. SMS is one of the most famous examples of this. When SMS was launched, the operators had no idea that it would be a mass-market success. But it will often be 'less', for example with respect to communication speeds on the Internet.

The next step is the difference between expected and perceived value. A relevant example could be that users, when buying new communication devices or services, may have all kinds of expectations with respect to their use of the new products. In reality, however, they will only use a fraction of what is actually offered, and the perceived value of the products is smaller than the expected value. But the perceived value can also be bigger than expected. An example is related to network effects, where users will have a tendency to concentrate on the immediate intrinsic values, while the extrinsic value of communication network offerings may be undervalued.

More details on the PN business models can be found in [27].

2.6.1.1 Users

In the case of PNs, the users are in the central position when discussing service design. The reason is obviously that if services are to be personal, the specificity

[5] A differentiation is now and then also made between literal and virtual network effects, where the term literal denotes that we are dealing with physical networks, while the term virtual means that the networks are non-physical such as, for instance, languages. In the context of this chapter it is, however, sufficient to differentiate between direct and indirect network effects.

of the user is essential. In a traditional mobile network, there are few services and only little differentiation between different categories of users. The differentiation mainly consists of different price packages, which are marketed to different user groups. However, with the technological developments, it is possible to develop more services and to differentiate, to a larger extent, between different service types. Furthermore, it will be necessary with a higher degree of differentiation between different user groups. PNs constitute an extreme example of this. In the case of PNs, service packages are customized and adapted (in principle) to the individual user.

This puts high requirements of the providers of services. Where, formerly, they have been offering more or less uniform services to the great mass of customers, service providers will have to adapt to a new and very heterogeneous environment. The requirements on the service delivery systems and the charging systems will, therefore, increase.

Furthermore, there is the issue of the differentiation between the users and the buyers. The users and the buyers are not necessarily the same. This will often be the case in the health care area, where the patients will be the users of the systems, while the buyers will be the health organisations, i.e. hospitals etc. It also applies in the cases of 'nomadic' workers where the employers will pay for the PNs, while the end-users will the employees. The reason for bringing in this issue in the context of a discussion on service design is that there may be a difference in the service design needs of the users and the buyers. This could be important to take note of for the providers of PNs.

2.6.1.2 Networks and Applications

The basic intrinsic value of PNs is – apart from the intrinsic value characterising all mobile services, i.e. mobility – the real personalisation of the package of services. Personalisation is also an intrinsic value of the present day mobile communication, as the terminals are more personalised than, for instance, traditional fixed line telephony. Each person has his/her personal terminal, and the users develop a personal interface on their terminals. However, when moving to PNs, it is not only the terminals, which are personalised, but also the services provided. The whole idea of PNs – seen from the service side – is that users have access to all relevant personal information and communication. This is the fundamental intrinsic value of personal services.

When looking at the network side of PNs, an important intrinsic value could be the efficiency of communications between close-by interacting PANs. Traditional mobile networks can also transfer files from one mobile terminal to another terminal. But, depending on the size of the files transferred, the price could be prohibitive. In the case of directly interconnecting PANs, the price could be zero or negligible – and the efficiency is thus translated into a low price for communication.

The question of extrinsic value is highly important is the case of PNs. With respect to the networking side, the number of nodes in a peer-to-peer based network of PANs is of crucial importance. The more PANs the better, as this will facilitate seamless communication without having the need to use the networks of commercial mobile network operators. The mass of interconnecting PANs will constitute a network which possibly can be used free of charge. This is a case of strong direct network effects.

The question of indirect network effects is less straightforward. As in all other networks, there can be indirect network effects related to information services: The larger the number of users, the potentially lower the price and the potentially higher also the use value of the services. But there are also indirect network effects related to all the different kinds of applications that will run on personal networks. Such applications will be complementary to the basic network offerings, and there will be a possibility for indirect network effects. An example is a 'digital business card' application, where business information is transferred from one terminal directly to another. This will only function if the users have the same 'digital business card' software on their systems.

PNs make use of various methods to establish connectivity with others. Because connectivity is an intrinsic part of PNs, it is important, in general, to consider the more important forms of connectivity methods available here.

The first method is that of peer-to-peer networking. Peer-to-peer gained a lot of interest in the Internet content business where peer-to-peer overlay networks are seen as a means of increasing the distribution of content over the Internet. It has also been seen as a way of increasing the efficiency of bandwidth resources to allow more users to access data or services simultaneously. Within PNs, peer-to-peer networking presents a way for users to connect to other users 'locally', without having to initially establish a connection with a service or access provider. Peer-to-peer networks, therefore, allow PN users to bypass the operator when there is no need to use their services and to create a user-to-user connection. Communication will then take place though this peer-to-peer network. Now, the intrinsic value of this is to be able to establish connectivity without the need to use an expensive operator network. The extrinsic value lies in the fact that as more users are connected to this network, the higher the number of users may be inter-connected. This is one of the most relevant extrinsic values of peer-to-peer networking.

The formation of PNs and PN Federations also deserve special mention. These are new concepts to networking, as they do not require the user to make a new connection every time he/she wishes to talk to someone or to share a file or information with someone. After the initial setup, PNs and PN Federations will ensure that users are constantly connected to their friends and colleagues either through an infrastructure based interconnecting structure or an ad-hoc based network. The main intrinsic value of PNs and PN-Federations are that they provide an 'always on' connectivity to the 'community' of the user. That is to say: users do not have to establish new connections when they need to connect to their friends or family as they are already are a part of the same PN, allowing communication to take place at any time.

2.6.2 *Business Model Design Elements*

2.6.2.1 Service Design

The main objective of the service design is to present 'value' to the end user. The provider intends and delivers a certain value proposition while the end user expects and perceives a value proposition. One other important issue on service design is the nature of the service or innovation. This can be categorised into two types: the first is a new version service, which is an evolution of an existing service to make it better, and the second is an entirely new service, a revolutionary service that is new in all aspects.

The concept of value is very important and has been described above in the conceptual framework part. The present section examines the service design aspects of business models for Personal Networks (PNs). This means that the section deals with the attributes of the services (network offerings and applications) that users meet – the intrinsic as well as extrinsic attributes.

Intrinsic Value

The intrinsic value of the PN is the being in a network that allows the users to access information, contact a friend/colleague/family when needed, and make use of the different services and applications in the PN to make their life simpler. Trust and security are fundamental elements of the PN and this may be considered an intrinsic value of the PN.

The intrinsic value of the PN Federation is being securely connected to other users for specific purposes. PN Federations have to be set up by the users or by another management entity. Intrinsically, PN Federations give value to the user by being able to contact or get information from other members of this PN federation.

The intrinsic value of file transfer/sharing services between different PNs is the possibility to share files and folders securely with other users. Because security and trust are inherent in the PN, this is also an intrinsic value of being able to transfer files and share files between different PNs.

Extrinsic Value

The interesting thing about the PN Federation is that its extrinsic value and intrinsic value are strongly related. Due to the nature of PN Federations, the intrinsic value, which is to have connectivity to others, and the extrinsic value is that others have connectivity to you and others in the PN federation. The network effect of being connected to the same PN Federation is a direct effect of the service. Having more users in your PN federation means that you have visibility to all these users, and information sharing may take place amongst you and other users in your PN Federation. The extrinsic value of the PN is that others are connected to you within the

PN. The PN consists of different PANs that are, in essence, interconnected to one another through a secure MAGNET infrastructure. One extrinsic value of the PN is the ability to share documents and information within the PN, and to make use of devices in the PN. The PN Federation are collections of PNs that belong to different users who share similar interests and have a reason to be federated. The extrinsic value of PN Federations is, therefore, similar to that of PNs.

2.6.2.2 Technology Design

Technical resources and capabilities are the components that the technical architecture is built with. But at the same time, the technical resources of the actors in the network impose requirements on the technical architecture and it has to work with those resources. The technical architecture encompasses the delivery of service as well as the connection of different actors to work together. Different performance measures are also part of the technology design such as the type of underlying network, the types of software, hardware and applications as well as personalisation of services.

Personal Networks are available in an environment with many heterogeneous communication technologies with different bandwidths, latencies and quality of connections. The devices are mobile which means that there are continuous changes in availability of devices and other communication infrastructures. An adaptation to changes is required on all levels. Moreover, the various devices have different computational capabilities, including mobile phones, PDAs, laptops and fixed servers.

Business Issues in the Technology Domain

In order to find a detailed description of a mobile operator's business model, the three-layer description developed by the MAGNET subproject for network architecture has been used. The three-level PN architecture consists of three abstraction levels: Connectivity, Networking and the Service Level. Each level has its own business model. The total business model for future mobile operators could be described as the aggregation of the business models of each level.

Going from the bottom up, the first level is the Connectivity Level, which can roughly be mapped onto OSI layers 1 and 2. Here, the devices are organized in Radio Domains (RD).

The Network Level, consisting of OSI layers 3, 4 and 5, is placed above the Connectivity Level. The P-PAN and the PN are defined at this level. In order to reflect the provision and usage of services in the P-PAN/PN concept, a Service Level is defined above the Network Level and fills the remaining OSI layers 6 and 7. It contains all the services offered on the nodes/devices in the Network Level.

The technology design is an intricate weave of different components from the access networks to the backbone infrastructure, from the applications and devices. All are related to the technology of the final product. Services have not been included as a part of the technology but will be held as a separate component but one that would contribute to the overall technology design.

It should also be mentioned that the technological architecture of the product is one that is the result of planning and investment from the different actors in the value chain. The technological architecture - because of investments and other costs involved - will generate costs to the value chain.

Important business issues that originate from the technology domain are security, Quality of Service, system integration, accessibility and management of user profiles.

Business Evaluation of the Technology Aspect

The Business Model will be affected by the need for using PN Federations, common resource utilisation capacity, personalisation/individualisation, security, trust, privacy, context awareness, service discovery, interconnection to other networks and implementation of constraints.

An Open Architecture with well defined interfaces will open up for more players in the value network and there will be an evident need for close partnership relations and partnership management on behalf of the actors.

PANs and PNs are likely to play a big role in the mobile operator's future service offering. But the traditional operator role could be threatened by major device manufacturers and content providers who will be able to offer independent terminal-based services from networking to applications and client software in order to provide a more comprehensive suite of services and a one-stop-shop option.

There is an ongoing technological research and development work among the device manufacturers that has resulted in numerous new devices hitting the market all the time. From the production of simple mobile phone, the device manufacturer has moved to produce handheld devices that are mini computers, phones, and personal devices all at the same time. Attractive design and simplicity of use are important design criteria, but as data services gain in popularity, the number of important applications is growing.

The telecoms industry grows and there is a lot of technology driven market changes like IMS, P2P, and PN with enhanced functionality. In order to deliver complete services there must be collaboration between a large numbers of market players. Also the complexities due to mobility regarding development of applications and services will require broader spectra of competencies.

There is a richness of terminals and devices but also a lack of useful and compatible applications, services, and content based on common standards. For one single player, it is not possible to create an end-to-end service between the demand and the supply side.

Partnerships and partnership management issues will grow in importance. Every partner has to have a profitable business model. Today there are high costs but low utilisation of the infrastructure, and big players will have greater opportunities for market differentiation.

2.6.2.3 Organisation Design

The organisation design is a description of the value network that is needed to realise a particular service offering. This network may consist of many different actors that have certain resources and capabilities, that when brought together, will create value for the customers and at the same time, realise their own strategies and goals.

In any value network, there are different degrees of resources and capabilities from different actors and they can be more or less powerful in this network. Structural partners are ones who provide the essential, non-substitutable assets. Contributing partners are those that provide services to meet the specific network requirements. Supporting partners are ones who provide substitutable, generic services to the network. Structural partners are theoretically better positioned to exert control over the network than supporting partners.

From a PN operator's point of view, examples of contributing partners would be the connecting infrastructure vendors and the mobile device manufacturers. These partners contribute to the specific network environment. Service providers and application providers would either be classified as contributing partners or as supporting partners if the role they play is a minor one.

Business issues in the organizational domain have to do with how the value network is organized and controlled and how the third parties and end-users are given access to network resources and capabilities. Network complexity will imply a need for many partnerships and some conductor will have to manage this partnership network.

Organisational Arrangements and Partnership Agreements

Personal Networks add a great deal of complexity to application and services development, which requires broader competencies and partnerships. Today provisioning of complete service solutions requires the collaboration of a large number of market players. There will probably be a richness of PN enabled devices, but a lack of useful applications, services and content.

This is the background for why players in the mobile markets are so interested in the creation of a sustainable network of partners. A sound and sustainable business model involving a network of partners requires that the model is profitable for each actor involved.

The resources of the PN operator will be further enhanced with partnership agreements. This has been a growing trend with data services where partnership agreements were made between the mobile operators and software developers, content developers and application developers for new data services and application on their mobile portal.

As the PN operator moves from being a pure network operator or facilitator to a service provider, the trend is to create partnerships with others to increase content as well as coverage (geographical). Partnership agreements and business relationships allow the PN service provider/network operator to offer bundled services such as PN with fixed, mobile and WiFi access as a package.

In the MAGNET Beyond project an extended personalization concept is presented that enables value networks of content providers, network providers, and service providers to offer personalized services to mobile users in a way that suits their individual needs at a specific place and time. Therefore, a new value network with different types of interactions between stakeholders will be needed in the new PN market. New networks will consist of many different actors that have certain resources and capabilities, that when brought together, will create value for the customers. It is important to point out that different roles may be taken on by the same actor, e.g. a Mobile Network Operator (MNO) may take on the role as a Service Provider, PN Operator, Network Operator at the same time. The possible roles of different stakeholders are in part described in Section 2.3.5.1 and in part presented below.

Identity Management Provider is a special Service Provider and will fulfill important functions as an authentication service provider and will build the bridge between different Service Providers and users. Identity Management Provider will be responsible to fulfil security requirements: privacy/anonymity: non-disclosure of personal information, identity information and anonymity support and can also act as a digital representative predicting the needs of a user, finding the relevant services, exchanging user information based upon the user's policies and making the service value-added before presenting it to the user.

Devices manufacturers are well-established stakeholders of the mobile value system and will provide hardware as well as software solutions. Devices manufacturers have access to the user because of the direct buying relationship. Therefore MAGNET Beyond products will be successful if the equipment manufactory manages to deliver product that meet the operator requirements. The key lies in delivering the performance promised at reasonable cost in a timely fashion. Standardization aspect will be a very important to reduce equipment and component costs through integration and economies of scale that in turn allow for mass production at lower cost. Devices manufacturers and content providers will be able to offer independent terminal based services from networking to applications and client software.

Future business models will increase the flexibility of roles and actors. The borders between traditional roles and administrative domains are blurring. The roles may change in the same active context implying a very flexible business model e.g. MNO may become service provider or content provider or retailer.

2.6.2.4 Finance Design

The finance design is a description of how financial arrangements between different actors in the network are made. The intention of this value network is to capture revenue or monetary value. The set of financial arrangements between the different actors includes how profit, investment, cost, risk and revenue sharing are arranged.

The tariff structure is part of this arrangement and it is worth mentioning because this is the most visible part of the finance design to the end user. Revenues come directly from the end user but there may be other forms of revenue coming from grants from the government or from advertisements. Investments and costs are related to the design choice made in the technology design. Investment sources provide capital to the network while cost sources generate costs for the network. Risks that occur within the other domains will incur financial consequences. How the network copes with these financial consequences from risks is part of the financial arrangements.

The result of the finance design is the set of financial arrangements between the actors in the value network in which the profit, investment, cost, risk and revenue sharing among the actors are arranged.

Descriptions of the various costs, the sources of revenue, as well as a description of the potential benefits of actors are very important in the financial design.

Cost Structure

The cost structure is a very important element of the finance design in the sense that it measures all the costs the firm incurs in order to create market and deliver value to its customers. In PN activities, there is an important potential for cost savings in the value creation process. The right use of new technologies in the PN environment opens up new opportunities for delivering new services and, therefore, additional value at reasonable costs.

When operators will implement new technologies to the network, they will be able to reduce CAPEX. Operators will be able to cut the costs resulting from new business processes, new organization, elimination of network elements, and reduction in network complexity. To keep CAPEX down, it will be necessary to share the network with other operators by leasing or renting capacity from other operators. In MAGNET Beyond, cost reductions could be achieved due to the sharing of common activities by different entities.

Cost Savings

PNs can provide large advantages in terms of cost savings, improved services to users, and new business opportunities. PNs will integrate different access networks (ad-hoc and infrastructure based networks) and will make it possible for mobile devices to connect to any access network or any other devices.

By deploying a heterogeneous wireless network, operators can adapt capacity to demand and thereby lower their capital and operational expenditures (CAPEX/OPEX).

It is clear that infrastructure cost savings is a strong incentive for new technology adoption. Peer-to-peer based networks can offer efficient means to implement various types of services while avoiding high investment and maintenance costs. The P2P model will provide better scalability, lower costs, more power and more efficient utilization of resources. Therefore, providers should consider and support the utilization of peer-to-peer networking, which may exploit the benefits of this emerging technology for increasing profits.

PNs enable a number of potential sources of cost savings, e.g.: Operators in PN will be able to keep operational cost down because all services will be provided on one common platform. They also may drive down costs through a gradual migration towards managed and hosted communication solutions. Such solutions represent an opportunity to manage all voice and data communications via a specialized supplier and eliminate costly premise-based equipment. Business relationships will allow the PN service provider/network operator to offer bundled services such as PN with fixed, mobile and WiFi access as a package, and minimize the cost.

Billing and Charging Structure

Charging and billing systems are complex and constitute a crucial part of telecom service providers' operations to recover financial investments in the infrastructure and generating profits for shareholders. Charging is the process where subscriber accounting information is retrieved for billing purposes, i.e. to write a bill according to a specific tariff and criteria. Billing will be a very important area for operators' business in PN.

In the new mobile network market it will be necessary to adapt and combine all of the charging and billing models into 'unified' flexible models which will cover the more diversified requirements of mobile charging and billing. The reason is that subscribers want a simple charging structure and receive only one bill. They also would like to receive micro payments included in the one bill. The more complicated the offer will be, the more consumers will not use services because they prefer certainty with respect to price schema. Users in PN will have a strong preference for simple pricing system [28].

New systems within PN will be characterised by a much more flexible and diverse charging method. Charging will be more focused on what service is provided, i.e. based on QoS requirements, security, user profile etc, and different types of bundling rates will be provided. The market will be heavily influenced by charging and the business cases will be linked to where the different players are located in the value chain. It is most likely that flat rate, prepaid and real time charging will dominate during the next few years.

For charging purpose, the data needs to be collected within a PN to enable charging and billing by third party service providers so that the cooperation and service composition can be achieved between all involved actors in provisioning PN services.

The IMS charging system might be a solution for different business models for IMS operators because it supports offline, online and flow-based charging. The main advantage for operators and end users of new charging models is the capability of charging based on session, event, volume or service.

Payment processing is no longer the exclusive domain of operators. Other parties, such as specialized billing companies, and mobile commerce platform vendors, have opportunities to get involved in this activity.

Revenue

There is no doubt that the providers' revenue models in PN will be composed of different revenue streams with the different pricing models. Revenue sharing arrangements with a large number of content and service providers will be an important component of new business models. It should be expected that no one revenue model will dominate, but rather a variety of service specific business and revenue models will exist. The operating revenue side will consist of revenue from the end user market and/or income from sold services to other actors. In new PN environments, the revenue will depend on the roles, the service offerings, the charging and the market share per service. MNOs can increase revenue by taking some percentages of each transaction. MNOs can also increase ARPU (Average Revenue Per User) by providing content.

Different models regarding revenue can be used in the PN concept. One is the one bill concept, where the mobile operator gathers the revenue and distributes it between service/content providers, and second is the multi bill concept, where the content/service providers manage their subscribers themselves. The model where the end-users' network providers are delivering all the different services and applications and are controlling the contact to the end-users is one of the solutions regarding business models. The operator controls the value chain, by billing the end-user and dividing the revenue within the value chain, e.g. with service providers, application providers, content aggregators etc.

Content provider-revenues may come from subscriptions fees, usage fees, syndication agreements and airtime revenue sharing. Content and other applications would be obtained through content providers and application service providers. Revenue sharing models would be in place between content provider and application provider and the mobile operator. Alternatively, instead of working with several or many content and application providers, the mobile operator could work with a content aggregator who provides consolidation services.

Application providers will earn revenue streams from sale of license fees, installation fees, and rental agreements for hosting, operation and maintenance services, and consulting services.

New revenue generation from the provision of special services that include services over and above the traditional voice and data services that are offered today will be a new avenue, which the mobile operator can assess. Revenue from the special services such as security services and multi access will be part of the PN service offering by mobile operators.

Pricing

The new pricing mechanisms should be used in order to maximise revenues of companies. Particularly the Internet and wireless technologies have an important impact on pricing and have created a whole new range of pricing possibilities.

The services need to be packaged differently for corporate users and consumers even though the basic applications are the same for both segments. Users may be attracted by multi-terminal and multi-subscription packages and pricing will play a key role. In new convergent environments, it has also become easier to compare prices not only by the users but also by content and service providers, which will probably conduct network operators to lower pricing. Creating links between service management and billing systems to ensure adequate pricing is an essential part of new business. The new pricing mechanisms enabled should be used in order to maximize revenues. For example, context providers could sell their content in several different ways. They could collect subscription fees from their private customers and demand fixed prices for content (articles, films, and sound) from their business customers.

More details on the business model concept can be found in [29].

2.7 Conclusions

The vision of MAGNET has been ambitious in the sense that the project aimed at specifying and demonstrating a future personal networks architecture, which will support most users in their communication needs in the future. This ambition covers technical challenges, which are far beyond 3G. On the user side, a main challenge is, that users seldom know exactly what they want and need in relation to technology, and when talking about the future, yet to be implemented technology it is even harder to imagine the needs and possibilities.

As illustrated above MAGNET has achieved to demonstrate technical solutions based on user requirements, but with build in flexibility and "safe" solutions in order to create solid results that are able to interact to shape a positive and preferred social environment and thereby presenting sustainable and innovative business cases and solutions.

User centricity is, however, only a direct key concept for part of MAGNET. MAGNET has included a large number of narrow technically areas which all make up a PN-architecture. Each of the technical areas in MAGNET is a research area in itself, and it is in many situations not at all needed to specifically address users or other persons in developing these. However, MAGNET has with its overall focus made clear, that users are important for the technical development process, and that user centricity has been a relevant and essential concept through the systems design, development and implementation.

It has further been demonstrated the user centric approach has positive economic implications. The business perspectives for companies offering PN services has been examined using a business model concept includes service design, technology design, organisation design, and finance design. It is concluded that PN activities provide important potentials for both users ad suppliers in the value creation process. The right use of new technologies in the PN environment opens up new opportunities for delivering new services and, therefore, additional value at reasonable costs.

An overall important lesson is that perhaps the biggest challenge in this multidisciplinary project has been the timing of user oriented input to the technical parts of MAGNET, as well as the common recognition to what and which data is needed.

References

1. S. Bødker, J. Greenbaum, M. Kyng, Setting the stage for design as action, in *Design at Work: Cooperative Design of Computer Systems*, ed. by J. Greenbaum, M. Kyng (Lawrence Erlbaum Associates, Hillsdate, NJ, 1991), pp. 139–154
2. M.J. Muller, PICTIVE – An exploration in participatory design. Paper presented at the Computer-Human Interaction Conference, Australia, 27 Apr to 2 May 1991
3. B. Garver, T. Dunne, E. Pacenti, Cultural probes. Interactions (1999)
4. N. Schultz, L. Sørensen, D. Saugstrup, Participatory design and creativity in development of information and communication technologies, in *Designing for Networked Communications. Strategies and Development*, ed. by S.B. Heilesen, S.S. Jensen (Idea Group Publishing, England, 2007)
5. E. Bergman, R. Haitani, Designing the PalmPilot: a conversation with Rob Haitani, in *Information Appliances* (Morgan Kaufmann, San Francisco, CA, 2000)
6. Draft user functionalities and interfaces of PN services (Low-Fi Prototyping), MAGNET Beyond Internal Report IR1.4.1 (Aug 2006), http://www.ist-magnet.org
7. Preliminary report: User centric scenarios for PNs of a valid architecture, MAGNET Deliverable D1.3.1a (Sept 2004), http://www.ist-magnet.org
8. Usability of PN services (low-fi prototyping), MAGNET Beyond Deliverable D1.4.1 (June 2007), http://www.ist-magnet.org
9. J.E. Bardam, J. Bunde-Pedersen, M. Soegaard, Support for activity based computing in a personal computing operating system, in CHI'06: *Proceedings from SIGCHI Conference on Human Factors in Computing Systems*, New York, 2006, pp. 211–220
10. The conceptual structure of user profiles, MAGNET Beyond deliverable D1.2.1 (Sept 2006), http://www.ist-magnet.org
11. Specification of user profile, identity and role management for PNs and integration to the PN platform, MAGNET Beyond Deliverable D4.3.2 (D1.2.2) (Mar 2007), http://www.ist-MAGNET.org/public + deliverables. Retrieved 15 May 2007
12. Human factors (HF); User profile management, ETSI Guide EG 202 325 v1.1.1 (2005), http://webapp.etsi.org/action/PU/20051018/eg_202325v010101p.pdf. Retrieved 15 May 2007
13. A.K. Dey, Providing architectural support for building context-aware applications PhD thesis, Georgia Institute of Technology, Atlanta, GA, Nov 2000
14. *Service requirement for the 3GPP Generic User Profile (GUP); Stage 1, (Release 6)*. 3GPP Technical Specification Document TS 22.240, Version 6.5.0 (Jan 2005); *Architecture, Stage 2, (Release 6)*, 3GPP Technical Specification Group Services and System Aspects TS23.240, Version 6.7.0 (Mar 2005); *Network, Stage 3, (Release 6)*, 3GPP Technical Specification Group Core Network and Terminals TS29.240; Version 6.1.0 (June 2005)
15. The Liberty Alliance Project, http://www.projectliberty.org/
16. J. Kögel, The Daidalos Virtual Identity Concept, Beträge zum 22. Treffen der VDE/ITG-Fachgruppe 5.2.4 Mobilität in IP-basierten Netzen, Darmstadt, 2007
17. B. Weyl, P. Brandao, A.F. Gomez Skarmeta, R. Marin Lopez, P. Mishra, H. Ziemek, C. Hauser, Protecting privacy of identities in federated operator environments, in *Proceedings of the 14th IST Mobile and Wireless Communications Summit*, Dresden, 2005
18. S. Grégoir, H. Verbandt, Alcatel's user-centric data repository and provisioning architecture. Alcatel Telecommunications Review, 4th quarter (2005), http://www.alcatel.com/com/en/appcontent/apl/T0512-User-Centric_DATA-EN_tcm172–521371635.pdf

19. The role of user profiles in PN Services and context awareness, MAGNET Beyond Deliverable D1.2.3 (June 2008), http://www.ist-magnet.org
20. The TV Anytime Forum, http://www.tv-anytime.org
21. H. Sharp, Y. Rogers, J. Preece, *Interaction Design* (Wiley, Chichester, England, 2007)
22. Usability evaluation of plans and schemes for low fidelity prototypes, MAGNET Beyond Internal Report IR1.4.2 (Dec 2006), http://www.ist-magnet.org
23. Usability testing of pilot services, MAGNET Beyond Deliverable D1.4.3 (June 2008), http://www.ist-magnet.org
24. A. Osterwalder, S.B. Lagha, Y. Pigneur, An ontology for developing e-business models, IN-FORGE. 'Ecole des HEC, 1015 Lausanne-Dorigny, Switzerland, DSIage 2002
25. E. Faber, P. Ballon, H. Bouwman, T. Haaker, O. Rietkerk, M. Steen, Designing business models for mobile ICT services. 16th Bled Electronic Commerce Conference eTransformation, Bled, Slovenia, 9–11 June 2003
26. P. Pedersen, L. Methlie, Exploring the relationship between mobile data services business models and end-user adoption, *IFIP – International Federation for Information Processing*, DOI 10.1007/b98978
27. Inclusion of models for competitive dynamics for PNs, MAGNET Beyond Deliverable D1.5.2 (Dec 2006), http://www.ist-magnet.org
28. R.R. Prasad, V.S. Kaldanis, *Interconnection and Billing Policies for Personal Networks* (Telenor Telektronikk, Jan 2007), pp. 26–33
29. A. Henten, V. Kaldanis, R. Roswall, I. Windekilde, Business models for Personal Networks. Third CICT Conference, Copenhagen, November 2007

Chapter 3
PN Networking

Ernö Kovacs, Luís Sánchez, Jorge Lanza, Jeroen Hoebeke, Marc Girod Genet, Martin Bauer, Rasmus L. Olsen, Majid Ghader, Henrik Thuvesson, and Luís Muñoz

3.1 Introduction

Despite the inaccuracy of long-term technology forecasts there seems to be a strong consensus that new technologies should be centred on the user, improving the quality of life and adapting to the individual, without the need to be aware of the technical details. The environment needs to become smarter, more responsive, and more accommodating to the needs of the people. Future technologies will provide context-aware services and will introduce new levels of personal comfort and safety. Personalisation and ubiquitous access to information and communication will be essential. Users will be able to create a personal profile that, according to the situation and moment, will allow them to access the most suitable means of communication and the most relevant information. These ideas can be found in visions for the future produced in various scenarios, such as WWRF's Book of Visions [1].

E. Kovacs (✉)
NEC Europe Ltd., Kurfürsten-Analge 36, Heidelberg 69115, Germany
e-mail: ernoe@nw.neclab.eu

L. Sánchez, J. Lanza, and L. Muñoz
Universidad de Cantabria, Spain

J. Hoebeke
Interuniversitair Micro-Elektronica Centrum vzw, Belgium

M.G. Genet
Groupe des Ecoles des Télécommunications – Institut National des Télécommunications, France

M. Bauer
NEC Europe Ltd., Germany

R.L. Olsen
Aalborg University, Denmark

M. Ghader
The University of Surrey, UK

H. Thuvesson
Telia-Sonera, Sweden

R. Prasad (ed.), *My Personal Adaptive Global NET (MAGNET)*, Signals and Communication Technology, DOI 10.1007/978-90-481-3437-3_3,

In the future, computation will be human-centred: it will enter the human world, handling our goals and needs and helping us to do more by doing less. Computation will be pervasive, like batteries, power sockets, and the oxygen in the air we breathe. Configurable generic devices, either handheld or embedded in the environment, will bring computation to us, whenever we need it and wherever we might be. As we interact with these "anonymous" devices, they will adopt our information personalities. They will respect our desires for privacy and security. Mobile users are demanding anywhere and anytime access to high-speed data real- and non-real time multimedia services from Next-Generation Wireless Systems (NGWS). New systems will boost our productivity. They will help us automate repetitive human tasks, control a wealth of physical devices in our environment, find the information we need (when we need it, without forcing our eyes to examine thousands of search-engine hits), and enable us to work together with other people through space and time.

3.1.1 Personal Networking Concept

Next Generation Wireless Systems should provide to the user access with a broad range of services in a transparent way, independently of user location by making the technology invisible and embedded in the natural surroundings. Reaching this goal requires efficient cooperation between heterogeneous networking technologies and different protocols. Wireless personal networks are an integral part of such an emerging heterogeneous infrastructure. It is highly desirable, and in fact required due to economical constraints, to incorporate the present wireless systems in building the new paradigm.

Take the concept of pervasive computing and combine it with strong user focus. The result is the idea of Personal Networks (PN) [2,3]. A PN (Fig. 3.1) is a collection of one's most private devices, referred to as personal devices/nodes, that forms a virtual network where collocated personal devices organize themselves in clusters, which are in turn interconnected via infrastructure-based networks, e.g., the Internet, an organisation's intranet, or via ad hoc networks such as another person's PN, a vehicle area network, or a home network. From a technical point of view, the PN is seen to consist of devices sharing a common trust relationship. Security and privacy are the fundamental properties of a PN, as well as its ability to self-organize, and adapt to mobility and changing network environments. PNs will support the users' professional and private activities, without being obtrusive and while safeguarding privacy and security [4].

The concept of a PN goes beyond the concept of a Personal Area Network (PAN). The latter refers to a space of small coverage (less than 10 m) around a person where ad-hoc communication occurs, e.g., using Bluetooth or IEEE 802.15.3. These are intended to interconnect portable and mobile computing devices such as PCs, Personal Digital Assistants (PDAs), peripherals, cell phones, and consumer electronics. PNs extend the local scope of PANs to a global one by addressing virtual personal

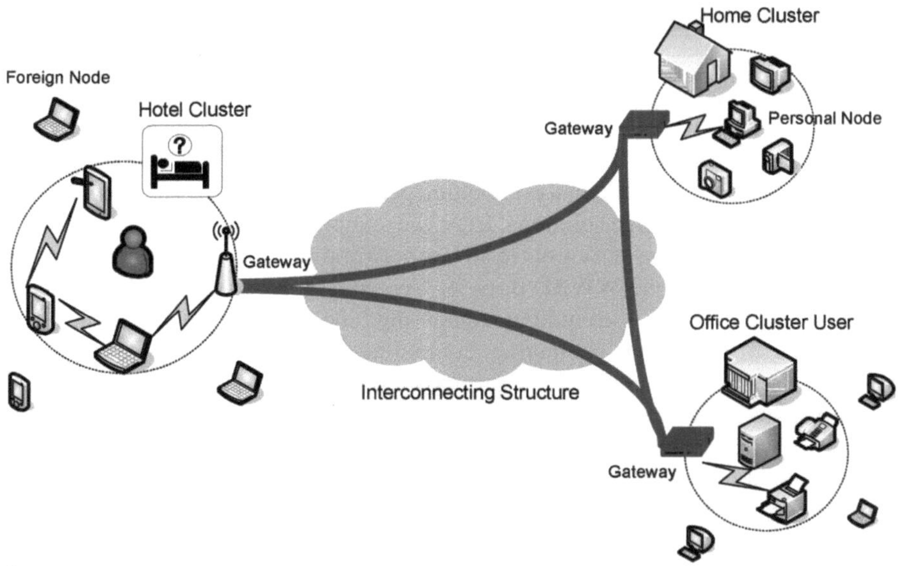

Fig. 3.1 Personal Network concept

environments that span a variety of infrastructure as well as ad-hoc networks. PNs are very much centred on a person and his/her needs. They will be dynamic in composition, configuration and connectivity depending on the time, place and circumstances, the resources required and the partners one wants to interact with. Besides the personalization and privacy requirements that are imposed on the Personal Networking paradigm, self-configuration and heterogeneity support are the main cornerstones for supporting this concept.

A PN is a person-centric network that provides the user with access to personal resources, services, and contents regardless of the location of the user. Nonetheless, personal communications cannot be restricted to the services provided by the devices the user owns. The possibility to interact with other user's PN has to be enabled in order to support the user in his/hers private and professional activities. It is beneficial to share personal resources, services, and content with others to achieve a common objective that would not be possible by a single PN. For instance, to get access to infrastructure networking facilities or to provide access to specific information, such as documents, pictures, movies, real-time images, and sensor information, PNs can federate into a group-oriented network. The PN federation (PN-F) is defined as a temporal, ad hoc, opportunity- or purpose-driven secure network of independent PNs [5]. The concept of PN Federations (PN-F) is even a more challenging one since the relations between users have to be managed and the security has to be reinforced in order to not open security holes while allowing authorized users to cooperate with you.

3.1.2 Comparison with Other Initiatives

Many technologies have been proposed in the area of personal and wireless communication, but there have been very few attempts to achieve a complete and integrated solution for all personal communication issues. In Annex A, we list some earlier and current work aimed at either analyzing future personal communication requirements or building such integrated solutions. Here, we summarize that.

Personal networks have its roots in ad hoc networking and user-focused research, such as identified by the WWRF Book of Visions [1]. When comparing personal networks to pervasive or ubiquitous computing, one major difference stands out. In the view of ubiquitous computing, computing devices are seen as commodity items that serve any user. They are meant to be shared by everybody. In personal networks, we are assumed to have our own devices, which form a network for us. This acknowledges the fact that we love to have our own devices and to personalize them by giving them their unique look and feel. Nevertheless, it must be said that much of the research that has taken place within the vision of ubiquitous computing and communication is very useful also for personal networks.

Ambient Networks (AN) [6] was an integrated project sponsored by the European Commission under the Information Society Technology (IST) priority under the Sixth Framework Programme. Its main objective was to create network solutions for mobile and wireless systems beyond 3G.

Most of the work carried out in this project concerns user devices and networks and their connections to access networks. The main idea behind this project was to make all of these networks into ambient networks (ANs). AN offers a fundamentally new vision based on the dynamic composition of these ANs to avoid adding to the growing patchwork of extensions to existing networks. Ambient Networks is more about the linkage between users' networks and infrastructure networks and between the different infrastructure networks than about the users' networks themselves. However, these links are still important for personal network communication in supporting infrastructure networking with QoS-support and reliability. Ambient Networks is therefore an important building block that can provide seamless infrastructure support to personal networks.

Power Aware Communications for Wireless Optimised Personal Area Network (PACWOMAN) [7] and Security for Heterogeneous Access in Mobile Applications and Networks (SHAMAN) were two other IST projects that started slightly ahead of IST MAGNET. PACWOMAN worked mainly on WPANs and ad hoc networking. The networking environment was divided into three distinctive spaces [8]. The first space was the Personal Area Network (PAN), where personal devices can communicate with each other. The second space was the Community Area Network (CAN), which consists of nearby PANs belonging to different people that wish to interact with each other. The last space was the Wide Area Network (WAN), which provides each of the PANs with connectivity to remote devices. IST SHAMAN [9] focused on providing a security architecture for PANs. The basis for their architecture was a trust model [10] that describes the basic security relations between different PAN devices (components in the SHAMAN terminology). Each device is owned by one user and that user determines, by means of security policies, who can use it. The security framework covers both local communication within a PAN and global access to the infrastructure. The work of both PACWOMAN and SHAMAN is highly relevant to personal networks and was partially used as a foundation for developing many of the concepts of personal networks.

Personal Distributed Environment (PDE) [11,12] has a very similar vision to personal networks and also goes further than just defining a vision. PDE is an attempt to define a concrete architecture and implement solutions that meets the vision. A PDE consists of a user's local and remote devices and services [13]. At the centre of the PDE is the so called PDE server. Each device in a PDE stays in contact with the PDE server to update its location, capabilities, and services. In this way, it is possible for a device in the PDE to use services on any other device in the PDE through the PDE server. The PDE assumes that each sub network already implements the necessary network and security solutions. These local mechanisms may differ between different sub networks and networking environments, but are kept unchanged. Instead, to make sure the PDE and its devices do not perform unauthorized tasks, a trust management system based on a trust engine that bridges the various trust and security systems in the various sub networks is proposed. PDE is, because of this, not a single homogeneous system, at least not when it comes to security and the sub networks.

The MyNet project [14] is a recently started project that is a collaboration between Nokia and MIT. They aim to study and develop a network architecture, tools and applications for simple, secure, personal overlay networks. The work is based on previous work within MIT [15, 16]. These projects stem from the peer-to-peer research community, but are still highly relevant as they focus on many overlapping areas with personal networks. Security, ease of use, and self-organization are goals also for this project. Hence, this project do not fully support ease-of-use self-organised networking, but do address security and trust as well as naming management.

The P2P Universal Computing Consortium (PUCC) [17] is a university and inter-industry cooperation of some Japanese universities and companies active in Japan, such as NEC, Toshiba, and NTT DoCoMo. The target for PUCC is to realize a seamless peer-to-peer (P2P) communications technology platform that enables the creation of ubiquitous services between networked devices. The initiative has been going on since December 2004, but until recently, very little has been published. The goal of PUCC is very similar to that of personal networks. With P2P overlays, they provide seamless communication between IP networks and non-IP networks such as home networks and sensor networks. A service platform provides seamless integration of services and other higher layer functionalities. However, the network layer is kept as is without any extra support.

There are numerous other projects that touch on the aspects of personal networks. From industry, we have, for example, Siemens' LifeWorks [18], which is a visionary concept of a unified communications experience for both business and private users. IBM defined and showcased a concept called Personal Mobile Hub (PMH) [19], which acts as a hub between a PAN and the infrastructure network. In the academic world, we have the work on personal networking by Robin Kravets' group at University of Illinois at Urbana-Champaign. Among the solutions they worked on, there is one called Mobile Grouped Device (MOPED) [20]. MOPED is a system that represents a person's set of personal devices as one entity towards the Internet using only one single Internet address. That address is given to a proxy node that is always available through the Internet. It is the task of the proxy to keep track of all the other personal devices and how they are connected to the Internet and to each other.

It is also worth mentioning that the Third Generation Partnership Project (3GPP) recently started to consider use-cases similar to personal networks in their drive towards All-IP networks (AIPN) [21]. In fact, they use the term "Personal Networks Management" for those use-cases, which involve a person with devices in different locations that are interconnected using 3GPP-networks as well as non-3GPP networks. Just recently, the Open Mobile Alliance (OMA) has also investigated the needs for a "Converged Personal Network Service". OMA will start a work item on this topic in their next meetings.

3.2 PN Architecture

In this section, an introduction to the PN architecture is given. It consists of three layers: the connectivity, the network and the service abstraction levels. After describing this architecture, we introduce the main concepts and terms that have been introduced during the development of the PN solution. The clarification of the terminology used will help on following the specification of the MAGNET solutions.

3.2.1 The Three Abstraction Levels View

As it is shown in Fig. 3.2 the architecture defined within MAGNET presents a layered view where three abstraction levels have been identified. This approach allows detaching the different requirements and challenges that need to be tackled on each of the different abstraction levels. Going from the bottom up, the first level is the *Connectivity Abstraction* level. Here the devices are organised in Radio Domains (RD). A Radio Domain is a set of Devices that have a common radio interface, a single Medium Access Control (MAC) mechanism and can communicate directly with each other. It is important to note that a node can belong to multiple RDs since it can be equipped with multiple access technologies interfaces.

The Network Abstraction level is placed above the connectivity abstraction level. The P-PAN and the PN are defined at this level. There are two types of Nodes and Devices in the network plane: Personal Nodes and Devices and Foreign Nodes and Devices. The Personal-PAN (P-PAN) is the set of Personal Nodes around the user. Further, a PN is an extension of the P-PAN as it is a collection of all "my active personal nodes", both remote and in the vicinity of the user. As in Fig. 3.2, the Personal Nodes outside of the P-PAN are grouped in Clusters such as: home

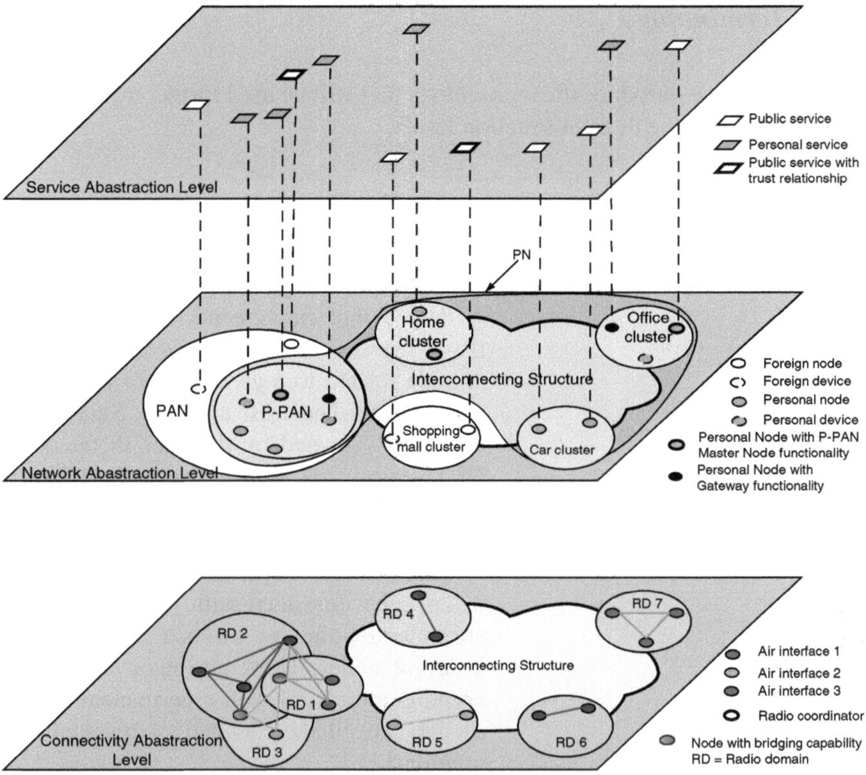

Fig. 3.2 The three abstraction levels view of a PN

Cluster, office Cluster, etc. The communication between different Clusters is done via the Interconnecting Structure (such as the Internet). The important point in this architecture is the strong focus around the long term trust concept which is used to make the distinction between Personal and Foreign Nodes and Devices. Only Nodes and Devices that are able to establish long term trust (i.e. Personal Nodes and Devices) can be part of the user's P-PAN/PN.

In order to reflect the provision and usage of services in the P-PAN/PN concept, a service abstraction level is defined above the network abstraction level. It contains all the services offered by the Nodes/Devices in the network abstraction level. Only these services are in practice visible to the user. Also less obtrusive services like name servers and service discovery protocols are part of this level. The services can be personal or public. Personal services are offered and used only by Personal Nodes in PN sense. This implies that these services can be used only if the long term Trust Relationship is established. On the other hand, the public services can be offered by Foreign Nodes to Personal or Foreign Nodes and from Personal Nodes to Foreign Nodes. The public services do not require a long-term Trust Relationship but many of them will require establishment of an ephemeral or short term Trust Relationship between the service provider and the user.

3.2.2 Terminology

In this section we introduce the terminology that will be used further throughout the chapters based on the three abstraction levels.

3.2.2.1 Common Terminology

Device	Any communicating entity
Node	A device that implements IPv6 [22] and/or IPv4 [23]
Personal Node	A node related to a given user or person with a pre-established trust attribute. Such a node is typically owned by the user in the MAGNET concept. However, any node exhibiting the trust attribute can be considered as a personal node. For instance an arbitrary node can be perceived as a personal node as long as it has been imprinted with the common trust attribute defining in essence a fully trusted group of nodes. These attributes are typically cryptographic keys with a permanent (as long as not cancelled, redefined or revoked) trust relationship
Personal Device	A device related to a given user or person with a pre-established trust attribute. These devices are typically owned by the user. However, any device exhibiting the trust attribute can be considered as a personal device. The same remarks as those for the personal nodes definition hold for devices
Private Personal Area Network	A Private Personal Area Network or P-PAN is a dynamic collection of personal nodes and devices around a person. The privacy in a P-PAN is guaranteed by mandating a mutual trust relationship between every node and device in a P-PAN. A P-PAN is often referred to as a personal bubble around a person
Personal Network	A Personal Network (PN) includes the P-PAN and a dynamic collection of remote personal nodes and devices in clusters that are connected to each other via Interconnecting Structures
Trust Relationship	Trust relationship is established when two parties communicate and determine with a measure of certainty each other's credentials to

	set up a secure communication channel using encryption mechanisms. When devices and nodes want to establish a secure communication channel, they build a trust relationship by whatever means possible
Imprinting	A procedure to bootstrap a trust relationship between two nodes that basically consists of an authenticated key exchange

3.2.2.2 Terms in Connectivity Abstraction Level

Radio Coordinator	Logic functionality responsible for medium access granting over a given radio technology
Radio Domain	A collection of nodes/devices with a common radio interface that are controlled by a single MAC mechanism (either centralised or distributed) and a single Radio Coordinator

3.2.2.3 Terms in Network Abstraction Level

Cluster	A network of personal devices and nodes located within a limited geographical area (such as a house or a car) which are connected to each other by one or more network technologies and characterised by a common trust relationship between each other. Nodes and devices in a cluster can become members of a P-PAN when a person with the P-PAN enters an area where the cluster nodes are located
Foreign Node	A node that is not personal and cannot be part of the PN. Foreign nodes can either be trusted or not trusted. Whenever trusted, they will typically have an ephemeral trust relationship with a node in a PN
Foreign Device	A device that is not a personal device and cannot be part of a PN. Foreign devices can either be trusted or not trusted
Interconnecting Structures	Public, private or shared wired, wireless or hybrid networks such as a UMTS network, the Internet, an intranet or an ad hoc network
Gateway Node	A Personal Node within a Cluster that enables connectivity to nodes and devices outside the Cluster

3.2.2.4 Terms in Service Abstraction Level

Service Management Node	The Service Management Node (SMN) is a selected P-PAN node responsible for a centralised service discovery within the P-PAN. SMN conducts also distributed (possibly peer to peer) local and remote service discovery with non P-PAN service discovery components or peers
Personal Service	Personal services are provided by personal nodes and devices and are available only to personal nodes and devices. This means that the service is accessible only after establishing a trust relation with the provider of the services
Public Service	Public services can be given by any device/node (both personal and foreign). In this case there will be services that can be accessible only after setting up an adequate authentication/ authorisation handshake (e.g., some bank service, payable printing service, etc.) or without requiring the establishment of a trust relationship with the provider of public services (public printer available for everybody)
Context	Information that characterizes a person, place or object. In that regard we talk in MAGNET about user, environment and network context. The context information is used for example to enable context-aware service discovery

3.2.3 PN Federation

While Personal Networking is focused on the communication between personal devices only, many communication patterns need to extend the boundaries of the Personal Network and involve the secure interaction of multiple people having common interests for various professional and private services. This motivates the concept of *PN Federations*. A PN Federation (PN-F) can be defined as a secure cooperation between different PNs, making selected service(s) and resource(s) available to selected receiver(s) for the purpose of achieving a common goal. In fact when devices belonging to different PNs need to communicate and/or share resources, a secure connection between involved devices will be established. Devices allow each other access to specific services as well as share resources to perform the common tasks. The main goal is to extend the PN solutions and architecture with necessary networking functionalities and group trust mechanisms to enable interactions between multiple PNs. More about the PN Federations is discussed in Section 3.5.

3.2.4 *Service and Context Management for PNs*

Service discovering is one of the most important steps for PNs to connect to secure personal services or foreign services. Because PNs are formed in the way to offer the user different services and they have to be discovered to be useful. In order to offer the user viewing, managing and accessing to all PN resources and services from anytime and anywhere, proper mechanism for service discovery, management and provision has been introduced. Users should be able to discover and use external services that are offered in their current environment. For management of the services within the MAGNET Beyond, a service management system is proposed, and is called MAGNET Service Management Platform (MSMP). The structure of the MSMP follows a centralised approach for the clusters. A Service Management Node (SMN) is elected and discovers and manages services at the P-PAN/cluster level and interacts with other SMNs at the PN level in a peer-to-peer fashion. The SMN is also responsible for discovering and advertising remote services. More about PN service management and MSMP is discussed in Section 3.5.3.

The context information and the services based on the context – which can be environmental, position, the network related – is an important aspect of PNs. Based on the context information, services are optimally offered, in the sense that context of the user matches context of the service. An example would be to offer the services nearby the user, which are also available instead of offering all potential service which may be far away or very busy. A dedicated Secure Context Management Framework (SCMF) provides the architecture and entities providing the functionality of gathering, communicating, processing, and storing relevant context information and to make it easily accessible to, e.g. the service discovery component, or other applications and services requiring context from the PN. More about the secure context management framework is discussed in Section 3.6.

3.3 Self-organization at Network Level

The solutions adopted for establishing secure communications within the PN at both connectivity and network level will be specified in this section. Starting from the automatic creation and maintenance of clusters of personal nodes and defining the approach taken to interconnect them across the available interconnecting infrastructures, this section will focus on the description of the mechanisms developed to deploy the secure overlay network.

3.3.1 *Establishing a Secure PN*

Before any specific description of the PN self-configuration mechanisms in the abstraction levels can be presented, a number of basic security notions and concepts must be introduced since privacy and security are the key features that rule the formation of the PN. The PN architecture relies on the notion of long term and short term trust relationships. The long term trust, which could also be perceived as permanent trust, is used to establish a strong security association or relationship between the nodes and devices of the PN. Long term secrets, in fact cryptographic keys, are used to form in essence the trust among the PN constituents, and especially the P-PAN/Cluster components.

These trust relationships are intended to be used between personal nodes owned by the same user. That is to say, the design is based on node ownership, which is a concept easily understood by end users. This is crucial since the end-user understanding of the trust relationship model influences the security of PNs. A lack of understanding of how this works and what consequences it has can jeopardize the security of that person's PN. Nevertheless, while the design is made with ownership in mind, there is nothing in the technical solution that will prevent a user to use the trust relationships in different way. Someone can create long-term trust relationships between nodes of a family for instance.

The long term trust keys are used as a basis to establish communications between PN nodes. The process of inserting a given secret in a device or node is referred to as imprinting a device [24]. The goal of imprinting is to exchange the pair-wise keys that will be used afterwards as the basis to derive the actual session keys used for protecting any communication between that particular pair of nodes.

Thereby, when introducing a new device to the PN, this device will be paired with at least one other device participating in PN and thus trusted by the other personal nodes. During this procedure the new device will securely exchange a long term pair-wise key with a personal node. This key will be referred to as the PN master key. As a result of the pairing procedure, the peers derive a long-term shared key that is subsequently used to secure the communication between them. Each device must store this information securely in the form of a device record. A peer record contains the following information: (1) Peer identifier – a unique identifier associated to the device; (2) PN key – the shared secret derived from the pairing process.

Opposite to other descriptions of cluster or Personal Area Network [25] that limit the concept to a matter of radio coverage (e.g. 10 m range), the concept of cluster proposed in this architecture stands on an opportunistic, distributed, multihop and proactive approach based on the trust relationships established between the cluster constituents. Further, it copes with the heterogeneity support, dynamic adaptation, infrastructureless environment survival and privacy requirements imposed by the P-PAN concept. Clusters are dynamic in nature. Nodes are switched off or become available as well as roam and show up in a different cluster. Clusters can split when a person takes some of the Nodes and leaves the rest behind. Likewise, clusters can be merged when a person arrives home and her P-PAN merges with the home cluster. Potentially, there is no limit on how large a cluster can grow, both in terms of number of nodes and hops. However, typically we expect clusters to have a small number of nodes and a limited geographical span, because of the way they will be deployed. In this sense, the clusters will be as large as possible (as long as a new personal node or device is reachable through a PAN air interface, the cluster will add a new wireless hop to its structure), adding new personal nodes and devices as soon as they appear in the cluster surroundings.

In order to form the PN and realise inter-Cluster communication over a fixed infrastructure, four requirements need to be fulfilled. First of all, the clusters need to have access to the fixed infrastructure through one or multiple Gateway Nodes (GW). Secondly, once access to the fixed infrastructure is available, the clusters need to be capable of locating each other. Thirdly, once they have located each other, they should establish tunnels between them. Last but not least, once the PN has been formed, it should be able to maintain itself in view of dynamics in the network. We will now discuss how these requirements lead to a conceptual PN architecture that relies on the concept of a PN Agent.

Connectivity between remote clusters can only be realised if they can locate each other. The PN Agent concept has been introduced to assist in this localisation and in the overall PN establishment as shown in Fig. 3.3. The PN Agent could be implemented as part of the user's fixed PN Cluster (e.g. the cluster of nodes around the user's home or office). It can also be implemented as a service under the control of service or network providers.

The PN Agent keeps track of each cluster GW point of attachment. Clusters that have connectivity to the infrastructure need to register themselves with the PN Agent. Based on this information, the PN Agent can inform the other registered clusters on the location of respective PN clusters. This information is indispensable for the creation of the tunnels between the remote clusters. The purpose of the tunnels is twofold. First, they provide a secure means for inter-Cluster communication

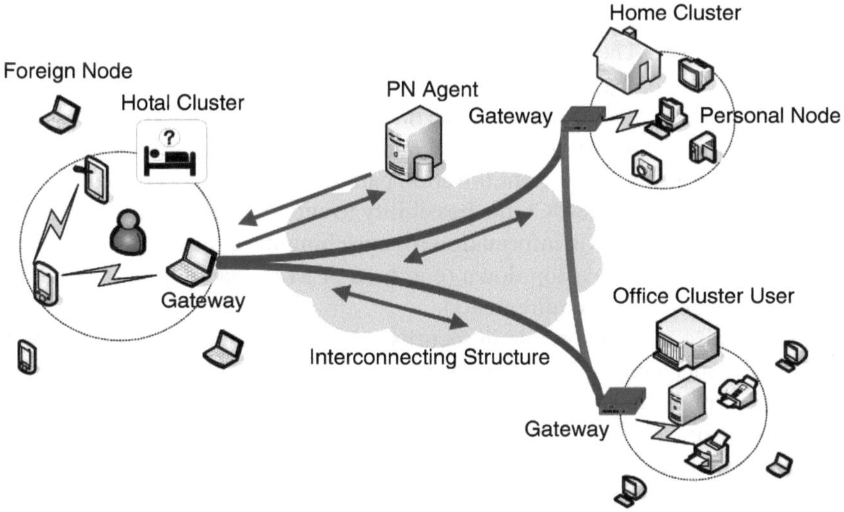

Fig. 3.3 PN architecture introducing the PN Agent

by shielding the intra-PN communication from the outside world. Secondly, these tunnels will be established and maintained dynamically, efficiently dealing with cluster mobility.

Establishing and maintaining these tunnels dynamically is based on the same concept since GW nodes keep their registration updated on the PN Agent and this one informs the others upon any change that occurs on the point of attachment of any of the registered GWs.

In addition, the PN Agent concept can be extended, meaning that it could provide additional functionalities such as naming, service discovery and foreign communication. The PN Agent offers a good entry point for PN to PN communication. The PN Agent should be considered as a concept rather than as a PN entity, since there may exist many different solutions to implement the PN Agent concept.

3.3.2 Universal Convergence Layer

The first step on achieving a self-configurable and automatically adaptable Personal Network is to solve the connectivity issues at cluster level by establishing a secure link between every pair of personal nodes. The main problems faced on this level are the heterogeneity, in terms of available wireless access technologies and personal devices capacity, and the provision of security over the unsecure wireless mean. Additionally, it must be possible to optimize the communications as well as supporting the backwards and forward compatibility.

The concept of isolating the upper-layers from underlying wireless technologies and thus providing real multi-mode can be achieved by introducing a Universal Convergence Layer. The UCL can be seen in a twofold approach. It mainly will act as an enabler for backward and forward compatibility by defining a common interface towards the network layer while managing several different wireless access technologies independently of their PHY and MAC layers. On the other hand, UCL can also enable the cross-layer optimisation paradigm. Its privileged location within the protocol stack gives the UCL the possibility to support the information flow both bottom-up (e.g. use of SNR information for enriching the decision process in an ad hoc routing algorithm) and top-down (e.g. tune of MAC parameters depending on the battery status or QoS requirements).

The UCL also plays a key role in security issues as an enabler for providing link layer security mechanisms that ensures data confidentiality and integrity, authenticity and non-repudiation.

The following sections will introduce the software architecture used for the UCL implementation as well as some concepts regarding the technological options followed to carry out the implementation work. It will also depict the different procedures and data flow of the packets through the UCL.

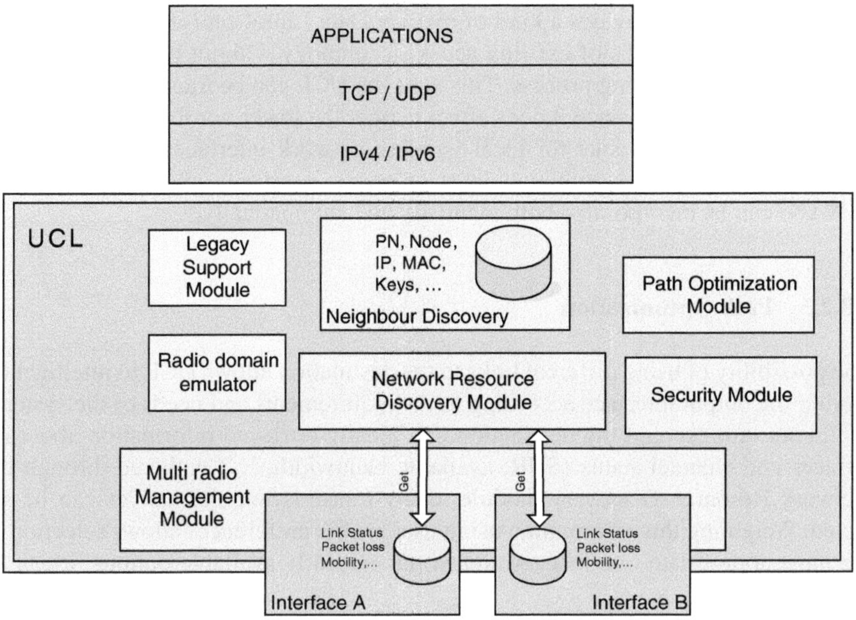

Fig. 3.4 Universal Convergence Layer high level architecture diagram

3.3.2.1 High-Level Architecture

Figure 3.4 presents the different building blocks (modules) forming the UCL. Each of these modules implements one of the basic functionalities offered by the UCL. Note that the proposed architecture aims at being highly scalable and thus it is based on a common skeleton to which different modules could be added. This modular approach allows adding and removing functionalities easily depending on the requirements and characteristics of the system it will run on.

3.3.2.2 Multi-radio Management

One of the main objectives of the UCL is to hide the complexity of the available air interfaces and to offer a unique interface to the upper layers. This module will handle this task by discovering and managing the different network resources (set them up, acquire statistics for feeding cross-layer optimization techniques, etc...).

UCL aims at masquerading multihoming by aggregating the different network interfaces (one per access technology the node is equipped with) on a single interface. By doing this, IP address of this unique interface become a valid identifier for that host thus alleviating the protocol stack from having to implement multihoming solutions on Layer 3 or 4.

Moreover, UCL provides a kind of overlay Data Link Control (DLC) layer that sets on top of the DLCs of existing access technology without having any impact on their standard working process. This way, the UCL can be transparently inserted into the protocol stack since it does affect neither the lower nor the upper layers.

On start-up, UCL looks for local wireless network interfaces and incorporate them under its control, although later on more interfaces (WPAN, WLAN or WWAN) can be incorporated both manually and automatically.

3.3.2.3 Path Optimization

The possibility of using different links to the destination allows UCL to intelligently modify the output interface according to the requirements and needs of the system.

Taking into account the destination and locally retrieved information about interfaces and channel status (SNR, available bandwidth, ...) gathered through the Network Resource Discovery module, many transmission alternatives can be selected. Weighting this information using user profile preferences allows selection of the most appropriate interface. Amongst the currently available options, it can be found:

- Traffic striping using at the same time all available transmission channels
- Use of the best link on the basis of SNR, bandwidth, packet loss, ... statistics retrieved

3.3.2.4 Neighbour Discovery

The secure cluster formation is based on long-term bilateral shared secrets which are the materialization of trust relationships shared between each pair of personal nodes. The long-term pair-wise secrets, in fact cryptographic keys (so-called K_{PN}), are used to form a strong security association between any pair of nodes that are part of the network. The neighbour discovery and authentication algorithm used on our system relies on the results of the imprinting procedure.

It is important to note some of the design assumptions that have been followed:

- Proactive approach for forming the cluster and for discovering the peers that become part of it has been selected.
- Node discovery is an issue that is resolved at connectivity level. Any node is aware of other nodes and/or devices within the same radio domain.
- The Neighbour Discovery module performs at link layer, so it is only retrieving information about the nodes at a one hop distance.

The main characteristic that rules when a node is inside or outside the PN is the long-term trust relationship established with the other nodes. In this sense, when two nodes meet and discover each other, they can leverage the shared secrets to verify their membership.

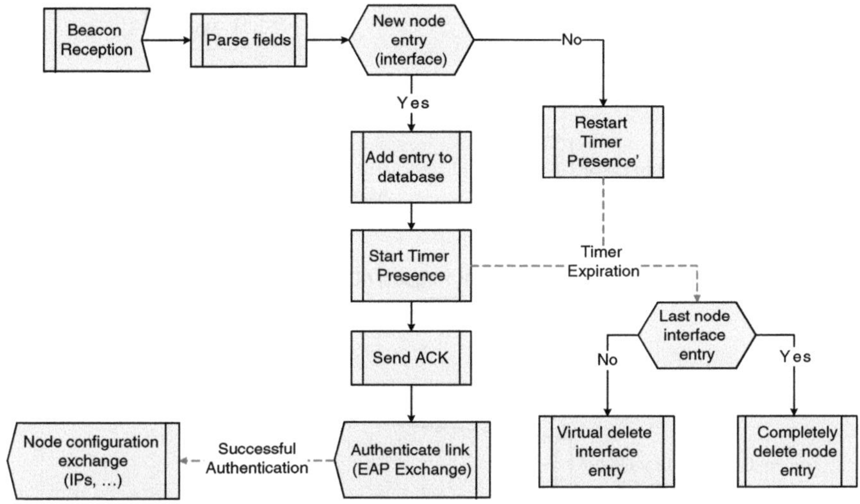

Fig. 3.5 Node discovery procedure flow diagram

To proactively discover neighbours, each node periodically broadcasts beacon messages advertising its presence. The periodicity of the beacons is to be designed depending on the dynamicity of the cluster. Context awareness techniques could be applied to set the inter-beacon time.

The proposed beacon structure is extensible in order to support future neighbour discovery features and may vary depending on the capabilities of the node. However it is mandatory that the node and PN identifiers are included since these are the indexes used for addressing the corresponding pre-established primary keys and therefore determining the trust relationship with peers.

Upon the reception of a beacon the procedure depicted on Fig. 3.5 is triggered. By parsing beacon payload fields, data such as node identifier or node name is retrieved and inserted or updated into the neighbours database. In addition to this, the MAC address and link layer interface the beacon has arrived from is collected.

For any new neighbour discovered (node plus network interface) an authentication procedure is triggered, so the peers catalogue each other. Successful authentication implies that a secure communication channel can be established between both nodes. It is then the time to securely exchange significant configuration information about the nodes, as private personal IP address.

3.3.2.5 Authentication and Security

Authentication

The first step in any communication is to establish a link layer channel. The neighbour discovery module, after detecting a new neighbour claiming to be one of these personal nodes (by the node and PN identifier included in its beacons), uses the

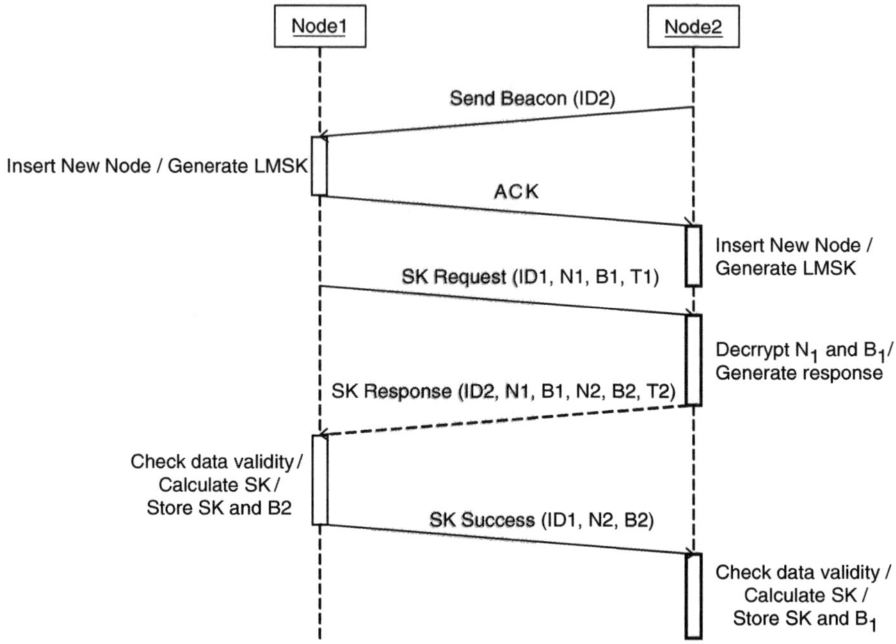

Fig. 3.6 Authentication plus Session and Broadcast keys exchange protocol

appropriate primary key to derive a session key that secure the newborn link layer channel. Obviously, the session key cannot be used for protecting the broadcast traffic because it is bilateral. Hence, each node has a broadcast key for encrypting the broadcast frames that is exchanged during the authentication process.

Figure 3.6 shows the four-way handshake used for authentication and link layer session key derivation. The following notations are used:

|	Concatenation
HMAC(key, data)	Hashing function
N_X	Nonce
B_X	Broadcast key
E(key, data)	Symmetric encryption

Symmetric encryption is done using Advanced Encryption Standard (AES) cryptographic algorithm with a key length of 256 bits.

1. Node 1 receives a beacon from Node 2
2. Node 1 sends EAP_request (E($LMSK_{1\_2}$, $N_1 \mid B_1 \mid T_1$))
3. Node 2 replies with EAP_response (E($LMSK_{1\_2}$, $N_1 \mid B_1 \mid N_2 \mid B_2 \mid T_2$))
4. Node 1 sends EAP_success (E($LMSK_{1\_2}$, $N_2 \mid B_2$)

where $LMSK_{1\_2}$ (Link Master Session Key) is calculated as HMAC_SHA_256 (K_{PN}, "$MAC_1 + MAC_2$").

Use of the MAC addresses of the candidate radios in the derivation function ensures that different pairs of hardware adaptors of a radio subsystem share different link keys even for the same pair of devices. This is particularly relevant in the presence of detachable wireless interface adaptors (USB or card based).

The SK_{1-2} (Session Key) is computed as HMAC SHA-256($LMSK_{1\_2}$, $N_1 \otimes N_2$) and is valid for T_2 seconds ($T_2 \leq T_1$). This procedure is run any time a new neighbour is discovered by a peer and whenever the derived session keys expire.

The actual authentication and session keys exchange procedure has been encapsulated using modified Extensible Authentication Protocol (EAP) where success messages are also authenticated.

Neighbour authenticity is assured if the session keys exchange is finished successfully.

Security

From a security perspective, one of the most important design goals of UCL is to make sure that use of a legacy, radio-specific security system does not cause any additional security vulnerabilities. In order to accomplish this, the UCL uses the session keys derived and exchanged in order to provide confidentiality, integrity and origin authentication through the encryption and signing of all the traffic exchanged between two neighbouring trusted nodes. Figure 3.7 shows how signature and encryption is applied over the payload of the MAC frame. This way a homogeneous security framework on top of the underlying heterogeneity is provided, avoiding using specific radio interfaces features. The communication between two collocated personal neighbours is protected through the encryption of the complete MAC frame payload (i.e. the complete IP datagram including the IP headers). The MAC header is not encrypted since the source and destination addresses would not be understood by the underlying technologies and transmission/reception would not be possible. Additionally a cryptographic signature is added to the packet in order to assure the integrity of the packet. These extra security features can only be applied in case both nodes are UCL enabled.

The communication architecture for PNs that we are considering is based on pair-wise trust relationships. Every pair of personal nodes shares a long-term trust relationship that is enforced when they communicate with each other. When two personal nodes meet they authenticate each other and exchange link level session

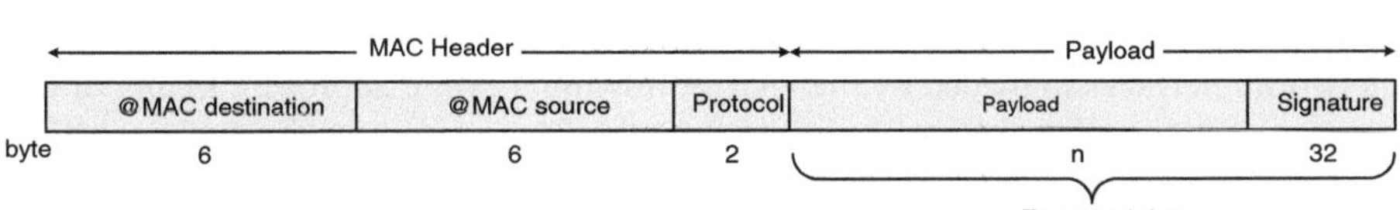

Fig. 3.7 Packet encryption format

keys, derived from the secret key they share, that are used to secure that particular link. This session keys are used to encrypt the IP datagram, using AES algorithm, and to securely sign the packet, using SHA-256. This way, only the counterpart neighbour is able to decrypt the information and verify the signature of the packet.

On multihop scenarios at cluster level, the end-to-end security is assured by securing each of the links of the communication. By definition, all the nodes in a cluster are personal, so the packet is protected by the security of each of the links that forms the end-to-end route. The counterpart is that the packet has to be encrypted and decrypted on every link of the route with the additional overhead that this implies.

3.3.2.6 UCL Data Flow

Once presented the components on the UCL architecture the flow of user data across the UCL will be presented in this section. Taking into account the information about the node's neighbourhood provided by the Neighbour Discovery module, the UCL focuses on enhancing the transmission and reception procedures by providing security and path optimization features in addition to the management of multiple interfaces.

UCL enables communication both with UCL enabled devices and legacy ones, assuring backwards compatibility and increasing the communication possibilities of a node. Hence, the UCL will not only deal with personal traffic but also with incoming and outgoing packets from/to non-personal nodes.

3.3.2.7 Downstream Data Flow: Transmission

UCL can be considered as an overlay Data Link Control layer atop all the different link layer interfaces the device has. In this sense, all the packets that are transmitted by the device go through the UCL. Figure 3.8 depicts the process followed by the packets follow on its way through the UCL. As the packet traverses the UCL, its type and destination is analyzed so that it can be redirected to the suitable network interface, adapted to support legacy operation, or protected with the appropriate security mechanism.

Packets arriving at the UCL transmission function might be of two types, signalling or data packets. When a packet arrives, it firstly has to be classified since it is going to be treated differently depending on its nature.

When a signalling packet is to be transmitted, it is firstly analyzed whether it is a broadcast or unicast packet. For broadcast ones, the packet is sent for each of the network interfaces managed by the UCL following a cyclic approach. In any of the two cases, the packet follows a similar process. There are three kind of signalling packets that are processed, Neighbour discovery (i.e. beacons, acknowledgement), Authentication packets (i.e. session key establishment and node configuration ones) and the legacy neighbour discovery ones (i.e. ARP and ICMPv6).

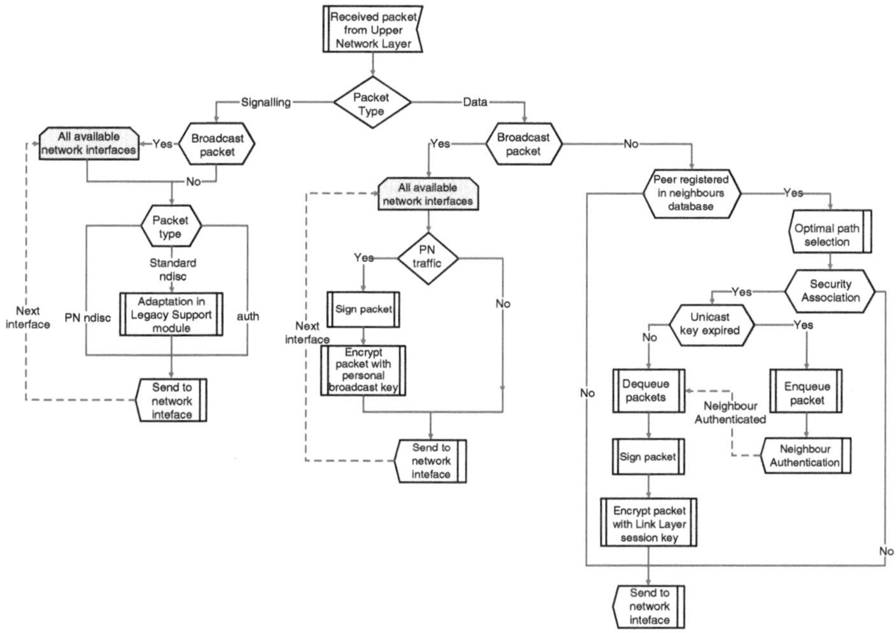

Fig. 3.8 UCL downstream data flow diagram

Data packets are also classified in broadcast and unicast. Broadcast data packets are also sent through all the node available network interfaces. Packets containing PN traffic are encrypted and signed using the node's broadcast key, so only other personal nodes will be able to decrypt and check the integrity of the packet. On the contrary, non-PN traffic packets are transmitted in clear.

For unicast packets, the destination MAC address is first checked. Packets addressed to nodes not registered on the neighbour's database, meaning that the destination node is non-UCL, are not encrypted and are sent in a legacy manner without using enhancements in packet transmission.

On the other hand, if the destination MAC address corresponds with one of the registered nodes, independently of whether it is personal or not, path optimization techniques are called. The main point is that the packet reaches its destination without suffering modification in the information it contains and following the more optimal path (quickest, lowest packet loss, less power consumer, etc.). Once the outgoing network interface has been selected on the Path Optimization module, the relationship with the peer node is checked. If destination corresponds with a personal node then a valid unicast session key (derived link layer session key from initial authentication process) bound to the output network interface is fetched and used to encrypt and sign the packet before sending it. Upon expiration of session key validity time, a new authentication process is triggered.

3.3.2.8 Upstream Data Flow: Reception

The scheme followed for dealing with incoming traffic is shown in Fig. 3.9. Traffic is classified depending on the identity of the source. Packet source MAC address is used as the index for searching in the neighbours' database.

Packets that arrive at the UCL from nodes that are not registered (mainly meaning that it is a non-UCL enabled device) are redirected to the upper layers after passing some sanity checks.

The process that follows traffic from registered nodes depends both on the ownership of the originator of the packet and on whether the packet is unicast or broadcast. Packets received from non-personal nodes are also checked in order to avoid impersonation attacks before accounting them and passing them to the higher layers. Packets coming from personal nodes are catalogued depending on

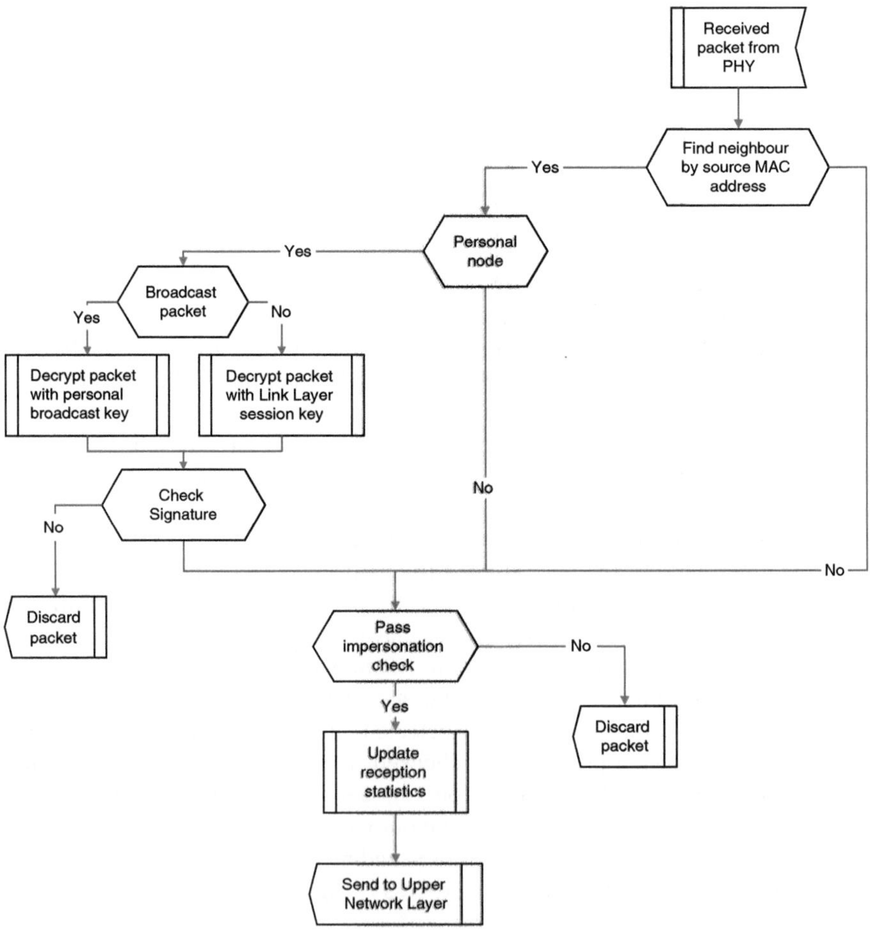

Fig. 3.9 UCL upstream data flow diagram

the dispersion. Unicast traffic is first decrypted using the corresponding link layer session key and then the integrity of the information is checked by comparing with the signature attached to the packet. Similar process is carried out for broadcast using the peer's broadcast key.

Once all security checks have been performed over the packet, the packet follows the standard path in the network stack. If any security check is not successfully passed, the packet is discarded. Impersonation check consists on the verification of the destination IP address. For non-personal nodes the only allowed destination IP address is one of the node's public addresses. This way it is assured that foreign nodes can only access to public services offered by this node and are not able to inject traffic in the personal cluster.

Before the packet leaves the UCL, link layer context information for the source node is updated.

3.3.2.9 Contribution to PN

As it has been already depicted throughout this book, a Personal Network consists on dynamic collection of personal (belonging to a user) and heterogeneous nodes and devices securely connected to each other, conforming what can be known as a user centric network.

Such kind of networks should provide the means to support heterogeneity, security and privacy as well as enable self-configuration and automatic adaptation to user context and needs. In this sense, UCL is one of the key enablers to make such a paradigm a reality.

UCL hides to the user the inherent complexity of dealing with multiple air interfaces, considering all of them as a unique one and enabling a seamless interworking between all the coexistent technologies in a transparent way to the user. Therefore the user will only have to worry about being connected, while forgetting all configuration and management of the different network interfaces. Besides that, UCL guarantees the confidentiality and integrity of the data transmitted, allowing the user to secure access and use of his/her devices without compromising any information and increasing its use experience. It also provides the necessary features to avoid access from not authenticated or non trusted nodes, acting as firewall.

UCL also adapts it capabilities based on user requirements, trying to always provide the best networking conditions and exports relevant network information so higher level applications can also adapt contents and operation to the network conditions.

3.3.3 The Network Overlay Approach

P-PAN, Cluster and PN are defined at the network level. This means that the common protocol layer used in both P-PAN (Cluster) and PN domains is the network layer. Clusters, including the P-PAN, can function independently as a network level group. The PN is an extended view intended to combine the P-PAN and the Clusters into a single Secure Personal Network. The "Secure" part of this name should be understood as Private (from the user's perspective) as well as robust and resistant to attacks from the outside. Considering that especially the P-PAN is expected to be located around the user and thus be mobile both geographically as well as logically in terms of network point of attachment, the PN should be maintained as clusters move around and change their point of attachment to the network. This, in combination with the common basic layer being the network layer, makes it clear that the mobility and security solutions for the PN should also be operating at that layer.

The basic approach taken to realize this PN concept was to implement the PN as a secure and self-organising overlay network consisting of all nodes that belong to the PN. This overlay network has its own private IP addressing space, creating a confined and private network in which personal nodes (PN nodes) can freely communicate with each other and on top of which a service discovery platform and PN applications can be deployed.

MAGNET explored the different possibilities for generating an addressing scheme (flat, hierarchical, etc.), because they were strongly related to the routing and mobility solution issues. The project adopted the PN-wide flat addressing scheme, where the user devices and nodes do not need to change the address when moving inside the PN. For the purpose of multicasting and broadcasting in the clusters and PN wide, special address formats are designed. Besides, an address configuration protocol with duplicate address detection allows PN nodes to automatically generate a unique PN IP address from the private IP addressing space assigned to the PN.

In order to establish IP connectivity within this overlay, routing functionality is needed. This functionality is provided by the PN Routing Protocol Module. The PN Routing Protocol Module provides a routing protocol that is capable of establishing paths between any two nodes in the PN. The protocol operates in a hierarchical fashion, thereby separating intra-cluster and inter-cluster routing. The routing protocol itself only deals with the establishment of paths used for forwarding unicast traffic. Next to this, the PN routing module also provides support for 1-hop, cluster-wide and PN-wide broadcasting through blind flooding and provides mechanisms for gateway selection.

3.3.3.1 Intra-cluster Routing Protocol

The intra-cluster routing protocol is a proactive ad hoc distance vector routing protocol. This protocol is a modified version of the Wireless Routing Protocol that has been adapted to meet the requirements of the PN environment. When the Neighbour Discovery module (see Sections 3.3.2.4 and 3.3.2.5) detects a new link or link break, the routing protocol is informed. Next, this new link or link break detection will trigger the exchange of intra-cluster routing protocol messages. These messages are one-hop broadcast messages that contain distance vector information and that are encrypted and forwarded by the UCL. Whenever such a routing update is broadcasted by a node, the sending node will request an acknowledgment from all neighbours for which the update message is intended, making the exchange of routing updates reliable. Upon reception of intra-cluster routing protocol updates, nodes will update their intra-cluster forwarding table. As a result, every node within a PN cluster will have an up-to-date path to every other node in the cluster.

Whenever a PN node wants to communicate with another PN node within the same cluster, its PN traffic will be sent to this intra-cluster forwarding table. This table will determine the next hop on the path to the destination node, after which the packet is handed over to the UCL for further encryption and forwarding. If an incoming packet is destined for the node itself, the packet is sent to that node where it will be processed by the higher layers.

3.3.3.2 PN Formation and Maintenance

Once the nodes have arranged themselves into clusters, for the PN to be formed, the different clusters have to establish tunnels between them. This phase completes the PN organization and maintenance and the main features that have to be assured are the network self-organization and self-healing (transparently to the intrinsic user dynamics) plus the assurance of user's communication security. The components that guarantee the aforementioned features to the PN formation and maintenance are described in following sub-sections.

In order to realize full PN connectivity, clusters at different geographical locations need to be interconnected through PN Gateway Nodes that have access to the Internet. A new PN entity called the PN Agent was designed and implemented for maintaining up to date the information of all the PN cluster attachment points. This PN Agent provides name registration/deregistration/discovery, publish subscribe and name resolution functions at PN and PN Federation level. During the PN formation process, the PN Gateway Nodes register themselves to the PN Agent (mainly in terms of attachment point to the Internet – public/private IP addresses and ports) and get, as registration response, the location information of the Cluster Gateway Nodes of all the remote PN Clusters. This remote PN Gateway information will be maintained up to date by the PN Agent through binding updates.

In addition, using the registration information provided by the PN Agent Client, the PN Gateway Node is now aware of the locations of other PN clusters and can use this information to establish tunnels to them.

3.3.3.3 PN Agent Framework

The PN Cluster information has to be maintained up to date and has to be made available to other PN Clusters and PN networking modules for setting up/establishing inter Cluster communication. Some type of agent, e.g. like a Home

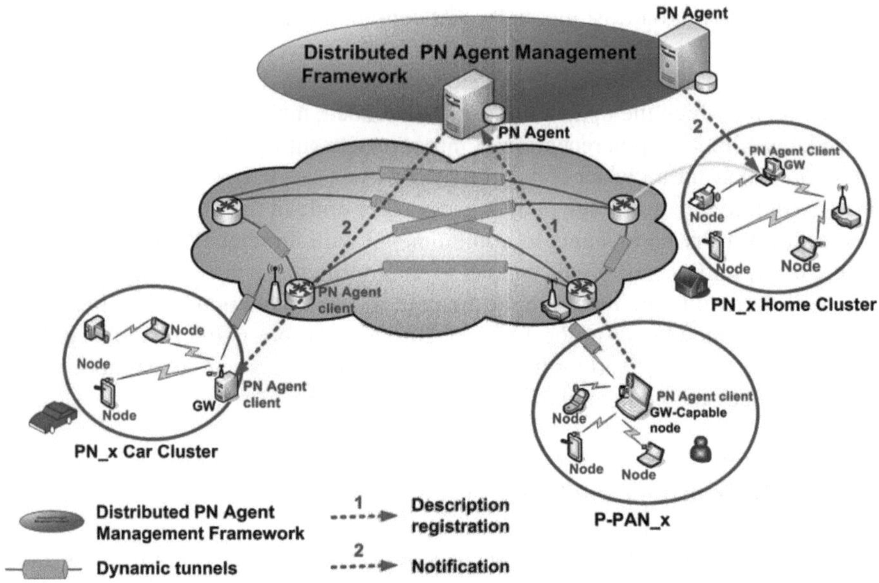

Fig. 3.10 PN Agent framework high level architecture

Agent or a PN-specific Agent, is usually required for that purpose. This has driven the introduction and the design of the PN Agent. The main role of this PN-Specific agent is to coordinate the Clusters and keep their location information up to date, including all their attachment points and IP addresses, in some kind of database. The proposed PN-dedicated solution is called PN Agent framework. Figure 3.10 introduces the main building blocks of the PN Agent framework which are mainly the following: a PN Agent Server (called PN Agent) and a PN Agent client. This figure also shows that the PN Agent can be either a centralized or a distributed functionality, including operation in P2P.

PN Agent

The PN Agent acts as distributed database (server) where all information related to the cluster locations, i.e. a short cluster description, is stored for a specific PN. So it implements a description repository and provides functionalities for publishing, removing or retrieving a description. This repository has also to be design for being distributed among Clusters in order to handle cluster mobility and the ad hoc case. Additional functionalities for resource publishing/notification/discovery are also implemented within the PN Agent in order to provide PN Cluster mobility support. This extra functionality will also allow, if necessary, the maintenance of the descriptions and the location of any PN fundamental components like, e.g. the

Service Management Node (SMN, see Section 3.5.3) or the Context Management Node (CMN, see Section 3.6).

Each PN needs to have at least one PN Agent available at anytime. Furthermore, the PN Agents need to have a well known identity, e.g. a fixed name, a public IP address or any identity that could be used by any PN component to interact to it.

Some of the PN Agent functionalities can be integrated in the naming system. This is one of the approaches used within the IST MAGNET Beyond research project.

The name resolvers in the system play one of the key roles of PN Agent by maintaining a name to address mapping database for PN networking purposes.

PN Agent Client

The PN Agent client provides all the functionalities that are necessary for interacting with the PN Agent. This includes the registration/deregistration of Cluster Gateway as well as the PN Agent notification/event handler. A PN Agent Client module will be implemented in any fundamental PN nodes that need to have their location information maintained up to date, e.g. Gateway-capable nodes, service Gateway/management nodes or context management nodes. A function in the gateway nodes that registers clusters when they have connectivity to the Interconnecting Structure and deregisters these clusters when they decide to disconnect from the Interconnecting Structure. A PN Agent Client module has then to be installed in any PN nodes that are Gateway-capable. When the PN clusters rely on a trusted interconnecting infrastructure via edge nodes (see Section 3.3.3.3)

Cluster Gateway Registration to the PN Agent

All the PN nodes that are Gateway-capable are provided with PN Agent Client functionalities and activate a PN Agent Client module for registering/deregistering their descriptions within the PN Agent framework. The Cluster Gateway description mainly contains the node name (serving as Gateway name), and all the node attachment points in terms of IP addresses. Therefore, using its PN Agent Client module, the Gateway-capable node of a Cluster can register its description to PN Agent and this is done through the sending of a description registration query to the PN Agent. In case the activated PN networking strategy is based on end to end tunnels between PN Gateways, the gateway-enable node directly registers its information to the PN Agent. When an Edge Node is used or involved in the PN networking, the registration/deregistration goes through the edge node to obtain the attachment point of the gateway to the interconnecting infrastructure. This obviously implies that the Edge Node registers its description to the PN Agent. The edge node IP address is in that case added to the description of the Gateway-enable node that is registered within the PN Agent framework. This is depicted in Fig. 3.11.

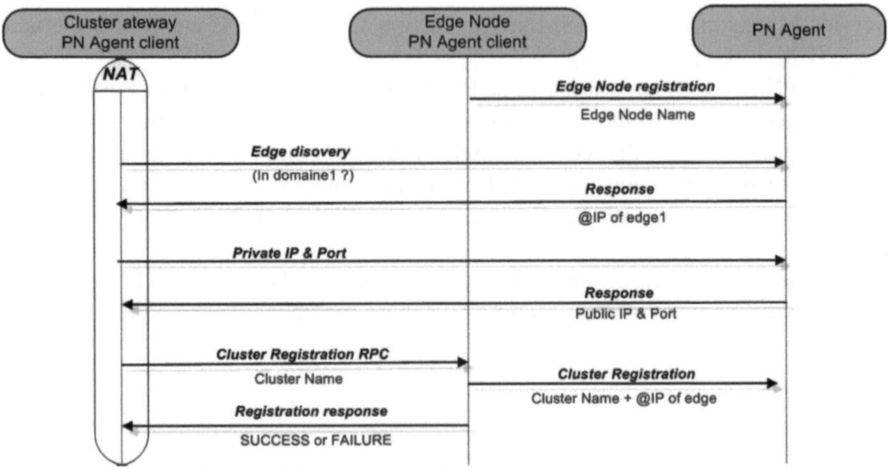

Fig. 3.11 Cluster registration procedure when an edge node is involved

An application layer NAT (ALLNAT) functionality has also to be implemented within the PN Agent client for handling the case where the Gateway-capable node is behind a NAT. Therefore, during the Gateway node registration procedure and through this ALLNAT module, the PN Agent client automatically detects the NAT presence and updates the Gateway-capable node description accordingly before sending any registration message to the PN Agent. This is also depicted in Fig. 3.11. A PN Agent receiving a cluster registration query first parses the query for retrieving the description of the new Cluster Gateway. It then registers this description into its repository and sends notification messages to the PN Agent clients of all the already registered PN Clusters Gateways. The PN Agent notification message mainly contains the attachment points of the new Gateway.

PN Agent as Cluster Mobility Support

Obviously, a Cluster on the move has to update its description registered within the PN Agent anytime its network environment is changing (new domain, new IP addresses, new attachment points ...). This can be handled within the PN Agent framework through the design of binding update mechanisms and dedicated messages allowing all the Cluster Gateways to update their new information into the PN Agent, with a workflow similar to the one already presented in Fig. 3.11 for the Cluster registration process.

If the cluster is disconnected without deregistration, the PN Agent deletes automatically the registration information since it does not receive more keep-alive messages from the cluster. As soon as the Cluster recovers its connectivity via one of its Gateways, this later will send an update message containing its new profile/description to the PN Agent via its embedded PN Agent Client. Upon the reception of a binding update mechanism, the PN Agent:

- Updates its repository with the new description
- Sends a notification message to all the others PN Cluster Gateways (i.e. their PN Agent clients) that are registered within its distributed repository

These described steps enable the PN Agent to keep information about the gateways and their attachment points to the infrastructure up to date. Not that these steps must be complemented with mobility management mechanisms to minimize latency in maintaining PN connectivity during cluster mobility. Any mobility management paradigm can be used to achieve mobility management. The PN Agent can also play the role of a location server provided it has been empowered with mobility management capability or a location server.

Edge Node Concept

The Edge Node, also called Edge Router or Access Router, is a powerful boundary node handling routing and forwarding functionalities and mainly providing one or several LANs with the connectivity to a backbone or an infrastructure network. It often belongs to a network provider but can also be held by private premises.

A traditional Edge Node lacks of flexibility and is not appropriate for highly dynamic environments like PNs. We rather envision an open and programmable Edge Node for supporting PNs and their very dynamic clusters. Indeed, Edge Nodes must be open and programmable with separate data, control and management planes in order to achieve the flexibility required by PN services in dynamic routing and forwarding as well as service adaptation [26]. This separation allows services to be deployed independently from any routing and data technology used in the PN.

The presence of open and programmable edge routing technology is foreseen as highly interesting for PN services since it can support a PN for naming and addressing, network overlay establishment, QoS, mobility and to some extend service discovery [26]. The Edge Node can thus assist in establishing tunnels and PN overlays, at run time, in order to achieve networking within the PN. In addition, Edge Nodes can support ad hoc routing between P-PAN and PN Clusters. If network overlays are set up, ad hoc routing algorithms and frameworks can enable dynamic connectivity within the PN over the network overlay. In the infrastructure case, this scenario obviously implies that a service level agreement (SLA) is in place with the providers to establish the overlays and to allow the Ad Hoc networking of the PN constituents. If providers support this open framework, networking can involve several personal networks and users and extend networking to PN Federations.

Edge Node functionalities and features can be deployed in part or entirely in the Cluster Gateways or in Edge Nodes actually in the PN Clusters themselves [26]. The PN user could also delegate some of his PN functionalities to Edge Nodes of completely trusted parties, like, e.g. his employer or even his network provider (a specific SLA and a trust relationship has to be somehow established in that case). The need for edge routers to support PN services is even more important for private premises. The edge routers belong in this case to the private premises owner (a campus, a hospital, an enterprise, a private site) that is willing to offer PN services support from edge routers. The routers would have the capabilities to establish dynamically and at run time tunnels for all active P-PANs in the private premise coverage area. Instead of putting the burden on the P-PANs, the provider routers can act on behalf of the P-PANs gateways and nodes and manage thousands of tunnels according to dynamic changes in the P-PANs and ambient environment and conditions. The presence of active or programmable intelligent routers in private premises can simplify the deployment of PN services and certainly take much control and computational burden away from P-PAN nodes. This can reduce significantly the complexity of the P-PAN nodes and allow distribution of intelligence with the private premises edge routers.

Edge Router Management

A generic and conceptual view of the overall management architecture envisioned is shown in Fig. 3.12. Management planes (partly centralised or fully distributed) supporting PN services is composed of a naming system (an alternate to names

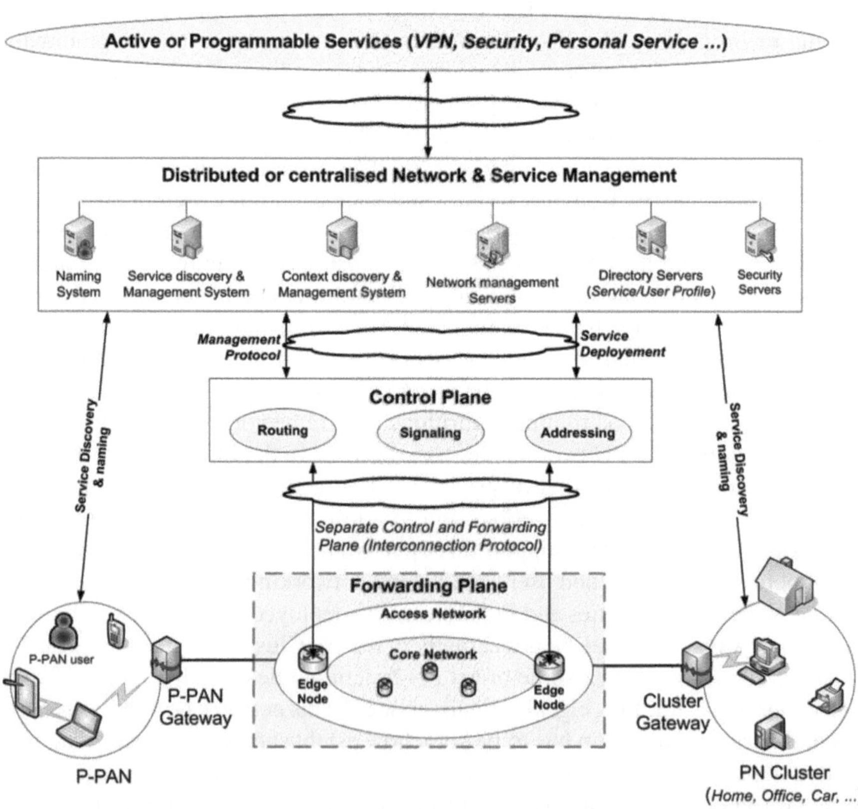

Fig. 3.12 Generic Management Plane for the support of PN services

could be identities), service and context discovery and management frameworks, distributed directories, security servers (AAA), and interact with mobility management paradigms, protocols and frameworks and network management servers. The services above the management plane can support PN requirements via open and programmable network architectures. The principle of separation allows active services to reside anywhere in the networks (P-PAN, clusters in PN and external networks) and control, locally or remotely, active or programmable routers in the infrastructure edges (if trusted somehow established) or in the private premises (inside the P-PAN or the clusters within the PN). For example, dynamic VPNs can be more easily established and become most importantly modifiable at run time. Further, such architectures provide high flexibility with respect to how services are triggered, controlled and deployed. This can happen via active packets capable of achieving coordinated discovery of edge routers with the management plane. This would assist the establishment of the dynamic VPNs for PN or PN Federations connectivity and services. The commands for the control of the edge routers can be achieved through the use of configuration rules (policies) following analysis of the packets flowing through the routers.

3.3.3.4 Dynamic Tunnelling

Using the PN Agent Client module, a node can register with the PN Agent. Upon successful registration this node will become a PN Gateway Node, meaning that this node is capable of providing connectivity to PN nodes in remote Clusters. Using the registration information provided by the PN Agent Client, the PN Gateway Node is now aware of the locations of other PN clusters and can use this information to establish tunnels to them. In case an Edge Router is used by the PN Gateway Node, the PN Gateway Node only needs to establish a tunnel to this Edge Router, since the Edge Router will take care of the establishment of all other tunnels to the remote clusters.

The Dynamic Tunnelling establishes these inter-cluster tunnels and stores all information related to these PN tunnels. It divides the establishment and maintenance in two different phases: a Tunnel Negotiation phase and a Tunnel Management and Enforcement phase. During the Tunnel Negotiation phase the tunnels are actually established. The information needed to establish these tunnels (IP addresses of the tunnel endpoints, PN prefix, the tunnel type (i.e. between which entities the tunnel is established), the tunnel maintenance type and the NAT information in case the requesting end point is behind a NAT) is provided by the PN Agent Client and passed to the module responsible for setting up new tunnels. This information is then, together with the negotiated keys, kept into a Tunnel Manager. From then onwards, the Tunnel Manager is responsible for maintaining and enforcing the tunnels. The information about the tunnels will be used to encrypt/encapsulate and decrypt/decapsulate packets sent to or coming from a tunnel using IPSec ESP in tunnel mode or IPSec over UDP in case a NAT box must be bypassed. Finally, when cluster deregistration or update is triggered explicitly, the action to remove or update tunnels is also passed from the PN Agent Client to the module responsible for managing the tunnels.

In this sense, as soon as a PN Gateway Node changes its point of attachment a new tunnel is negotiated where one of the endpoints of the tunnel changes from the previous tunnel.

3.3.3.5 Inter-cluster Routing

As already explained, when a PN node successfully registers this node becomes a PN Gateway Node, which is capable of providing connectivity to PN nodes in remote clusters. Intra-cluster routing protocol will then propagate this PN Gateway information within the cluster. As a result, all nodes in the cluster will have an overview of all available PN Gateway Nodes and this information is stored in a PN Gateway Selection table.

In order to enable IP communication between the nodes in remote Clusters, GW nodes should be able to exchange routing information over these tunnels. To this end, the intra-cluster routing protocol has been extended with an inter-cluster routing module that allows both proactive and reactive inter-cluster routing. Inter-cluster forwarding is not based on next hop information anymore, but on the unique tunnel identifiers of the dynamically established tunnels as it is shown in Fig. 3.13. The end result, after the exchange of routing information over these tunnels, is full inter-cluster connectivity within the PN IP addressing space, allowing secure communication between every pair of PN nodes.

When the intra-cluster forwarding table is not able to forward a PN packet because the destination node is in a remote PN cluster, the packet is sent to this PN Gateway Selection Table. If this node is not a PN Gateway Node, the packet is forwarded to the selected gateway (advanced gateway selection mechanisms using context information are possible). If this node is a PN Gateway Node, the packet

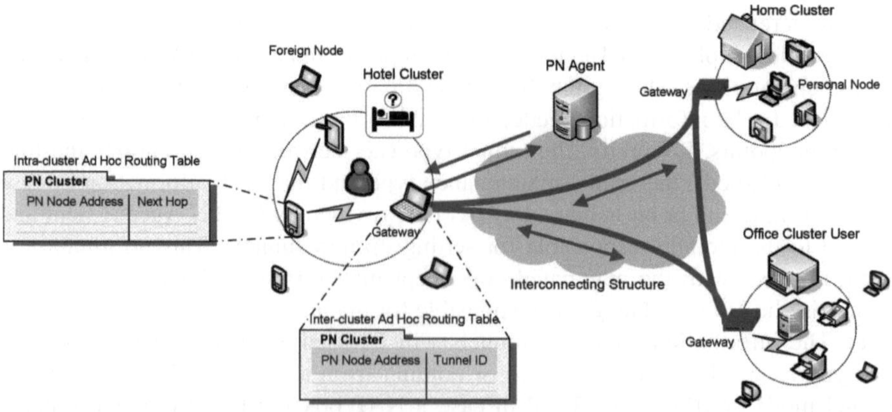

Fig. 3.13 PN Agent registration, dynamic tunnelling and PN routing

can be forwarded to the remote cluster where the destination is located using the inter-cluster forwarding table.

This inter-cluster forwarding table will have (a) all routes to nodes in remote clusters when proactive inter-cluster routing is used (b) a reactively established route to the nodes in remote clusters with which nodes in the cluster are communicating with when reactive inter-cluster routing is used (c) a default entry to an Edge Router when an Edge Router is used. Inter-cluster forwarding is based on tunnel identifiers: the tunnel identifier of the tunnel that need to be used in order to reach the remote destination is retrieved from the inter-cluster forwarding table and is then used by the Tunnelling Module to encrypt and encapsulate the packet. If no route exists, an ICMP error message is sent.

Since the intra-cluster routing protocol is proactive, every PN Gateway Node will have in its intra-cluster forwarding table an overview of all PN nodes that are in the same cluster. When proactive inter-cluster routing is used, the list of addresses in this intra-cluster forwarding table is exchanged with other PN Gateway Nodes. In case an Edge Router is used, this information is sent to the Edge Router, which will store it and further propagate it to the remote clusters. Upon reception, this information together with the identifier of the tunnel over which the information has been received is used to update the inter-cluster forwarding table, resulting in a route to all PN nodes in the PN Gateway Nodes (if they do not use an Edge Router) or Edge Routers. When reactive inter-cluster routing is used and in the PN Gateway Node a route to a remote PN node is needed, a route request will be sent to the remote clusters. In case an Edge Router is used, the packet will be forwarded immediately to the Edge Router using the default route and the Edge Router will take care of the reactive route establishment. Upon reception of this request, a PN Gateway Node or Edge Router can immediately check if the destination node is in its cluster or not and can send back a route reply, thereby establishing a bidirectional communication path in the inter-cluster forwarding tables.

3.4 PN-Aware Service Management

PN concept is introduced for allowing a user to be, as permanent as possible, able to access all his/her personal devices and resources, regardless of their cluster attachment and location. Obviously, resources also comprise services and even a physical resource can be viewed as a service, to some extent, and can then be managed in a similar way.

Therefore, a service publishing and management environment is required for PN environments [27]; however, designing a PN-oriented service architecture is not easy, since for that purpose, all the PN-specific characteristics and constraints must be taken into consideration. Some of the PN constraints impacting on the service architecture are summarized below:

- Security and privacy of the personal data and services have to be guaranteed
- Heterogeneity of networks environments, terminals, services and applications imposes to design generic modules that also address service interworking

- Cluster mobility/PN user on the move, i.e. PN service mobility, has to be taken into account
- Ubiquity; as for PN networking architecture, PN service architecture has to support/handle both the infrastructure mode and the ad hoc case, i.e. when the PN has no connectivity to any infrastructure network
- A PN is user centric, which obviously implies that the proposes service publishing and discovery mechanisms have to take user profile/preferences into account and have to be context-aware
- The proposed service management architecture has to be portable as far as possible in order to be carried out in embedded devices. A lot of those devices will, e.g. be used in the P-PAN

3.4.1 Service Life Cycle Management

From initial idea to the service termination, a service goes through several stages. This process is called service life cycle. Steps involved in the service life cycle management (Fig. 3.14) are Initial Idea Stimulation, Service Planning and Definition Initial idea stimulation, Service Development planning and definition, Service Deployment development, Service Packaging deployment, Service Monitoring packaging, Service monitoring and Maintenance and Service Evolution and Withdrawal. Main goal with service life cycle is to minimize time-to-market and integration cost. A brief description of stages mentioned above is presented below:

- Initial idea stimulation is the first step of creating new services. Based on market needs analysis, new ideas are evaluated.
- Service planning and definition is the stage where opportunities for new services are further defined. Further service creation depends on commercial feasibility.
- Service development includes implementing and testing the applications. Also specifications describing requirements of the new service, design, implementation and tests are presented at this stage.
- Service deployment is the final stage before offering the service to the customers. At this stage the service is installed on service provider environment, tested and activated. Service enablers offered by third party providers are also handled at this stage.
- Service packaging is the stage where the service is offered to the customer. Service features and the billing condition, as well as commercial packages are defined. Packaging of services offered by third party providers are also handled at this stage.
- Service monitoring and maintenance is the most important stage of service life cycle. At this stage the service has been tested and offered for use to the customers. In order to keep the service at the maintenance stage some requirements has to be fulfilled. It has to be possible to update and modify the service without

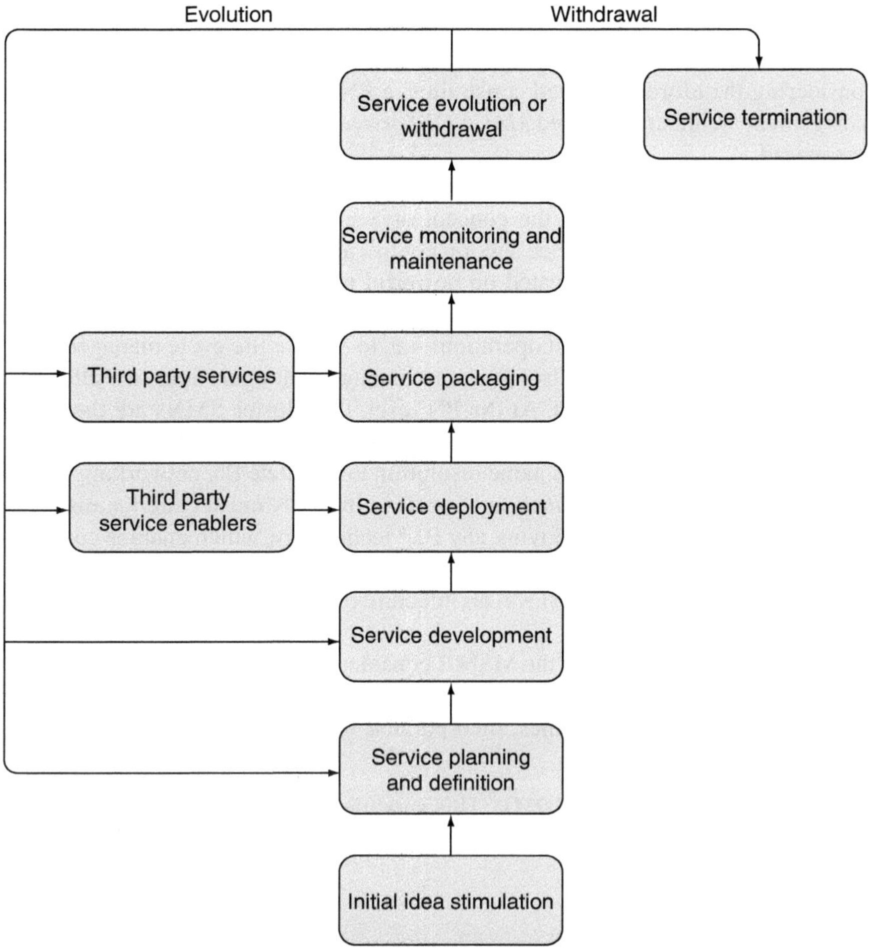

Fig. 3.14 Service life cycle management

interrupting ongoing sessions. Furthermore, in order to keep continuous evolution of the service the system should support different interfaces, components and applications. When a service is shut down it should be possible for service provider to make sure that there are no users subscribed for that service and that no other services depend on this service.

- Service evolution and withdrawal is the final stage of the service life cycle. At this point it has to be decided whether service is going to be further developed or terminated. If it is decided to completely terminate the service a proper process of dealing with subscribers, services that are dependent of terminated service etc. most be done.

3.4.2 MAGNET Service Management Platform

Considering the aforementioned constraints, a PN-oriented service publishing and management architecture, called *MAGNET Service Management Protocol (MSMP)*, is proposed.

At the Cluster level, the requirement for supporting Cluster service gateway functionality leads to introducing the concept of Service Management Node (SMN), foreseen for Cluster-wide service session control and management. The SMN functionalities are enabled or activated on powerful nodes within Clusters, capable of handling tasks and transactions related to secure and context-aware service publishing/discovery and management operations, i.e. to service life cycle management.

For the PN-wide service discovery and management operations, a P2P service overlay approach was decided. At the PN level, the Cluster SMNs are the natural candidates for participation in service overlay. A P2P overlay of SMN nodes located typically in clusters guarantees name resolution to facilitate PN networking and implements a service locating function to achieve inter-PN cluster service discovery. This overlay can be built employing any P2P technology, which enables communication between the Cluster SMNs (acting as super peers).

Figure 3.15 depicts the high level architecture of the MAGNET Service Management Protocol.

The internal architecture of the MSMP, considers the aforementioned constraints. Figure 3.16 depicts the proposed architecture. Different functionalities are supported by a variety of software modules, incorporated within an SMN. The modules are briefly explained below.

Service Discovery Module (SDM). This module acts as the core of the service discovery system. It is responsible for all discovery process operations, such

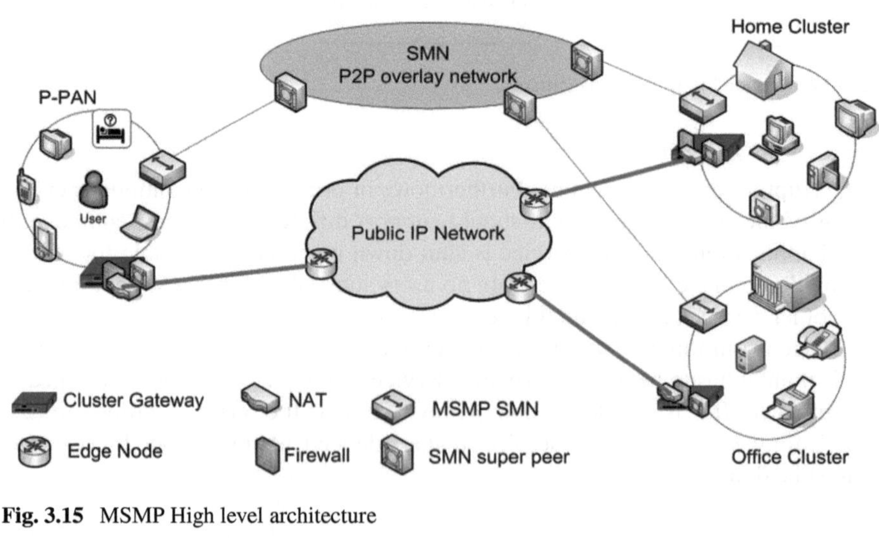

Fig. 3.15 MSMP High level architecture

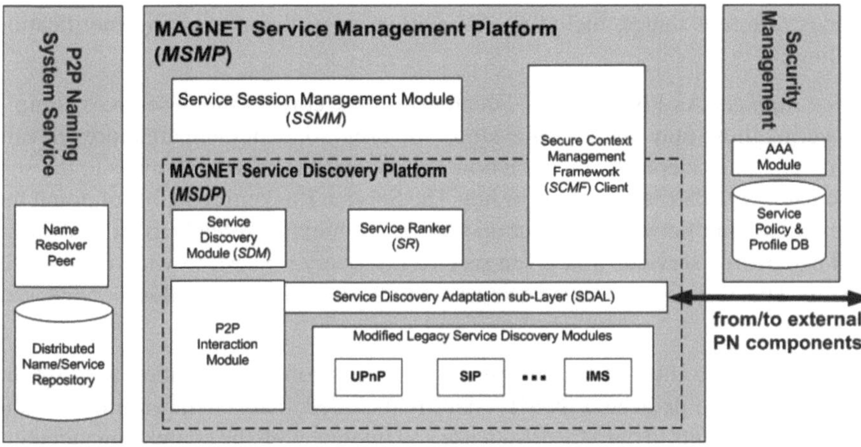

Fig. 3.16 MSMP internal architecture

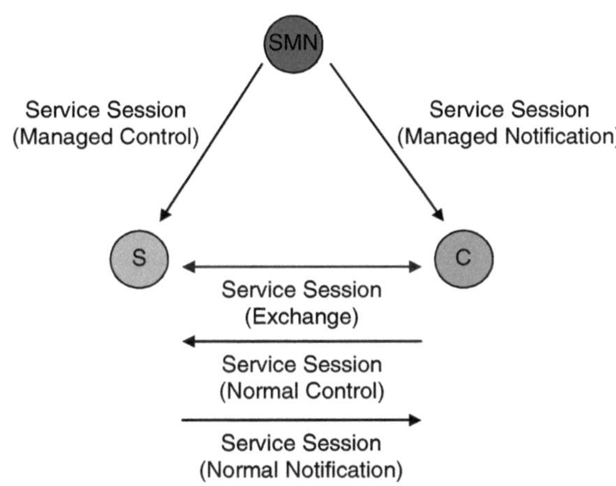

Fig. 3.17 SMN acting as an intermediary node between clients and servers

as accepting registration of the advertised services, replying the service discovery requests made by the clients, and interacting with other SMNs within the P-PAN/cluster (e.g. individual SMN of radio domains) to compile all the available services in the corresponding network.

Service Session Management Module. The existing discovery protocols do not usually provide proper tools for management of the service session. The enhancement on the existing legacy protocols is that the SMN can be employed as a broker of the service and be used as the service manager. Service sessions are established through the SMN. Control and Notification messages all are re-directed to be manageable by the SMN. Figure 3.17 presents the idea of SMN acting as an intermediary, monitoring (by spying) and controlling (by interfering) node. Service provisioning includes different stages, including description, control and eventing (notification handling).

Service Ranker. As PN's and PN-Federations may contain many services, simply discovering those may not be sufficient for the user. Some of them are more relevant than others, i.e. it depends much on how the context of the user matches to the service context whether it is relevant or not. The Service Ranker is capable of doing the context matching between the user and service context, and leads to an evaluation of all discovered services in a given service discovery request, to what degree it is relevant. The evaluation is based on a set of rules specific to individual service, see, e.g. [28, 29].

SCMF Client. To interact and access context and profile information used by the Service Ranker, a dedicated SCMF client is included. This client ensures that the internal components in the MSMP can use and interact with the context management framework.

P2P Naming System Service. The P2P Naming System Service is designed for handling the distributed service repository and the wide-area service discovery operations among Cluster SMN peers of the PN SMN service overlay already depicted in Fig. 3.15.

Modified Legacy Service Discovery Modules. This SMN sub-block, already depicted in Fig. 3.15, includes all the SMN lower layer modules that are designed for interacting with legacy service discovery framework and external service frameworks, such as, e.g. UPnP, Bluetooth SDP, SIP-based services and IMS. This provides the Cluster SMN with service interworking functionalities.

Security Management. The Security Management is designed for handling all the secure service discovery and management operations. It mainly provides service clients (SMN clients) for authentication and service access control.

Service Discovery Adaptation sub-Layer (SDAL). The SDAL acts as a convergence layer that links the SMN lower layers components, i.e. the Modified Legacy Service Discovery Modules, to:

- The distributed service repository of the P2P Naming System Service, through its P2P Interaction Module
- The Service Session Management Module
- The SDM and the SR for insuring context-aware service publishing and discovery
- The AA server Module for insuring the secure service publishing and discovery operations, and vice versa

The SDAL is also provided with a dedicated communication interface that allows any PN components (e.g. like applications) to interact with the SMN for service description publishing and discovery purposes.

3.4.3 PN Interactions with External Service Frameworks

The PN concept extends the use of a handheld terminal or client to a larger network; the entire PN can be seen as a big user terminal that can be contacted by an external network. The internal nodes in a PN are hidden from all external nodes (peers) [30].

An interesting aspect of this concept is enabling access to an external service framework, from a node inside the PN. Examples of such an access could be web surfing, internet banking, remote login, etc. A backend server external to PN is contacted by an internal node within the PN, through a gateway node and other foreseen entities, such as *Network Address Translation* (NAT) boxes and firewalls. There are also cases that an internal PN node should be discovered and contacted from an external node, to provide a shared service to the outside world, or to receive and take an external call. Figure 3.18 illustrates the concept of establishment of service sessions between the PN nodes and external nodes.

The HTTP traffic shown in blue is an example of outbound traffic initiated from a node inside the PN. A possible approach would consist of calling the external server using a URL, the name is resolved by the naming system of the PN, and the IP address of the external node is used for making the HTTP request. NAT is carried out at the gateway node, and the external server will be eventually contacted. In return, the backend server replies to the HTTP request, and again at the gateway node (acting as a network address external to the internal translator), the reply will be forwarded towards the requesting node.

The VoIP traffic, as an inbound traffic shown in red, is initiated by an external node. The external IP-Phone, actually calls the internal peer, however, that external node sees the whole PN as an entity. The only visible node from outside world is the PN agent, which holds the addresses of the gateway nodes within the PN. The PN agent determines the corresponding gateway node, which enables the destination peer to be contacted. The gateway node is contacted for taking the call, and then automatically forwards all inbound calls to a dedicated entity, which is called *Service Gateway Node* (SGN). The difference between the SGN and the gateway node (contacted earlier) is that the SGN is intelligent in terms of finding the most

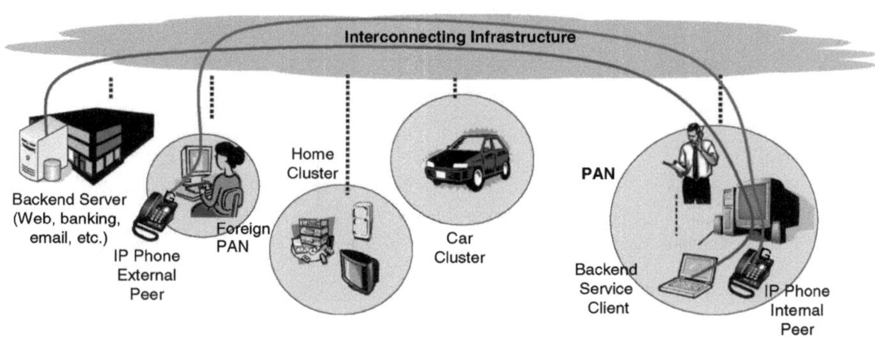

Fig. 3.18 External IP phone session and web surfing enabled within a PN

appropriate node in the PN (relying on the context information and capabilities of the PN devices). However, the gateway node only functions at the network layer, which forwards all calls to a pre-determined SGN. The SGN acts as a proxy for the destination node, and caches the specification (obtained from the MSMP) of the best node (at any time) potentially able to take the call. When the SGN receives the call, forwards it to the most appropriate internal node (already known in the cache) for taking the call. There the call is taken and an acknowledgement is sent to the initiating node through the gateway node and NAT, and finally the service session will be virtually established between the SGN and the external peer, whilst the actual service session end-points (internal and external peers) will eventually communicate with each other.

Service Gateway Node (SGN) in for IMS. For interaction with IMS calls, almost the same statements are valid. The PN Service Gateway Node is the entity that interacts with the outside world and inside the PN with other PN components. It acts as a firewall, with an embedded NAT, and can manage *Service Name Translation*. The interaction between MSMP (as an IMS client) and IMS core is provided via G_m interface. This interface is shown in Fig. 3.19. (The firewall icon is used for the SGN to stress these expected functions from this node).

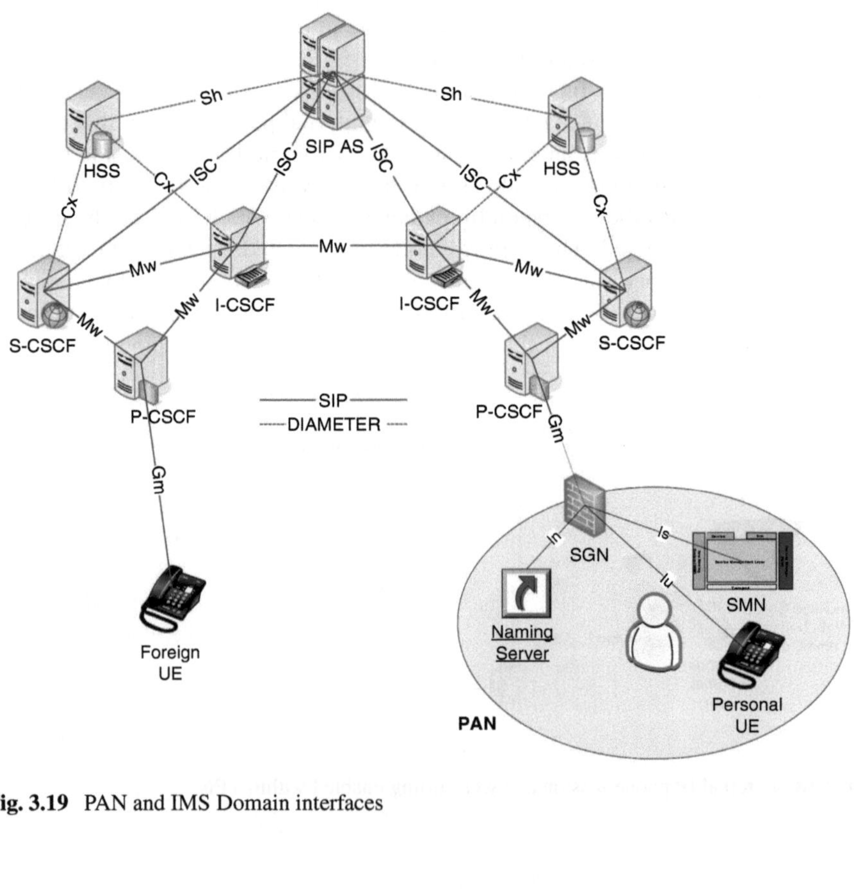

Fig. 3.19 PAN and IMS Domain interfaces

In order to interact with the IMS system, the SGN is considered as a User Equipment (UE) proxy (although the end node inside the PN is the actual endpoint UE). The interfaces depicted in Fig. 3.19 are described as follows.

- **G_m**: This is the normal traditional interface between the SGN and the other parts of the IMS system, namely the *Proxy Call Session Control Function* (P-CSCF). This interface is based on SIP.
- **I_s**: This interface is used in case an invite message arrives to the SGN. The MSMP is contacted for retrieving the list of UE-capable nodes. These services (as IMS clients) are already registered with the MSMP. MSMP, with the help of context information, provides the URL of the end user UE device to take the call.
- **I_n**: This interface is used to resolve the destination address of the called device, i.e. PN UE.
- **I_u**: This interface must be used for forwarding the signalling to the destined UE. This interface is equal to the Gm.

3.4.4 Charging and Billing

Three basic business models have been outlined in MAGNET Beyond, namely a self-organised model, a service-oriented model and a combination model.

The self-organised model is one where no financial exchange takes place, for example PAN resources are local and belong to the user, or two users connecting to each other's devices using Bluetooth P2P, and they share services and resources freely. It is also possible that when a user connects to the WiFi network but does not have to pay for this service (it may already be paid for by his company, belongs to a friend or may be paid for by an advertiser and sponsor). The self-organised business model is therefore one that is formed based on its own actions and is independent of any external chargeable resources.

The service-oriented model is one where a financial transaction takes place based on chargeable resources, e.g. a payment by the user to the WiFi Service provider in exchange for connecting to the Internet. This model involves often many business partners each delivering their 'bit' of the final PN service, partners that are interested in providing services for users and charging these services. This in itself will be a challenge service-wise, technologically, organizationally, and financially. One of these challenges is the issue of simple and transparent billing for customers. This includes understanding costs in advance to get full cost control across different access and device technologies, geographical locations, PNs domains etc to support user-centric PN and PN-F communications.

The combination model would encompass both earlier models where a self-organised and a service-oriented model exist. This would probably be the most common case in a PN, where different types of communication will take place, either through a network operator's or service provider's connection or through a personal peer-to-peer connection. Ad-hoc networks may exist in any of the combinations.

In general, there are large already-done investments in charging and billing infrastructure. Protocols have been specified in IETF and have been extended by 3GPP and 3GPP2 for bearer-level charging. This includes charging embedded in different domains (e.g. Packet Switched), services (e.g. SMS, MMS) and in subsystems (e.g. IMS). OMA has defined charging flows and data definitions for a couple bearer independent services enablers for application-level charging (e.g. SIMPLE IM). Seamless interoperability of IP services has been specified by GSM Association (GSMA) and so on. Charging will not come from scratch. This implies a need to specify how existing infrastructures can be exposed and re-used in MB's context, architecture and services/applications and eventually point out gaps based on the new MB scenarios, requirements and technologies. These gaps are candidates for future standards.

However, charging that is very important in any commercial applications/services are also very deployment and service provider specific, so even if existing infrastructures and mechanisms can be exposed and re-used in MAGNET Beyond's context, there is often a need to decide specific chargeable events, triggers, information flow etc from case to case. Which ones do you want to use?

But even if no standardised charging are defined, any implementation can still trigger charging events if it so decides, but these features have to be defined and implemented for the specific deployment. They will not be available as standard features by default. Instead of addressing a full charging/billing model, MAGNET Beyond implementation can provide some hooks for charging, whereby a service provider may be able to implement a charging. On the other hand, if standardised charging triggers are defined to the most likely business models, any developers are free to quickly choose which ones to use or not.

So this section can just describe some general advises and guidelines. OMA Best Practises document [41] introduces charging concepts, terminology and things to consider including testing considerations for charging specification development. This guideline can probably serve MAGNET Beyond scope as well to propose some steps and recommendations to specify/generate charging for different parties, see in Table 3.1.

3.5 Collaboration Between Users

While Personal Networking is focused on the communication between personal devices only, many communication patterns need to extend the boundaries of the Personal Network and involve the secure interaction of multiple people having common interests for various professional and private services, introducing the concept of PN Federations. A PN Federation (PN-F) can be defined as a secure cooperation between different PNs, making selected service(s) and resource(s) available to selected receiver(s) for the purpose of achieving a common goal. In fact when devices belonging to different PNs need to communicate and/or share resources, a secure connection between involved devices will be established. Devices allow each other

Table 3.1 Proposed steps for clarifying charging concept based on OMA charging best practises [41]

	Steps	Example in MAGNET Beyond scope
1.	Identify chargeable events	Identify potentially chargeable events for the relevant MB application or service (not all events will be used). Understand the charging needs of PN and/or PN federations
2.	Identify which entities/functions to trigger charging requests	Identify which MB entities/functions that trigger the charging requests, e.g. different MB servers with controlling functions
3.	Identify when to trigger charging requests	Identify when to trigger, e.g. before service delivery has started, during delivery, after the delivery has been competed
4	Identify information to be included in the charging events	Identify information needed in the charging events, e.g. service identifier, type of action, data volume, level of quality. A MB service may require new charging data elements to carry such MB service specific information and information exchange between the internal entities in PN or PN federations

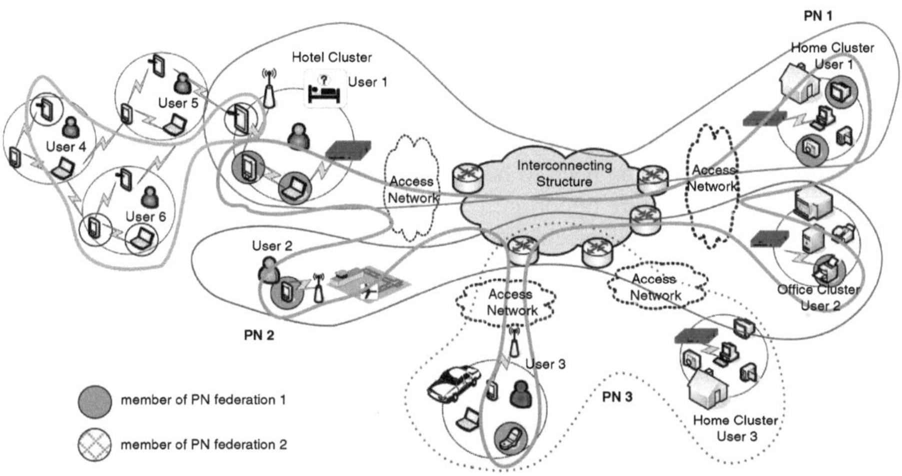

Fig. 3.20 Illustration of Ad hoc based versus Infrastructure based federations

access to specific services as well as share resources to perform the common tasks. The main goal is to extend the PN solutions and architecture with necessary networking functionalities and group trust mechanisms to enable interactions between multiple PNs.

In [35], the concept of a PN federation is illustrated, together with the underlying devices that participate in the federation. Based on how the cooperation between the devices of different people is realized in order to establish the federation, we can discriminate between *Infrastructure-based* and *Ad Hoc-based PN federation*. In Fig. 3.20, these two different PN federations are illustrated. The first PN federation

(PN federation 1) is established between devices that are all connected to an infrastructure network – either directly or via some other devices belonging to the same federation. In this federation, support functionality available in or through the fixed infrastructure can be used to assist in the PN federation definition and establishment. This can be compared to the PN Agent introduced in the Personal Network architecture. In the second PN federation, the federation is formed in the absence of a fixed infrastructure. As no infrastructure is accessible, the definition and establishment of the federation need to be done in a distributed ad hoc fashion, having implications on the solutions that need to be developed to realize PN federations. This type of federation is called an Ad Hoc PN federation and will mostly occur when nearby users collaborate within a federation and will impose different requirements on the networking solutions. Of course, hybrid federations that are a combination of these two types are also possible. We can also classify PN federations based on a number of other characteristics. First of all, depending on the way the federations are initiated, we can discriminate between *purpose driven* PN federations and *opportunity driven* PN federations. Purpose driven means that the formation of the federation is explicitly requested or defined beforehand, whereas opportunity driven means that the federation is formed spontaneously when interesting circumstances to do so arise. In both cases, and especially in the second case, context information can play an important role. Next, depending on the lifetime of the federation, we can make the distinction between very *short-lived* federations and *longer term* federations. This distinction will have its implications on the complexity of the solutions to establish the federation. In the case of short-lived federations, solutions to setup and manage the federation need to be lightweight and simple. Longer term federations open up much more opportunities to introduce more complex and powerful management and definition mechanisms.

3.5.1 Automatic, Profile Controlled Establishment of PN-F

For this PN-F concept, a PN-F life cycle has been derived illustrating the different phases in the life-time of a PN-F. Figure 3.21 shows this PN-F life cycle. In the following subsections we will describe in more detail the different components and phases.

3.5.1.1 PN-F Profile and PN-F Participation Profile

In order to be able to create trustworthy PN federations, rules are needed that determine who is (or can become) a member of the federation and how. We refer to this as membership management. When in a federation, a member needs to define which resources are available to other members as well as who is able to setup or update these rules and profiles. Let us refer to this as resource management. Based on this, we have identified two different profiles, a PN-F profile, which is a profile common

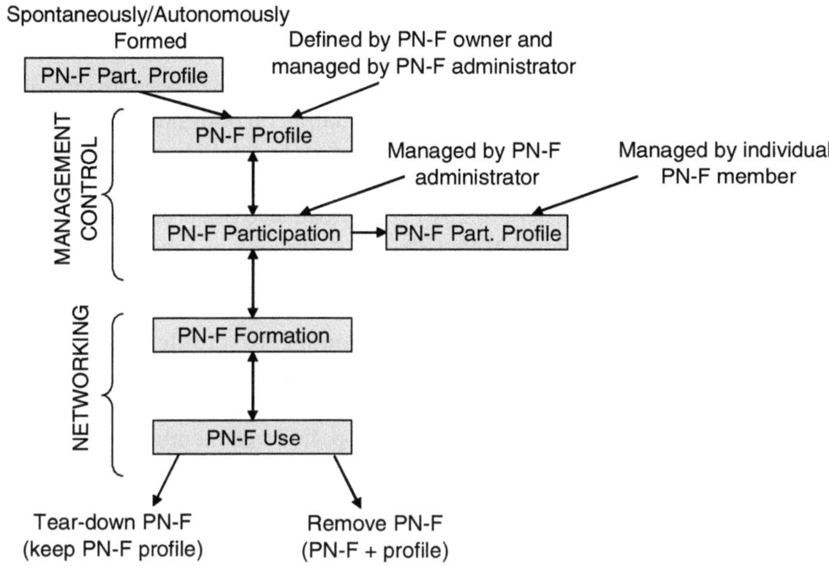

Fig. 3.21 PN-F life cycle

to the federation and individual PN-F participation profiles, which are bound to the individual members. The former is used for the federation's membership management whereas the latter is used for resource management of individual members.

The PN-F participation profile can be specific or generic. A specific PN-F participation profile defines for an existing PN-F the resources and services the member wants to make available to that PN-F. A generic PN-F participation profile defines user interests and requirements related to participating in or setting up new federations and the resources a user wants to make available in case a PN-F is formed based on this profile.

The PN-F profile contains the following policies, rules, agreements common to the PN-F. First of all, the PN-F needs to have an owner. The owner is the one that manages the PN-F Profile. The owner can define a list of administrators, who can have read and/or write access to the PN-F profile in addition to the owner.

The policies of the PN-F determine how the memberships are managed. The members of the federation can be defined explicitly. Alternatively, rules can be defined to dictate how new members can be added to the PN-F.

Further, the above PN-F profile contains global information, i.e. relevant for the members of the federation, which needs to be securely stored and accessible by all members. For infrastructured federations, storage can be done centralized or distributed in each PN participating in the federation. For ad hoc federations, storage needs to be completely distributed. Of course, as the profile can only be modified by specific people, strong and efficient security solutions that verify, protect and enforce the rules defined therein and their authentication are needed. In addition, updates to the profile need to be propagated to all involved parties and a lifetime could be assigned to the profile.

3.5.1.2 PN-F Participation

The participation phase is the process of building up the group of participants, establishing secure communication channels and negotiating on both sides the conditions for joining the federation. In order to make it possible for PNs to join a PN-F, a PN-F owner can publish the new PN-F into a search function database or invite or notify other PNs to join the federation. A PN user can also use a search and browse functionality to find interesting PN-Fs (PN-F descriptions or tags) in categorized databases.

These mechanisms make it possible for PNs to join a PN-F. Adding new members to the PN-F will involve the updating of the common PN-F profile and the creation of an individual PN-F participation profile for the newly joined PN.

During the lifetime of the PN-F it is possible that some characteristics of the PN-F change. Therefore upon any change in the policies or members, an update or redefinition of the PN-F has to be carried out. This does not imply that the current PN-F is completely terminated, but a secure renegotiation of the PN-F parameters has to be carried out again. Several circumstances may cause the modification of the characteristics of a PN-F and major renegotiations (update of members, exchange of new group key, etc.).

3.5.1.3 PN-F Formation

The next phase in the PN-F cycle is the PN-F formation. Once the PN-F profile has been created and members have fixed their PN-F participation profile, the federation can be established (according to any of the formation policies defined in the PN-F profile) at the network level, offering secure communication between the different PN-F members.

3.5.1.4 PN-F Use

After the formation phase follows the PN-F use phase consisting of secure service access and service provisioning of shared services according to the PN-F participation profiles of the PN-F members. It should be possible to rate shared devices/resources/services. The shared object rate gives a good estimation about the quality of a shared object.

3.5.1.5 PN-F Termination

A PN-F member can decide at any time to remove his/her devices/resources/services from sharing in an individual PN-F by updating the individual PN-F Participation Profile. The PN-F will end the related recourse and/or service sharing and optionally notify the related PN-F members.

A PN-F member may decide to quit from the PN-F. Whenever this happens, the common PN-F profile is updated and all members are informed by the creator or an administrator of the PN-F who are the ones in care of the management of the PN-F. A new secure relationship is then established between the remaining nodes, avoiding that the former member can still make use of the federation.

The PN-F creator or administrator can decide to ban an individual PN-F member from the PN-F. If the creator detects some irregularities in the PN-F such as for instance a member no longer fulfilling the requirements (e.g. behaving on a selfish way not sharing any resources), he might decide to kick him/her out. As in the previous case, the PN-F Profile is updated (the members' list specifically), the rest of members notified and a new group key exchanged.

Finally, a PN-F creator or authority can completely or temporary close down the PN-F. All the resource and service sharing will terminate and the PN-F users will be notified. Nevertheless, the common PN-F Profile will not be cancelled if the closure is only temporal so it can be reused afterwards. When the closure is permanent, the PN-F profile will be cancelled and the PN-F stops its existence.

3.5.2 Joining the PN Federations

The PN-F architecture (Fig. 3.22) introduces the Federation Manager as the entity, which manages the participation of a PN in PN Federations and the various resulting PN-F profiles. The Federation Manager that is responsible for the creation of the PN-F, manages the PN-F profile for the whole PN-F while the Federation Manager of each member manages its own PN-F participation profile. This profile management consists of creation, storage, updating and distribution. To allow authentication, a Certification Authority, CA, is required. To this end, the Personal Network Directory Server (PNDS) is introduced as a trusted third party that will issue the personal certificates. However, in an ad-hoc PN-F this is not always possible and solutions such as the use of a Proximity Authenticated Channel could be used and extended to allow ad hoc PN-Fs. In the following sections, the different components of the architecture and their interactions will be further discussed according to the life cycle of a PN-F.

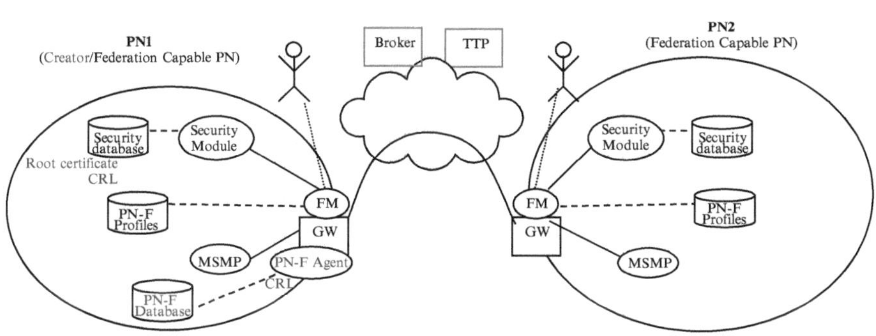

Fig. 3.22 PN-F architecture

3.5.2.1 PN-F Participation and Management

A PN engaged in a federation can have one of two roles: creator or participant. The PN-F Creator generates a PN-F Profile containing the main details of the PN-F and stores it. The federation profile is a data structure unique for each federation. The public part used for announcements contains the federation ID, name, short description. To search or identify the creator of that federation, his nickname and PN ID are provided, as well as a X.509 certificate from the directory service PNDS (see next paragraph). The rules for joining and for other federation management decisions are delivered in the federation profile in form of semantic policies written in the Notation3 language. The public part of the PN-F Profile is made public and candidates (i.e. other PNs) go on a dialogue with the creator to see whether they are allowed to enter on the PN-F or not. For infrastructure-based federations, a central directory component, the PN directory server (PNDS) stores the PN-F announcements. Interested parties can search this directory according to keys such as the topic, and eventually select a federation to join. In an ad-hoc federation, two devices that belong to different PNs and are in the wireless range of each other interact from the bottom to the upper protocol layers to form a federation.

According to the formation protocol, the private part of the federation profile is disclosed only after a secure (encrypted) channel is established between creator and participant. In this part we can find:

- More policy rules about starting and stopping a federation.
- Information about the current members that have joined the federation.
- Address of the PN-F agent that is the seed of the overlay (per federation) on which the participants advertise their resources and services.

In order to proceed with the next step in the PN-F participation phase, the PN-F Creator and potential PN-F members (i.e. other PNs) need to be able to authenticate each other and to establish a security association that can be used to secure all ensuing communication. A new PN component, called Personal Network Directory Service, is also introduced as the identity provider (i.e. trusted third party entity). The PNDS, operated by a service provider, acts as a Certificate Authority (CA) providing X509 certificates which associate public key with a particular user. The PNDS certificates are leveraged by CPFP to establish bilateral trust relationships between the PNs that are afterwards enforced each time the two PNs communicate under the auspices of any federation.

After this authentication and security association step, the PN-F member can actually join the PN-F. A PN-F participation profile that lists the services that the new PN-F member will make available within the PN-F is created and stored in the SCMF. At this stage, each member knows in which PN-Fs she/he participates, which other PNs are currently member of the PN-F and, optionally, what services are made available by these members. This information can in any case be retrieved through a PN-F wide service discovery mechanism since the MSMP implementation has been extended to support also this feature.

The implementation of the FM participation protocol makes use of a protocol state machine. The rules for going into the in-use state (active federation) for example can be flexibly formulated based on the number of participants, time, or the presence of certain participants.

3.5.2.2 PN-F Network Overlay Formation

Similarly to the PN case, the concept of a network overlay has been selected to realize secure PN-F communication to enable all PN nodes of the PN-F members to become part of the PN-F overlay. In order to separate the internal PN communication from any PN-F communication, every PN-F will also have its own PN-F addressing space (defined in the PN-F profile) and every involved node will obtain a unique PN-F IP address within this addressing space. In a similar way to the PN, the PN-F overlay will be established.

As it was the case on the PN self-organization solution, for infrastructure based PN-Fs, the location of all other clusters of the PN-F members needs to be discovered in order to form the overlay. Since this information is stored in the PN Agent, it can be retrieved by contacting each PN-F member's PN Agent. Authentication and the establishment of security associations will be firstly carried out using certifications issued by the PNDS. Next, dynamic tunnels will be established between all involved clusters. Within the overlay, a hierarchical routing protocol (separating intra-cluster and inter-cluster routing) is running that provides end-to-end routes between all nodes in the PN-F overlay. Finally, nodes will get an overview of the available services within PN-F through service announcements or requests based on the content (policies, etc.) of the PN-F participation profile.

For ad hoc PN-Fs, neighbouring clusters of other PNs are discovered through their beacons announcing their presence. A secure association with neighbour PNs is established, exchanging a pair-wise primary master key which will be used as the seed for deriving link level session ones. After a secure link is guaranteed, PN-F routing information is transmitted and a PN-F cluster is formed. Interconnection between PN clusters at different locations requires support of the infrastructure to establish the end-to-end paths within the network overlay. Therefore the procedure to be followed is similar to the already depicted above for infrastructure based PN-F overlays (Fig. 3.23).

3.5.3 PN-F Service Management

The implemented service overlay designed at the PN level, can be extended to the PN-F level. A new service gateway named PN-F Agent is introduced for playing the role of the PN-F service overlay node. The PN-F Agent implements some of the SMN functionalities but it is exclusively dedicated for storing and discovering PN-F resources and services.

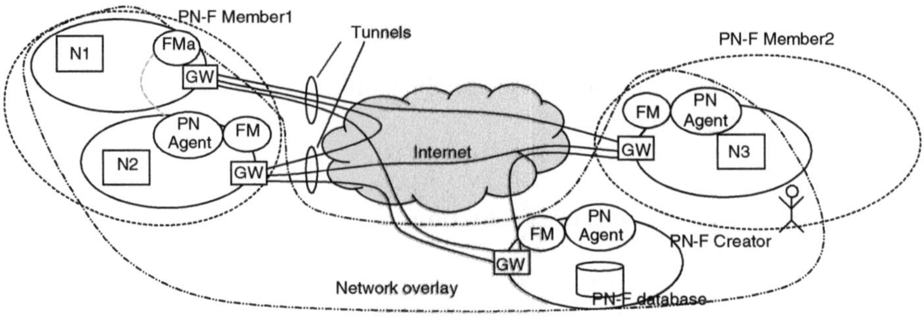

Fig. 3.23 PN-F network overlay

The proposed PN-F service framework establishes P2P service overlays dedicated to PN Federations (one per PN-F). The PN-F Agent serves as a super-peer within a PN-F service overlay. It can be viewed as a PN SMN dedicated for managing PN-F level service information publishing and discovery. The PN-F participants publish/register, update and discover the information on their shared services within the PN-F through their PN-F Agent by relying on an intentional name format providing the needed service descriptions.

Figure 3.24 depicts the high level architecture of both PN and PN-F service overlay solutions.

3.6 PN Context Management

Personal networks provide the unique opportunity to adapt and personalize applications, services and the whole networking environment to the current needs of the user, while at the same time protecting the security and privacy of the user. User preferences and context information can be gathered, processed and used within the secure environment of the personal network. It will only be provided to other users in PN Federations or external infrastructures if the user explicitly allows it. This enables a level of personalization and context-awareness that would not be possible in public infrastructures due to privacy concerns of the user.

To support this, we introduced the Secure Context Management Framework (SCMF) for Personal Networks that decouples the applications from the context sources, making it possible to share context between applications and allowing applications to seamlessly work in different environments. In Personal Networks the access to a fixed infrastructure cannot be guaranteed due to changing connectivity and availability of network resources. Also, limitations in bandwidth and battery power require reducing the communication overhead as much as possible, so context information should only be exchanged if required. Therefore, we designed a context management framework that takes the special requirements of a Personal Network into account [31, 32].

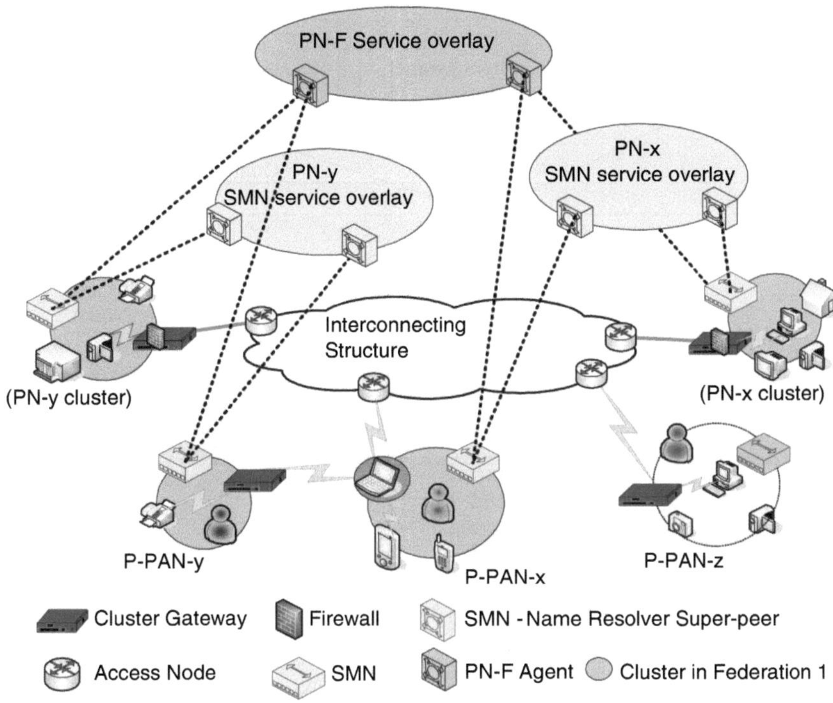

Fig. 3.24 Service management architecture on PN-F scenario

The main functionalities of the SCMF are retrieving, processing, storing, exchanging, and providing context and user profile information. The SCMF consists of Context Agents running on each node in the Personal Network. Figure 3.25 provides a high-level view of such a Context Agent and the components interacting with it.

Applications and services can be designed using a standardized interface (Context Access Layer – CAL) for accessing context and user profile information through their local Context Agent. In the same way, data sources can be implemented using another standardized interface (Data Source Abstraction Layer –DSAL) for providing context data. The SCMF takes care of everything in between, including processing, storing, distributing, and access control of available context and user profile information.

In summary, the following advantages are obtained by the Secure Context Management Framework:

- A developer writes all his/her applications against this common interface.
- A developer does not have to know anything about the specifics and internals of the context sources.
- A developer/user can replace sources (e.g., use completely different sensors with different protocols) easily, as long as they provide the same type of context information in the end.

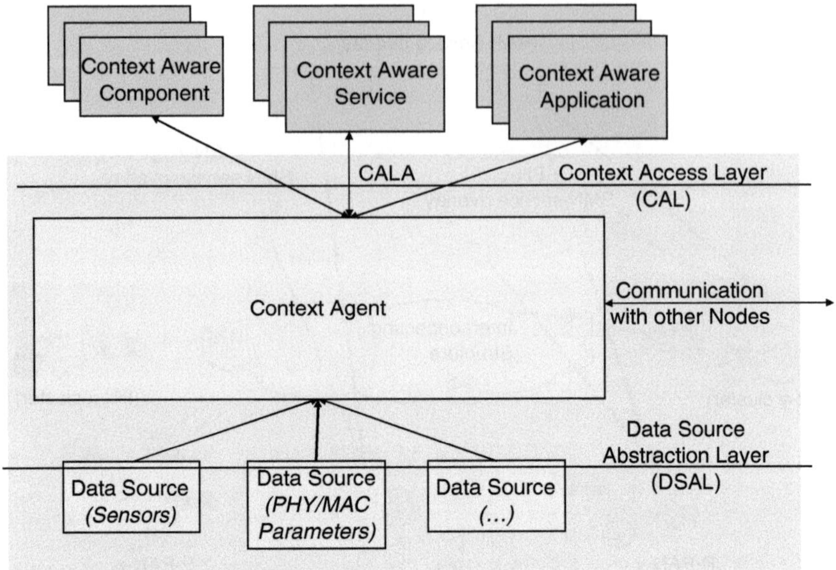

Fig. 3.25 High level view of a Context Agent and interaction with other components

- A developer can reuse context processing components as they operate on the common model, they do not have to be adapted to different sensors/ representations etc.
- A developer does not have to know the exact distribution of context information and context sources; this is made transparent by the SCMF. However, the developer can influence the access to context information through the use of scopes, which is explained in Section 3.6.3.

The remainder of this chapter is organised as follows: In Section 3.6.1 an overview is given how the Context Agents form the SCMF in a PN and how they interact in the PN-F case. Then the context modelling is explained in Section 3.6.2 and the Context Access Language (CALA) in Section 3.6.3. Together they show how the SCMF supports Context-aware components. Implementation related aspects of the SCMF are presented in Chapter 7.

3.6.1 Network Organisation and Distribution of Context Information

In Fig. 3.26 an example scenario is shown on how Context Agents may be distributed in the Personal Network to form the SCMF. The framework allows as mentioned efficient access to context information distributed in the PN, which means that all agents cooperate in this process.

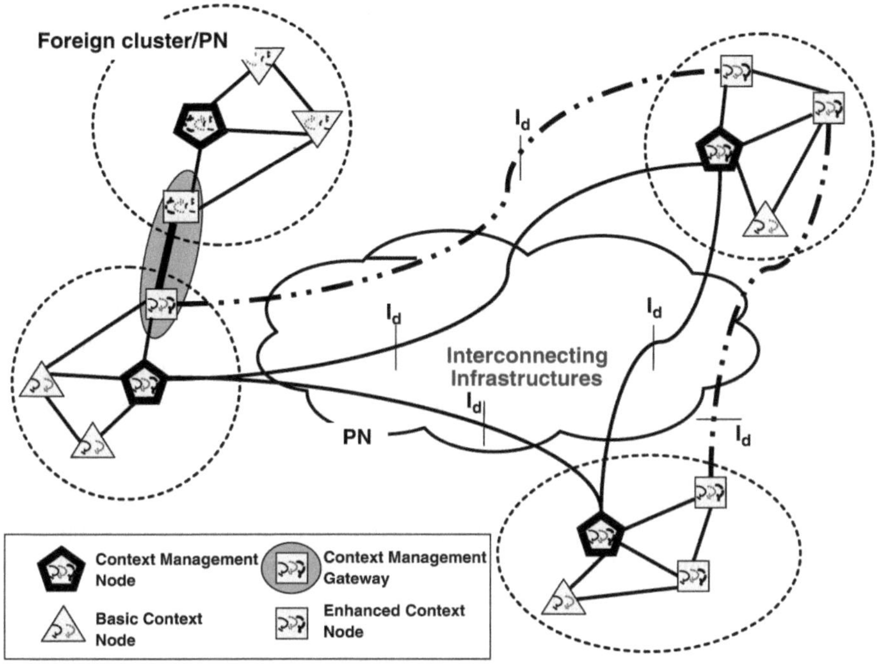

Fig. 3.26 Overview of network structure of SCMF specific entities

Context Agents may, however, be configured according to the processing capability of the device which it resides on, e.g. low end mobile phone may offer different storage and sensing capabilities than a high end laptop, thus the Context Agent may be configured as a basic or enhanced Context Agent. Within a single cluster one enhanced Context Agent is selected a dedicated role, namely a Context Management Node (CMN). The CMN has index information regarding all the information available on nodes in the cluster. Context Agents also interact with their peers in other clusters to handle PN-wide requests. This gives a hierarchical structure, enabling scalable access to context information. Access to the information may happen directly between involved nodes, though, as to minimize the delay for obtaining context information remotely. Context may also be shared among different PN's through dedicated Context Management Gateways (CMG), which ensures that privacy enforcement at PN-Federation level is ensured.

3.6.2 *Context Modelling*

In this section we present our approach to modelling context and user profile information. Context-aware components need access to all information pertaining to the aspects of the user's situation according to which they should adapt their behaviour. This means we need to model aspects pertaining directly to the user, but also to his environment, e.g. (other) people, objects, places, devices, services, networks, etc.

A context model has to provide a suitable semantic definition for this information. It also has to allow providing efficient access to the information. Another essential aspect of the context model is its extensibility, i.e., if new aspects have to be modelled to support new scenarios this has to be easily possible, for example, without having to change interfaces.

Ontologies allow the formal definition of concepts and properties that allow the modelling of relations between instances of the concepts. Thus ontologies can be used to define a common vocabulary with a well-defined semantics for sharing data between different components, applications and services. The use of ontologies also paves the way to ontology-based reasoning, for which a number of different reasoners are readily available.

The concepts in the ontology can be organized in a class hierarchy, which may also provide the basis for an object-oriented internal representation, if this is desired for the implementation. Therefore, we decided to use an ontology as the basis for modelling context information in MAGNET Beyond, using OWL, and more specifically OWL-DL for staying decidable.

Figure 3.27 shows the core concepts of the MAGNET Beyond Integrated Ontology. The underlying idea is to define a hierarchy of entity types, facilitating a type-based access to context and user profile information. Its top level concept is the *MagnetEntity*. The *MagnetEntity* concept introduces the property *hasIdentifier*. Any entity that can be uniquely identified using an identifier can thus be modelled as a *MagnetEntity*. Based on the unique identifier an index can be built that provides

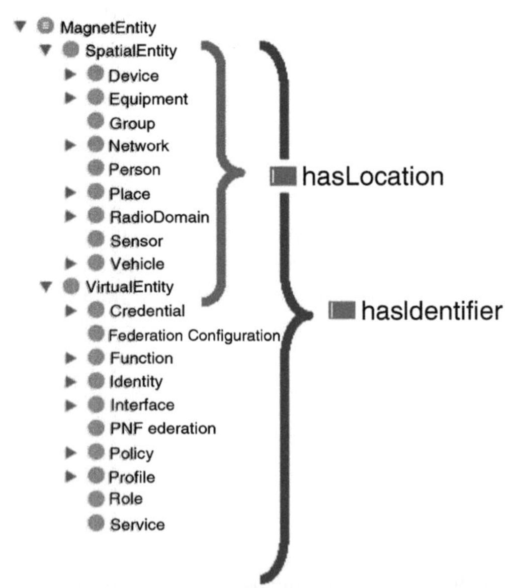

Fig. 3.27 Core part of the MAGNET Beyond Integrated Ontology

the basis for efficiently accessing context information in all cases in which the specific entity is known.

The *MagnetEntity* concept has two subconcepts, the *SpatialEntity* and the *VirtualEntity*. The *SpatialEntity* concept introduces the *hasLocation* property. All physical entities have a geographical location and for some other concepts it may also make sense to define a location. For example, a wireless network may have a spatial extent or the location of a group may be the aggregated location of its members. Taking the *hasLocation* property (whenever available) and the type information, a spatial index structure can be built to allow efficient type-based access to context information with a location scope, e.g., if all networks are to be found that cover a certain area. The *VirtualEntity* concept comprises all types of entities that are not associated with a geographical location. The attributes of MAGNET entities are modelled as properties in the ontology.

In addition to the context information itself, we need to provide meta information, including at least the following kind of meta information:

- *Confidence*
  Expresses the grade of confidence that a piece of context information is true, e.g., that a person is at a specific geographic location. The confidence is given as a value from the interval (0.0, 1.0] [33].

- *Accuracy*
  Expresses how precise a certain value, typically a measurement from a sensor is, e.g., the temperature is 10.5°C with an accuracy of ±0.2°C.

- *Creation Time*
  Expresses the moment in time when the information was created, i.e., the value was measured or the profile information was added.

- *Validity Interval*
  Expresses the period of time for which the information is considered to be valid.

Not every kind of meta information will make sense for all types of context or user profile information, e.g., accuracy typically makes sense with measured values like sensor values only.

3.6.3 Context Access Language

The Context Access Language (CALA) is used at the interface that allows components to access information from the SCMF. There are three different types of interactions, two for retrieving information from the SCMF, one for modifying information:

- Synchronous retrieval: query/response
- Asynchronous retrieval: subscription/notification
- Synchronous modification: insert/update/delete

For synchronous query/response-based interaction the following CALA parameters can be specified:

- *Selector.* The entities of interest and the attributes to be returned are selected. There are two general options. The entity is already known, in which case it can be selected by providing its unique identifier; or only the type of entity is known. For example, information about all currently available networks is requested. So the type is used as the selector. In this case special attention has to be paid to the scope of the query. The attributes that are to be retrieved also have to be listed. In case the attribute refers to another MAGNET entity, only a reference with its unique ID is provided. The special attribute *ALL* allows to retrieve all available attributes for the selected MAGNET entities.
- *Restrictions.* By specifying restrictions, the MAGNET entities to be returned can be restricted based on the values of one or more attributes. The following operators can be used in restrictions:

  - Comparison operators on simple data types (attribute < comp > value)
  - Composition Operators for combining restrictions

  In the future, special operators on complex data types may also be supported.

- *Scope.* The scope of the query restricts where the SCMF looks for the requested information. In contrast to restrictions which are used for filtering after the results have been gathered, the scope is used before gathering the results, restricting the places where to look for information. The following scopes are considered relevant: *Network Domain* (for PNs: *node*, *cluster*, *PN*, *federation* and *external*), *Physical Location*, and *Time,* to access history information.
- *Options.* The Option field is for providing additional information and extensions to the basic functionality.

For asynchronous subscription/notification based interaction a subscription condition has to be specified in addition to the parameters explained above:

- *Subscription Condition.* The general options for the condition are: *Notify once* (e.g., for cases in which retrieving context information takes a long time), *Notify on change* (with different options: any change, an absolute threshold, or a threshold relative to the last reported measurement), and *Notify periodically* (for which a time period is required to be specified).

An example for the XML-representation of a CALA query and a response can be found in Section 7.2.11.2.

3.6.4 Further Reading

The material presented here is only the most important part of the whole SCMF. Much more material and many useful details have been worked on and published in deliverables and scientific papers. For the complete overview of the framework, the deliverables [34, 35] provides much more reading material. In addition to this, several papers has been published ranging from scenarios and requirement analysis in [31], analytic modelling of access strategies in [33, 36, 38] which focuses on reliability aspects of context information. Some of this work has in fact lead to ways of selecting appropriate access strategies, as dedicated PhD works show in [37]. In particular for caching strategies, works in [37, 38] show that there is much reliability to be gained by selecting appropriate timings when caching or when selecting update time intervals. It was also shown in [40] how information on the reliability metric can be used to ensure a high quality of context information if the Context Agent is aware of certain meta information, mainly related to network delay and information dynamics. This work has also been used for evaluation of how context aware applications are influenced. An example with context aware service discovery is given in [39] which also proposed methodologies to increase application reliability based on the reliability of the used context information by using estimated values/processed information in conjunction with accessed information. Finally, context aware service discovery has previously been evaluated experimentally in [29] showing the network and timing overhead associated to the context aware behaviour of service discovery. In fact, all the above sources alone provide in detail and useful knowledge about context management systems, and reveal many challenges and the difficulties in achieving context aware systems.

3.7 Conclusions

Personal Networking is an exciting new concept permitting user to combine his own personal devices to form a unique network for his own personal use. He can form this network by imprinting new devices into the network, by connecting his different clusters wherever he is, and by letting his PN interact with his environment through context sensing as well as through interactions with external systems.

In this chapter, we explained the technologies that can make PN secure, safe, and easy to use. The introduced security mechanisms permit the establishment of secure communication between the nodes and the clusters. A later chapter will describe the PN security mechanisms as well as more sophisticated aspects of PN security in more details. Then, we introduced the connectivity mechanisms that permit to use many different (wireless) networks. The introduced Universal Convergence Layer is an important aspect of an advanced wireless protocol stack and a blueprint for future mobile devices. Based on the advanced connectivity layer, MAGNET Beyond has evaluated advanced forms of networking for clusters, PN and PN-Federations. The solutions found have been evaluated and compared (as can be seen in Chapter 7). Interactions with the environment of the users are performed on the sensor level using the highly advanced context management system. That self-organizing system provides various forms of personal information to all nodes in the system. It is the storage system for distributed user profiles and can interact with the external systems. This interaction with external systems has been shown for HTTP and SIP-based services.

The Personal Network was extended to a new form of user-to-user networking with the concept of Personal Network Federation. Using that concepts, PNs form the base of interactions between. With the security, networking and automatic adaptation concepts introduced, PN-Fs offer a very versatile and powerful method for creating new services. The mobility of the users as well as his changing needs are taking into account using the dynamic establishment of PN-F based on the conditions and rules contained in the PN-F profiles.

The self-organization of the services and the context management show that PN/PN-F are truly designed with the end-user in mind. Services can be easily found and automatically used. New sensor and context sources are automatically included into the context processing and therefore used in the applications. The systems adapt without user intervention to the changes in connectivity, networking, and federation.

References

1. Wireless World Research Forum, Book of Visions (2001) online, http://www.wireless-world-research.org/index.php?id=107
2. I. Niemegeers, S.H. de Groot, From personal area networks to personal networks: A user oriented approach. J Wireless Pers. Commun. **22**, 175–186 (2002)
3. I. Niemegeers, S.M. Heemstra de Groot, Research issues in ad-hoc distributed personal networking. Wireless Pers. Commun. **26**(2–3), 149–167 (2003)
4. E. Gustafsson, A. Jonsson, Always best connected. IEEE Wireless Commun. **10**(1), 49–55 (2003)
5. I. Niemegeers, S. Heemstra de Groot, FEDNETS: Context-aware ad-hoc network federations. Wireless Pers. Commun. (Springer) **33**(3–4), 305–318 (2005)
6. Ambient Networks (AN), http://www.ambient-networks.org/
7. IST PACWOMAN – Power aware communications for wireless optimised personal area networks, http://www.imec.be/pacwoman/Welcome.shtml
8. F. Louagie, L. Muñoz, S. Kyriazakos, Paving the way for the fourth generation: A new family of wireless personal area networks. In the 12th IST Mobile and Wireless Communications Summit, Aveiro, Portugal, June 2003
9. IST-2000–25350 SHAMAN, D13 – Final technical report – results, specifications and conclusions, 30 Nov 2002
10. C. Gehrmann, T. Kuhn, K. Nyberg, P. Windirsch, Trust model, communication and configuration security for Personal Area Networks. In the 11th IST Mobile and Wireless Telecommunications Summit, Thessaloniki, Greece, 16–19 June 2002
11. J. Dunlop, R.C. Atkinson, J.M. Irvine, D. Pearce, Personal distributed environment for future mobile systems. In the 12th IST Mobile and Wireless Communication Summit, Aveiro, Portugal, 15–18 June 2003
12. J. Dunlop, The concept of a personal distributed environment for wireless service delivery. NEXWAY White Paper, June 2004
13. S. Schwiderski-Grosche, A. Tomlinson, D.B. Pearce, Towards the secure initialisation of a personal distributed environment. Technical Report RHUL-MA-2005–09, Department of Mathematics, Royal Holloway, University of London, 20 July 2005, http://www.rhul.ac.uk/mathematics/techreports
14. MyNet, http://projects.csail.mit.edu/nrcc/mynet-uia.php

15. F. Kaashoek, R. Morris, User-relative names for globally connected personal devices. In the 5th International Workshop on Peer-to-Peer Systems (IPTPS'06), Santa Barbara, CA, Feb 2006
16. B. Ford, Unmanaged Internet Protocol: Taming the edge network management crisis. In the Second Workshop on Hot Topics in Networks (HotNets-II), Cambridge, MA, Nov 2003
17. Universal Computing Consortium (PUCC), http://www.pucc.jp/
18. The Siemens LifeWorks Concept, White Paper (2008), http://www.siemensenterprise.com/attachments/2gip/LifeWorksWhitePaper.pdf. Accessed Mar 2008
19. D. Husemann, C. Narayanaswa, M. Nidd, Personal mobile hub. In the Eighth IEEE International Symposium on Wearable Computers (ISWC'04), Arlington, VA, 31 Oct 2004 to 3 Nov 2004
20. R. Kravets, C. Carter, L. Magalhaes, A cooperative approach to user mobility. ACM Comput. Commun. Rev., **31**(5), 57–69 (Oct 2001)
21. Third Generation Partnership Project (3GPP), Service requirements for Personal Network Management (PNM) – Stage 1. Technical Specification, 3GPP TS 22.259 V8.3.0 (2006–06), Mar 2007
22. S. Deering, R. Hinden, Internet Protocol,Version 6(IPv6) Specification. IETF RFC 2460, Dec 1998
23. R. Braden, Requirements for internet hosts-communication layers. IETF RFC 1122, 1989
24. IST-507102 MAGNET, Deliverable D4.3.2, Final version of the Network-Level Security Architecture Specification, S. Mirzdeh et al., Mar 2005
25. IETF Mobile Ad hoc NETworks (MANET) working group, http://www.ietf.org/html.charters/manet-charter.html
26. W. Louati, D. Zeghlache, Network based virtual personal overlay networks using programmable virtual routers. IEEE Commun. Mag. (Special issue, Self organization in networks today), **43**(8), 86–94 (Aug 2005)
27. E. Kohler et al., The click modular router, ACM Trans. Comp. Sys., **18**(3), 263–97 (Aug 2000)
28. M. Ghader, R.L. Olsen, M. Giro-Genet, R. Tafazolli, Service management platform for personal networks. 14th IST Mobile and Wireless Communications Summit, Dresden, Germany, 19–22 June 2005
29. R.L. Olsen, A. Nickelsen, J. Nielsen, H.P. Schwefel, M. Bauer, Experimental analysis of the influence of context awareness on service discovery in PNs, in *Proceedings of the IST Summit 2006*, Greece, 2006
30. E. Kovacs, D. Kraft, A. Cimmino, S. Bessler, M. Ghader, L. Gavrilovska, Personal networks as distributed clients for IMS. ICT-MobileSummit 2008, Stockholm, Sweden, 10–12 June 2008
31. M. Bauer, R.L. Olsen, L. Sanchez, et al., Context management framework for MAGNET Beyond. Accepted for Workshop on Capturing Context and Context Aware Systems and Platforms, IST Mobile and Wireless Communications summit, Myconos, Greece, 2006
32. L. Sanchez, J. Lanza, M. Bauer, R.L. Olsen, M. Girod Genet, A generic context management framework for personal networking environments. Accepted for Workshop on Personalized Networks, Third Annual International Conference on Mobile and Ubiquitous Systems, San Jose, CA, 2006
33. R.L. Olsen, H.-P. Schwefel, M.B. Hansen, Quantitative analysis of access strategies to remote information in network services. Globecom06, San Fransisco, CA, Nov–Dec 2006
34. IST-027396, Deliverable D2.3.1, Specification of PN networking and security components, M. Jacobsson et al., Dec 2006
35. IST-027396, Deliverable D2.3.2. PN secure networking frameworks, solutions and performance, M. Jacobsson et al., June 2008
36. M.B. Hansen, H.-P. Schwefel, R.L. Olsen, Probabilistic models for access strategies to dynamic information elements, to appear in Performance Evaluation, Elsevier
37. R.L. Olsen, Enhancement of wide-area service discovery using dynamic context information, Ph.D. dsertation thesis, Aalborg University, Jan 2008, ISBN: 87–92078–37–0
38. H.P. Schwefel, M.B. Hansen, R.L. Olsen, Adaptive Caching strategies for Context Management systems, invited paper for PIMRC'07

39. R.L. Olsen, H.P. Schwefel, M. Bauer, Influence of unreliable information on Context Aware Service Discovery. Third Workshop on Context Aware Proactive Systems, Guildford, United Kingdom, June 2007
40. R.L. Olsen, H.-P. Schwefel, Determination of context value in multiple context source scenarios for Context Management systems, in *Proceedings of WPMC'07*, Jaipur, India
41. Open Mobile Alliance (OMA), Charging specification best practices. Approved Version 1.0, 25 Mar 2008

Chapter 4
PAN-Optimized Air Interfaces

Dirk Dahlhaus, Thomas Hunziker, Spyridon Vassilaras, Hamed Al-Raweshidy, and Mauro De Sanctis

4.1 Introduction

For the design of air interfaces (AIs) being suitable for typical WPAN application scenarios, it is important to consider the overall objective of MAGNET Beyond, namely to design, develop, demonstrate and validate the concept of a flexible Personal Network (PN) that supports resource-efficient, robust, ubiquitous personal services in a secure, heterogeneous networking environment for mobile users. As a consequence, two PAN-optimized AI solutions, one for high and one for low data rate applications, have been envisaged. The high data rate (HDR) PAN applications will be enabled by a multi-carrier spread spectrum (MC-SS) air-interface solution and a MAC layer scheme utilizing IEEE 802.15.3. For low data rate (LDR) applications, a low-power, low-complexity frequency modulation based UWB (FM-UWB) air-interface solution and a MAC layer based on IEEE 802.15.4 is proposed. A so-called Universal Convergence Layer (UCL) sits on top of the both AIs and is in charge of interfacing the LDR and HDR MAC layers with higher layer protocols. The structure of selected air interfaces is depicted schematically in Fig. 4.1.

In order to showcase typical applications supported by the LDR and HDR AIs, some baseline scenarios are presented. A prerequisite is that a PAN with heterogeneous (HDR and LDR) air interfaces has been established. The PAN may have the structure as in Fig. 4.2.

Examples for usage of the AIs are the following:

1. **Showing Video on Screen**

    - A video is stored on the internet tablet.

D. Dahlhaus (✉) and T. Hunziker
University of Kassel, Wilhelmshöher Allee 73, Kassel 34121, Germany
e-mail: dahlhaus@uni-kassel.de

S. Vassilaras
Intracom/Athens Information Technology, Greece

H. Al-Raweshidy
Brunel University, UK

M. De Sanctis
University of Rome "Tor Vergata", Italy

R. Prasad (ed.), *My Personal Adaptive Global NET (MAGNET)*, Signals and Communication Technology, DOI 10.1007/978-90-481-3437-3_4,

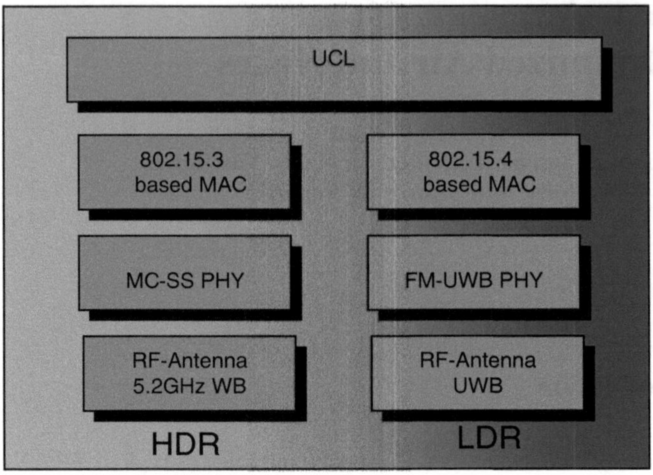

Fig. 4.1 Structure of MAGNET Beyond air interfaces

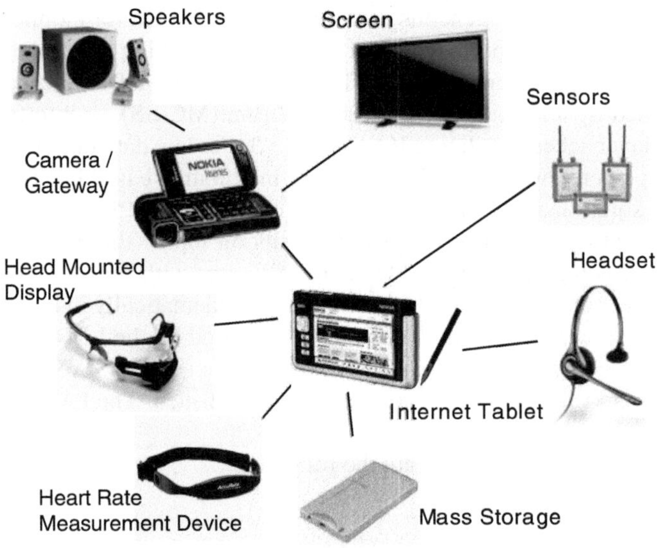

Fig. 4.2 Potential structure of PAN

- The user wants to present the movie on a big screen.
- The video is shown on the screen (streaming).

2. **Play HiFi Audio on Remote Speakers**

- Audio is stored on the internet tablet.
- The user wants to hear the audio files with the speakers in the room.
- The audio files are streamed to the speakers.

3. **High Speed access to Mass Storage**
   - Large data files are stored on the mass storage.
   - The user accesses the files on the mass storage and works with them.
   - Some large files are copied to the internet tablet.
4. **Exchanging data between mobile devices of different PANs**
   - Two people having their own PAN running meet and want to exchange data.
   - A direct connection between both PANs is established.
   - Data (e.g. a movie) is transferred from one user's device to another user's device.
5. **Personal Medical Care**
   - A person wears several body mounted sensors as shown in Fig. 4.3.
   - A connection to a mobile gateway device is established that is able to read and monitor the data measured by the sensors.
   - The mobile gateway may establish a connection to the infrastructure network to compare measurements with a database.

A summary of baseline scenarios is shown in Table 4.1. The table indicates the use of LDR and HDR AIs as suited for the application at hand.

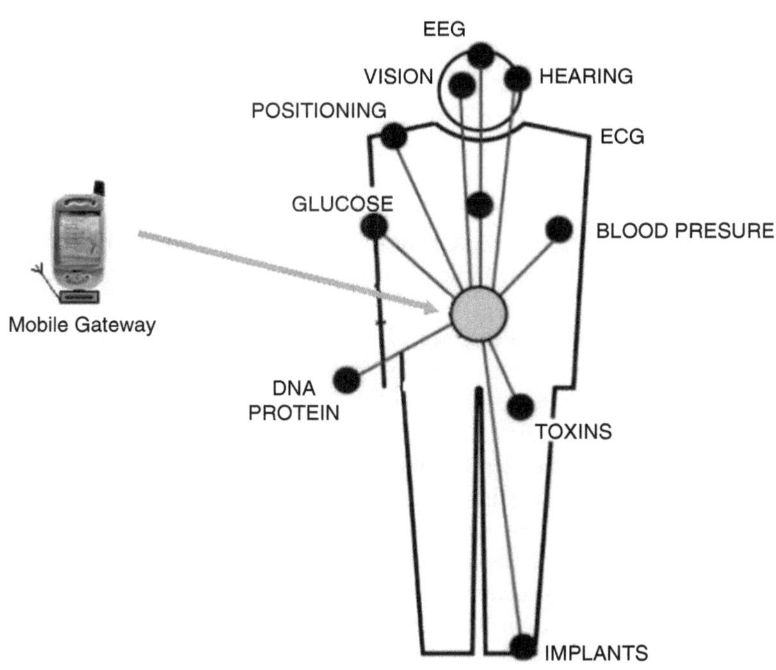

Fig. 4.3 Example of medical care scenario

Table 4.1 Baseline scenarios

Scenario	Range (m)	Mobility (m/s)	Frequency (GHz)	PHY	Bit rate
Showing video on screen	<10	<1	5.2	MC-SS	HDR < 130 Mbps
Play HiFi audio on remote speakers	<10	<1	5.2	MC-SS	HDR < 130 Mbps
High speed access to mass storage	<10	<1	5.2	MC-SS	HDR < 130 Mbps
Exchanging data between mobile devices of different PANs	<10	<3	5.2	MC-SS	HDR < 130 Mbps
Personal medical care	<2	<1	UWB	FM-UWB	LDR < 100 kbps

4.2 Air Interface Description

In this section the MAGNET LDR PHY and the MAC are briefly described. Complete specifications of the PHY and the MAC can be found in [27].

4.2.1 Low Data Rate Transmission with FM-UWB Modulation

The LDR air interface targets short range applications up to 10 m distance and is based upon a robust radio with very low start-up time (<1 ms). The transceiver architecture has been selected to optimise performance and minimise power consumption, meaning at least one order of magnitude lower than conventional mobile handset RF circuitry.

The main focus of the prototyping is a 125 kbps (raw bit rate) system exploiting mainly TDMA (IEEE 802.15.4 MAC) but also, RF and sub-carrier FDMA for multiple access at the physical layer. Table 4.2 summarises the LDR radio characteristics according to the specifications.

RF centre frequency, bandwidth and TX power are dictated by UWB regulations. Switching times and latency are switching times of the hardware components and the group delay occurring in the various filters in the transmitter and receiver. While the current consumption values are estimations for an optimised IC version in appropriate IC technology, Chapter 6 contains the values of the implemented AI.

The target prototype will feature a star topology, in which a base station controls communications with remote nodes in a point to multipoint fashion meaning that no real dynamic routing will be needed so the overall latency remains low. Table 4.3 presents the target user specifications.

The effective data rate equals half the raw data rate due to the use of Manchester encoding.

Ideally power consumption should be determined by the auto discharge time of the battery rather than the radio power consumption. Typical lithium 50 mAh capacity button cell has an auto discharge time equal to 10 years.

Table 4.2 FM-UWB radio characteristics

RF centre frequency	High band (HB)	4.5 GHz
	Low band (LB)	6.4–8.7 GHz
RF bandwidth		500 MHz
RF output power		−14 dBm
Sub-carrier frequency		1–2 MHz
Sub-carrier modulation		FSK, $\beta = 1$
Raw bit rate		≤125 kbps
Receiver sensitivity		−80 dBm
TX, RX switching time		≤10 μs
Latency (at PHY level)		<1 ms @ 100 kbps
RX synchronisation time		<50 bits
Current consumption RX		7 mA
Current consumption TX		4 mA

Table 4.3 LDR target user specification

Effective data rate	<62.5 kbps
Latency (PHY + MAC)	<10 ms[a]
Range	<10 m
BER	$<10^{-3}$
Antenna pattern	Omni-directional
Autonomy	Auto discharge time of battery
Security	128 AES[b]
Cost	1 $/device
Size	USB stick – credit card size

At data rate of 100 kbps and a message of 400 bits sent every minute this means 2,102 s of transmission time per year. With a sensor plus transmit power consumption equal to 5 mA, this means 10,000 mAs = 2.8 mAh per year which is lower than the 5 mAh auto discharge.

4.2.1.1 PHY Layer

This section gives a description of the FM-UWB transceiver. Both transmitter and receiver architecture(s) are described.

[a]The latencies can be further minimised if you forego the beacon environment and are willing to risk potential interference from accidental data interference from accidental data collision with other sensors on the network. Data latency can also affect battery life so for a truly low power sensor network it has to be as low as possible. For simple star networks (few clients, one network coordinator) latencies in the order of few ms can be expected.

[b]This level of security as dictated in IEEE 802.15.4 is not implemented in the Magnet Beyond prototype.

Transmitter Architecture

Figure 4.4 shows the block diagram of the FM-UWB transmitter that implements double FM with a low modulation index digital FSK followed by high modulation index analogue FM creating a constant-envelope UWB signal [1].

The transmitter consists of a 1–2 MHz sub-carrier oscillator generating a triangular signal that is FSK modulated by the transmit data. This sub-carrier signal modulates the RF VCO, yielding a constant-envelope UWB signal with a flat power spectral density and steep spectral roll-off. Figure 4.5 shows the data d(t), the sub-carrier m(t) and the UWB V(t) signals in the time domain for a data transition at $t = 0$ and sub-carrier frequency of 1 MHz; the centre frequency of the UWB signal V(t) was chosen to be 10 MHz for the sake of visibility. In a real FM-UWB system the centre frequency would be 4.5 GHz or higher.

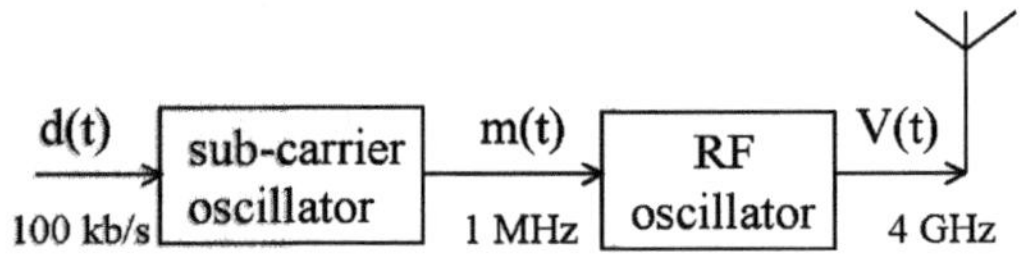

Fig. 4.4 UWB transmitter block diagram

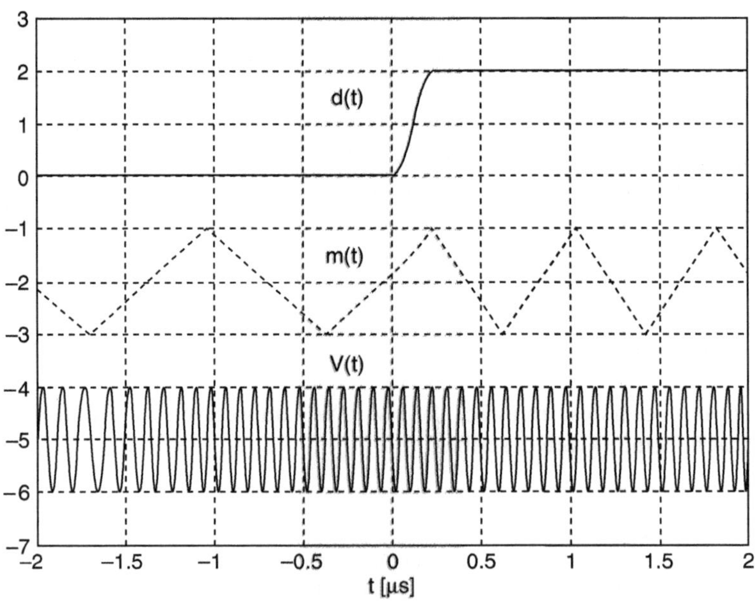

Fig. 4.5 Time domain view of data d(t), sub-carrier m(t) and UWB signal V(t)

Sub-carrier Generation

The sub-carrier generation ($f_{SUB} = 1–2$ MHz) includes the following functionality:

- Data pre-filtering
- Sub-carrier oscillator and FSK modulator

Figure 4.6 shows the block diagram of the complete sub-carrier generation system. The sub-carrier generation ($f_{SUB} = 1–2$ MHz) is implemented by a Direct Digital Synthesiser (DDS) operating at a clock frequency of 20 MHz. The DDS approach offers both precision and flexibility. Sub-carrier frequencies and sub-carrier frequency deviation can be easily modified.

The raw data is Manchester encoded before it enters the DDS to ease clock recovery in the receiver.

Table 4.4 presents the DDS characteristics.

The DDS digital output word that represents the instantaneous phase of the signal is converted into a triangular wave in a DAC. Generation of a triangular wave does not need a look-up table as would be required for the generation of a sine wave. The DAC output signal is next lowpass filtered by interpolation inside the DAC [2] to attenuate aliasing components. This interpolation corresponds to a second order analogue lowpass filter with cut-off frequency of 10 MHz. The filtered DDS signal is passed on to the RF VCO. By using a multiplying DAC the amplitude of the sub-carrier signal and as a result the RF deviation of the FM-UWB output signal can be adjusted. Table 4.5 shows the sub-carrier frequencies for the four-user 100 kbit/s FM-UWB system with sub-carrier frequencies between 1 and 2 MHz that has been the focus of the prototyping.

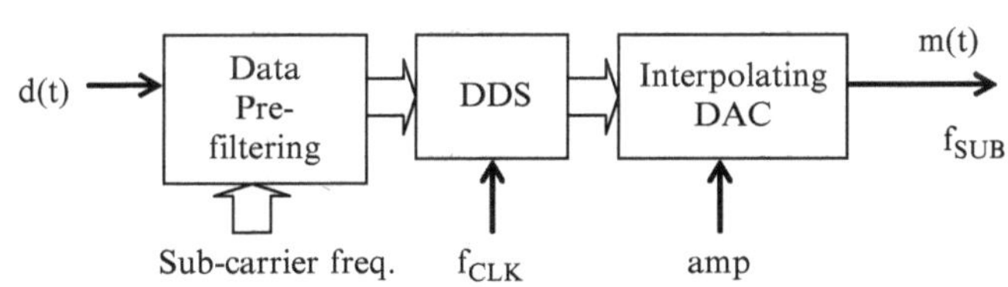

Fig. 4.6 Block diagram of transmitter DDS for sub-carrier generation

Table 4.4 DDS characteristics

Clock frequency	20 MHz
Phase resolution	16 bits
Amplitude resolution	10 bits
Output frequency	1–2 MHz
Frequency resolution	305 Hz
Digital data pre-filtering	Gaussian, BT = 0.7

Table 4.5 Sub-carrier frequencies used in prototype

Sub-carrier	Sub-carrier frequency (MHz)
1	1.00
2	1.25
3	1.50
4	1.75

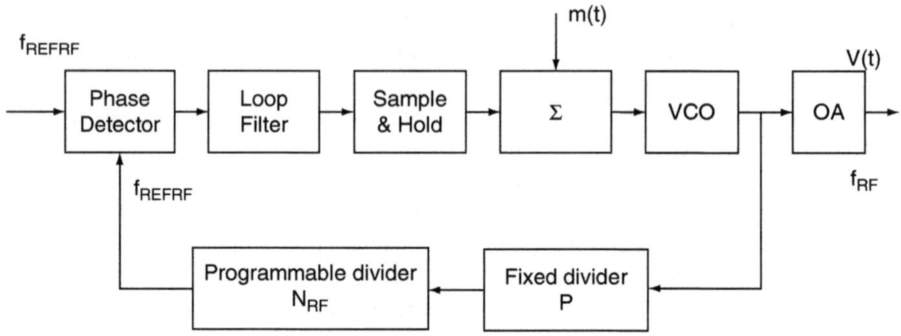

Fig. 4.7 Block diagram of RF signal generation

RF UWB Signal Generation

Figure 4.7 shows the block diagram of the RF signal generation based upon a free-running RF VCO that is regularly calibrated by a PLL frequency synthesiser. This frequency synthesiser ensures the correct centre frequency of the UWB signal. As it does not operate continuously, it does not really impact transmitter power consumption. Modulation of the RF VCO with the sub-carrier signal occurs in open loop mode with the sample-and-hold circuit in hold mode and the frequency synthesiser switched off.

The VCO output signal is the FM-UWB signal which is fed to the output amplifier (OA) providing the appropriate RF output level. The VCO output signal is also fed into a fixed ratio prescaler ($P = 128$ for LB, 256 for HB) that reduces the high VCO output frequency (4.5 or 6–9 GHz) to a lower frequency (35 or 23–35 MHz) compatible with the programmable divider hardware. The RF centre frequency is directly related to the division number N_{RF} of the programmable divider by

$$f_{RF} = N_{RF} P f_{REF} \tag{4.1}$$

With a reference frequency $f_{REF} = 250$ kHz, the centre frequency of the UWB signal will have a resolution of 32 MHz for LB and 64 MHz for HB. Table 4.6 shows the division numbers for the three channels in the lower UWB band between 3.1 and 5 GHz.

Table 4.6 Transmitter division numbers and resulting RF centre frequencies

Channel	NRF	RF centre frequency (MHz)
L1	141	4,512
H1	100	6,400
H2	109	6,976
H3	118	7,552
H4	127	8,128
H5	136	8,704

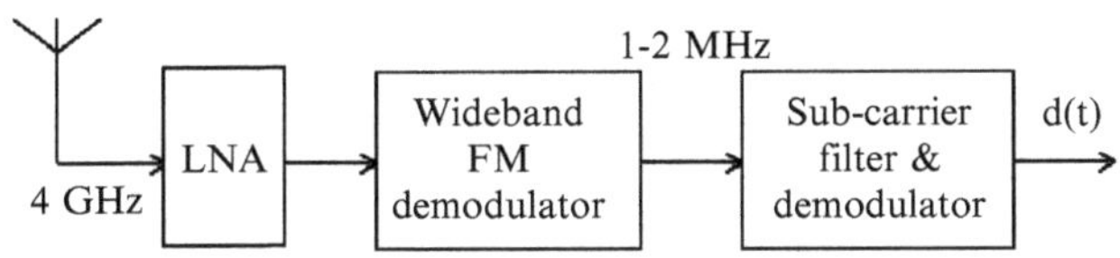

Fig. 4.8 Zero-conversion receiver architecture

One can imagine different frequency allocation schemes than the one shown above. FM-UWB is very flexible in this respect and one can trade bandwidth against receiver processing gain. E.g., a single user with higher RF bandwidth (up to 3 GHz bandwidth) may occupy channel H_3.

Receiver Architecture

Since the transmitter uses double FM modulation, the receiver needs to perform two FM demodulations; one at RF and another one at the sub-carrier frequencies. In the simplest and most low power receiver architecture, the receiver demodulates the FM-UWB signal without frequency translation (i.e., no mixing). Thus, no local oscillator or carrier synchronisation is required. Figure 4.8 shows the zero-conversion receiver block diagram.

The receiver in its most basic form comprises a Low Noise Amplifier (LNA), a wideband FM demodulator, low-frequency sub-carrier filter and amplification stages, and sub-carrier demodulator. A wideband FM demodulator (not preceded by limiting) is the key component of the FM-UWB receiver. The absence of carrier synchronisation allows rapid synchronisation (<1 ms) in ad-hoc networks. In addition, the hardware implementation for FM-UWB is potentially very low power and low cost compared to other wireless schemes because of relaxed oscillator phase noise requirements. Multiple users can share the same RF carrier and distinguish themselves by using different sub-carrier frequencies. Multiple-access interference sets a limit on the number of users [3].

Wideband FM Demodulator

Figure 4.9 presents the wideband FM demodulator based upon a delay line and multiplier.

The relation between the input frequency and the demodulator output voltage for the delay line demodulator is given by

$$V_{FMDEMOD}(f) = \frac{A_1^2}{2} \cos\left(N\frac{\pi}{2}\frac{f}{f_c}\right) \tag{4.2}$$

and shown in Fig. 4.10. The delay time τ is chosen equal to an odd multiple of a quarter period T for the carrier frequency f_c of the FM-UWB signal, i.e.

$$\tau = N\frac{T}{4} = \frac{N}{4f_c} \tag{4.3}$$

with $N = 1, 3, 5$.

The FM demodulator overdrive O is defined as

$$O = \frac{2\Delta f}{B_{DEM}} = N\frac{\Delta f}{f_c} \tag{4.4}$$

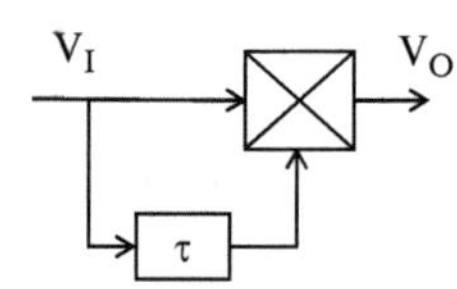

Fig. 4.9 Delay line FM demodulator

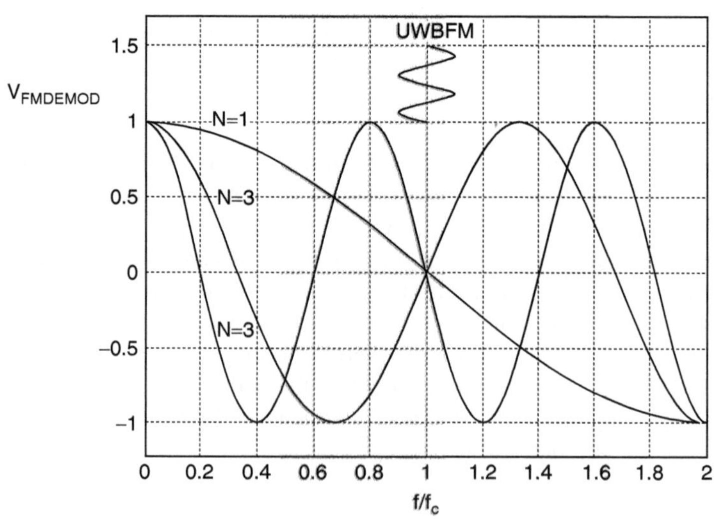

Fig. 4.10 Relation between normalised delay line demodulator input frequency and normalised output voltage for various values of N

One possible choice of the wideband FM demodulator delay time is to choose a value for N that allows for demodulation of the complete lower part of the UWB spectrum (3.1–5 GHz) without retuning of the FM demodulator. This requires $N = 3$ and yields a bandwidth B_{DEM} equal to

$$B_{DEM} = \frac{2}{N} f_c = 2.67\ \text{GHz for a centre frequency}\ f_c = 4\ \text{GHz}.$$

The price to be paid for such a flexible wideband FM demodulator is a sensitivity penalty P at RF equal to

$$\mathrm{P} = -10\ \log 10\ \mathrm{O}\ [\mathrm{dB}] \tag{4.5}$$

For a 1 GHz wide FM-UWB signal the penalty is 4.3 dB, for a 500 MHz wide FM-UWB signal (the minimum required bandwidth), the penalty equals 7.3 dB.

This sensitivity penalty can be avoided by tailoring the bandwidth of the FM demodulator to the UWB signal bandwidth, i.e. by choosing the demodulator bandwidth equal to the bandwidth of the FM-UWB signal, resulting in an overdrive $\mathrm{O} = 1$. Tuning of the demodulator centre frequency is required in that case. Figure 4.11 shows the demodulator bandwidth as a function of the delay time τ normalised to $1/4\mathrm{f_c}$.

It is worthwhile to go for the highest possible overdrive in order to maximise receiver sensitivity. E.g., for a 500 MHz bandwidth FM-UWB signal, N can be increased up to 15 to obtain an overdrive $\mathrm{O} = 1$.

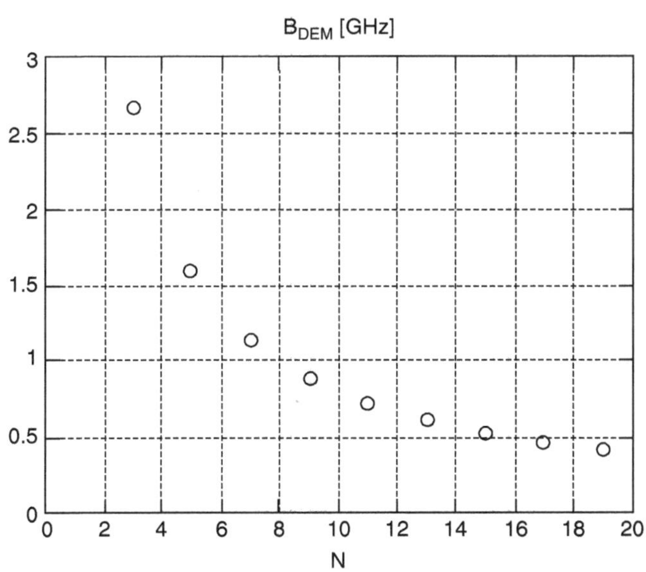

Fig. 4.11 Demodulator bandwidth as a function of delay time ($\mathrm{N} = 4\mathrm{f_c}\tau$)

In reality it is not so straightforward to implement a real delay line on an integrated circuit. A tunable delay is even more challenging if not impossible to realise. The next sub-section proposes the use of an alternative approach using the group delay of a resonator to implement the delay.

Optimised Delay Circuit Implementation

A straightforward approach from a realisation point of view would be to implement the delay not by a delay line but rather as a group delay associated with e.g., a resonator. The most commonly used resonators in harmonic oscillators or frequency selective amplifiers have a bandpass transfer function H described by two complex poles and one real zero. With $p = j\omega$, their transfer function in the frequency domain is given by

$$H = H_0 \frac{\frac{\omega_0}{Q} p}{p^2 + \frac{\omega_0}{Q} p + \omega_0^2} = \frac{H_0}{1 + jQ\frac{\omega}{\omega_0} - jQ\frac{\omega_0}{\omega}} \tag{4.6}$$

By introducing a variable named detuning ν defined as

$$\nu = \frac{\omega}{\omega_0} - \frac{\omega_0}{\omega} \approx 2\frac{\omega - \omega_0}{\omega_0} = 2\frac{\Delta\omega}{\omega_0} \tag{4.7}$$

the resonator transfer can be rewritten as

$$H = \frac{H_0}{1 + jQ\nu} \tag{4.8}$$

This second order resonator is fully characterised by its resonant frequency ω_0, quality factor Q and the maximum value of its transfer function H_0. The dimension of H may be, e.g., that of impedance for a parallel resonant circuit as shown in Fig. 4.12.

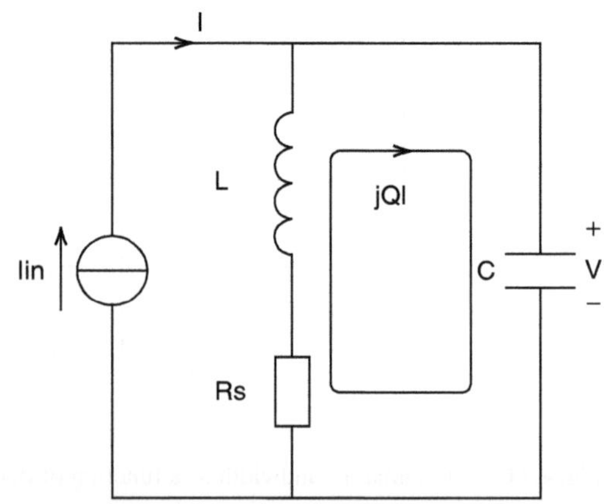

Fig. 4.12 Parallel resonant circuit

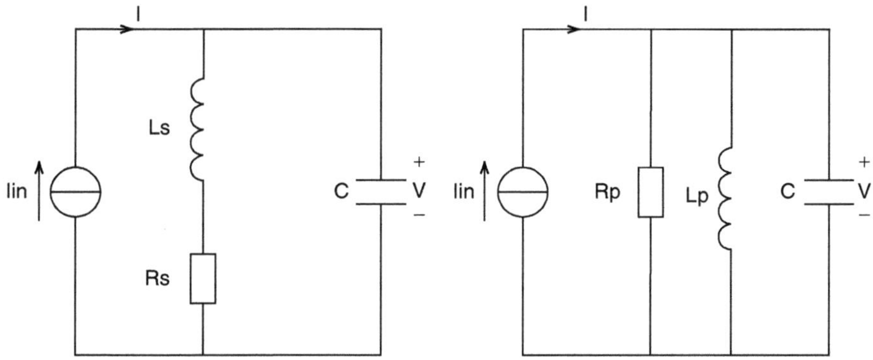

Fig. 4.13 Equivalent circuits for parallel resonant circuit near ω0

For frequencies $\omega \approx \omega_0$, the series connection of inductance L_s and series loss resistance R_s can be replaced by an inductance L_p in parallel with an equivalent parallel loss resistance R_p as shown in Fig. 4.13.

For the parallel resonant circuit, the internal current flowing through L_p and C is a factor Q larger than the external current. The values for the inductance L_p and loss resistance R_p are

$$\begin{aligned} L_p &= \frac{1+Q^2}{Q^2} L_s \approx L_s \\ R_p &= \left(1+Q^2\right) R_s \approx Q^2 R_s \end{aligned} \tag{4.9}$$

The impedance of the parallel resonant circuit is given by

$$\begin{aligned} Z = \frac{V}{I} &= \frac{R_P}{1 + j\omega R_p C - j\omega \frac{R_p}{\omega L}} \\ &= \frac{R_p}{1 + jQ\frac{\omega}{\omega_0} - jQ\frac{\omega_0}{\omega}} = \frac{R_p}{1 + jQ\nu} \end{aligned} \tag{4.10}$$

Its magnitude and phase are given by

$$\begin{aligned} |Z| &= \frac{R_p}{\sqrt{1+Q^2\nu^2}} \\ \arg(Z) &= -a\tan(Q\nu) \approx -Q\nu = -\frac{2Q\Delta\omega}{\omega_0} \end{aligned} \tag{4.11}$$

The resonant frequency and quality factor are given by

$$\begin{aligned} \omega_0 &= \frac{1}{\sqrt{LC}} \\ Q &= \frac{\omega_0 L}{R_s} = \frac{R_p}{\omega_0 L} = R_p\sqrt{\frac{C}{L}} \end{aligned} \tag{4.12}$$

The group delay τ_g equals the frequency derivative of the phase

$$\tau_g = -\frac{\partial}{\partial\omega}\arg(Z) = \frac{2Q}{\omega_0} = \frac{Q}{\pi f_0} \tag{4.13}$$

The resonator group delay τ_g equals the delay τ required in the delay line demodulator and $f_0 = f_c$

$$\tau_g = \tau = \frac{N}{4f_0} \text{ with } N = 1, 3, 5, \ldots \text{ which yields for the quality factor Q}$$

$$Q = \frac{\pi}{4}N \approx 0.79N \tag{4.14}$$

The quality factor of the resonator can be varied continuously resulting in a continuous tuning of the wideband FM demodulator bandwidth.

There are two additional details that need to be taken into account when we go for circuit implementation:

1. The phase shift at the centre frequency $f_c = f_0$ needs to be equal to $\pi/2$ (90°). This means that an additional phase shift of $\pi/2$ is required in order to be able to use the parallel resonant circuit in the wideband FM demodulator. This can be accomplished by, e.g., a differentiator circuit.
2. The magnitude of the impedance of the parallel resonant circuit and the differentiator are not flat over the full bandwidth of the demodulator, which will introduce some distortion in the demodulator output signal. These imperfections that show up as harmonic distortion in the demodulated signals (the sub-carriers) are quite acceptable. Harmonic distortion is present anyway, as a result of multipath (see Section 4.3).

Figure 4.14 shows how the concept may be translated into a circuit realization.

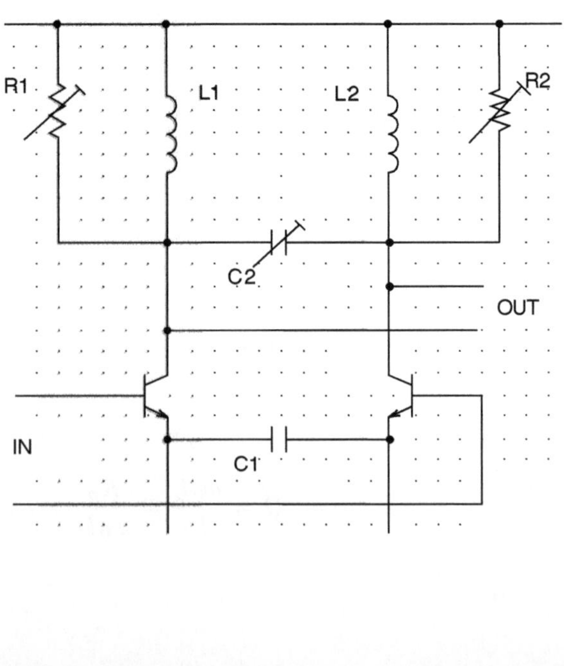

Fig. 4.14 Possible implementation of variable delay circuit

The differentiation is taken care of by the capacitively degenerated differential pair. L_1, L_2, C_2, R_1 and R_2 constitute the parallel resonant circuit. The centre frequency can be tuned by varying C_2 (on-chip varactor) whereas the delay can be tuned by varying R_1 and R_2 (MOS resistors). Another characteristic of this circuit is that its gain is proportional to the quality factor.

Receiver Sub-carrier Processing

A direct-conversion architecture with baseband low-pass filters and a baseband FSK demodulator alleviates the filter requirements. Figure 4.15 shows a corresponding architecture.

Multiple-Access Interference

In addition to TDMA and RF centre frequency FDMA techniques, FM-UWB technology exploits sub-carrier FDMA as possible access scheme. Individual users with a common RF centre frequency distinguish themselves by different sub-carrier frequencies. Using a FM demodulator without hard-limiting allows for simultaneous demodulation of multiple FM signals at the same centre frequency.

We will now investigate the multiple-access interference occurring in the FM-UWB system using this sub-carrier FDMA access scheme. Figure 4.16 shows the case where the wideband FM demodulator input equals the sum of N UWB signals $\{V_1(t), V_2(t), \ldots, V_N(t)\}$.

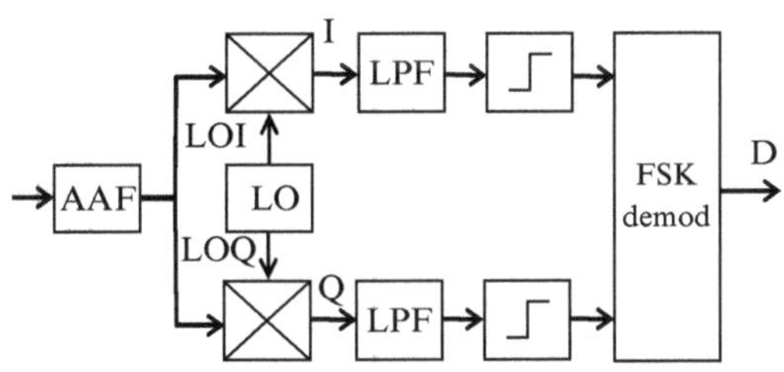

Fig. 4.15 Receiver sub-carrier processing with anti-aliasing filtering (AAF)

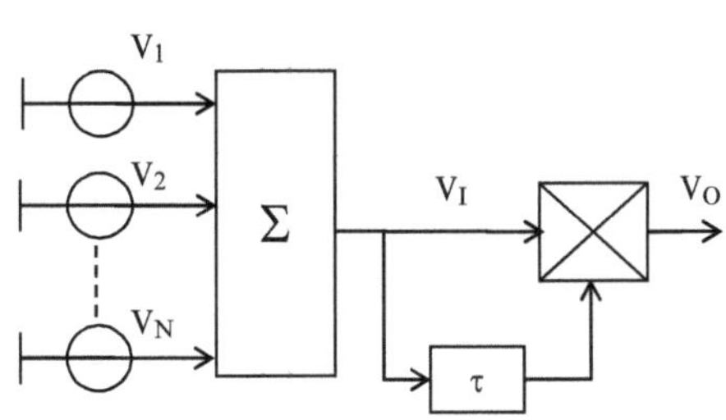

Fig. 4.16 Wideband FM demodulator with N FM-UWB input signals

The output signal V_O of the wideband FM demodulator contains N^2 terms:

- N terms of the form $V_i(t)V_j(t-\tau)$, $(i = 1, 2,.., N)$ constituting the sum of the N sub-carrier signals
- N(N-1) terms of the form $V_i(t)V_j(t-\tau)$, $(i = 1, 2,.., N; j = 1, 2,.., N, j \neq i)$ constituting the multiple-access interference residue

We assume that all N users have the same centre frequency f_c, the same deviation Δf, yet each of them has its particular and unique sub-carrier frequency f_{SUBi}. The sub-carrier waveform is triangular resulting in a flat power spectral density of the UWB signal. Two cases are of particular interest in a multi-user environment:

- 2 FM-UWB signal of unequal power $(S1 \neq S2)$
- N FM-UWB signals of equal signal power $(S1 = S2 = S3 = \cdots SN)$

In particular, the two-user case is often encountered where a strong interferer represented by signal V_2 causes interference to the desired signal V_1.

4.2.1.2 MAC Layer

The LDR MAC for MAGNET Beyond is based on IEEE 802.15.4 standard. This section describes the MAGNET Beyond LDR MAC specifications.

Operational Overview

The IEEE 802.15.4 specifications provide the option of operating a network in a number of topologies viz. star, peer-to-peer and cluster tree. Multiple topologies make the protocol versatile enough to cater to different applications and scalabilities. The basic building block of an 802.15.4 network in Fig. 4.17 is a PAN.

A PAN is started and maintained by a coordinator. The coordinator is responsible for beaconing, device associations, regulating channel access, etc. Two different types of devices can participate in a LR-WPAN:

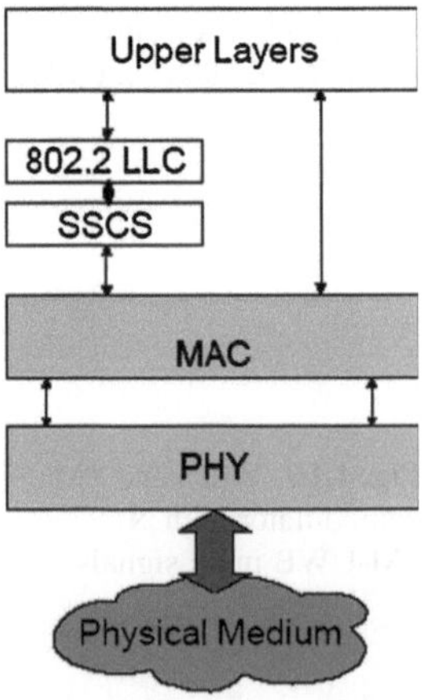

Fig. 4.17 IEEE 802.15.4 Architecture

- Full function devices (FFDs)
- Reduced function devices (RFDs)

An FFD can operate serving as PAN coordinator unlike an RFD. An FFD can talk to RFDs or other FFDs whereas an RFD can communicate only with an FFD. An RFD implies a very simple implementation and is intended for extremely simple applications (e.g., light switch).

The network has the option of operating in a beacon enabled mode or a non-beacon mode. In beacon enabled mode, channel access is facilitated by superframes. Each superframe is bounded by a beacon from the coordinator. Any device that wants to communicate in the PAN should synchronise with the beacon. In non-beacon enabled mode, the devices specifically ask for a beacon from the coordinator in order to carry out any transmission in the PAN.

Channel can be accessed either by CSMA/CA or TDMA mechanisms. The superframe is divided into a mandatory contention access period (CAP) and an optional contention free period (CFP). In the CAP a slotted CSMA/CA medium access scheme is followed while CFP is TDMA based. All devices that require dedicated time slots for data transmission can request the coordinator for guaranteed time slot (GTS) allocation.

Supported Network Topologies (Star, Peer to Peer, Cluster)

IEEE 802.15.4 provides the option for the three operational topologies in Fig. 4.18.

Star Topology

All the member devices communicate only with the coordinator and not among themselves. All star networks operate independent of the other star networks operating in the region. This can be achieved by choosing a unique PAN ID.

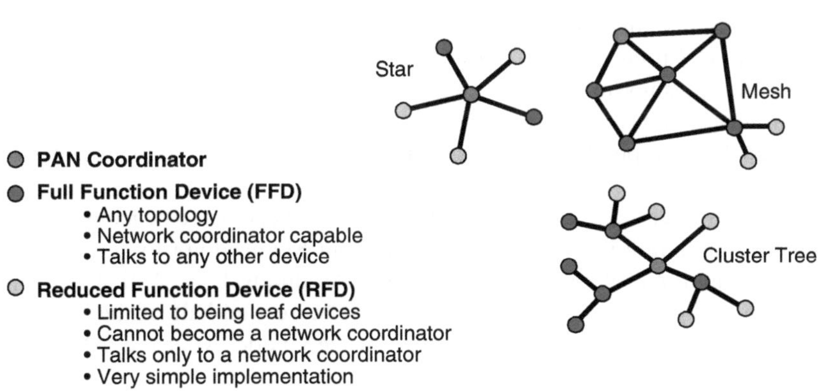

Fig. 4.18 IEEE 802.15.4 network topologies

Peer-to-Peer (Mesh) Topology

This network topology provides the option of communication between peers within a PAN. Since, only FFDs can communicate among themselves, this topology is more conducive for situations wherein the member devices are FFDs. One device will be nominated as the PAN coordinator, for instance, by virtue of being the first device to communicate on the channel. Further network structures can be constructed out of the peer-to-peer topology and may impose topological restrictions on the formation of the network.

Cluster Tree

Cluster tree topology can be viewed as an example of peer to peer networking in which most of the devices are FFDs. An RFD may connect to a cluster tree network as a leave node at the end of a branch, because it may only associate with one FFD at a time. Any of the FFDs may act as a coordinator and provide synchronisation services to other devices or other coordinators. Only one of these coordinators can be the overall PAN coordinator, which may have greater computational resources than any other device in the PAN. The PAN coordinator forms the first cluster by establishing itself as the cluster head (CLH) with a cluster identifier (CID) of zero, choosing an unused PAN identifier, and broadcasting beacon frames to neighbouring devices. A candidate device receiving a beacon frame may request to join the network at the CLH.

Superframe Structure

The MAC spec defines superframe as a chunk of 16 equally sized slots. Superframe consists of a CAP (Contention Access Period) followed by CFP (Contention Free Period). Nodes use CSMA to get medium access during CAP. Each CFP consists of up to seven GTS (Guaranteed Time Slots) to various nodes. A GTS may occupy more than one slot. Information about GTS for various nodes is mentioned in the beacon. All GTS transmissions must end before the start of the beacon transmission. All the transmission in CAP must end before the start of CFP and start of beacon transmission. Acknowledgement for a packet is optional and the requirement for an ACK is specified in a data packet. Figure 4.19 illustrates the superframe structure.

BI (Beacon Interval) is the duration between two successive beacons and SD is the superframe duration. Generally $SD = BI$, but SD need not to be equal to BI. BI can be a multiple of SD i.e. $BI = k * SD$. Assuming $k = 2$, first half of the superframe will be active with CAP and CFP. The second half of the superframe will be inactive and the PAN coordinator is in sleep mode. After the SD elapses, it wakes up and the same cycle goes on.

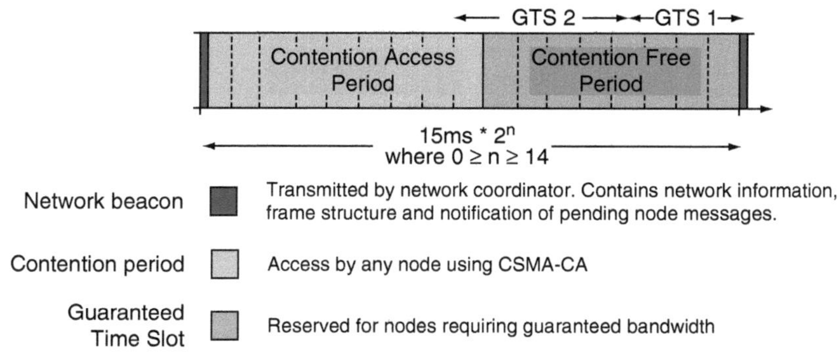

Fig. 4.19 Superframe structure

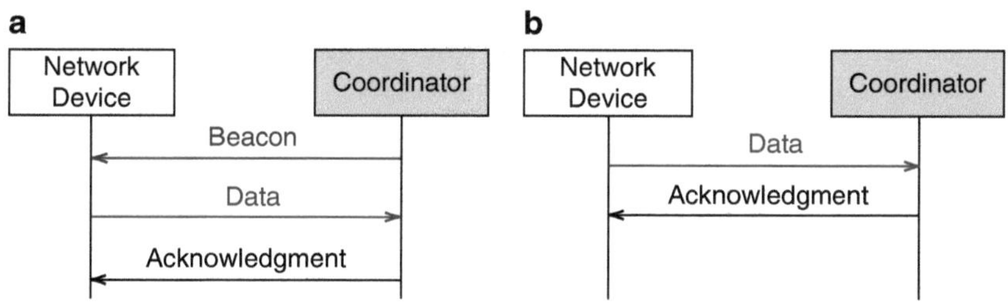

Fig. 4.20 Direct data transmission in (**a**) beacon enabled mode (**b**) non-beacon enabled mode

In IEEE 802.15.4 all GTS transmission must end before the start of the beacon transmission. All the transmissions in CAP must end before the start of CFP and start of beacon transmission. This is done to save power of RFDs so that they can wake up at regular intervals for beacons without having to wait.

Data Transfer Models

The mechanisms for each transfer type depend on whether the network supports the transmission of beacons. A beacon-enabled network is used for supporting low-latency devices, such as PC peripherals. However, the beacon is still required for network association.

Three types of data transfer transactions exist.

1. *Direct data transmission* (Fig. 4.20): data are transferred from a device (Tx) to a coordinator (Rx).
   In a *beacon-enabled* network if a device wishes to transfer data to a coordinator:
   - First the device listens for the network beacon.
   - When the beacon is found the device synchronises to the superframe structure.
   - Then the device transmits its data frame to the coordinator (i.e. using slotted CSMA-CA).

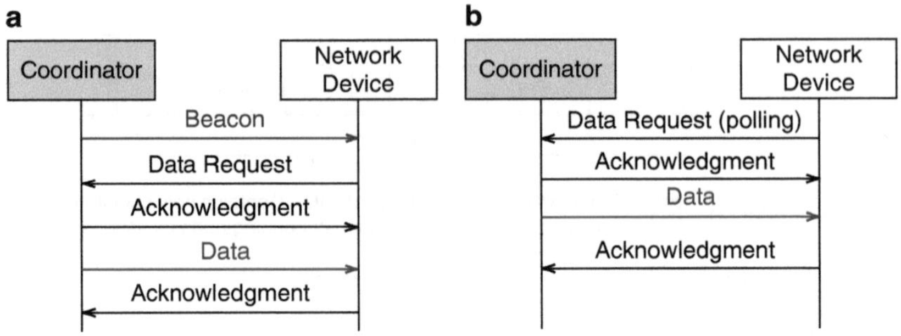

Fig. 4.21 Indirect data transmission in (**a**) beacon enabled mode (**b**) non-beacon enabled mode

The coordinator sends an acknowledgement frame upon successful reception
In *non-beacon-enabled* network, when a device wishes to transfer data to the coordinator, it does so using unslotted CSMA-CA.

2. *Indirect data transmission* (Fig. 4.21): data are transferred from a coordinator (Tx) to a device (Rx).
   In a *beacon-enabled* network, when the coordinator wishes to transfer data to a device:

   - The coordinator indicates in the network beacon that a data message is pending.
   - The device periodically listens to the network beacon.
   - The device transmits a MAC command requesting the data using slotted CSMA-CA.
   - The coordinator sends an acknowledgement frame upon successful reception of the data request.
   - The coordinator sends the pending data frame using slotted CSMA-CA.
   - The device sends an acknowledgement frame upon successful reception of the data.

   The message is removed from the list of pending messages in the beacon
   In a *non-beacon-enabled* network if a coordinator wishes to transfer data to a device:

   - The coordinator stores the data.
   - The device sends a request to its coordinator using unslotted CSMA-CA.
   - If data are pending the coordinator transmits the data frame to the device, using unslotted CSMA-CA.
   - If data are not pending, the coordinator transmits a data frame with a zero-length payload which indicates that no data were pending.
   - The device sends an acknowledgement frame upon successful reception of the data.

3. *Transmission between two peer devices*: in star topology only direct or indirect data transmissions are possible, because data may be exchanged only between

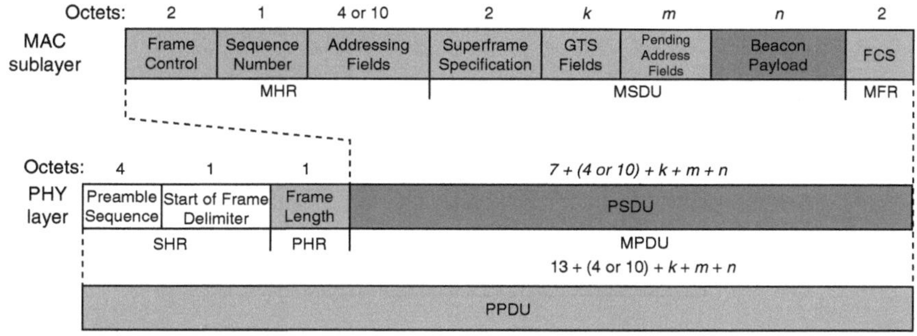

Fig. 4.22 Beacon frame

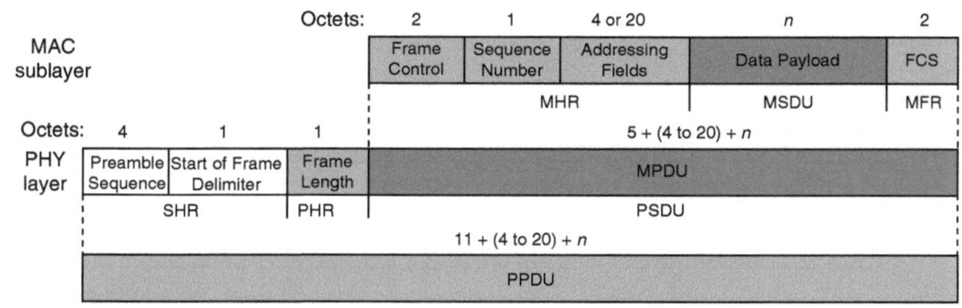

Fig. 4.23 Data frame

the coordinator and a device. In a peer-to-peer topology data may be exchanged between any two devices on the network; consequently all three data transfer models are allowed.

Frame Structures

Four types of frame structures are defined by the MAC standard viz.

- Beacon frame (Fig. 4.22): Used by the coordinator.
- Data frame (Fig. 4.23): Used for all transfer of data.
- Acknowledgement Frame (Fig. 4.24): Used to acknowledge reception of command and data frames.
- MAC Command Frame (Fig. 4.25): Used for handling all MAC peer entity control transfers.

MAC Functional Description

Channel Access

Depending on the network configuration, an LR-WPAN may use one of two channel access mechanisms. In a beacon-enabled network with superframes, a slotted carrier

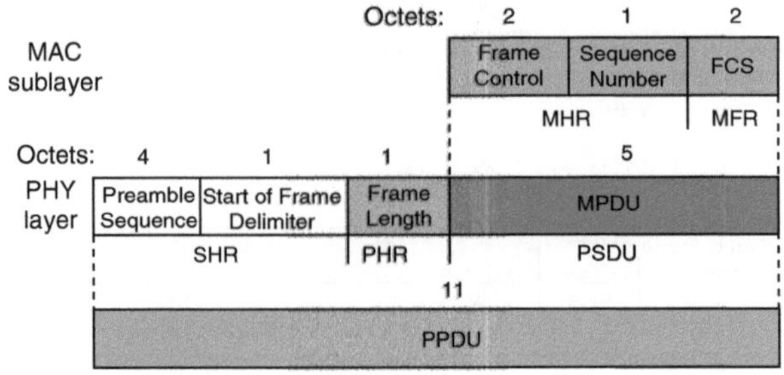

Fig. 4.24 Acknowledgement frame

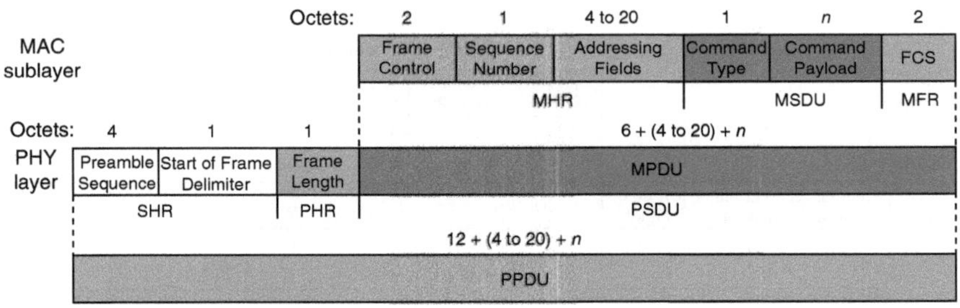

Fig. 4.25 MAC command frame

sense multiple access with collision avoidance (CSMA-CA) mechanism is used. In networks without beacons, unslotted or standard CSMA-CA is used. The MAGNET LDR WPAN implements the IEEE 802.15.4 MAC layer in beacon-enabled mode only. In this mode, the channel access mechanism is a slotted CSMA-CA, where the back-off slots are aligned with the start of the beacon transmission. Each time a device wishes to transmit data frames during the CAP, it shall locate the boundary of the next back-off slot and then wait for a random number of back-off slots. If the channel is busy, following this random back-off, the device shall wait for another random number of back-off slots before trying to access the channel again. If the channel is idle, the device can begin transmitting on the next available back-off slot boundary. Acknowledgement and beacon frames shall be sent without using a CSMA-CA mechanism.

Super-Frame Structure

The IEEE 802.15.4 MAC standard allows the optional use of a superframe structure. This option is used for the LDR WPAN in MAGNET. The superframe is bounded by network beacons and is divided into 16 equally sized slots. The beacon frame is sent in the first slot of each superframe. The beacons are used to synchronise the attached devices, to identify the PAN and to describe the structure of superframes.

In beacon mode the superframe can have an active and an inactive portion. During the inactive portion, the coordinator shall not interact with its PAN and may enter a low-power mode. The active portion consists of CAP and CFP. Any device wishing to communicate during the CAP shall compete with other devices using a slotted CSMA-CA mechanism. On the other hand, the CFP contains guaranteed time slots (GTSs), which always appear at the end of the active superframe starting at a slot boundary immediately following the CAP. The PAN coordinator may allocate up to seven of these GTSs and a GTS can occupy more than one slot period.

The duration of different portions of the superframe are described by the values of *macBeaconOrder* and *macSuperFrameOrder*. *macBeaconOrder* describes the interval at which the coordinator shall transmit its beacon frames. The beacon interval, BI, is related to the *macBeaconOrder*, BO, as follows: BI $=$ *aBaseSuperFrameDuration* $*2^{\mathrm{BO}}$, for $0 \leq \mathrm{SO} \leq \mathrm{BO} \leq 14$. The superframe is ignored if BO $= 15$.

The value of *macSuperFrameOrder* describes the length of the active portion of the superframe, which includes the beacon frame. The superframe duration, SD, is related to *macSuperFrameOrder*, SO, as follows: SD $=$ *aBaseSuperFrameDuration* $*2^{\mathrm{SO}}$ symbols, for $0 \leq \mathrm{SO} \leq 14$. If SO $= 15$, the superframe should not remain active after the beacon.

The active portion of each superframe is divided into a *aNumSuperFrameSlots* equally spaced slots of duration $2^{\mathrm{SO}} *$ *aBaseSlotDuration* and is composed of three parts: a beacon, a CAP and CFP. The beacon is transmitted at the start of slot 0 without the use of CSMA.

An example superframe structure is shown in Fig. 4.26.

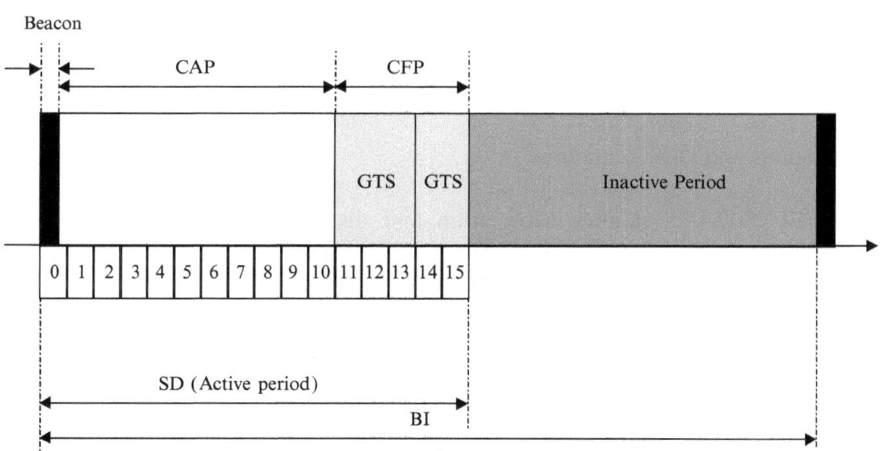

Fig. 4.26 The superframe structure and relationship between CAP, CFP, SD, and BI

CAP

The CAP starts immediately after the beacon and completes before the beginning of the CFP on a superframe slot boundary. The CAP shall be at least *aMinCAPLength* symbols unless additional space is needed to temporarily accommodate the increase in the beacon frame length to perform GTS maintenance. All frames except acknowledgement frames or any data frame that immediately follows the acknowledgement of a data request command that are transmitted in the CAP shall use slotted CSMA-CA to access the channel. A transmission in the CAP shall be complete one IFS (InterFrameSpace) period before the end of the CAP. If this is not possible, the sender defers its transmission until the CAP of the following superframe.

CFP

The CFP shall start on a slot boundary immediately following the CAP and extends to the end of the active portion of the superframe. The length of the CFP is determined by the total length of all of the combined GTSs. No transmissions within the CFP shall use a CSMA-CA mechanism. A device transmitting in the CFP shall ensure that its transmissions are complete one IFS period before the end of its GTS.

IFS

IFS time is the amount of time necessary to process the received packet by the PHY. Transmitted frames shall be followed by an IFS period. The length of IFS depends on the size of the frame that has just been transmitted. Frames of up to *aMaxSIFSFrameSize* in length shall be followed by a SIFS (ShortIFS) whereas frames of greater length shall be followed by a LIFS (LongIFS).

Associations and Disassociations

The IEEE 802.15.4 defines association and disassociation functions for self-configuration in its MAC sublayer. The coordinator is responsible for association and disassociation of devices in the PAN. To associate with a coordinator, a device will perform channel scan to find the existing coordinators. There are three kinds of scans in the 802.15.4:

- Energy scan (ED)
- Passive scan
- Active scan

The Energy scan measures the energy level of each channel to select suitable channel to be used. The Passive channel scan in which no beacon request frame is sent and the Active channel scan in which a beacon request frame is sent locate a suitable coordinator.

After successful completion of the scan procedure, the results of the channel scan are used to choose a suitable PAN. A device shall attempt to associate only with a PAN that is currently allowing association. The device stores all required information for the selected PAN: channel, PAN identifier (ID), the beacon order (BO) and superframe order (SO). The last three parameters are acquired from the beacon of the selected coordinator found while scanning. The algorithm used in the selection of which PAN to associate with from the list of PAN descriptors returned from the channel scan is beyond the scope of the standard.

Following the selection of a PAN with which to associate, the next higher layers request that MLME configures the *phyCurrentChannel* to the appropriate logical channel on which to associate, *macPANId* to the identifier of the PAN with which to associate and *macCoordExtendedAddress* or *macCoordShortAddress* to the address of the coordinator with which it associates.

An unassociated device shall initiate the association procedure by sending an associate request command to the coordinator of an existing PAN. If the association request command is received correctly, the coordinator shall send an acknowledgement. This acknowledgement, however, does not mean that the device has associated. The coordinator needs time to determine whether the current resources available on the PAN are enough to allow another device to associate. This decision should be made within *aResponseWaitTime* symbols. If sufficient resources are available, the coordinator shall allocate a short address to the device and generate an association response command containing the new address and a status indicating the successful association. If there are not enough resources, the coordinator shall generate an association response command containing a status indicating failure. This response is sent to the device using indirect transmission, i.e., the association response command frame shall be added to the list of pending transactions stored on the coordinator and extracted at the discretion of the device.

On the other side, the device, after getting the acknowledgement frame, waits for the response for *aResponseWaitTime* symbols. It either checks the beacons in the beacon-enabled network or extracts the association response command from the coordinator after *aResponseWaitTime* symbols. On reception of association response command, the device shall send an acknowledgement. If the association is successful, the device shall store the address of the coordinator with which it has associated.

When a coordinator wants one of its associated devices to leave the PAN, it shall send the disassociation notification command to the device using indirect transmission. Upon reception of the packet, the device should send the acknowledgement frame. Even if the ack is not received, the coordinator shall consider the device disassociated.

If an associated device wants to leave the PAN, it shall send a disassociation notification command to the coordinator. Upon reception, the coordinator sends ack. Even if the ack is not received, the device shall consider itself disassociated.

An associated device shall disassociate itself by removing all references to the PAN. A coordinator shall disassociate a device by removing all references to that device.

Synchronisation

In a beacon enabled network, devices shall be permitted to acquire synchronisation only with beacons containing the PAN identifier specified in *macPANId*. If tracking is specified in the MLMESYNC.request primitive, the device shall attempt to acquire the beacon and keep track of it by regular and timely activation of its receiver. It shall enable its receiver at a time prior to the next expected beacon frame transmission, i.e. just before the known start of the next superframe. If tracking is not specified, the device shall attempt to acquire the beacon only once.

To acquire beacon synchronisation, a device shall enable its receiver and search for at most *aBaseSuperframeDuration* $*(2^n + 1)$ symbols, where n is the *macBeaconOrder*. If a beacon frame containing the current PAN identifier of the device is not received, the MLME shall repeat the search. Once the number of missed beacons reached *aMaxLostBeacons*, the MLME notifies the next upper layer by issuing MLME-SYNC-LOSS.indication with a reason BEACON-LOSS. The MLME shall timestamp each received beacon frame at the same symbol boundary within each frame, the location of which is implementation specific. The symbol boundary shall be chosen to be the same as that used in the timestamp of the outgoing beacon frame, stored in *macBeaconTxTime*.

Transmission

The MAC is responsible to create a header for the data that has to be transmitted. Each packet is identified by the packet number, *macDSN* for all the frames and *macBSN* for beacons.

In order to transmit a data or a MAC command frame or a beacon, the MAC sublayer shall copy the value of *macDSN* or *macBSN* into the sequence number field of the MHR of the outgoing frame and then increment it by one. The source address field shall contain the address of the device sending the frame. If the device has been allocated a short address, it shall use that address in preference to its 64 bit extended address. If the source address field is not present, the originator of the frame shall be assumed to be a PAN coordinator and the destination address shall contain the address of the recipient. The destination address shall contain the intended recipient of the frame, which may be either a 16 bit short address or a 64 bit extended address. If the destination address field is not present, the recipient of the frame shall be assumed to be the PAN coordinator. The destination and source address may be in different PANs, which are identified by the PAN identifier fields.

In beacon-enabled PANs, the transmitting device shall attempt to find the beacon before transmitting. If it cannot find the beacon, it shall use unslotted CSMA-CA. Once the beacon is found, it transmits in the appropriate portion of the superframe. Transmissions in the CAP shall use slotted CSMA-CA and those in GTS shall not use CSMA-CA.

Reception

Reception is important in terms of energy consumption. Each device may choose whether the MAC sublayer is to enable its receiver during idle periods. During these idle periods, the MAC sublayer shall still service transceiver task requests from the next higher layer. On completion of each transceiver task, the MAC sublayer shall request that the PHY enables or disables its receiver, depending on whether *macRxOnWhenIdle* is set to TRUE or FALSE, respectively. In MAGNET LDR MAC, that supports only beacon enabled mode, the value of *macRxOnWhenIdle* shall be considered only during idle periods of the CAP.

Due to the nature of radio communications, a device with its receiver enabled will be able to receive and decode transmissions from all devices complying with IEEE 802.15.4–2003 that are currently operating on the same channel and are in its Personal Operating Space (POS), along with interference from other sources. The MAC sublayer shall, therefore, be able to filter incoming frames and present only the frames that are of interest to the upper layers.

Upon reception of packets, the MAC sublayer shall discard all its received frames that do not contain a correct value in their FCS field in the MFR.

The next level of filtering shall be dependent on whether the MAC sublayer is currently operating in promiscuous mode. In promiscuous mode, the MAC sublayer shall pass all frames received after the first filter directly to the upper layers without applying any more filtering. The MAC sublayer shall be in promiscuous mode if *macPromiscuousMode* is set to TRUE.

If the MAC sublayer is not in promiscuous mode, it shall accept only frames that satisfy all of the following requirements:

- The frame type subfield of the frame control field shall not contain an illegal frame type.
- If the frame type indicates that the frame is a beacon frame, the source PAN identifier shall match *macPANId* unless *macPANId* is equal to 0xffff, in which case the beacon frame shall be accepted regardless of the source PAN identifier.
- If a destination PAN identifier is included in the frame, it shall match *macPANId* or shall be the broadcast PAN identifier ($0 \times$ ffff).
- If a short destination address is included in the frame, it shall match either *macShortAddress* or the broadcast address ($0 \times$ ffff). Otherwise, if an extended destination address is included in the frame, it shall match *aExtendedAddress*.
- If only source addressing fields are included in a data or MAC command frame, the frame shall be accepted only if the device is a PAN coordinator and the source PAN identifier matches *macPANId*.

If any of the requirements listed above are not satisfied, the MAC sublayer shall discard the incoming frame. If all of the requirements listed above are satisfied, the frame shall be considered valid and processed further.

Acknowledgement

An important function of the MAC is confirming successful reception of a received frame. The specifications provide two acknowledgement policies for ensuring that the data exchange is reliable: No Acknowledgement and Acknowledgement.

A data or MAC command frame shall be sent with the acknowledgement request subfield of its frame control field set appropriately for the frame and successful reception and validation are confirmed with an acknowledgement.

If the receiving device is unable to handle the incoming message for any reason, the receipt is not acknowledged. The frame control field indicates whether or not an acknowledgement is expected. The acknowledgement frame is sent immediately after successful validation of the received frame. Beacon frames sent by a PAN coordinator and acknowledgement frames are never acknowledged.

A frame transmitted with its acknowledgement request subfield set to 0 (No Acknowledgement) shall not be acknowledged by its intended recipient. The originating device shall assume that the transmission of the frame was successful.

A frame transmitted with the acknowledgement request subfield of its frame control field set to 1 (Acknowledgement) shall be acknowledged by the recipient. If the intended recipient correctly receives the frame, it shall generate and send an acknowledgement frame containing the same DSN (Data Sequence Number) from the data or MAC command frame that is being acknowledged.

The transmission of an acknowledgement frame in the CFP shall commence *aTurnaroundTime* symbols after the reception of the last symbol of the data or MAC command frame. The transmission of an acknowledgement frame in the CAP shall commence at a back-off slot boundary.

GTS

A GTS allows a device to operate on the channel within a portion of the superframe that is dedicated exclusively to that device. A device shall attempt to allocate and use a GTS only if it is currently tracking the beacons. A GTS shall be allocated only by the PAN coordinator and it shall be used only for communications between the PAN coordinator and a device. A single GTS can extend over one or more superframe slots. The PAN coordinator may allocate up to seven GTSs at the same time, provided there is enough capacity in the superframe.

A GTS shall be allocated before use, with the PAN coordinator deciding whether to allocate a GTS based on the requirements of the GTS request and the current available capacity in the superframe. GTS shall be allocated on a first-come-first-serve basis and all GTSs shall be placed contiguously at the end of the superframe and after the CAP. Each GTS shall be de-allocated when the GTS is no longer required, and a GTS can be deallocated at any time at the discretion of the PAN coordinator or by the device that originally requested the GTSs. A device that has been allocated GTS may also operate in the CAP.

The management of the GTSs shall be undertaken by the PAN coordinator only. For each GTS, the PAN coordinator shall be able to store its starting slot, length, direction and associated device address.

The GTS direction is specified as either transmit or receive. Each device may request one transmit GTS and/or one receive GTS. For each allocated GTS, the device shall be able to store its starting slot, length and direction. If a device has been allocated a receive GTS, it shall enable its receiver for the entirety of the GTS. In the same way, a PAN coordinator shall enable its receiver for the entirety of the GTS if a device has been allocated a transmit GTS.

A device is instructed to request the allocation of a new GTS through the GTS request command, with GTS characteristics (e.g. direction, length) set according to the requirements of the intended application. On receipt of this command, the PAN coordinator shall send an acknowledgement frame. Following the ack transmission, the PAN coordinator shall first check if there is available capacity in the current superframe based on the remaining length of the CAP and the desired length of the requested GTS. The superframe shall have available capacity if the maximum number of GTSs has not been reached and allocating a GTS of the desired length would not reduce the length of the CAP to less than *aMinCAPLength*. The PAN coordinator shall make its decision within *aGTSDescPersistenceTime* superframes. On receipt of the ack from the coordinator, the device shall continue to track the beacons and wait for at most *aGTSDescPersistenceTime* superframes. If no relevant GTS descriptor is received in the superframe during this period of time, the MAC sublayer of the device shall notify the next upper layer of failure to obtain the requested GTS.

When the coordinator determines whether capacity is available for the requested GTS, it shall generate a GTS descriptor with the requested specifications and the short address of the requesting device. This descriptor indicates the length and the start of the GTS in the superframe and notifies the next upper layer of the new GTS allocation. If there was not sufficient capacity to allocate the requested GTS, the start slot shall be set to 0 and the length to the largest GTS length that can currently be supported. This GTS descriptor shall remain in the beacon frame for *aGTSPersistenceTime* superframes. On receipt of the beacon frame, the device shall process the descriptor and notify the next upper layer of the success.

In the same way, the MAC of a device is instructed to request the deallocation of an existing GTS through the MLME-GTS request primitive using the characteristics of the GTS the MLME wishes to deallocate. From this point on, the GTS to be deallocated shall not be used by the device. To request the deallocation of an existing GTS, the MLME shall send the GTS request command to the PAN coordinator. Upon successful reception, the PAN coordinator sends an ACK to the device. The PAN coordinator then deallocates the GTS whose characteristic in the packet matches those in its allocation. The PAN coordinator shall also ensure that any gaps occurring in the CFP, appearing due to the deallocation of a GTS, are removed to maximise the length of the CAP.

The MLME of the PAN coordinator shall also attempt to detect when a device has stopped using a GTS using the following rules: For a transmit frame GTS, the

MLME of the PAN coordinator shall assume that the device is no longer using the GTS if a data frame is not received for at least $2n$ superframes. For receive GTSs, the MLME of the PAN coordinator shall assume that the device is no longer using its GTS if an acknowledgement frame is not received for at least $2n$ superframes. The value of n is equal to $2^{8-macBeaconOrder}$ if $0 \leq macBeaconOrder \leq 8$ and 1 if $9 \leq \text{macBeaconOrder} \leq 14$.

CAP

The PAN coordinator shall preserve the minimum CAP length of *aMinCAPLength* and take preventative action if the minimum CAP is not satisfied. However, an exception shall be allowed for the accommodation of the temporary increase in the beacon frame length needed to perform GTS maintenance. If preventative action becomes necessary, the action chosen is left up to the implementation, but may include one or more of the following:

- Limiting the number of pending addresses included in the beacon.
- Not including a payload field in the beacon frame.
- Deallocating one or more of the GTSs.

Frame Security

The MAC sublayer is responsible for providing security services on specified incoming and outgoing frames when requested to do so by the higher layers. IEEE 802.15.4–2003 supports the following security services:

- Access control is a security service that provides the ability for a device to select the other devices with which it is willing to communicate. In this standard, if the access control service is provided, a device shall maintain a list of devices in its ACL (Access Control List) from which it expects to receive frames.
- Data encryption is a security service that uses a symmetric cipher to protect data from being read by parties without the cryptographic key. Data may be encrypted using a key shared by a group of devices (typically stored as the default key) or using a key shared between two peers (typically stored in an individual ACL entry). In this standard, data encryption may be provided on beacon payloads, command payloads, and data payloads.
- Frame integrity is a security service that uses a message integrity code (MIC) to protect data from being modified by parties without the cryptographic key. It further provides assurance that data came from a party with the cryptographic key. In this standard, integrity may be provided on data frames, beacon frames, and command frames. The key used to provide frame integrity may be shared by a group of devices (typically stored as the default key) or by two peers (typically stored in an individual ACL entry).

- Sequential freshness is a security service that uses an ordered sequence of inputs to reject frames that have been replayed. When a frame is received, the freshness value is compared with the last known freshness value. If the freshness value is newer than the last known value, the check has passed, and the freshness value is updated to the new value. If the freshness value is not newer than the last known freshness value, the check has failed. This service provides evidence that the received data are newer than the last data received by that device, but it does not provide a strict sense of time.

The protocol also provides the following security modes [4]:

- Unsecured mode
- ACL mode
- Secured mode

When using ACL security, a device maintains a list of devices from which it expects to receive communications. When the MAC layer receives a frame, it notifies the upper layer that the frame was received and also indicates whether the sender of the frame is in its ACL.

Devices operating in secured mode may provide any of the security services defined above. When using cryptographic security, the MAC layer uses the Advanced Encryption Standard (AES).

Summary of Differences Between MAGNET LDR-MAC and IEEE 802.15.4 MAC

In previous sections we have described the MAGNET LDR-MAC characteristics that are based on IEEE 802.15.4. From the analysis, we observe that the differences between the two approaches are mainly on channel access mechanisms.

Depending on network configuration, the IEEE 802.15.4 MAC may use one of two channel access mechanisms. In a beacon-enabled network with superframes, a slotted carrier sense multiple access with collision avoidance (CSMA-CA) mechanism is used. In networks without beacons, unslotted or standard CSMA-CA is used. MAGNET approach is instead based uniquely on beacon-enabled mode with all the associated characteristics

Medium Access Mechanisms

Multiple users can be accommodated in a number of ways in the FM-UWB system depending on the required Quality of Service (QoS):

- RF FDMA for the highest QoS (no multiple-access interference)
- IEEE 802.15.4 MAC (TDMA) for standard applications
- Proprietary MAC (TDMA) for sensor networks, e.g., WISEMAC
- Sub-carrier FDMA for ultra low power applications

The frame structure of the transmitted signal corresponds to the option chosen. For standard applications, this will be the IEEE 802.15.4 frame structure. The IEEE 802.15.4 MAC will be the focus of the prototyping.

Frequency Allocation for RF FDMA

Following the draft ECC Decision on Devices using UWB technologies in the bands below 10.6 GHz it was decided to use the following frequency bands for FM-UWB.

- Low band (LB) 4.2–4.8 GHz
- High band (HB) 6.0–9.0 GHz

Table 4.7 presents the channel centre frequencies. The single LB frequency L_1 is a multiple of 32 MHz and the 5 HB frequencies are multiples of 64 MHz for HB.

Figure 4.27 shows the spectral density of the six FM-UWB signals corresponding to six different users, following the frequency allocation of Table 4.2. Each of them has a signal power of -15 dBm and a deviation Δf of 250 MHz. Multiple-access interference between the five non-overlapping FM-UWB signals is non-existent resulting in the highest QoS.

Table 4.7 Example of FM-UWB channel centre frequencies

Channel	RF centre frequency (MHz)
L1	$141 \times 32 = 4{,}512$
H1	$100 \times 64 = 6{,}400$
H2	$109 \times 64 = 6{,}976$
H3	$118 \times 64 = 7{,}552$
H4	$127 \times 64 = 8{,}128$
H5	$136 \times 64 = 8{,}704$

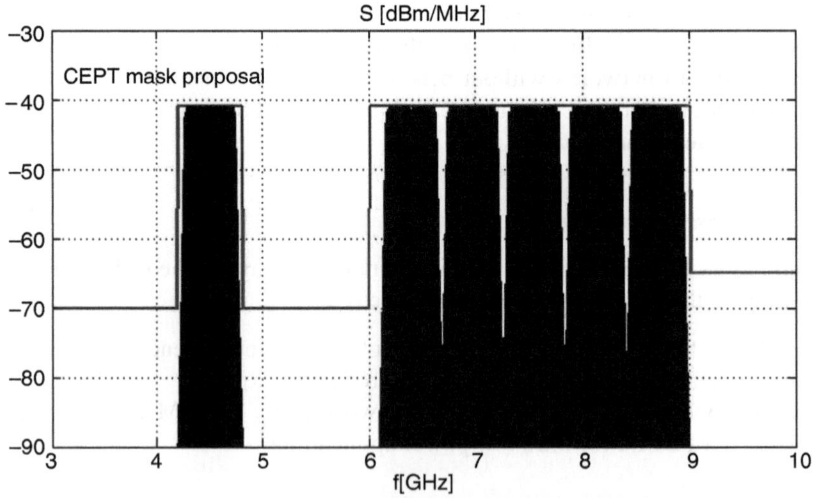

Fig. 4.27 Spectral density of the L1 H1-H5 FM-UWB signals spaced 576 MHz apart

Sub-carrier FDMA Frequency Allocation

In addition to classical TDMA and RF FDMA techniques, FM-UWB has the unique possibility to also exploit sub-carrier FDMA. Individual users with a common RF centre frequency distinguish themselves by their sub-carrier frequencies. The receiver processing gain creates a margin for strong interferers and also determines the number of users that can be in the same piconet.

The prototyping of the *sub-carrier FDMA* is done in the sub-carrier frequency range of 1–2 MHz. Four 100 kbit/s users are accommodated as shown in Table 4.5.

Duplex Timing

The LDR system works in half duplex mode, so the radio is either in standby, transmit or receive mode. The receiver requires a certain time before its output data are valid. Receiver synchronisation time is limited by the bit synchronisation process. This time is bit rate dependent and typically in the order of 25 bits. Synchronisation occurs during the preamble. Since it is not possible to start synchronising on a signal that is not yet there, this delay cannot be avoided.

4.2.2 High Data Rate Transmission with MC-SS Modulation

The MAGNET high data rate air interface utilizes a multi-carrier spread-spectrum (MC-SS) PHY layer along with a MAC layer that is based on IEEE 802.15.3. In this section the MAGNET HDR PHY and the MAC are briefly described. Complete specifications of the PHY and the MAC can be found in [27].

4.2.2.1 PHY Layer

The PHY layer is designed to operate in 5.2 GHz bands allocated to wireless access systems. The preferred bandwidth is 40 MHz; however an alternative bandwidth of 20 MHz has been defined to ensure compliance with worldwide regulations. The MC-SS system is based on OFDM-TDMA together with spreading in frequency domain for multi-code transmission. Transmission links in a PAN are separated in time domain due to TDMA approach, and each transmission link occupies the entire bandwidth in a given time duration. Figure 4.28 shows the overall structure of the PHY layer.

First, bits received from the MAC layer are channel encoded, punctured, and interleaved. The mapping block maps a certain number of coded bits to one complex modulation symbol. Then spreading and multi code transmission is done over complex modulated symbols. Null subcarriers for guard bands are added in the OFDM framing block. In the OFDM modulation block the IDFT operation is carried out

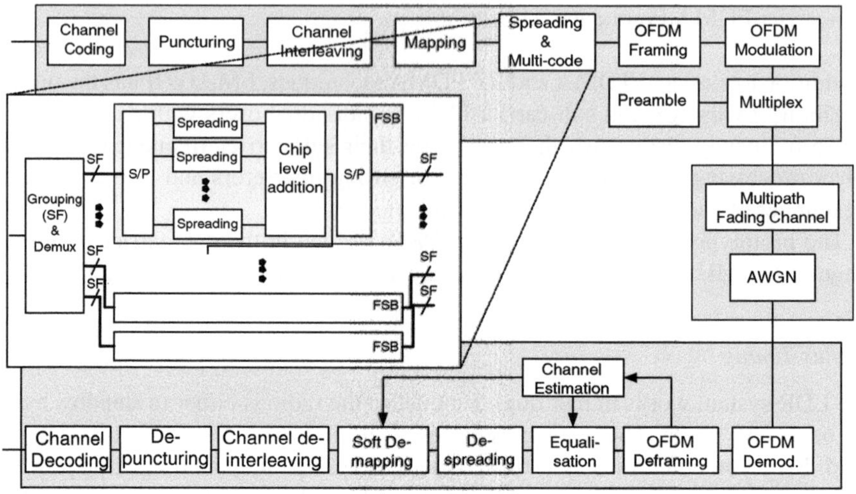

Fig. 4.28 Block Diagram of MC-SS Physical Layer [27]

and the cyclic prefix is added in time domain. The receiver encompasses the inverse operations of the transmitter detection and channel decoding plus channel estimation and equalisation modules.

Basic System Parameters

The basic system parameters for 20 MHz as well as 40 MHz are listed in Tables 4.8 and 4.9. The 40 MHz system supports data rates from 28.87 Mb/s to 129.29 Mb/s by applying different modulation schemes (QPSK, 16-QAM and 64-QAM) and code rates (1/2, 2/3, 3/4).

The FFT size in the 40 MHz version is 256, with 192 data carriers, 19 pilot and 45 guard carriers. The system applies spreading in frequency domain with a spreading factor fixed to 8. The number of transmitted codes per spreading block can be varied between zero and eight.

Frame Structure

The frame structure is depicted in Fig. 4.29.

The spreading allows for increased flexibility and adaptability by varying the used encoding scheme with respect to the required data rate and channel conditions. Moreover, spreading in frequency domain increases robustness against narrow band interferers.

The PHY frame format is illustrated in Fig. 4.30.

Table 4.8 Data rate in Mbit/s and modulation and coding scheme in full-load configuration

Modulation	40 MHz			20 MHz		
code rate	QPSK	16QAM	64QAM	QPSK	16QAM	64QAM
1/2	28.87	57.74	86.62	14.435	28.87	43.31
2/3	38.50	76.99	115.49	19.25	38.50	57.75
3/4	43.31	86.62	129.92	26.66	43.31	64.96

Table 4.9 OFDM system parameters

	40 MHz	20 MHz	
Carrier frequency	5.20	5.20	GHz
Sampling frequency Fsig	40	20	MHz
FFT size N_{FFT}	256	128	
Total subcarriers	256	128	
Subcarriers for guard band No	45	23	
Subcarriers for pilot N_{pilot}	19	9	
Subcarriers for data N_{pm}	192	96	
Percentage of guard band	17.578	17.969	%
Subcarrier spacing Δ_{FFT}	156.250	156.250	kHz
Occupied signal bandwidth Bw	33.13	16.56	MHz
Number of time samples per data symbol	256	128	Samples
Samples for guard interval	10	5	Samples
Samples for total OFDM burst	266	133	Samples
Maximum delay spread	0.213	0.213	μs
Sample duration in time	0.025	0.050	μs
Length of data interval in time	6.40	6.40	μs
Length of guard interval in time Tcp	0.250	0.250	μs
Length of total OFDM interval in time TFFT	6.65	6.65	μs
Percentage of guard interval	3.91	3.91	%
Channel coding	Convolution code		
Generator polynomial	$g1 = 133,\ g2 = 171$		
Tail		6	Bits
spreading factor	8	8	Chips
Maximum velocity	3	3	km/h
Maximum Doppler spread	14.4	14.4	Hz
Coherence time $= 9/(16\pi f_D)$	12.4	12.4	Ms

The PHY layer attaches the PHY header to the MAC header, calculates the HCS (Header Check Sequence) over the combined PHY and MAC headers, and appends the HCS to the end of the MAC header. The PHY preamble, which is used for PHY synchronisation, channel estimation and RF Impairment is sent first, followed by the PHY header, MAC header, HCS and header tail bits, followed by the frame payload, the FCS, the stuff bits (SB), if necessary, and finally the zero tail bits.

Channel Encoding

The basic channel coding scheme of MC-SS physical layer is a convolutional encoder of coding rate 1/2 used by NASA. The convolutional encoder shall have a

Fig. 4.29 MC-SS Frame Structure

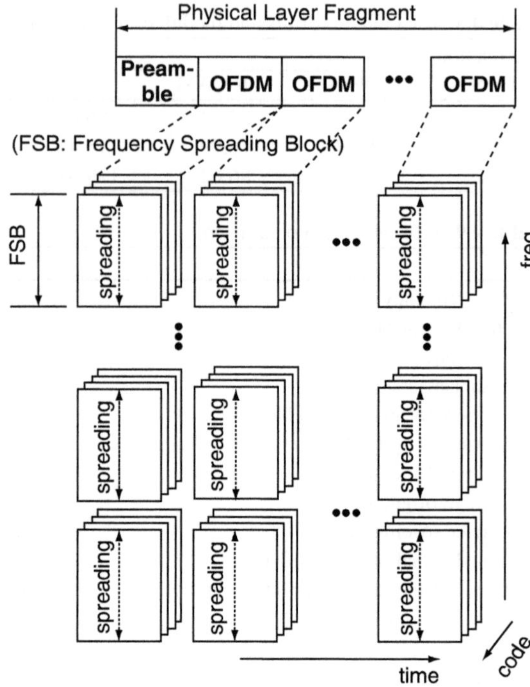

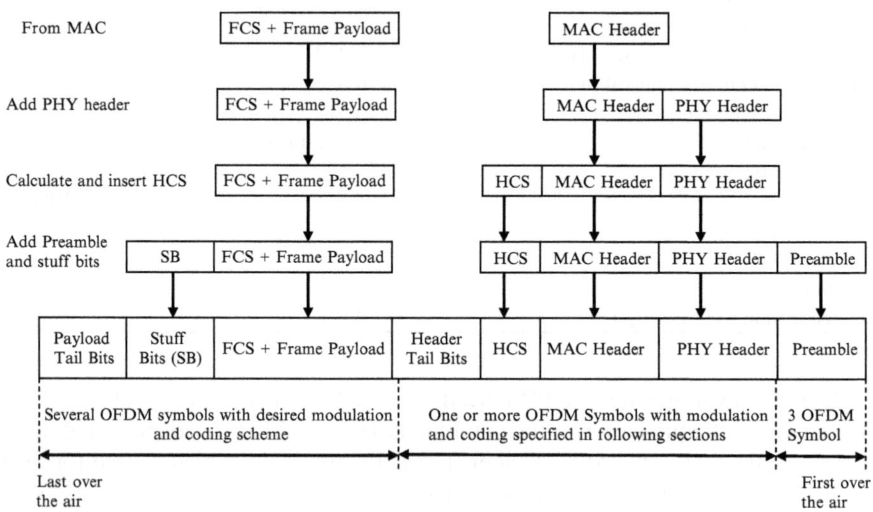

Fig. 4.30 PHY Frame formatting

constraint length equal to seven (6 memory elements, K = 7) and shall use the generator polynomials, $G_0 = 133$ oct. and $G_1 = 171$ oct. 12 bits are added as tail bits at the end of an encoded frame.

Table 4.10 Puncturing pattern

	Code rate 2/3	Code rate 3/4
Puncturing pattern	$\begin{Bmatrix} 1\ 1\ 0 \\ 1\ 0\ 1 \end{Bmatrix}$	$\begin{Bmatrix} \vert 1\ 1\ 0 \\ \vert 1\ 0\ 1 \end{Bmatrix}$

Puncturing

The higher code rates 2/3 and 3/4 are provided from a half rate convolutional code by puncturing specified bits according to a predefined pattern. Puncturing pattern for rate 2/3 and 3/4 are given in Table 4.10. This pattern defines the positions of the bits that have to be punctured. For each input bit, if the corresponding puncturing value is false (0) then the bit is punctured. If the value is true (1), the input is copied to the output.

Interleaver

Encoded bits are interleaved by an interleaver at the bit-level rather than at symbol-level. The interleaver permutes the encoded bits within one OFDM symbol. The following algorithm is foreseen as interleaving pattern:

The first permutation, is defined by the rule:

$$i = (L_{\mathrm{CBPS}}/16)\ (k \bmod 16) + \mathrm{floor}(k/16),\ k = 0,\ 1, \ldots,\ L_{\mathrm{CBPS}} - 1$$

where L_{CBPS} denotes the number of coded bits per OFDM symbol. The function floor(.) denotes the largest integer not exceeding the parameter, and mod is the integer modulo operator.

The second permutation is defined by the rule:

$$j = s \times \mathrm{floor}(i/s) + (i + L_{CBPS} - \mathrm{floor}(16 \times i/L_{CBPS})) \bmod s$$

The value of s is determined by the number of coded bits per sub-carrier, L_{CBPS}, according to $s = \max(L_{BPSC}/2, 1)$, where L_{BPSC} means the number of coded bits per subcarrier. Consequently, the interleaving pattern is given by pattern$[k] = j,\ k = 0, 1, \ldots, L_{CBPS} - 1$.

Mapping

QPSK, 16-QAM or 64-QAM with Gray encoding have been considered as possible modulation schemes. Normalization is used to ensure the same average power for all mappings. Normalization factors are indicated in Table 4.11.

Table 4.11 Mapping schemes

Modulation	Number of coded bits per symbol	Normalizing constants (kmod)
QPSK	2	$1/\sqrt{2}$
16-QAM	4	$1/\sqrt{10}$
64-QAM	6	$1/\sqrt{42}$

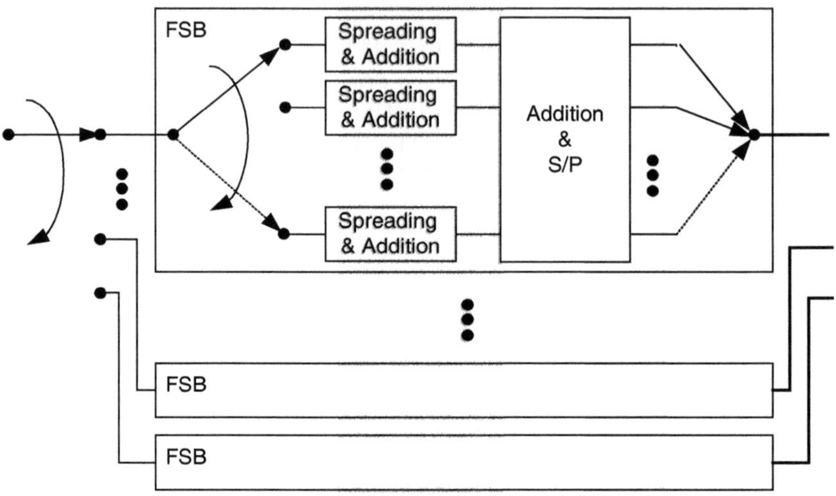

Fig. 4.31 Spreading and multi-code transmission

Spreading, Multi-code Transmission and Modulation

The spreading and multi-code transmission block de-multiplexes input modulated symbols, spreads each de-multiplexed modulated symbol with orthogonal spreading code, added in chip-level, and serial-to-parallel converted which results in multi-code transmission in frequency domain. Note that MC-SS cluster considers the case where the full set of spreading codes are utilized (full rate multi-code transmission) and the IFFT length is an integer multiple of the spreading factor, so several frequency-spreading blocks (FSB) can be transmitted within one OFDM symbol.

Figure 4.31 shows the detailed operation of spreading and multi-code transmission block in FSB.

A spreading factor of 8 is resulting in 12 FSBs per OFDM symbol using 20 MHz bandwidth, and 24 FSBs per OFDM symbol with 40 MHz, respectively. Modulated symbols are spread by a Walsh-Hadamard code with a length of 8 chips. OFDM-modulation is done by a standard IDFT with a length of 256 carriers.

4.2.2.2 MAC Layer

The MAC scheme developed in IEEE 802.15.3 makes use of a piconet structure, which is controlled by a piconet coordinator (PNC). All components in the piconet are devices (DEVs), and one of them is required to perform the role of the PNC. Every piconet is started, maintained and terminated by a PNC. The main responsibility of the PNC is to coordinate the channel access mechanism which regulates the wireless medium access by the members of the piconet. All control information is sent out via beacons from the PNC to the devices. The member devices can communicate with each other in ad-hoc fashion either by contending for the channel via CSMA/CA or by getting a dedicated time slot allocated in a TDMA fashion. A potential piconet structure is depicted in Fig. 4.32.

The time axis is divided into superframes, which consist of a beacon frame transmitted by the PNC, a contention access period (CAP), and a contention free period (CFP). Figure 4.33 shows a typical superframe structure.

The beacon frame is transmitted by the PNC at the beginning of each super frame. The CAP period is provided for non-real time data transmissions, and uses CSMA/CA for the medium access. The CFP adopts a standard TDMA mechanism, and allocates guaranteed time slots (GTSs) each device. The start time and duration of each GTS are determined by the PNC.

The MAC also provides mechanisms for co-existence of multiple piconets as well as range extension. In case of fully independent piconets a neighbouring piconet can be established. The neighbouring piconet operates independently in dedicated time slots, which are reserved by the parent piconet. Data exchange between both networks is not possible. A child piconet is established to increase the range or to shift computation power from one network to the other. In case of a child piconet the PNC of the child is member of the parent piconet. Consequently, the child PNC can exchange data with any other device in the parent piconet. A parent, child and neighbour piconet is depicted in Fig. 4.34.

Some MAC functionalities are briefly explained in the following sections.

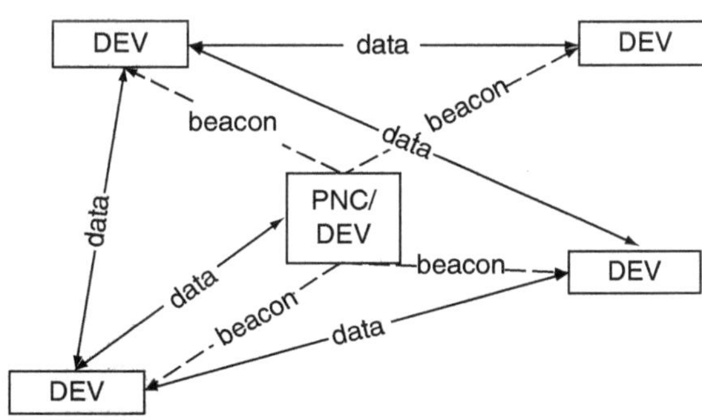

Fig. 4.32 IEEE 802.15.3 piconet

Beacon	Contention Access Period (CAP)	Channel Time Allocation Period (CTAP)						
		MCTA 1	MCTA 2	CTA 1	CTA 2	•••	CTA n-1	CTA n

Fig. 4.33 Superframe structure

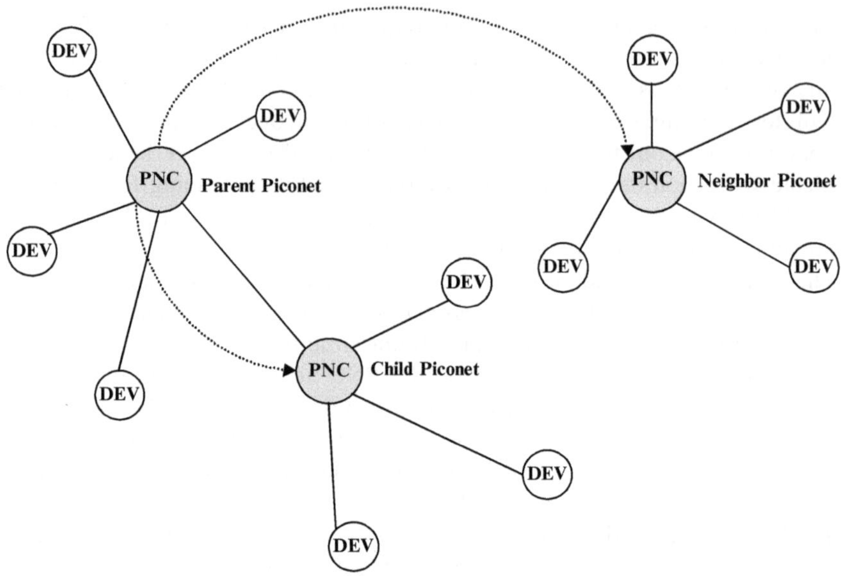

Fig. 4.34 Child and neighboring piconets

Starting, Maintaining and Stopping Piconets

To start a piconet, a device that is capable of acting as the PNC scans the available channels to find one that is not being used; it starts the piconet by sending the beacon. The device becomes the PNC that provides the basic timing for the piconet with the beacon.

The PNC is responsible to regulate associations, handle channel time requests, maintain synchronization among devices in power save modes, regulate transmission power in the piconet, etc. There can be more than one PNC capable devices in a piconet. The standard provides the option of PNC handover depending on the capabilities of the devices. The PNC can change the superframe parameters depending on the requirements.

In IEEE 802.15.3, three types of piconets are defined:

- Independent piconet: A piconet with no dependent piconets and no parent piconets.
- Parent piconet: A piconet that has one or more dependent piconets.
- Dependent piconet: A piconet that requires a time allocation in another piconet, called the parent piconet, and is synchronized with the parent piconet's timing.

There are two types of dependent piconets: Child piconet and Neighbor piconets. Child piconet is a dependent piconet where the PNC is a member of a parent piconet, and is used for range extension or co-existence of IEEE 802.15.3 compliant networks, whereas neighbor piconet is a dependent piconet where the PNC is not a member of a parent piconet, and facilitates co-existence with networks operating with other wireless protocols.

Scanning

All DEVs shall use passive scanning to detect an active piconet. That is, DEVs shall be in receiving mode for a period of time in a channel to look for beacon frames from a PNC.

DEVs search for piconets by traversing through all the channels. The result of a scan shall include information on any parent, child, or IEEE 802.15.3 neighbor, piconets that were detected. This provides a complete inventory of each channel.

The channel is selected by passive scanning of available channels. Advertising is done by sending out beacons that contain the signatures of the piconet. Devices that can listen to these network beacons send association requests to the PNC. The PNC then based on the resources accepts the association or rejects it.

Channel Access

The PNC coordinates channel access by sending a beacon that marks the start of a superframe. The channel time is divided into superframes, and each superframe beginning with a beacon. The superframe is composed of three major parts: the beacon, the CAP and the CTAP.

The CAP period is mainly used to exchange management information with the PNC and for non-real time data transmissions. The CAP uses carrier sense multiple access/collision avoidance (CSMA/CA) for the medium access. CTA is used for handling isochronous streams that have high QoS requirements. The CTAP adopts a standard TDMA mechanism and allocates guaranteed time slots (GTSs). The start time and duration of each GTS are determined by PNC according to the devices requests and announced in the beacon interval of the superframe.

There can be proprietary algorithms for channel time allocation. Devices in the piconet synchronize with the PNC after receiving the beacon and then transmit in either the CAP or CTAP. To transmit in the CTAP, a device requests CTAs from the PNC by specifying the time units required in order to support its relevant real-time streams. If the PNC grants the device's request, the device then has exclusive rights to transmit during its CTA slots. The device requesting for a time slot in the CTA must ensure that it asks for a time duration which is required for the complete transfer of its data taking into account the inter frame spacing, guard time and the ACK policy in use. All devices other than the ones involved in a flow shall not transmit in the allocated time slot for that flow.

Synchronization

The PNC maintains synchronization of the piconet and all the devices should be synchronized with the PNC's clock. In addition, child or neighbor PNCs shall synchronize their piconet's time usage to the parent PNC's beacon and their CTA. The beacon sent at the beginning of every superframe contains the information necessary to time-synchronize the DEVs in the piconet.

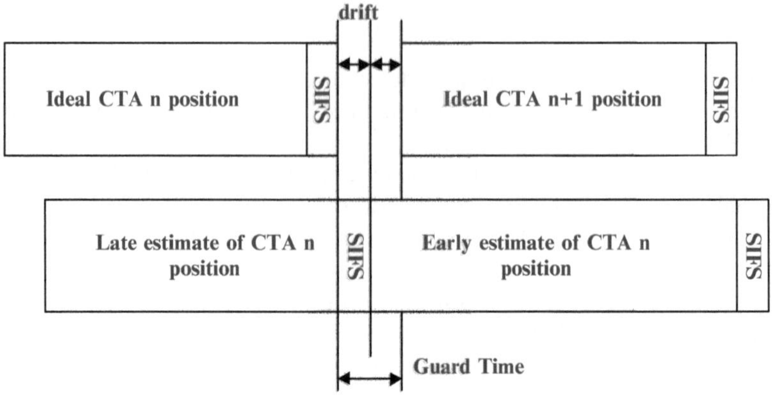

Fig. 4.35 Guard time

Figure 4.35 shows the use of guard time intervals between two CTAs so as to account for the drift in the clocks, which may cause a wrong estimate of CTA location resulting in loss of information. The guard time is the time between the end of one CTA and the start of the next CTA. Including SIFS as part of CTAs and allocating guard time between CTAs ensures that transmissions are spaced by at least a SIFS. The required guard time depends on the maximum drift between a DEV's local time and the ideal time. This drift is a function of the time elapsed since a synchronizing reference event.

Channel Time Management

The MAC supports two kinds of data streams: isochronous and asynchronous. A device can support more than one type of stream depending on the application it is designed to support. The channel time management involves creation, modification and termination of isochronous data streams and the reservation and termination of asynchronous channel time for the exchange of asynchronous data.

Acknowledgement and Retransmission

The MAC layer adopts the ACK and retransmission mechanism to provide a reliable communication for higher layer.

The IEEE 802.15.3 MAC defines three types of ACK policies:

- No ACK (No-ACK)
- Immediate ACK (Imm-ACK)
- Delayed ACK (Dly-ACK)

When using the No-ACK policy, the destination shall not acknowledge the received frame. The two successive frames are separated by minimum interframe space (MIFS). The No-ACK policy is appropriate for frames that do not require guaranteed delivery. The Imm-ACK policy provides an acknowledgement process in which each data frame is individually ACKed following the reception of the frame. The Dly-ACK policy allows the source DEV to send a burst of frames without the intervening ACK frames. When necessary, the source DEV adds Dly-ACK request information to a data frame's MAC header. Once the destination DEV receives this frame which includes request information, it will send the Dly-ACK frame which acknowledges those correctly received frames in current burst. The source shall not start or resume the next burst transmission until a Dly-ACK frame is received. These frames which are not ACKed should be retransmitted in the next burst. During the CAP, retransmissions shall follow the backoff rules. During CTAs within the CTAP when an Imm-ACK or Dly-ACK is expected, but is not received during a RIFS (RetransmissionIFS), the source DEV shall start the retransmission of the frame (or new frame if the failed frame's retransmission limit has been met) after the end of RIFS as long as there is enough channel time remaining in the CTA for the entire frame exchange.

A DEV determines the number of times a frame is retried before the DEV discards that frame. If the DEV gives up on a fragment of an MSDU/MCDU, the DEV shall discard all MPDUs of that MSDU/MCDU.

Fragmentation and Defragmentation

The MAC Protocol Data Unit (MPDU) is the MAC data frame that will be sent over the PHY. It is fragmented from the MAC Service Data Unit (MSDU) which is passed from an upper layer. Fragmentation may be performed at the transmitting device on each MSDU. In addition, certain commands, i.e. MCDUs, may be fragmented. All the MPDUs from the same MSDU have the same size except the last MPDU, which may be shorter.

Once the MSDU/MCDU is fragmented and a transmission attempted, it shall not be refragmented. Each fragment shall be sent with the Last Fragment Number field set to the highest fragment number of the current MSDU/MCDU, which is one less than the total number of fragments of the current MSDU/MCDU. The first fragment shall be sent with the Fragment Number field set to zero. Each subsequent fragment shall be sent with the Fragment Number field incremented by one. The MSDU/MCDU shall be completely reassembled in the correct order before delivering it to the frame convergence sub layer (FCSL). Any MSDU/MCDU with missing fragments shall be discarded. The receiver shall not deliver an MSDU/MCDU to the FCSL until all of the fragments have been obtained. The receiving DEV may discard the fragments of an MSDU/MCDU if it is not completely received within a timeout determined by the receiving device.

4.3 Performance Comparison of MB AIs with Existing Technologies

The MAGNET HDR air interface has been benchmarked with the WiMedia UWB solution. It has been found that though both PHY layers make use of OFDM they differ significantly. Bandwidth, data rates, spreading techniques and carrier frequencies are completely different. PHY layer benchmarking has proven that the MC-SS system provides much higher radio coverage than WiMedia system if we consider maximum transmit powers. However, WiMedia provides much higher data rates for very short ranges. Whereas the WiMedia system can achieve up to 480 Mb/s in very short ranges, the MAGNET system can only transmit up to 129.92 Mb/s. Further differences can be found in the number of non-interfering piconets. Due to the current regulations in Europe only four to five orthogonal frequency channels are available to host non-interfering WiMedia systems using a transmit power of $-41.3\,\mathrm{dBm/MHz}$. In contrast, nine orthogonal channels are available for MC-SS systems. Consequently, more orthogonal systems can operate close to each other.

The MAC layers differ mainly in the network structure. Whereas MAGNET utilizes a centralized control structure, WiMedia is fully distributed. This has advantages in case of mobility as any device can leave the piconet without affecting the overall performance. If the piconet controller leaves the IEEE 802.15.3 net, a new piconet controller needs to be found, which will cause an interruption. The MAGNET MAC, however, is more flexible in time domain. The length of a superframe as well as of the channel time allocation periods can be adapted according to the needs. In WiMedia the length of a superframe and of the medium access slots are fixed. Similarities can be found in supported types of traffic, ACK policies and security aspects.

FM-UWB has been benchmarked with ZigBee, Bluetooth and WiBree. All of the four systems target on short range low data rate radio communications. Physical layer simulations show that the BER performance of FM-UWB in the multipath channel with interference is significantly better than Bluetooth. FM-UWB is also expected to perform better than ZigBee and WiBree in the robustness to the fading channel and interference due to the large bandwidth and the elaborated air-interface. Among the four systems, FM-UWB, ZigBee and WiBree have similar throughput and range. Bluetooth can obtain higher data rate and larger coverage range, but the cost is higher transmission power.

In the MAC layer, several scenarios have been simulated to evaluate the performance of LDR-UWB and Bluetooth systems. The results reveal that depending on certain scenarios and conditions the performance of LDR-UWB is better than that of Bluetooth. In general, LDR-UWB system has lower mean delay than that of Bluetooth. Meanwhile the Block Error Rates of the two systems are comparable. It indicates that the performance at the MAC layer of the LDR-UWB is compatible to Bluetooth. For further information on the performance of the implemented MAGNET Beyond AIs, read [44] for a detailed performance comparison of the MAGNET Beyond AIs with the aforementioned legacy systems.

4.4 Advanced Topics in the Design of LDR and HDR AIS

The previous sections have addressed the design and performance of the LDR and HDR AIs which have been implemented in the project according to the description in Chapter 6. In the following, certain advanced topics of the AI design are addressed which are not part of the demonstrator, but are foreseen to be implemented in future versions of the AIs. Some of the aforementioned topics are generic in that they can be used also with other AIs, while others are specific to the AIs selected in the MB project.

Section 4.4.1 focuses on advanced techniques for increasing the spectrum efficiency and mitigating the effects of interference in the MAGNET HDR wireless communication link. Section 4.4.2 describes extensions of IEEE 802.15.3 for inter-PAN communication and Section 4.4.3 addresses interference issues arising among LDR and HDR AIs being located close to each other, potentially within the same device.

4.4.1 Interference Mitigation and Spectrum Efficiency

A wide range of PHY, MAC and cross-layer techniques that can independently or jointly improve the performance of the HDR air interface have been investigated in the MAGNET Beyond project. Here, two of these techniques are presented in greater detail: Adaptive Modulation and Coding (AMC) is presented in Sections 4.4.1.1 and 4.4.1.2 describes an amplify-and-forward approach for implementing a cooperative diversity scheme.

4.4.1.1 AMC for Real Time or Streaming Media Transmission

The QoS requirements for streaming media transmission are substantially different than those of data transmission. When transmitting streaming media, packets received after a maximum allowable delay D_{max} are useless because the time for displaying their content has already passed. Therefore, retransmissions of erroneously received packets are only possible in the time limits imposed by the maximum delay restriction. In HDR WPANs were the Round-Trip Time of packets is lower than the maximum allowable delay a certain number of retransmissions for each packet is permissible. On the other hand, a small fraction of received bits can be incorrectly received without noticeable degradation of the medium quality. Thus the QoS constraint should take the form: "probability that a packet (or bit) is not received correctly at the receiver within D_{max} from its arrival time at the transmitter is less than a QoS parameter P_{QoS}".

Such soft or statistical QoS guarantees are well established and studied in the context of wired networks. In broadband wired networks BER is practically negligible ($<10^{-9}$) and the only cause of errors is packet loss (or excessive delay) due

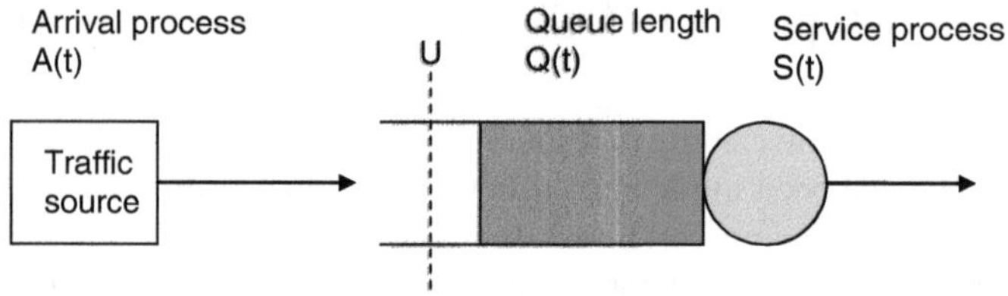

Fig. 4.36 Single queue model for defining the effective bandwidth of a traffic generating source

to congestion. Statistical QoS provisioning in such networks was made possible by the extensive work on *effective bandwidths* [5, 6]. Briefly, the effective bandwidth of a traffic generating source is defined as the minimum constant service capacity $b(P_{QoS}, U)$ that attains a buffer overflow probability $P_{loss} < P_{QoS}$, where U is the buffer size (Fig. 4.36). Thus, by performing Call Admission Control (CAC), the network can guarantee a buffer overflow probability less than P_{QoS} if the effective bandwidth of the aggregate arrival process of all admitted calls is less than the available capacity c of the communication link. Note that while increasing the buffer size results in an (exponential) decrease of the overflow probability, all traffic arriving to a queue with length $> U$ will experience a queuing delay $> U/c$. Therefore for a given maximum allowable delay D_{max} and maximum service capacity c, it is useless to have a buffer with size greater than $U = c \cdot D_{max}$. In such a setting, D_{max} and P_{QoS} are the QoS parameters whose values are negotiated between the application/user and the network/service provider.

Calculating effective bandwidths and buffer overflow probabilities for traffic sources with a variety of arrival process statistics has been the subject of extensive research work. Since for most of the arrival processes of practical interest obtaining analytical solutions for the buffer overflow probability is intractable, Large Deviations (LD) techniques [7] have been applied to obtain asymptotic approximations of such probabilities. With the risk of oversimplification, such LD approximation results are of the form [8]

$$P[Q(t) > U] \approx \alpha \cdot e^{-\theta \cdot U}, \tag{4.15}$$

where α and θ depend on the statistics of the arrival process and the service capacity c (but not on U). Selecting an appropriate stochastic model for an observed arrival process and estimating the model's parameters based on a sample realization of this process is another important practical issue for applying LD approximations in real life CAC [9].

Unlike wired links, wireless links exhibit variations in the channel quality which in the physical layer can be measured by the SINR at the receiver. By using different modulation and coding (M&C) modes different BER vs. transmission rate curves can be achieved. Furthermore, BERs are commonly much larger than in wired networks. Thus, the probability of incorrectly received frames is not negligible. Consequently, the AMC policy should optimize the tradeoff between higher data rate and BER which will result to a lower delay violation probability and lower data rate which will translate to lower BER, but higher delay violation probability.

Recently, many researchers have studied this problem proposing a range of optimization criteria for AMC with QoS guarantees for streaming media transmission. In a series of papers [10–12], the authors proposed to derive the AMC policy which maximizes link capacity under an average *Packet Error Rate* (PER) (denoted by P_0) constraint imposed for every M&C mode. The overflow probability (denoted by P_d) is then calculated based on Markovian models for the service capacity process, Poisson arrivals and exact numerical solutions for calculating the probability of overflow with a finite buffer. Consequently, the cumulative packet loss rate is obtained as $\xi = 1 - (1 - P_d)(1 - P_0)$. By performing an optimization over the range of all possible P_0, the best achievable ξ is determined. However, exact solutions for queues with Markovian arrival and service processes suffer from an exploding state space which results to excessive computational load and aggregating numerical errors. As an alternative, LD approximations have been employed in [13] to introduce a dual notion to the effective bandwidth called *effective capacity* that captures the varying capacity of wireless links. Building on the work in [13], an LD approximation approach is used in [14–16] to calculate the maximum delay violation probability after specifying the AMC policy using a transmission rate optimization criterion under a BER constraint.

Although decoupling the two loss metrics P_d and P_0 results in simpler optimization problems, solving the full optimization problem of finding the AMC policy which minimizes the total packet loss ξ (without imposing an equal average PER in all transmission modes) will produce the truly optimal AMC policy. In MAGNET Beyond, it has been proposed to minimize ξ (or more rigorously an upper bound for ξ) as the AMC criterion for streaming media transmission. A significant gain in total packet loss compared to existing AMC schemes is achieved by this improved algorithm as demonstrated in this section. Intuitively, the expected gain will be obtained by attaining – for each value of the SNR – the optimal tradeoff between a lower P_d (which translates to a higher transmission rate and hence to a higher P_0) and a lower P_0 (and higher P_d). In Section 4.4.1.2, we will assume that erroneously received packets are not retransmitted. Combining our AMC algorithm with ARQ is also addressed in Section 4.4.1.2.

In the following two sections, we present and evaluate the proposed AMC algorithm based on the same channel models (i.e., Nakagami fading model and derived Markovian time correlation model) used in the literature and MATLAB based simulations. This will make our results generic enough to be useful for a variety of wireless communication systems and will allow for comparisons to existing optimization schemes and algorithms. In Section 4.4.1.4, we shift our focus to the MAGNET MC-SS radio channel and M&C modes and explain how we can adapt our AMC algorithm to this particular system.

4.4.1.2 Generic AMC for real time media transmission without ARQ

As mentioned in the previous section we adopt the same generic system model as in [11] (with a few minor modifications), which is summarized in this section.

Table 4.12 Transmission modes with convolutionally coded modulation

	Mode 1	Mode 2	Mode 3	Mode 4	Mode 5
Modulation	BPSK	QPSK	QPSK	16-QAM	64-QAM
Coding rate R_c	1/2	1/2	3/4	3/4	3/4
R_n (bits/sym.)	0.50	1.00	1.50	3.00	4.50
a_n	274.7229	90.2514	67.6181	53.3987	35.3508
g_n	7.9932	3.4998	1.6883	0.3756	0.0900
γ_{pn} (dB)	−1.5331	1.0942	3.9722	10.2488	15.9784

According to this model, a discrete time arrival process is serviced by a buffered wireless link with a single-transmit and a single-receive antenna. The buffer can accommodate an infinite queue which operates in a first-in-first-out (FIFO) mode (as in Fig. 4.36 with $U = \infty$). While the model used in [11] assumes a finite buffer, it can be argued that in most cases a sufficiently large buffer can be afforded so that excessive delay rather than buffer overflow will be the preeminent cause of queuing originated packet loss.

At the wireless link, multiple transmission M&C modes are available. More specifically, the convolutionally encoded M_n-ary rectangular or square QAM (dubbed TM2 in [11]), adopted from the IEEE 802.11a and HIPERLAN/2 standards, is employed. The 5 M&C modes of this scheme are listed in Table 4.12, in a rate ascending order.

Although in this section we focus on the specific M&C scheme, the proposed AMC algorithms can be applied to other M&C schemes in a similar manner.

Data are transmitted frame by frame, where each frame contains a fixed number of information symbols N_{ds} and overhead symbols N_{os}. Information symbols are transmitted at the selected M&C mode while overhead symbols are always transmitted at a base mode (usually Mode 1). Given a fixed symbol rate, the frame duration (T_f seconds) is constant, and represents the *time slot* for the discrete time arrival and service processes. For convenience, each packet contains a fixed number of bits $N_b = N_{ds} \cdot R_1$. Thus, when transmitting with M&C mode n (at a rate of R_n bits/symbol), exactly $r_B(n) = R_n/R_1$ packets per time-slot (frame) can be serviced (transmitted).

The AMC algorithm selects the M&C mode to be used for transmission during each time slot based on the measured SNR at the receiver and communicates the selection back to the transmitter. Therefore, the statistics of the service process depend on the stochastic characteristics of the fading channel, the transmission rate achieved by each M&C mode and the AMC algorithm. The employed channel fading model is based on the following set of assumptions:

A1: The wireless channel quality remains constant per frame, but is allowed to vary from frame to frame. This corresponds to a block fading channel model, which is suitable for slowly varying channels. AMC is thus implemented on a frame-by-frame basis (time slot granularity).

A2: Perfect channel state information (CSI) is available at the receiver relying on training-based channel estimation. The corresponding mode selection is fed back to the transmitter without error and latency.

A3: Error detection based on cyclic redundancy check (CRC) is perfect, i.e., sufficiently powerful error detection CRC codes are used.

A4: If a packet is received incorrectly at the client after error detection, we declare packet loss which is equivalent to complete loss of all data bits contained in the packet.

For fading channels adhering to A1, the channel quality is captured by a single parameter, namely the *instantaneous* received SNR γ. Since the channel varies from frame to frame, we adopt the general Nakagami m model to describe γ statistically. The received SNR γ per frame is thus a random variable with a Gamma probability density function

$$p_\gamma(\gamma) = \frac{m^m \gamma^{m-1}}{\bar{\gamma}^m \Gamma(m)} \exp\left(-\frac{m\gamma}{\bar{\gamma}}\right), \tag{4.16}$$

where $\bar{\gamma} := E\{\gamma\}$ is the *average* received SNR, $\Gamma(m) := \int_0^\infty t^{m-1} e^{-t} dt$ is the Gamma function and m is the Nakagami fading parameter ($m \geq 1/2$). This model includes the Rayleigh channel when $m = 1$. A one-to-one mapping between the Rician factor and the Nakagami fading parameter m allows Rician channels to be well approximated by Nakagami-m channels. This channel model is suitable for flat-fading channels as well as frequency-selective fading channels encountered with OFDM systems.

The wireless channel parameters $\bar{\gamma}$ and m are determined by the hardware parameters of wireless equipment and the operational environment (e.g., the transmitter output power P_t, the receiver noise power P_N, the antenna loss L_i), and by the propagation conditions of radio waves (e.g., the distance between transmitter and receiver d, the carrier frequency f_c and the path-loss model). We also use the following PHY abstraction and modulation and coding modes following [11] (with minor modifications):

Let N denote the total number of transmission modes available ($N = 5$ for the considered M&C schemes). As in [11] (see also [17]), we assume constant power transmission, and partition the entire SNR range in $N + 1$ non-overlapping consecutive intervals, with boundary points denoted as $\{\gamma_n\}_{n=0}^{N+1}$. The AMC policy is to transmit at M&C mode n, when $\gamma \in [\gamma_n, \gamma_{n+1})$. To avoid deep channel fades, no data are sent when $\gamma_0 \leq \gamma < \gamma_1$, which corresponds to the mode $n = 0$ with rate $R_0 = 0$ (bits/symbol). The design objective of AMC is to determine the boundary points $\{\gamma_n\}_{n=0}^{N+1}$. For simplicity, we approximate the instantaneous packet error rate (PER) as:

$$PER_n(\gamma) \approx \begin{cases} 1, & \text{if } 0 < \gamma < \gamma_{pn}, \\ a_n \exp(-g_n \gamma), & \text{if } \gamma \geq \gamma_{pn}, \end{cases} \tag{4.17}$$

where n is the mode index, γ is the received SNR, and the mode-dependent parameters a_n, g_n, and γ_{pn} are obtained by fitting (4.17) to the exact PER. With packet length $N_b = 1,080$, the fitting parameters for TM2 are provided in Table 4.12. Based on (4.16), transmission mode n will be chosen with probability [17]

$$\Pr(n) = \int_{\gamma_n}^{\gamma_{n+1}} p_\gamma(\gamma) d\gamma = \frac{\Gamma(m, m\gamma_n/\bar{\gamma}) - \Gamma(m, m\gamma_{n+1}/\bar{\gamma})}{\Gamma(m)} \tag{4.18}$$

where $\Gamma(m,x) := \int_x^\infty t^{m-1}e^{-t}dt$ is the complementary incomplete Gamma function. Let $\overline{\mathrm{PER}}_n$ denote the average PER corresponding to mode n. In practice, we have $\gamma_n > \gamma_{pn}$, which implies that $\overline{\mathrm{PER}}_n$ can be obtained in closed-form as [17]

$$\begin{aligned}\overline{\mathrm{PER}}_n &= \frac{1}{\Pr(n)}\int_{\gamma_n}^{\gamma_{n+1}} a_n \exp(-g_n\gamma)p_\gamma(\gamma)d\gamma \\ &= \frac{1}{\Pr(n)}\frac{a_n}{\Gamma(m)}\left(\frac{m}{\bar{\gamma}}\right)^m \frac{\Gamma(m,b_n\gamma_n)-\Gamma(m,b_n\gamma_{n+1})}{(b_n)^m},\end{aligned}$$

where $b_n := m/\bar{\gamma} + g_n$. The average PER of the AMC scheme can then be approximated as the ratio of the average number of packets in error over the average number of transmitted packets:

$$\overline{\mathrm{PER}} \approx \frac{\sum_{n=1}^{N} r_B(n)\Pr(n)\overline{\mathrm{PER}}_n}{\sum_{n=1}^{N} r_B(n)\Pr(n)}. \tag{4.19}$$

In [11] the thresholds $\{\gamma_n\}_{n=0}^{N+1}$, are determined so that a prescribed PER P_0 is achieved for each transmission mode, i.e., $\overline{\mathrm{PER}}_n = P_0$, which naturally leads to $\overline{\mathrm{PER}} = P_0$. Once these thresholds have been selected, queuing analysis is performed to obtain the buffer overflow probability $\mathrm{P_d}$ in the case of a finite buffer. The overall loss probability ξ is then calculated as $\xi = 1-(1-\mathrm{P_d})(1-\mathrm{P_0})$ which is proven to hold (although a packet being successfully transmitted is not independent to not suffering a buffer overflow) because $\overline{\mathrm{PER}}_n = P_0$ for all n [11]. In order to obtain an improved ξ the following optimization is performed:

$$\min_{P_0} \xi\,(P_0) \quad \text{s.t.} \quad \overline{\mathrm{PER}}_n = P_0. \tag{4.20}$$

The detailed queuing analysis is based on the following Finite State Markov Chain (FSMC) channel model:

The SNR region $[\gamma_n,\ \gamma_{n+1})$ corresponding to transmission mode n constitutes the channel state indexed by n. Assuming slow fading conditions so that transitions happen only between adjacent states, the probability of transition exceeding two consecutive states is zero [11], i.e.,

$$P_{l,n} = 0, \qquad |l-n| \geq 2.$$

The adjacent-state transition probability can be determined by:

$$\begin{aligned}P_{n,n+1} &= \frac{N_{n+1}T_f}{\Pr(n)}, &&\text{if } n = 0,\dots,N-1,\\ P_{n,n-1} &= \frac{N_nT_f}{\Pr(n)}, &&\text{if } n = 0,\dots,N,\end{aligned}$$

where N_n is the cross-rate of mode n (either upward or downward), which can be estimated as

$$N_n = \sqrt{2\pi \frac{m\gamma_n}{\bar{\gamma}}} \frac{f_d}{\Gamma(m)} \left(\frac{m\gamma_n}{\bar{\gamma}}\right)^{m-1} \exp\left(-\frac{m\gamma_n}{\bar{\gamma}}\right),$$

where f_d denotes the mobility-induced Doppler spread. The probability of staying at the same state n is:

$$P_{n,n} = \begin{cases} 1 - P_{n,n+1} - P_{n,n-1}, & \text{if } \ 0 < n < N, \\ 1 - P_{0,1}, & \text{if } \ n = 0, \\ 1 - P_{N,N-1}, & \text{if } \ n = N, \end{cases}$$

so that the transition matrix of the FSMC is a banded matrix denoted by

$$\mathbf{P}_c = \left[P_{i,j}\right]_{(N+1)\times(N+1)} \tag{4.21}$$

Although a banded channel transition matrix is adopted for simplicity, the ensuing results apply to general channel transition matrices.

The described FSMC channel model combined with the AMC policy dictates a dynamic, rather than deterministic, service process S(t) for the queue. In particular, if at time slot t the channel is at state n, the number of packets that can be transmitted at time slot t is $S(t) = r_B(n) = R_n/R_1$. Therefore the service process S(t) is modeled as an FSMC with transition matrix $\mathbf{P}_c$ given in (4.21). The number of packets arriving at the queue during each time slot is modeled as a Markovian arrival process. This is a more general arrival process compared to the Poisson arrivals adopted in [11]. It can be used to model the time correlations observed in VBR (Variable BitRate) traffic generated by streaming media sources (e.g., MPEG encoded video) [9]. We propose to select the AMC policy (i.e., the thresholds $\{\gamma_n\}_{n=0}^{N+1}$) that minimizes the cumulative packet loss rate ξ without imposing a PER constraint as in (4.20). Note that by breaking this constraint it is no longer straightforward to calculate ξ. Thus, we resort to the upper bound $\xi \leq P_d + P_0$ and try to optimize this sum of probabilities (which we will denote by ξ').[1] To this end, a general method of computing the buffer overflow probability P_d is required. For a fixed service capacity c and maximum delay constraint D_{max}, the buffer size can be appropriately dimensioned and therefore buffer overflow probability and maximum delay violation probability are equivalent. However, for variable service capacity we should distinguish between these two quantities and take P_d to denote the maximum delay violation probability which is the probability of interest (provided that the buffer is large enough to prevent overflow).

[1] $\xi = P_d + P_0 - \Pr(\text{pckt transmission error} \cap \text{pckt excessively delayed})$ therefore $\max(P_d, P_0) \leq \xi \leq P_d + P_0$. For small probabilities P_d and P_0 this bound is quite tight, e.g., for $P_d = P_0 = 10^{-3}$ we have $10^{-3} \leq \xi \leq 2 \cdot 10^{-3}$ whereas for $P_d = 10^{-4}$ and $P_0 = 10^{-3}$ we have $10^{-3} \leq \xi \leq 1.1 \cdot 10^{-3}$.

Large deviations approximation of P_d

We suggest following an LD approximation approach to calculate the maximum delay violation probability along the lines of [14]. Briefly, the approximation used in [14] is obtained according to the following steps:

1. According to the statistical characteristics of the arrival A(t) and service S(t) processes, find the limiting log moment generating functions $\Lambda_A(\theta)$ and $\Lambda_S(\theta)$ of the two processes. Determine the solution θ^* of the equation $\Lambda_A(\theta) = -\Lambda_S(-\theta) = \delta$.
2. For any desired delay bound D_{max}, the delay bound violation probability can be derived as

$$\Pr\{\text{Delay} > D_{max}\} \approx e^{-\theta^* \delta D_{max}} \tag{4.22}$$

Note that the limiting log moment generating function of a FSMC process X(t) with transition probabilities matrix

$$P_X = \{p_X(i,j)\}_{i,j=1}^{M}$$

and number of arrivals (or departures) that are a deterministic function of the current state Y_t, i.e.,

$$X(t) = f_X(Y_t)$$

can be calculated analytically as [7]

$$\Lambda_X(\theta) = \log \rho \left[\Pi(P_X, \theta)\right],$$

where ρ is the Perron-Frobenious eigenvalue or the matrix $\mathbf{\Pi}(P_X, \theta)$ which has elements

$$\pi(i,j) = p_X(i,j) \cdot e^{\theta \cdot f_X(j)}.$$

This calculation of $\Lambda_X(\theta)$ can be easily extended to the case where $f_X(Y_t)$ is a random function of the state Y_t (see [7]).

Although the LD approximation in (4.22) captures the asymptotes of the delay bound violation probability for $D_{max} \to \infty$, it can be refined by heuristics that strive to compute a multiplying constant β such that

$$\Pr\{\text{Delay} > D_{max}\} \approx \beta \cdot e^{-\theta^* \delta D_{max}}. \tag{4.23}$$

For instance [18] and [19] discuss a very effective heuristic for approximating the multiplicative constant α in the buffer overflow probability:

$$\Pr\{Q(t) \geq U\} \approx a \cdot e^{-\theta^* U}. \tag{4.24}$$

Using the fact that the mean of a non-negative random variable is equal to the integral of its complementary distribution, we can see that

$$\mathbf{E}[Q(t)] = \int_0^\infty \Pr\{Q(t) \geq U\} dU \approx \int_0^\infty a \cdot e^{-\theta^* U} dU = \frac{\alpha}{\theta^*},$$

which leads to the approximation

$$\alpha \approx \theta^* \mathbf{E}[Q(t)]. \tag{4.25}$$

Thus, in order to find the refining constant α, we need to estimate the expectation of the queue length. The idea is that it is much easier to obtain a reliable estimate of the mean queue length through simulation than an estimate of small tail probabilities. An alternative method for calculating the refining constant α without any need for simulation is

$$\alpha \approx \frac{\mathbf{E}[A(t)]}{\mathbf{E}[S(t)]}. \tag{4.26}$$

Calculating α by (4.26) is much faster, but, in general, less accurate than using (4.25). Note that Pr{Delay > D_{max}} in (4.23) and the Pr{Q(t) ≥ U} in (4.24) express the probability that at a given timeslot one or more packets in the buffer are expected to violate the delay bound or exceed the threshold U respectively. Instead what is needed for calculating ξ' is the probability P_d that a given packet will violate the delay bound. In order to arrive to a refined approximation for P_d, we assume that no packet is dropped even if it is expected to violate the delay bound and consider a fluid queuing model instead of a packet queuing model (both to facilitate the mathematical derivation).

We start by deriving a refinement to the approximation of the fraction of overflowing fluid when the service process is constant (S(t) = c). Let us denote by F(u) the probability that an unconstrained queue does not exceed a given length u, i.e., F(u) = Pr[Q(t) ≤ u], by G(u) the probability that an infinite queue exceeds a given length u, i.e., G(u) = Pr[Q(t) ≥ u] and by f(u) the probability density function (PDF) of the queue length, that is, $F(u) = \int_0^u f(x)dx$. Then, the average amount of overflowing fluid per time slot can be expressed as

$$\int_U^{U+c} f(u)(u-U)du + \int_{U+c}^{\infty} f(u)c\,du =$$
$$\int_U^{\infty} f(u)(u-U)du - \int_{U+c}^{\infty} f(u)(u-U-c)du =$$
$$\int_U^{\infty} G(u)du - \int_{U+c}^{\infty} G(u)du \approx$$
$$\int_U^{\infty} a\cdot \mathrm{e}^{-\theta^* u}\mathrm{du} - \int_{U+c}^{\infty} a\cdot \mathrm{e}^{-\theta^* u}\mathrm{du} = \frac{\alpha}{\theta^*}\left[\mathrm{e}^{-\theta^* U} - \mathrm{e}^{-\theta^*(U+c)}\right].$$

Hence, the fraction of lost fluid is

$$P_{lf} \approx \frac{\alpha}{\theta^* \mathbf{E}[A(t)]}\left[\mathrm{e}^{-\theta^* U} - \mathrm{e}^{-\theta^*(U+c)}\right],$$

which by using the previously obtained approximation $\alpha \approx \theta^* \mathbf{E}[Q(t)]$ becomes

$$P_{lf} \approx \frac{\mathbf{E}[Q(t)]}{\mathbf{E}[A(t)]}\left[e^{-\theta^* U} - e^{-\theta^*(U+c)}\right].$$

Now for an independent and identically distributed (i.i.d.) service process with $\Pr\{S(t) = c_i\} = p_i$, we can use the same argument to arrive to the following expression:

$$P_{fl}^{(1)} \approx \frac{\mathbf{E}[Q(t)]}{\mathbf{E}[A(t)]}\left[e^{-\theta^* U} - \sum_i p_i \cdot e^{-\theta^*(U+c_i)}\right].$$

In case of a correlated service process (such as the FSMC used herein) the above expression is no longer rigorously correct, but it can be used in lack of a better approximation with p_i being the marginal probability of service capacity c_i. Another heuristic would be to use the mean service rate $\mathbf{E}[S(t)]$ or the effective rate δ in place of the fixed rate c to arrive to the approximations

$$P_{fl}^{(2)} \approx \frac{\mathbf{E}[Q(t)]}{\mathbf{E}[A(t)]}\left[e^{-\theta^* U}\left(1 - e^{-\theta^* \mathbf{E}[S(t)]}\right)\right]$$

and

$$P_{fl}^{(3)} \approx \frac{\mathbf{E}[Q(t)]}{\mathbf{E}[A(t)]}\left[e^{-\theta^* U}\left(1 - e^{-\theta^* \delta}\right)\right]. \tag{4.27}$$

In order to evaluate the three refined approximations suggested above, simulation experiments have been conducted with known Markovian service and arrival processes and the results have been compared to the values obtained by the three approximations. In Table 4.14, we report these results for the sample values to arrival and service process parameters shown in Table 4.13 and for varying buffer sizes which result to a probability range of three orders of magnitude. The banded probability matrix for the service process is in accordance with the FSMC channel fading models described earlier.

Table 4.13 Parameters of arrival and service processes used in comparing the 3 refined approximations of P_{fl}

	Arrival process	Service process
Number of states	M = 5	M = 6
Arrival/service rate per state in Mbps	$\mathbf{r}_A$ = [0.2 0.4 0.6 0.85 1.1]	$\mathbf{r}_S$ = [0 0.35 0.7 1.05 2.1 3.15]
State transition probabilities matrix	$\begin{bmatrix} 0.35 & 0.3 & 0.2 & 0.1 & 0.05 \\ 0.2 & 0.3 & 0.2 & 0.18 & 0.12 \\ 0.14 & 0.22 & 0.3 & 0.21 & 0.13 \\ 0.14 & 0.21 & 0.22 & 0.28 & 0.15 \\ 0.12 & 0.23 & 0.25 & 0.22 & 0.18 \end{bmatrix}$	$\begin{bmatrix} 0.649 & 0.351 & 0 & 0 & 0 & 0 \\ 0.0224 & 0.909 & 0.0686 & 0 & 0 & 0 \\ 0 & 0.0703 & 0.8687 & 0.061 & 0 & 0 \\ 0 & 0 & 0.1521 & 0.7273 & 0.1206 & 0 \\ 0 & 0 & 0 & 0.1315 & 0.7823 & 0.0862 \\ 0 & 0 & 0 & 0 & 0.0709 & 0.9291 \end{bmatrix}$

Table 4.14 Comparison of the four refined approximations of fluid loss probability with simulation results (Markovian arrival and service processes)

U (in Mbits)	P_{fl} simulation	$P_{fl}^{(1)}$	$P_{fl}^{(2)}$	$P_{fl}^{(3)}$	$P_{fl}^{(4)}$
50	9.43×10^{-3}	1.69×10^{-2}	1.75×10^{-2}	9.47×10^{-3}	8.32×10^{-3}
75	1.20×10^{-3}	2.15×10^{-3}	2.22×10^{-3}	1.20×10^{-3}	1.05×10^{-3}
100	1.51×10^{-4}	2.72×10^{-4}	2.81×10^{-4}	1.52×10^{-4}	1.34×10^{-4}
125	1.89×10^{-5}	3.45×10^{-5}	3.57×10^{-5}	1.93×10^{-5}	1.69×10^{-5}
150	2.37×10^{-6}	4.38×10^{-6}	4.52×10^{-6}	2.45×10^{-6}	2.15×10^{-6}

Clearly, the third refined approximation provides the best (and extremely good) approximation of the fluid loss probability estimated by simulation. In case that the time constraints for computing the fluid loss probability $P_{fl}^{(3)}$ are prohibitive for an adequate estimation of **E**[Q(t)], the simpler approximation of α obtained by (4.26) can be used instead to obtain

$$P_{fl}^{(4)} \approx \frac{1}{\theta^* \mathbf{E}[S(t)]} \left[\mathrm{e}^{-\theta^* U} (1 - \mathrm{e}^{-\theta^* \delta}) \right].$$

As an illustration of the limited accuracy of the above simpler approximation, the corresponding values of $p_{fl}^{(4)}$ are reported in the last column of Table 4.14.

All the above refined approximations can be used in case of delay violation probabilities instead of buffer overflow probabilities by substituting δD_{max} for U. Therefore, the suggested refined approximation for P_d is (cf. (4.27))

$$P_d \approx \frac{\mathbf{E}[Q(t)]}{\mathbf{E}[A(t)]} \left[\mathrm{e}^{-\theta^* \delta D_{\max}} (1 - \mathrm{e}^{-\theta^* \delta}) \right] \tag{4.28}$$

Note that all the approximations discussed so far concern infinite buffers where no fluid is thrown away because of buffer overflow (or because it will violate the delay constraint). In case of finite buffers, it has been shown that the asymptotic part of the LD approximation is identical to the infinite buffer case. However, the multiplicative constant should be different since the respective probability measure for finite buffer is obviously smaller than the one for infinite buffer. Unfortunately, there is no easy way to estimate the multiplicative constant for the finite buffer case. This is an open issue for future research. In case of an appropriately dimensioned buffer, where buffer overflows are not an issue (as is our working assumption), packet loss is due to excessive delays. In general, a packet that arrives to the front of the queue and gets ready to be transmitted could be dropped if it has already violated its maximum allowable delay. However, our refined approximation works only if all packets are transmitted. In practice this assumption holds in case of multihop transmissions where the end-to-end delay depends on the delays in subsequent hops and thus a packet should be transmitted even if it has violated the delay constraint at the current hop.

Once we have an expression for P_d we can formulate the AMC policy as the solution to the following optimization problem:

$$\begin{aligned} &\mathit{min} \quad \xi' = P_d + P_0 \\ &\mathit{decision\ variables:}\ \{\gamma_n\}_{n=0}^{N+1} \end{aligned} \tag{4.29}$$

where P_0 is given by (4.19) and P_d by (4.28). Clearly this is a non-linear optimization problem as both P_0 and P_d are non-linear functions of the decision variables. Due to the complexity of these functions it is rather impractical to obtain closed form solutions of this optimization problem so a numerical solution is preferred. Note that **E**[Q(t)] is dependent on the decision variables as well. Since estimating **E**[Q(t)] via simulation at each iteration of the optimization algorithm would be too expensive computationally, a two level iteration approach is suggested:

Step 1: For the initialization vector $\left\{\gamma_n^{(0)}\right\}_{n=0}^{N+1}$ estimate the value of **E**[Q(t)].

Step 2: Numerically solve the optimization problem in (4.29) for the fixed value of **E**[Q(t)].

Step 3: Re-estimate the value of **E**[Q(t)] at the new vector obtained in the previous step. Return to Step 2. Iterate until two subsequent iterations provide values ξ' that differ less than some small value ε (stopping criterion).

Evaluation of the Proposed AMC Algorithm

In this section, we demonstrate the benefits of the proposed AMC policy over the previously suggested policy of maximizing ξ under an equal $\overline{\mathrm{PER}}_n$ constraint. We experiment with an arrival process described by a FSMC which can capture time correlations commonly observed in VBR traffic such as MPEG coded streaming video. The specific values used in these scenarios are shown in Table 4.15. In

Table 4.15 Parameter values used for the evaluation of the proposed AMC policy

Parameter	Value	
Number of states M	5	
Arrival rates per state $\mathbf{r}_A$	[0.6 1.2 1.8 3.2 4.1] packets/time slot	
Transition probabilities matrix	Same as arrival process state transition probabilities matrix shown in Table 4.15	
Service rates $\mathbf{r}_B$	[0 1 2 3 6 9] packets/time slot	
Time slot duration T_f	2 ms	
Doppler spread f_d	$0.02/T_f$	
Nakagami parameter m	1	
Maximum allowable delay D_{max}	100 time slots = 0.2 s	
	Figs. 4.37 and 4.38	Figs. 4.39 and 4.40
Average SNR $\bar{\gamma}$	12–17 dB	12 dB
Reduction to all arrival rates	0 dB	0–6 dB

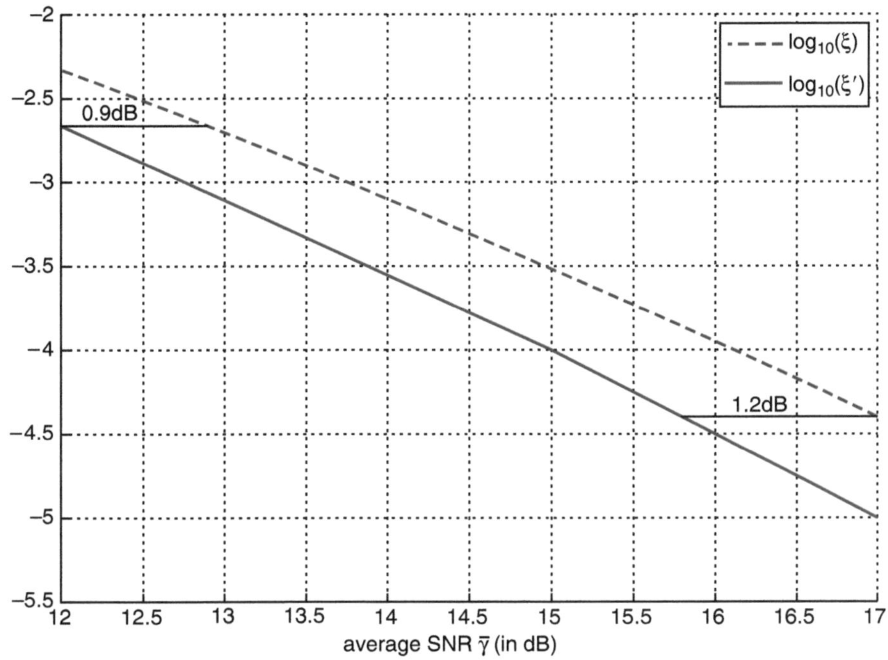

Fig. 4.37 Attained values of ξ and ξ' for a wide range of average SNR $\bar{\gamma}$

all cases we obtain the value of the total loss rate ξ under the equal $\overline{\text{PER}_n}$ constraint policy by solving (4.20) and the value of the upper bound ξ' to the total loss rate under unconstrained $\overline{\text{PER}_n}$ by solving (4.29).

In Fig. 4.37, we plot base 10 logarithms of the attained values of ξ and ξ' for average SNR $\bar{\gamma}$ ranging from 12 to 17 dB.

Figure 4.38 shows the ratio ξ/ξ' in the same SNR range. It can be seen that by employing the proposed AMC algorithm the total loss is reduced by a factor of 2–4 times. This gain in probability of loss can be translated into a gain in required transmit power in order to achieve the same probability of loss. As shown in Fig. 4.37, an increase to the average SNR of approximately 1 dB (more precisely 0.9–1.2 dB) is required in order to achieve equal loss rate, i.e., $\xi(\bar{\gamma}+1) \approx \xi'(\bar{\gamma})$. In Figs. 4.39 and 4.40, we keep $\bar{\gamma}$ constant at 12 dB and multiply the arrival rates rA shown in Table 4.15 by a factor w $\approx$ 1, 0.794, 0.631, 0.5, 0.398, 0.316, 0.25.

Figure 4.39 is a plot of log 10(ξ) and log 10(ξ') and Fig. 4.40 a plot of ξ/ξ' both as a function of $-10 \cdot \log 10(\text{w}) = 0, 1, 2, 3, 4, 5, 6$. Figure 4.40 reveals that the total loss is reduced by a factor growing from 2 to 22 times (for decreasing arrival rates) by employing the proposed AMC algorithm. This gain in probability of loss can be translated to improved arrival or source compression rates: by applying the proposed AMC algorithm we can afford arrival rates rA that are increased by 0.85 to 2.45 dB (horizontal distance between the two lines in Fig. 4.39) and still obtain equal loss rates to the ones achieved with the $\overline{\text{PER}_n}$ constrained policy. In other

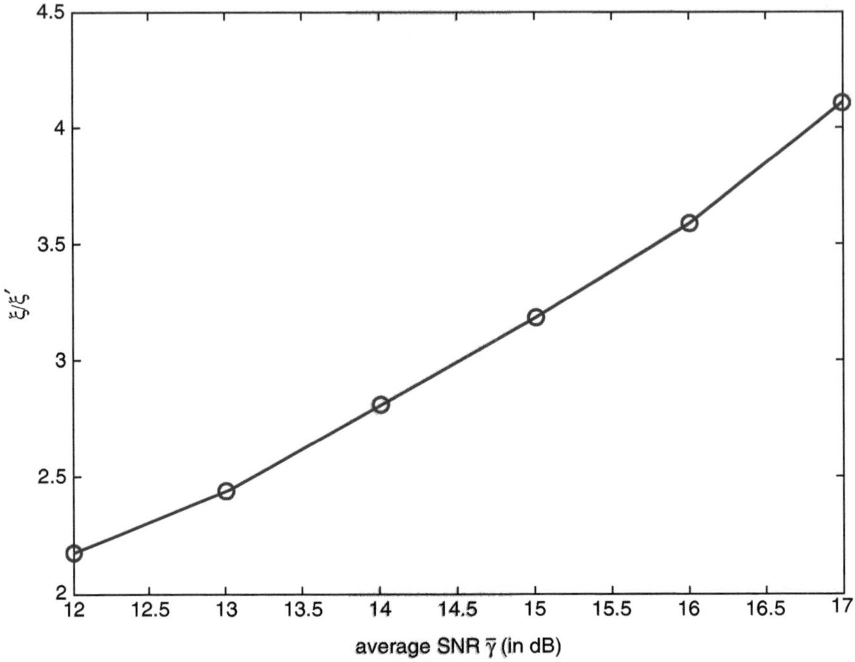

Fig. 4.38 Attained ratio ξ/ξ' for the values shown in Fig. 4.37

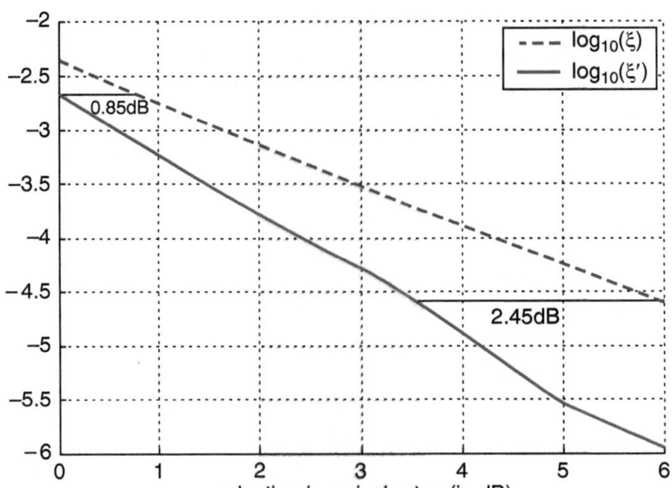

Fig. 4.39 Attained values of ξ and ξ' when reducing all arrival rates rA by the same factor

words, the proposed algorithm can accommodate arrival rates that are increased by a factor of 1.22–1.76 in this example (and obviously growing even larger for smaller arrival rates).

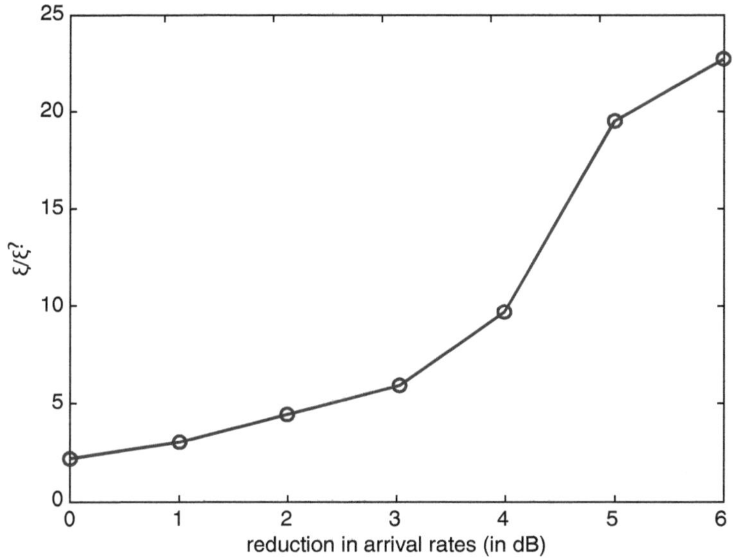

Fig. 4.40 Attained ratio ξ/ξ' for the values shown in Fig. 4.39

4.4.1.3 Generic AMC for real time media transmission with ARQ

In the previous section we have assumed that erroneously received packets are not retransmitted. When the RTT (i.e., the time from the moment a data packet is sent until the moment the associated ACK is received) is relatively small compared to the maximum allowable delay, it makes sense to retransmit corrupted packets. There are many different ARQ schemes that can be used for packet retransmission. On one extreme, every packet is retransmitted as many times it takes until it is correctly received. Other policies might put an upper bound to the number of retransmissions, retransmit with a certain probability or use hybrid ARQ, i.e., to send incremental information instead of retransmitting the whole packet with each subsequent retransmission.

In all cases, using retransmissions reduces or even eliminates the residual probability of packet loss due to incorrect reception at the cost of increasing the average time of packet transmission (and thus the probability of excessive delay). In principle, the analysis of the previous section can be extended to the case of ARQ by substituting the residual PER (which we will denote by P_r) for the PER P_0. In other words we still need to determine the AMC policy that minimizes the overall packet loss rate ξ (or ξ' when using our policy) under the specific ARQ scheme used. The difficulty, however, lies in the fact that depending on the ARQ scheme used, P_r and the service capacity might not be easily obtainable or might be dependent on the queue length (in which case LD analysis cannot be applied). More specifically, let us consider a number of ARQ policies and explore whether they are amenable to LD analysis:

ARQ policy 1: *Retransmit each packet as many times it takes to get correctly received.* In this case the residual PER is always zero. The service capacity $r_{B1}(n)$ at state n, i.e., the number of packets that can be cleared from the queue when using M&C mode n and ARQ policy 1, is now a random variable depending on whether packet transmissions are successful or not. If all transmissions are successful then $r_{B1}(n) = r_B(n)$. More generally, the service capacity $r_{B1}(n)$ is equal to x, where x is an integer satisfying $0 \leq x \leq r_B(n)$, with probability

$$p_i(n,x) \approx \binom{r_B(n)}{r_B(n)-x} \overline{PER}_i^{r_B(n)-x}[1-\overline{PER}_i]^x, \tag{4.30}$$

where we make the simplifying approximation that the PER experienced by all packets transmitted with M&C mode i is approximately equal to $\overline{PER}_i$. Since in this case $P_r = 0$, the overall packet loss probability is equal to P_d which can be computed as in the previous section with the service capacity at a given state of the underlying Markov process being a random function of this state. In fact, this is the only ARQ policy that is easily amenable to LD analysis and for which we have applied our AMC algorithm to investigate its performance. However, this policy has the obvious drawback that it will result in unnecessary retransmissions of packets in case that an incorrectly received packet has already violated the maximum allowable delay. By retransmitting this packet until it is correctly received, not only we will never get this packet to its destination on time, but we also delay subsequent packets in the queue risking to lose them as well due to maximum delay violation. To avoid this, the following ARQ policies can be employed:

ARQ policy 2: *Retransmit corrupted frames only if the current queue length Q(t) is less than a predetermined threshold.* Obviously this policy is not amenable to LD analysis because the service rate depends on the current queue size. Furthermore, calculating the residual probability of packet error is not an easy task.

ARQ policy 3: *Retransmit a corrupted frame up to a maximum number of times.* This is a simple ARQ policy very commonly used in the literature. Albeit its simplicity, calculating the residual PER and service capacity under this policy is rather intractable since the retransmissions of a given packet might span more than one frame (potentially at different channel states).

ARQ policy 4: *Retransmit a corrupted frame with a predefined probability* $p_{rtx}(i)$. Under this policy it is easy to calculate the service capacity distribution (as with policy 1). However, calculating the residual PER is as hard as under ARQ policy 3.

In the rest of this section we present results that illustrate the achievable packet loss probability under our AMC algorithm for ARQ policy 1 and compare it with the case of no packet retransmission. In Fig. 4.41, we plot the base 10 logarithm of the overall probability in both cases for an SNR ranging from 12 up to 17 dB.

The system parameters used are shown in Fig. 4.16. Note that the ARQ policy achieves much lower overall probability of loss (which in this case is equal to P_d) because the D_{max} used in this example is equal to 100 time slots. In Fig. 4.42 we examine the effect of varying the maximum allowable delay (for SNR $=$ 17 dB)

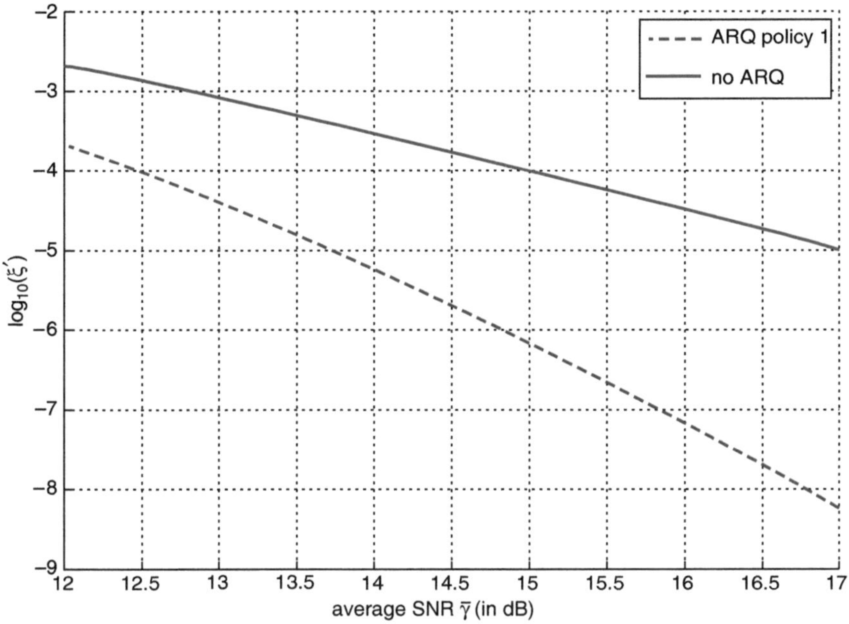

Fig. 4.41 Comparison of attained overall packet loss with and without retransmissions (D_{max} = 100 time slots)

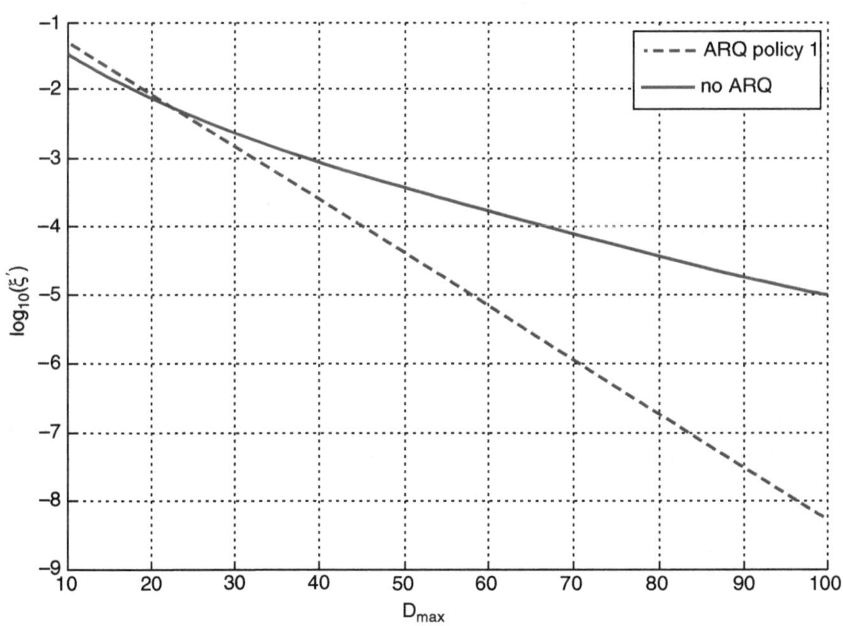

Fig. 4.42 Comparison of attained overall packet loss with and without retransmissions (average SNR = 17 dB)

and observe that for smaller D_{max} retransmissions are not resulting in a lower loss probability.

Obviously, an integrated AMC policy with ARQ can select between a retransmission and no retransmission scheme in order to minimize the overall packet loss probability.

4.4.1.4 Adaptation of generic AMC Schemes to the MAGNET HDR AI

The AMC policy introduced in the previous sections was developed for a Nakagami fading channel with Markovian time correlation and using the 5 M&C modes of Table 4.12. Let us now consider a MAGNET MC-SS link and seek to adapt our AMC algorithm to work with the 9 M&C modes shown in Table 4.8 and a fading stochastic model that better describes a WPAN channel. For instance, a channel model assuming a log-normal marginal distribution and an AR time correlation of the received SNR can be employed. In this case, (4.16) becomes

$$p_g(g) = \frac{1}{\sigma\sqrt{2\pi}} e^{-\frac{(g-\mu)^2}{2\sigma^2}},$$

where g is the received SNR per frame, measured in dB and μ and σ are the mean and standard deviation of g. Hence,

$$\Pr(n) = \int_{g_n}^{g_{n+1}} p_g(g)dg = P_g(g_{n+1}) - P_g(g_n) = \frac{1}{2}\left(erf\left(\frac{g_{n+1}-\mu}{\sigma\sqrt{2}}\right) - erf\left(\frac{g_n-\mu}{\sigma\sqrt{2}}\right)\right) \quad . \tag{4.31}$$

The payload BER at mode n as a function of g is obtained by the associated Look-Up Table and linear interpolation. The PER(g) is then derived as a function of the payload BER(g) using an approximation for the PER which is calculated according to the simplified (assuming independent bit errors) formula

$$\text{PER} = 1 - (1 - \text{BER}_{\text{hdr}})^{\text{H}} \cdot (1 - \text{BER}_{\text{payload}})^{\text{D}} \tag{4.32}$$

where H and D are the number of bits in the header (PHY&MAC) and payload, resp., and BER_{hdr} and $\text{BER}_{\text{payload}}$ are the corresponding BERs (recall that header and payload are transmitted with potentially different M&C modes, hence experience different BER). Here, H, D, and header BER are assumed constant. We can then calculate the average PER per mode n by numerical integration of the following integral:

$$\overline{\text{PER}}_n = \frac{1}{\Pr(n)} \int_{g_n}^{g_{n+1}} \text{PER}(g) p_g(g) dg. \tag{4.33}$$

The average PER, P_0, can be then derived using (4.19).

As far as the calculation of P_d is concerned, the essential difference lies in the calculation of the limiting log moment generating function $\Lambda_S(\theta)$ of the service process S(t). S(t) is not a Markovian process any more, but a process driven by the AR process defined by

$$u(t) = \sum_{i=0}^{I-1} a(i)\, u(t-i) + e(t) \tag{4.34}$$

where $u(t)$ is the simulated log-power at time t and the $a(i)$'s are the filter coefficients assumed known. The innovation $e(t)$ is a sample from a zero mean white Gaussian process.

However, S(t) is not AR because the SNR value u(t) is mapped to a number of transmitted frames value S(t) based on the AMC policy that maps a given SNR to a M&C mode. This interval mapping results in a PDF for the number of transmitted frames that is obviously not normal and hence S(t) is not AR. For this reason, although the limiting log moment generating function of an AR process can be obtained in an analytical form, the limiting log moment generating function of S(t) cannot be derived analytically and needs to be calculated numerically (using a sufficiently large value of n) based on the definition

$$\Lambda_S(\theta) = \lim_{n\to\infty} \frac{1}{n} log\mathbf{E}\left[e^{\theta[S(0)+S(1)+\cdots+S(n)]}\right]. \tag{4.35}$$

Obviously, this calculation is more computationally expensive than any closed form derivation of the limiting log moment generating function. Since this calculation needs to be repeated many times when solving the optimization problem in (4.29) obtaining the optimum AMC policy in this case will be more time consuming than in the Markovian service process case.

4.4.1.5 Amplify-and-Forwarding Cooperative Transmission

In this section a new approach for interference mitigation using cooperative transmission is proposed. The idea of improving the performance of wireless networks using cooperative schemes has recently attracted much interest. A number of methods have been proposed letting single antenna transceivers exhibit performance characteristics of MIMO links. These methods are known as user cooperation diversity [20], cooperative diversity [21–23] or cooperative communication [24].

In [21] cooperative protocols, such as amplify-and-forward and decode-and-forward, are developed and analyzed for the use in ad-hoc networks. The performance of the protocols is characterized in terms of outage events. The authors came to the conclusion that the amplify-and-forward protocol provides powerful benefits using distributed antennas. The work of [22] is also concerned with ad-hoc networks. Høst-Madsen defines the upper and lower capacity bounds for a 4-node cooperative diversity network. His results are numerical solutions showing that in general the gain from receiver cooperation is significant, while the gain from transmitter cooperation is more limited. In contrast to other approaches where relay stations are assumed to have no own information to send, [20] analyses the case where the relay stations have own information to send. It is shown that user cooperation is also beneficial in such scenarios resulting in significant gains over non-cooperative transmission. The difficulty of this approach is the need for a more complex receiver, as the device now has to be able to detect the signals of the other devices during the own transmission. The authors are concentrating in their work on cooperative transmission in cellular networks.

Optimizing the outage probability is aimed at in [25]. The authors show that the optimal selection of a single relay link among the set of multiple amplify-and-forward relay candidates has a better performance than simultaneous relay transmissions from more than one device. Their approach demonstrates the need for an intelligent scheduling among amplify-and-forward relay candidates.

Another approach for optimizing the outage probability using amplify-and-forward relaying is discussed in [26]. Kraidy and his co-authors are using turbo codes and compare rotated and unrotated turbo-coded schemes, demonstrating that both methods perform close to their corresponding outage limits.

For all devices which cannot afford multiple antennas, like WPAN devices, these cooperative transmission schemes are interesting. As MAGNET's HDR air interface is working in the unlicensed 5 GHz ISM band, the devices are exposed to cochannel interference (CCI) from other systems, such as WLANs following the IEEE 802.11a standard. Receiving devices can react to outages due to CCI by using retransmission requests. The use of ACKs in the IEEE 802.15.3 MAC as used in the MAGNET HDR mode is discussed in Section 4.2. However, frame retransmission may be unfavorable in the view of latency, since typically retransmissions requested by non-acknowledgements involve subsequent superframes. A further disadvantage of the existing ARQ scheme is the inefficient use of the energy reserves of battery powered devices. In case a device is short of energy, it would be helpful if another device could carry out the necessary retransmission of a certain frame. These issues are addressed by the cooperative protocols discussed in the following.

System Model

We consider a WPAN composed of M devices $\mathrm{D}_1, \cdots, \mathrm{D}_M$, which exchange information over a common channel on the basis of a TDMA scheme. Data is sent in frames of fixed length directly from peer to peer. Each frame fits into a time slot of a superframe. Additionally, there is occasional CCI from devices I_1, $\mathrm{I}_2, \ldots$ associated with other wireless systems in the neighborhood. Different time slots may be affected by the additive signals from different numbers of interferers. A signal originating from a certain interferer and coinciding with a certain time slot is in the following also referred to as a frame.

The devices transmit with power P_s, and the power of the additive noise emerging in the WPAN receiver front ends equals P_n. The channels between any two WPAN devices as well as those between an external interferer and a WPAN device are assumed frequency nonselective – as a result of the relatively short distances – and time-invariant over a superframe. The random attenuation in the channel between the two WPAN devices D_m and D_l is denoted as $h(D_m, D_l)$. We assume Rayleigh fading where $h(D_m, D_l) \sim \mathsf{CN}(0, \mu_{m,l})$, i.e., $h(D_m, D_l)$ is zero-mean circularly symmetric complex Gaussian distributed with variance $\mu_{m,l} > 0$. The attenuation $h(I_m, D_l)$ between an interferer I_m and D_l may follow any continuous distribution, and the different attenuations are assumed independent.

Besides of transmitting a regular frame, a WPAN device can relay the amplified signal observed in a previous time slot of the same superframe. A certain frame may thus be received in multiple time slots subject to different attenuations. We focus on a superframe or part of such and describe a scenario for which we define a slot allocation matrix (SAM) $\mathbf{T}$ with elements from $\{0,1\}$. The rows of the $(N \times K)$ matrix $\mathbf{T} = [\mathbf{t}_1 \cdots \mathbf{t}_k]$ relate to the N time slots, while the columns represent the K appearing frames (numbered $1, \cdots, K$). A'1' at the nth position of the column vector $\mathbf{t}_k$ indicates that the kth frame appears in the nth slot, possibly as a result of an amplify-and-forward. Corresponding to $\mathbf{T}$ the $(N \times K)$ matrix $\mathbf{A}_l = [\mathbf{a}_{l,1} \; \cdots \; \mathbf{a}_{l,K}]$ displays the complex valued attenuations of the frames as observed by the device D_l. The positions of the non-zero entries in $\mathbf{A}_l$ are in line with those in $\mathbf{T}$ with probability one. If, for example, device D_m transmits the kth frame in the nth slot, the nth element of $\mathbf{a}_{l,k}$ equals $h(D_m, D_l)$. Additionally, the N-dimensional vector $\mathbf{v}_l$ contains the powers of the composite additive noise per slot encountered by the device D_l. Every entry in $\mathbf{v}_l$ contains P_n (accounting for the front end of D_l) plus possibly the power of the amplified noise emerging in the front ends of relaying devices.

In the example depicted in Fig. 4.43, an intended frame transmission from D_1 to D_2 in time slot 1 coincides with interference from I_1. The device D_3 observes the composite signal and relays it in slot 2. The right hand side of the figure shows the corresponding $\mathbf{T}$, $\mathbf{A}_2$, and $\mathbf{v}_2$. Note that β_3 represents the factor by which the observed signal is amplified in D_3 in order to meet the transmit power constraint. To avoid requiring the devices to simultaneously transmit and receive, and to simplify the discussion on the diversity order, we make the following assumptions:

- Within a superframe a device has the role of either transmitting data frames, receiving and decoding frames, or being available for assisting with an amplify-and-forward procedure.

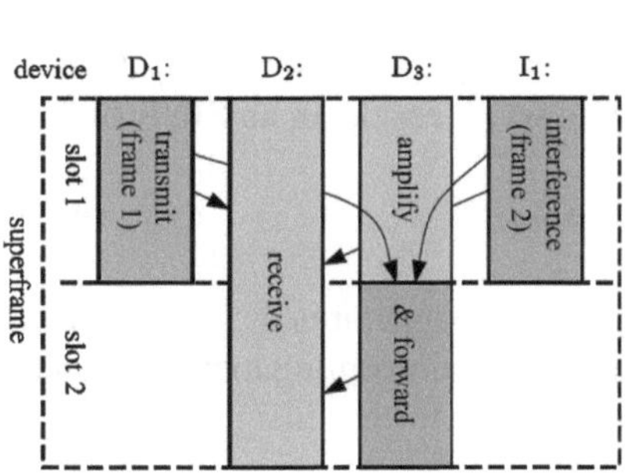

$$\mathbf{T} = \begin{bmatrix} 1 & 1 \\ 1 & 1 \end{bmatrix}$$

$$\mathbf{A}_2 = \begin{bmatrix} h(D_1, D_2) & h(I_1, D_2) \\ h(D_1, D_3)\beta_3 h(D_3, D_2) & h(I_1, D_3)\beta_3 h(D_3, D_2) \end{bmatrix}$$

$$\beta_3 = (|h(D_1, D_3)|^2 + |h(I_1, D_3)|^2 + P_n/P_s)^{-\frac{1}{2}}$$

$$\mathbf{v}_2 = \begin{bmatrix} P_n \\ P_n + |\beta_3 h(D_3, D_2)|^2 P_n \end{bmatrix}$$

Fig. 4.43 Example of a 2-slot superframe allocation and corresponding SAM, T A2 and v2

- A data transmitting device sends a certain frame in no more than one time slot of a superframe.
- We rule out recursive relaying, i.e., forwarding of signals which include relayed signals.
- An idle device D_m only relays a frame if the amplification factor β_m is within certain bounds.

Optimal Linear Signal Combining

Suppose the device D_l wants to decode the kth frame, which appears in one or multiple time slots with attenuations given by $\mathbf{a}_{l,k}$. The signals observed within the N time slots can be arranged such that they form an N-dimensional array signal, having the form of a sequence of $N \times 1$-sample vectors. Then, prior to the decoding, an appropriate combining of the array signal components mitigates the non-desired signal components, similar to a beamforming in a multi-antenna receiver.

Let us assume an ideal front end and regard the K signals emitted by the WPAN devices and interferers as well as the front end noise as independent zero-mean random processes. The contribution by the ith frame to any sample vector can be represented as a zero-mean random vector with covariance matrix $P_\mathrm{s}\mathbf{a}_{l,i}\mathbf{a}_{l,i}{}^\mathrm{H}$, and the noise component as a zero-mean random vector with covariance matrix $\mathrm{Diag}(\mathbf{v}_l)$, where $(\cdot)^\mathrm{H}$ denotes Hermitian transposition and $\mathrm{Diag}(\mathbf{v}_l)$ the diagonal matrix composed from $\mathbf{v}_l$, respectively. The interference-plus-noise covariance matrix thus reads

$$\mathbf{R}_{l,k} = P_\mathrm{s} \sum_{\substack{i=1 \\ i \neq k}}^{K} \mathbf{a}_{l,i}\mathbf{a}_{l,i}^\mathrm{H} + \mathrm{Diag}\,(\mathbf{v}_l) \tag{4.36}$$

As well known from array signal processing theory, the so-called MVDR (minimal variance distortionless response)-beamformer [28] defined as

$$\mathbf{w}_{l,k} = \frac{\mathbf{R}_{l,k}^{-1}\mathbf{a}_{l,k}}{\mathbf{a}_{l,k}^\mathrm{H}\mathbf{R}_{l,k}^{-1}\mathbf{a}_{l,k}}. \tag{4.37}$$

achieves the maximal SINR among all linear combining schemes $\mathbf{w}^\mathrm{H} \in C^N$ under the constraint $\mathbf{w}^\mathrm{H}\mathbf{a}_{l,k} = 1$. A signal combining by $\mathbf{w}_{l,k}{}^\mathrm{H}$ results in an SINR of

$$\gamma_{l,k} = P_\mathrm{s}\mathbf{a}_{l,k}^\mathrm{H}\mathbf{R}_{l,k}^{-1}\mathbf{a}_{l,k}. \tag{4.38}$$

Note that computing $\mathbf{w}_{l,k}{}^\mathrm{H}$ requires knowledge of the signatures $\mathbf{a}_{l,1}, \ldots, \mathbf{a}_{l,K}$ as well as $\mathbf{v}_l$. In the absence of this knowledge adaptive beamforming methods can be applied which can achieve an SINR of almost $\gamma_{l,k}$. For further insight we refer to the literature on adaptive beamforming.

Diversity Order

Successful decoding of the kth frame by device D_l requires that the SINR $\gamma_{l,k}$ attains a given threshold $\gamma_{\min}$. The event $\{\gamma_{l,k} < \gamma_{\min}\}$ is referred to as an outage, and necessitates a frame retransmission. Diversity is generally known to reduce outage probabilities. Diversity is achieved in the considered system by sending a frame in multiple time slots (via relays), and impaired by the incidence of interference. The diversity order associated with the kth frame is defined as [21]

$$d_{l,k} = \lim_{\gamma \to \infty} \frac{-\log \Pr\{\gamma_{l,k} < \gamma_{\min}\}}{\log \gamma} \tag{4.39}$$

where $\gamma = P_s/P_n$ represents the transmit signal-to-noise power ratio (SNR). A lower bound for $d_{l,k}$ is formulated in the proposition below. We let $\mathcal{I}\mathbf{t}$ denote the index set defining the positions of the 1's in the vector $\mathbf{t}$, and $|\mathcal{S}|$ the cardinality of the set $\mathcal{S}$.

Proposition 1: The diversity order $d_{l,k}$ fulfills $d_{l,k} \geq \varrho_k$, where [29]

$$\varrho_k = |\mathcal{I}\mathbf{t}_k| - |\{i \in \{1, \ldots, K\} : \mathcal{I}\mathbf{t}_i \cap \mathcal{I}\mathbf{t}_k \neq \oslash\}| + 1. \tag{4.40}$$

We note that the lower bound is independent of the device index and expressed on the basis of the SAM. Unlike the channel attenuation, the SAM is unique over the WPAN devices. Idle devices may use the SAM for computing the minimum diversity order associated with a certain frame, and use this number as a decision variable for possible relaying.

Relaying Policies

A network coordinator typically informs the WPAN devices about the time slot allocation through a control channel. A number of unallocated time slots, preferably towards the end of a superframe, may be reserved for relaying purposes. Devices not involved in a regular frame transmission can sense the channel, store the observed signals and along the way determine the numbers of interfering signals. If appropriate, one of the reserved slots can then be used for relaying a previously stored signal.

Provided that an idle device can figure out the SAM describing the past part of a superframe promptly before a reserved slot, different policies are conceivable for deciding if and how to assist with an amplify-and-forward procedure. Assuming there are L intra-WPAN frames in the past superframe part, numbered $1, \ldots, L$, and $K-L$ frames from interferers, an idle device can compute $\varrho_1, \ldots, \varrho_L$ and proceed using the following relaying policies.

Policy 1: Calculate first $\varrho_{\min} = \min\{\varrho_1, \ldots, \varrho_L\}$. If $\varrho_{\min} < 1$ then randomly choose an index from the index set $\{i \in \{1, \ldots, L\} : \varrho_i = \varrho_{\min}\}$, say k, and relay the observed signal within the slot with the index $\min \mathcal{I}\mathbf{t}_k$. If $\varrho_{\min} \geq 1$, remain idle.

Policy 2: Calculate $\varrho_{\min} = \min\{\varrho_1, \ldots, \varrho_L\}$. Randomly choose an index from the index set $\{i \in \{1, \ldots, L\} : \varrho_i = \varrho_{\min}\}$, say k, and relay the observed signal within the slot with the index $\min \mathcal{I}\mathbf{t}_k$.

Policy 3: Calculate $\varrho_{\max} = \max\{i \in \{\varrho_1, \ldots, \varrho_L\} : \varrho_i < 1\}$. Randomly choose an index from the index set $\{i \in \{1, \ldots, L\} : \varrho_i = \varrho_{\max}\}$, say k, and relay the observed signal within the slot with the index $\min \mathcal{I}\mathbf{t}_k$.

In words, policy 1 aims at avoiding that a frame is subject to a diversity order below 1, whereas in policy 2 all unused time slots are exploited for increasing diversity orders. Relaying policies 1 and 2 use the frame with lowest diversity order for relaying, while policy 3 relays frames with the highest diversity order below 1.

Numerical Results

In this section we employ numerical methods to evaluate the achievable outage probability reductions by the two relaying policies. A large number of superframes are generated in the simulations under the following assumptions. Every superframe comprises N_{SF} time slots. The number of intra-WPAN frames for the ith superframe is given as $Z_i + R_{i-1}$, where Z_i is subject to a Poisson distribution with mean ν_{frames}. If the number of frames exceeds N_{SF}, all time slots of the superframe are allocated for regular transmissions and the surplus $R_i = Z_i + R_{i-1} - N_{\text{SF}}$ frames are carried over to the next superframe, whereas otherwise $R_i = 0$. Hence, the average number of intra-WPAN frames per superframe equals ν_{frames}. These are followed by an average $N_{\text{SF}} - \nu_{\text{frames}}$ time slots available for relaying. The simulation starts with $R_0 = 0$. Additionally, the nth time slot of the ith superframe contains $Y_{i,n}$ frames originating from external interferers. The numbers $Y_{1,1}, Y_{1,2}, \ldots, Y_{1,N\text{SF}}, Y_{2,1}, \ldots$ are generated independently subject to a Poisson distribution with mean ν_{interf}.

Every frame is originally transmitted from a different device. There are $N_{\text{SF}} + 1$ WPAN devices involved in every superframe: one per transmitted intra-WPAN frame, one receiver aiming at decoding these frames, and one idle device per time slot reserved for relaying. The channel attenuations $h(\mathrm{D}_m, \mathrm{D}_l)$ and $h(\mathrm{I}_m, \mathrm{D}_l)$ for all possible device pairs are independently generated subject to $\mathcal{CN}(0, 1)$, and they are independently generated for every simulated superframe.

Immediately before each of the time slots reserved for relaying, an idle device decides if and which of the stored signals to amplify-and-forward on the basis of one of the relaying policies. Finally, the SINRs in (4.38) resulting from optimal signal combining at the receiver are evaluated and the outage events where SINRs do not achieve $\gamma_{\min}$ identified.

Figure 4.44 show the obtained outage probabilities versus the amount of interference ν_{interf} for superframes with $N_{\text{SF}} = 8$ time slots, comprising on average 2, 4, and 6 intra-WPAN frames, $\gamma = P_{\text{s}}/P_{\text{n}}$ equal to 20 dB, and $\gamma_{\min}$ equal to 10 dB. The outage probabilities are shown for employing relaying policies 1 and 2 as well as for no relaying.

We note that without relaying the outage performance does not depend on the intra-PAN traffic load, hence, the respective outage probabilities are similar

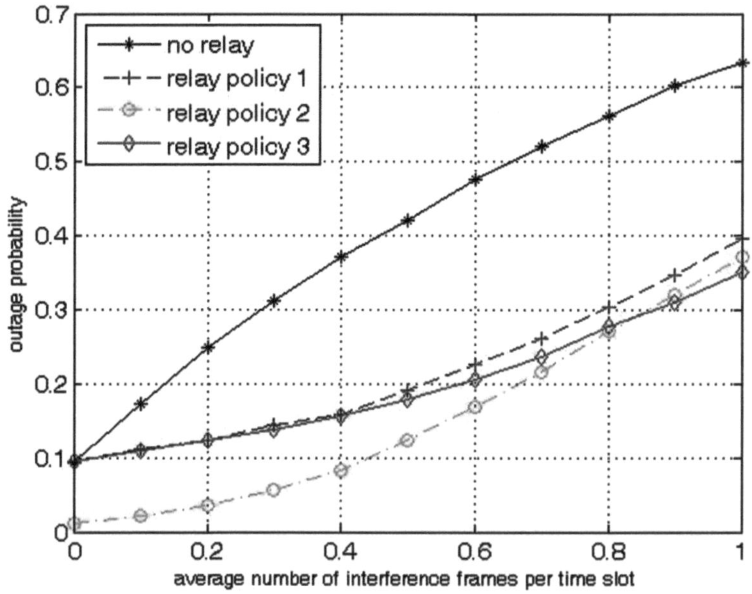

Fig. 4.44 Outage probability versus ν_{interf} for superframes comprising $N_{SF} = 8$ time slots for an average $\nu_{\text{frames}} = 2$ intra-WPAN frames

in all three figures. Applying policy 1, i.e., relaying frames with diversity order lower bounds below 1, significantly reduces outages in the presence of CCI. The improvements are larger at low intra-WPAN traffic load. At an average of two intra-PAN frames per superframe, i.e., a traffic load of 25% and 05 interfering frames per time slot on average, relaying policy 1 reduces the outage probability by more than 50%. Relaying policy 2 makes full use of the free time slots for increasing diversity orders, thereby reducing outages even in the absence of CCI since outages can also result from deep channel fades. In presence of CCI, policy 2 achieves larger outage probability reductions than policy 1. The performance of relaying policy 3 is similar to policy 1 in scenarios with less than 0.4 interfering frames per time slot. With increasing presence of CCI policy 3 outperforms the outage reduction of policies 1 and 2. As frames with lower diversity order need more relay time slots for a successful transmission, there is a greater probability of failure due to additional interference. Hence, it is wise in the case of frequent CCI to prioritize frames with diversity orders close to 1.

Figures 4.47–4.49 show results for similar scenarios except for having NSF = 16 time slots in every superframe. The traffic loads are again 25% (Fig. 4.47), 50% (Fig. 4.48), and 75% (Fig. 4.49). We note that for a given traffic load the increased number of time slots per superframe lets the three relaying policies become even more beneficial (Figs. 4.45–4.49).

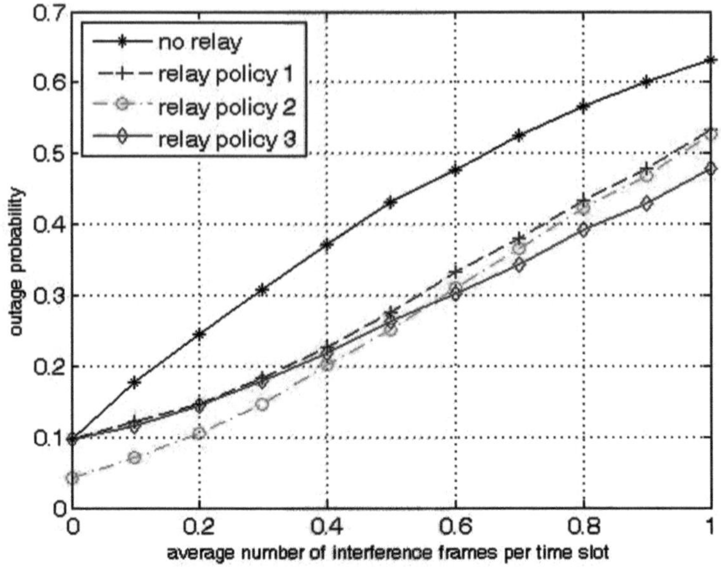

Fig. 4.45 Outage probability versus ν_{interf} for superframes comprising $N_{SF} = 8$ time slots for an average $\nu_{frames} = 4$ intra-WPAN frames

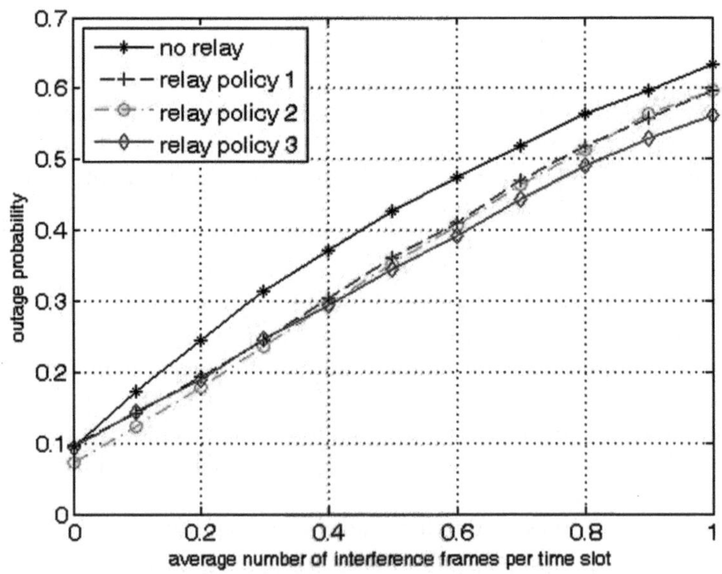

Fig. 4.46 Outage probability versus ν_{interf} for superframes comprising $N_{SF} = 8$ time slots for an average $\nu_{frames} = 6$ intra-WPAN frames

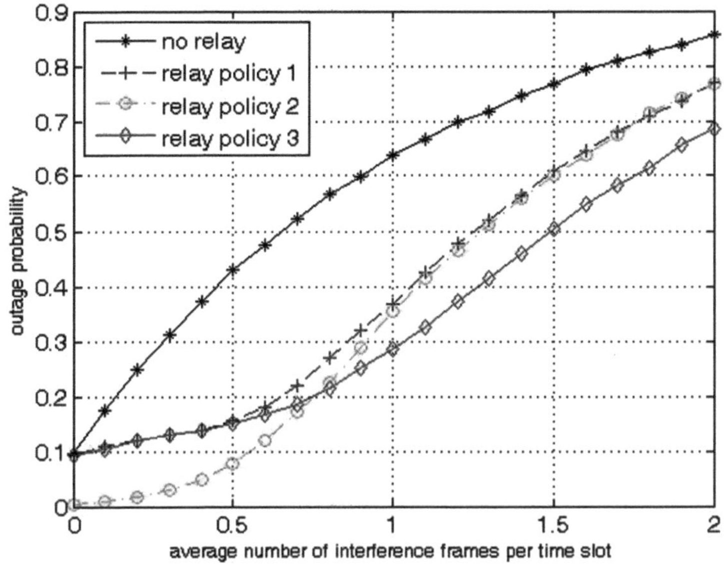

Fig. 4.47 Outage probability versus ν_{interf} for superframes comprising $N_{SF} = 16$ time slots for an average $\nu_{\mathrm{frames}} = 4$ intra-WPAN frames

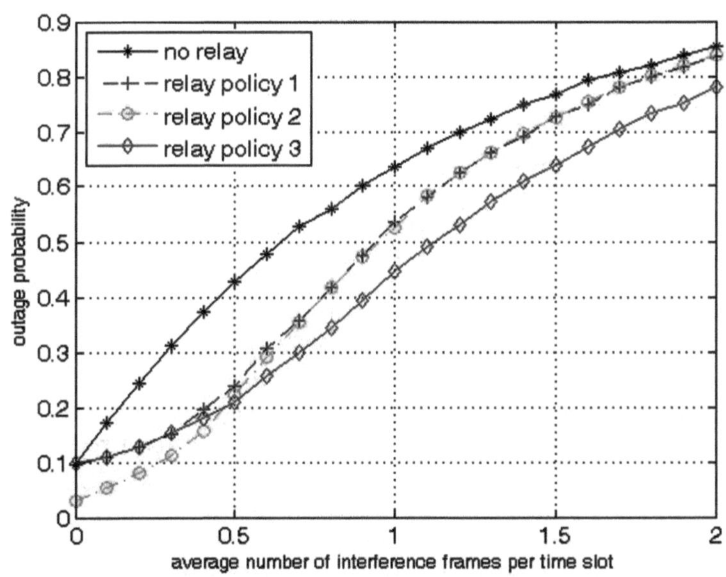

Fig. 4.48 Outage probability versus ν_{interf} for superframes comprising $N_{SF} = 16$ time slots for an average $\nu_{\mathrm{frames}} = 8$ intra-WPAN frames

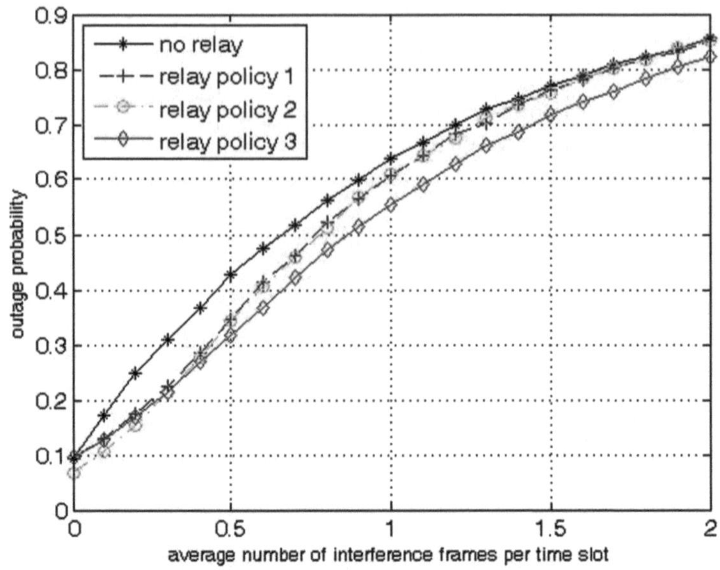

Fig. 4.49 Outage probability versus ν_{interf} for superframes comprising $N_{SF} = 16$ time slots for an average $\nu_{\text{frames}} = 12$ intra-WPAN frames

4.4.2 PAN-to-PAN Communication

WPANs span a limited operating space consisting of low cost and low power devices. There are different standards for WPANs depending on the type of air interface supported e.g. Bluetooth, UWB and the data rate (low with extended range in IEEE 802.15.4 and high with limited range in IEEE 802.15.3). As mentioned before, the PNC in a high data rate WPANs sends periodic beacons to provide timing information for the member devices. To communicate with devices in a piconet and access its resources, a device has to associate with the PNC by sending an association request command. The PNC then allocates a Device ID (DEVID) to the device upon successful association and informs the other member devices about the newly associated device and its advertised capabilities. The communication takes place in a superframe which consists of a beacon, a CAP and a CTAP (Fig. 4.50).

The CAP is CSMA/CA based and it can be used for sending association requests and other commands if allowed by the PNC. Small amounts of asynchronous data can also be sent in the CAP by the member devices. If the devices require channel time on a periodic basis, they can request the PNC for channel time in the CTAP which is TDMA based. The CTA request can be for either a sub-rate allocation (CTAs in alternate superframes) or a super rate allocation (multiple CTAs in the same superframe) depending on the type of traffic the device wants to send and its constraints, e.g. frame interarrival time. Since the communication in a piconet is single hop, to extend the range of a piconet IEEE 802.15.3 [30] allows the formation of child piconets which are dependent on channel time from the established parent

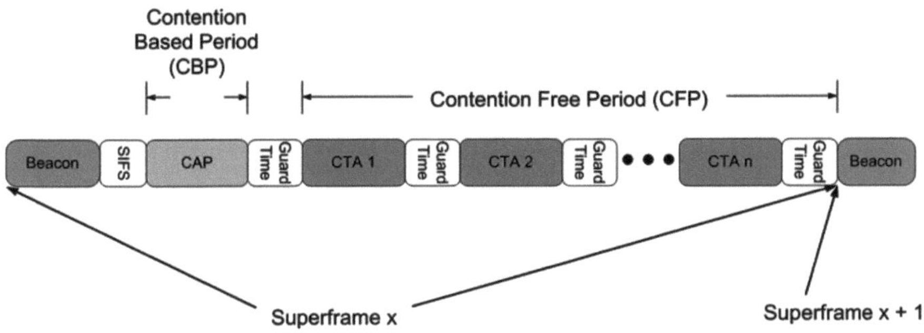

Fig. 4.50 IEEE 802.15.3 superframe

piconet. The devices in the parent piconet and the parent PNC can communicate with the child PNC and vice versa. The limitation in this extention is that the devices in the child piconet cannot communicate with the devices in the parent piconet and the parent PNC. Since the devices communicate with each other through a single-hop link, in case of bad channel conditions, the devices have to reduce their data rates. If the CTA rate allocation is the same, then by lowering the data rates, the duration of each CTA has to be increased which reduces the number of devices that can be supported by the superframe. This is because with the increased duration of CTAs for devices having more than one CTA per superframe shall occupy more time slots and hence lesser capacity for other devices. If most of the devices are suffering due to bad channel conditions, then the PNC has an option to change the channel. If two devices are at two extreme ends of the piconet, then they have to transmit at a relatively higher transmission power to keep the required SNR. Transmitting at a higher power results in increased energy utilization which is not beneficial for energy constrained devices.

4.4.2.1 Parent-Child Communication Model

The high data rate WPAN standard IEEE 802.15.3 defines the transmitter data rates (∂_T) of 11, 22, 33, 44 and 55 Mbps. The beacon and MAC headers of the frames are sent at the base rate of 22 Mbps and the rest of the payload at any of the desired values of ∂_T. Since the CTAP is TDMA based, it is not possible to achieve the defined ∂_{T}. Therefore the throughput achieved at the MAC layer is always much less than the transmitter data rate at the physical layer. If a device wants to send small amounts of asynchronous data in a single CTA, then it can transmit and achieve the defined data rates. For isochronous transmission, the requirement is to allocate more than one CTA per superframe (depending on the tolerable inter arrival delay) for the device. The achievable actual data rate (∂_A) is always less than the ∂_T and depends on certain factors such as number of CTAs allocated to the device, number of frames sent in each CTA and the time duration of each CTA (which depends on the required data rate). The number of devices in a piconet influences the decision of the PNC to allocate a particular number of CTAs to a device to ensure fair allocation.

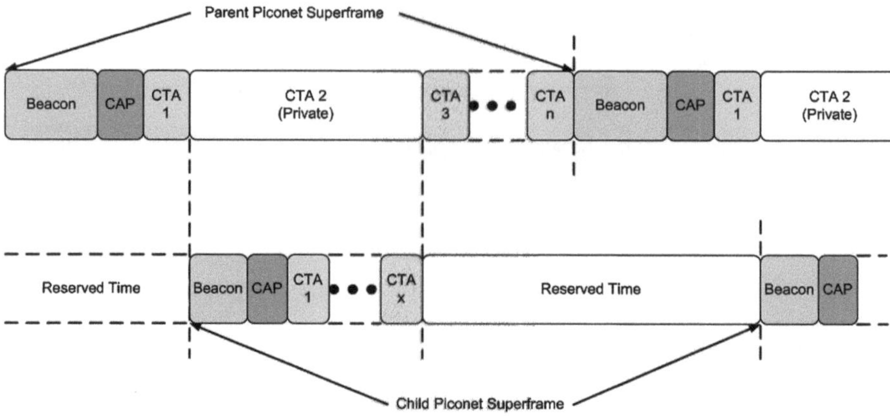

Fig. 4.51 Child superframe time allocation

Theoretically there can be 256 devices supported by the PNC in a piconet. Since some of the DEVIDs are reserved for special purposes, the maximum number of devices that a single PNC can support, as allowed by IEEE 802.15.3, is 243. The practical number of devices that a single PNC can support is, however, much lower than 243 if multimedia transfers are taking place between most of the devices. The increased number of devices also imposes additional processing overhead on the PNC. To resolve the processing burden and extend the range of piconet, IEEE 802.15.3 allows the formation of child piconets which are dependent on the established parent piconet. Though the administration of the child piconets is done autonomously by a child PNC, the channel time is provided by the parent PNC from its transmitted superframe through a private CTA. It can be seen in Fig. 4.51 that the time period in the superframe of a child piconet after the private CTA is reserved till the start of a subsequent private CTA in another superframe of the parent piconet.

This is to keep synchronized with the time allocated by the parent PNC to the child PNC. Figure 4.51 also indicates an issue related to the child superframe about the allocation of super rate CTAs for isochronous streams with strict delay constraints. If the reserved time after the private CTA allocated to the child PNC exceeds the maximum tolerable delay for most of the real time applications, then it is not possible for the child piconet to support them. If the reserved time is decreased by increasing the time allocation of the private CTA, it can disturb the CTAs in the parent piconet.

4.4.2.2 Scheduling Problems in the Parent-Child Model

As explained before, it is difficult in a parent-child relationship to maintain QoS for certain multi-media applications, especially voice. The issue is further aggravated if the formation of a child piconet within a child piconet is considered. Figure 4.52 shows such a formation which has a two level hierarchy.

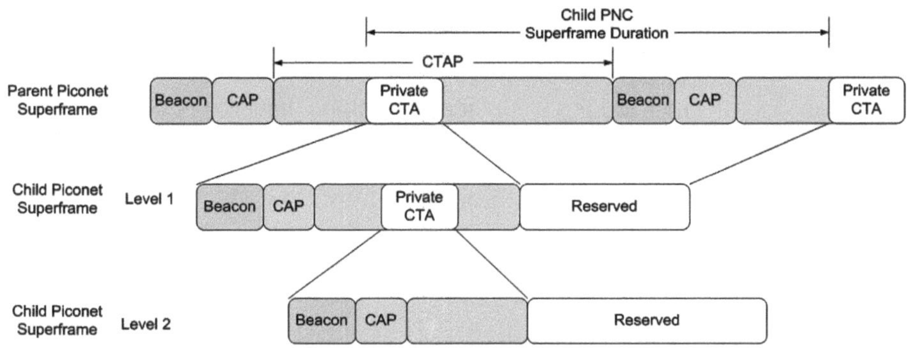

Fig. 4.52 Time allocation for hierarchical child piconets

If η_{PS} is the duration of the parent piconet superframe and η_{C1S} is the duration of the superframe for the level 1 child piconet, then $\eta_{C1S} = \eta_{PCTA1} + \eta_{RSVD} = \eta_{PS}$ and $\eta_{PCTA1} = k_1\, \eta_{PS}$ for $0 < k_1 < 1$, where η_{PCTA1} is the duration of a private CTA allocated to the level 1 child PNC and η_{RSVD} is the duration of the reserved time in η_{C1S} for η_{PS} till the start of a next successive private CTA for the level 1 child PNC. Similarly for the level 2 child piconet which is formed within the level 1 child piconet, we have $\eta_{C2S} = k_2\, \eta_{PCTA1} + \eta_{RSVD1} = \eta_{PS}$ and $\eta_{PCTA2} = k_2 \eta_{PCTA1}$ for $0 < k_2 < 1$, where η_{C2S} is the duration of level 2 child piconet superframe and η_{PCTA2} is the duration of a private CTA allocated to the level 2 child PNC by the level 1 child PNC. The value of k_2 is less than 1 to indicate that $\eta_{PCTA2} < \eta_{PCTA1}$. The number of super rate CTAs allocated to a device which is sending a real-time traffic depends on the maximum tolerable delay and jitter for that particular traffic type and the required data rate ∂_R. Since the super rate CTAs are evenly spread throughout the superframe, the duration of a private CTA allocated to a child PNC is a significant factor to determine if the parent and child piconet can support a particular real time traffic type with specific requirements of maximum tolerable delay and jitter. If X_{MTD} denotes the value of maximum tolerable delay and jitter for a particular real time traffic type, then the superframe can be split into logical partitions to make time allocations easier. The smallest partition size is taken to be equivalent of the strictest requirement for delay and jitter, which according to [31] is for voice (<10 ms). If $X_{MTD(min)}$ denotes the minimum compulsory logical partition size for the superframe, then $X_{MTD} = n X_{MTD(min)}$, where n can be any positive integer. The value of n shall be set to 1 to indicate voice applications. The number of logical partitions is given by $N_p = [\eta_{PS} - (\tau_{PNC} + \eta_{CAP})]/X_{MTD(min)}$, where τ_{PNC} is the beacon overhead and η_{CAP} is the CAP duration. The expression $\eta_{PS} - (\tau_{PNC} + \eta_{CAP})$ gives us the CTAP duration which is divided by $X_{MTD(min)}$ into a number of partitions. If the value of $X_{MTD(min)}$ is taken to be 8ms, then the superframe is split into 8 partitions, each of approximately 8ms. Once the superframe is partitioned, the time can be allocated much more easily for real time applications keeping the boundaries of logical partitions in view. However, the time allocation for a private CTA should be done very carefully as it can have a significant effect on isochronous streams with super rate CTA allocations.

Table 4.16 Important parameters of HDR WPANs for HRT (high-rate transmission), MRT (medium-rate transmission) and LRT (low-rate transmission)

Parameters	Values IEEE 802.15.3	IEEE 802.15.3a	IEEE 802.15.3c	Units
Superframe duration	0–65,535	0–65,535	128,000	μs
CAP duration	0–65,535	0–65,535	0–65535	μs
SIFS	10	10	2.5	μs
MIFS	2	2	0.05	μs
Supported data rates (Mbps)	11, 22, 33, 44, 55	55, 80, 110, 160, 200, 320, 480	>2,000–4,679 (HRT) =100–2,000 (MRT) <100 (HRT)	Mbps
Fragment (MPDU) size including FCS	64, 256, 512, 1024, 1280, 1536, 1792, 2048, 4024 (802.15.3a), 65535 (802.15.3c)			octets

If $\eta_{RSVD} > X_{MTD(min)}$ and $\eta_{PCTA1} > X_{MTD(min)}$, then both the parent piconet and the child piconet cannot support voice applications as required. If $\eta_{RSVD} < X_{MTD(min)}$ and $\eta_{PCTA1} > X_{MTD(min)}$, the child piconet can support voice applications, but the parent piconet cannot. In order for both the parent piconet and child piconet to support voice applications (since they have the strictest upper limit on tolerable delay and jitter) the following two conditions must be true, namely $\eta_{RSVD} < X_{MTD(min)}$ and $\eta_{PCTA1} < X_{MTD(min)}$. It can be shown that the above two conditions cannot be true at the same time since $\eta_{RSVD}<X_{MTD(min)}$ and $\eta_{PS}=\eta_{CIS}$, with $\eta_{PCTA1} + \eta_{RSVD}=\eta_{PS}$. If η_{RSVD} is assumed to be equal to $X_{MTD(min)}$, then $\eta_{PCTA1}=\eta_{PS}-X_{MTD(min)}$. This means that $\eta_{PCTA1} = N_p - 1$ and thus takes the major portion of the parent superframe. Therefore, the parent piconet cannot support voice applications. The same theory can be applied to other traffic types as well. This shows us that since the level 1 child piconet cannot support voice applications, there is no possibility for a level 2 child piconet or above to support multimedia applications. The increase or decrease in ∂_T determines the length of the CTA required to send a particular type of traffic. With higher values of ∂_T, the overhead per CTA increases, but the capacity of superframe also increases due to the reduced size of CTAs required by devices. IEEE 802.15.3a [30] defines an alternate physical layer based on UWB to achieve much higher data rates (cf. Table 4.16) using the same MAC layer of IEEE 802.15.3. Even higher data rates are proposed in IEEE 802.15.3c [32] in Gbps for the 60 GHz frequency band. Although by using much higher data rates, the capacity of the superframe is increased and much smaller CTA durations can be used using frame aggregation, the spacing of the super rate CTAs depending on the factor $X_{MTD(min)}$ does not change.

4.4.2.3 Inter-PAN Communication Model

The Inter-PAN Communication process has to address the following issues:

- Seamless merging of two or more piconets
- Seamless splitting of two or more piconets

- User's PAN device identity is not lost during the merging and splitting process
- All devices in each and every piconet are able to communicate with one another directly, provided that they are in the transmission range of each other
- The modifications should take into consideration the MAC reserved fields that are available in the IEEE 802.15.3 standard
- The scheduling issues proposed in case of the Parent-Child Model should be resolved so that the QoS for real time applications is not affected.

When two piconets merge for the purpose of inter-PAN communication, the intra-PAN piconet association and communication should not be disrupted. When the piconet splits, inter-PAN communication should end tidily and not abruptly, and intra-PAN communication should not be disrupted. Each piconet that merges and splits must be able to maintain its current association with its own piconet PNC. All devices in the Inter-PAN communication should be able to communicate with one another directly; however, channel access can be monitored by the PNC. The transmission range of devices limits the range in which a device can communicate between piconets. This document does not provide a solution to enable device communication between merged piconets that are out of range from one another. Lastly, all modifications to support inter-PAN communication will take into consideration the reserved fields in the IEEE 802.15.3 MAC layer only. The proposed modification would be appended onto the child piconet that is already part of the standard. The following sub-sections address the Inter-PAN communication issues.

Inter-PAN Communication

This process is first initiated by the discovery of an existing piconet through the MLME-SCAN.request primitive as shown in Fig. 4.53.

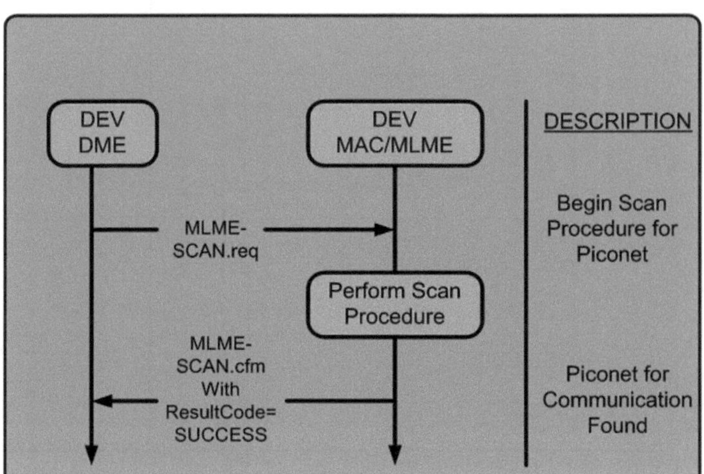

Fig. 4.53 Piconet scan initialization

This passive scanning request is carried out by the PNC or a device in a piconet. The PNC may allocate a CTA such that there is unallocated channel in the CTAP which provides quiet time for the PNC or a device to scan channels for other IEEE 802.15.3 piconets. When the PNC carries out piconet scanning, it goes into a silent mode where it shall cease piconet beacon transmission for one or more beacon intervals, but it is not allowed to suspend beacon transmission for more than twice *aMinChannelScan* [30]. If the device is scanning, then the PNC will make request to the device using the *MLME-REMOTE-SCAN.request.*

When the desired piconet for communication is found, the DME of the PNC will initiate the *MLME-SYNC.request* and receive an *MLME-SYNC.confirm.* Once completed, the PNC can begin associating itself with the new piconet using the *MLME-ASSOCIATE.request* primitive. Since the PNC is about to associate with a new piconet, it is referred as a device to the other PNC for descriptive purposes. The association process between the device (PNC of an established piconet) and PNC is illustrated in Fig. 4.54.

For the purpose of descriptive explanation an Inter-PAN device that has associated with another piconet shall be referred to as *inter-device*. As soon as the device (PNC) of one piconet associates itself with the PNC of the different piconet, it can request the formation of a dependent child piconet. This process is triggered by

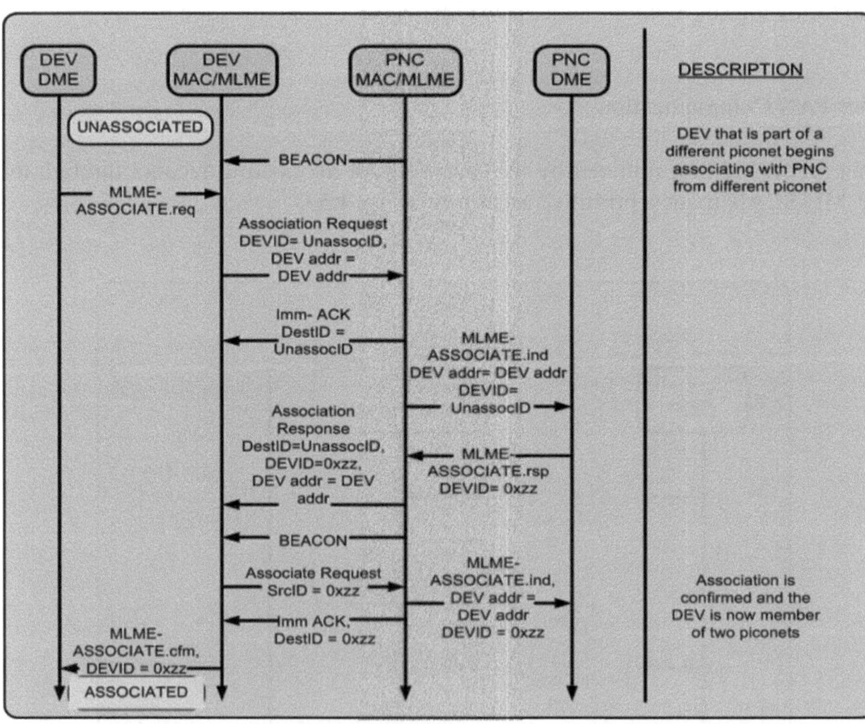

Fig. 4.54 Association procedure

the MAC layer using the *MLME-START-DEPENDANT.request* primitive. The device shall send a Channel Time Request command to request a pseudo-static private CTA. In a private CTA, the SrcID and the DestID are identical. The device shall set the SrcID and TrgtID fields in the Channel Time Request command to the DEVID of the originating device, the Stream Index field to zero and the PM CTRq Type field to ACTIVE. The PNC will then recognize that this is a request for a child piconet. If the PNC rejects the formation of a child PNC for any reason such as insufficient channel time or unable to allocate a pseudo-static CTA, it shall send a Channel Time Response command with the Reason Code field set to '*request denied*'. In this case, inter-PAN communication is not possible and the device should dissociate from the current piconet and return to its own piconet. If the device receives a private CTA from the PNC, the device DME configures the child PNC parameters using the *MLME-START-DEPENDANT.request* and confirms primitives.

Before the child PNC can begin transmitting its own beacon, it should return to its existing piconet channel and initiate moving the current channel to the newly allocated child piconet channel. The PNC will broadcast the *Piconet Parameter Change* Information Element with the change type set to CHANNEL in its current channel via its beacon for *NbrOfChangeBeacons* consecutive beacons. The Piconet Parameter Change IE shall contain the channel index of the new channel to which the PNC will be moving the piconet, and the *Change Beacon Number* field that contains the beacon number of the first beacon with a beacon number equal to Change Beacon Number field in the previous Piconet Parameter Change IEs. The device receiving this message shall change from the current channel to the new channel before the first expected beacon on the new channel. The devices shall not transmit on the new channel until a beacon has been correctly received on the new channel. To enable every device in the child piconet and parent piconet to communicate with one another, all members of the child and parent piconet should associate with one another.

A new command frame called *Inter-PAN Association Request* is created for this purpose and the process is illustrated in Fig. 4.55.

The command frame is sent by either the child or parent PNC or both PNC to its members. This new command frame has a type value '011' which indicates that it is a command frame. The PNID is set to PNID of the originating piconet. The SrcID is set to the PNC's DEVID and the DestID is either set to BcastID if the PNC requires all its members to Inter-PAN Associate or to individual DEVID if requires only a specific device to associate. The ACK policy bit is set to '01', i.e. Immediate Acknowledgement (Imm-ACK). The Inter-PAN Association Request MAC Frame payload will have the following fields:

- Inter PAN BSID (6–32 octets)
- Inter PAN PNC Address (8 octets).

The Inter PAN BSID and Inter PAN PNC Address are set to the target piconet address that the PNC requires its devices to associate. Upon receiving this command, the device(s) will begin listening for the specific beacon with the PAN BSID and PNC Address and begin the process of associating with the new piconet. The association process is similar to that described in the standard. Once the association

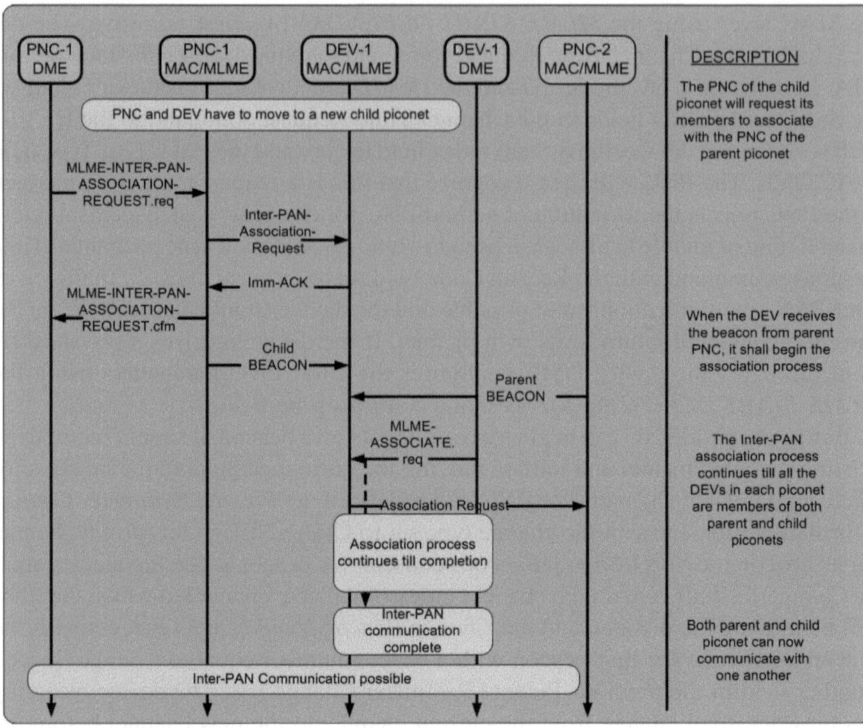

Fig. 4.55 Inter-PAN association procedure

process is successful, the devices which are member of both piconets can now communicate with one another using similar protocols defined in the standard. If the piconet to which a device is associating does not support Inter-PAN Communication, a new Reason Code is created within the Association Response message called *Inter-PAN Communication not supported.* The Reason Code will use one of the reserved fields that are available in the Association Response fields.

This process of disassociation is an extension to the manner in which a dependent piconet ends its relationship with the PNC. Since devices in each piconet are potentially associated to more than one piconet, modifications are necessary so that both piconets split seamlessly. Either the child or parent PNC should send a new command frame, *Piconet-Splitting-Request* to the PNC of the other inter-PAN requesting to split from one another. This new command frame has a type value '011' which indicates that it is a Command frame. This process is described in Fig. 4.56.

The PNID is set to PNID of the originating piconet. The SrcID is set to the PNC's DEVID of the originating piconet and the DestID is set to PNC's DEVID of the destination piconet. The ACK policy bit is set to '01' Immediate Acknowledgement (Imm-ACK). The MAC Frame payload is empty. Upon receiving Piconet Splitting Request command frame, both PNCs should begin informing their devices to disassociate themselves from the inter-PAN associated piconets. The new command frame is called *Force-Inter-PAN-Disassociation-Request* (cf. Fig. 4.57).

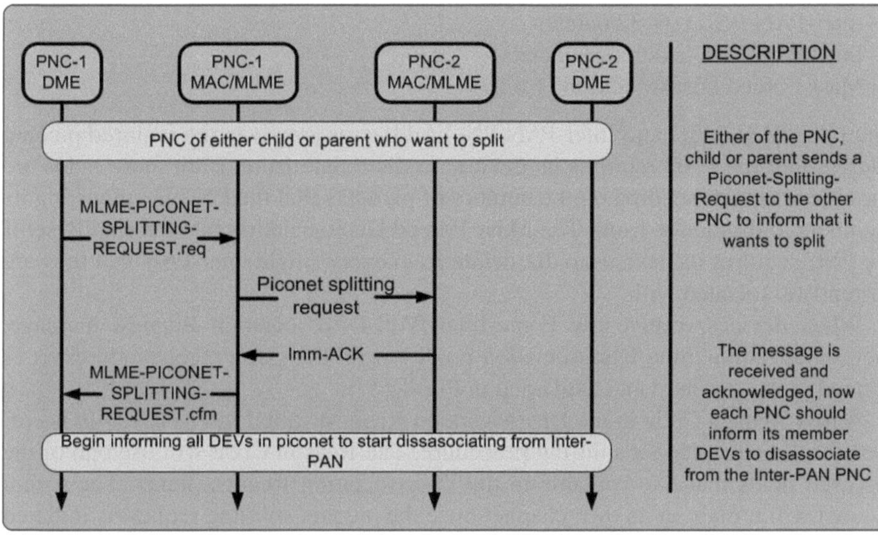

Fig. 4.56 Piconet splitting procedure

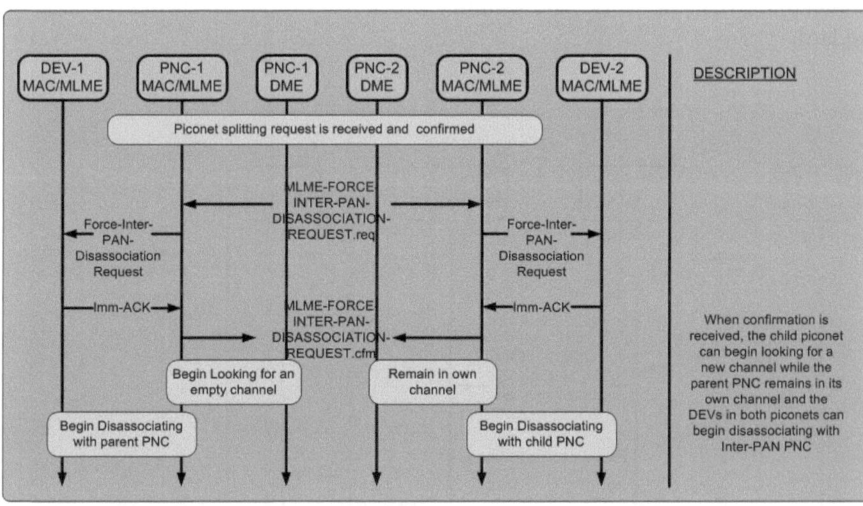

Fig. 4.57 Forced inter-PAN disassociation

The term *forced* is used since it is the PNC requesting its devices to dissociate instead of the devices. This new command frame has a type value ‘011’ which indicates that it is a Command frame. The PNID is set to PNID of the originating piconet. The SrcID is set to the PNC’s DEVID of the originating piconet and the DestID is set to BcastID of its member piconet. The ACK policy bit is set to ‘01’ Immediate Acknowledgement (Imm-ACK). The MAC Command frame will have the following fields.

- Inter-PAN BSID (6–32 octets)
- Inter-PAN PNC Address (8 octets)
- Mass Forced Disassociation (1 bit).

The Inter-PAN BSID and Inter-PAN PNC Addresses are set to the required piconet address that the PNC requires its devices to dissociate from. Both these fields are variable in size depending on the number of piconets that the PNC is requesting its members to dissociate from. The Mass Forced Disassociation bit is normally set if the PNC requires its devices to dissociate from every single inter-PAN that they are currently associated with.

When devices receive the Force-Inter-PAN-Disassociation Request message, they should initiate the Disassociation process with the given piconet addresses as defined in the standard and explained in Fig. 4.58.

A new Reason Code in the Disassociation Request called *Inter-PAN Split* is created for the new piconet splitting procedure. The Reason Code will use one of the reserved fields that are available in the Disassociation Request fields. The parent piconet will remain in its own channel once the piconet splitting request is initiated while all child piconets would have to shut down or move to a different channel. If the child piconet decides to maintain its piconet, it shall begin scanning for a new channel to move its network. The scanning process is similar to that described in the standard.

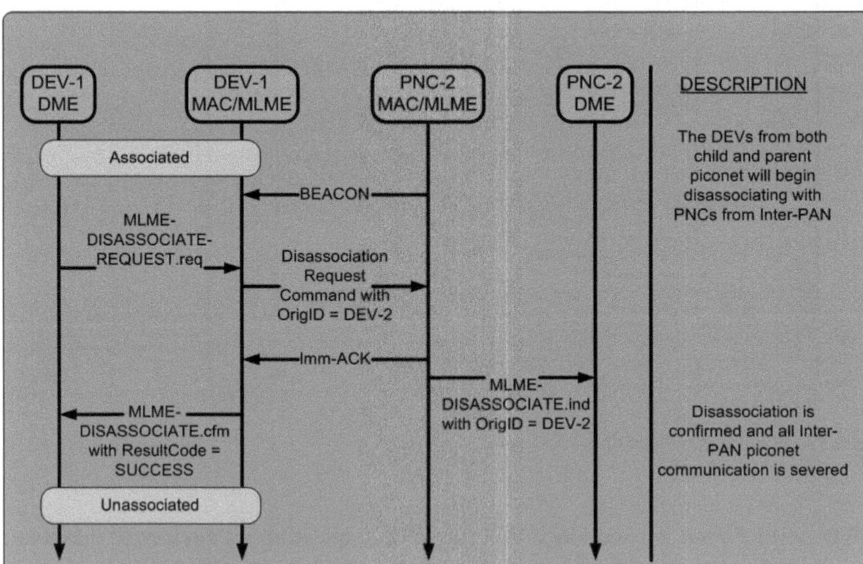

Fig. 4.58 Disassociation process

Parent-Child Scheduling Solution

Although there have been some scheduling algorithms proposed for HDR WPANs like [33], [34] and [35], the focus has been mainly on supporting VBR MPEG streams and a lot of details regarding the superframe utilization efficiency and the detailed parameters to support QoS have been missed to some extent. Reference [36] presents a hierarchical superframe structure, but it is not applicable in the inter-PAN communication scenario whereas [37] proposes an application aware MAC scheme which does not consider the requirements of [30] to support QoS.

When one of the PNCs associates with the parent PNC as the child PNC, it sends a channel time request command to the parent PNC for channel time. The source DEVID and the destination DEVID are the same in the channel time request command so that the parent PNC can determine that it is a request for the private CTA from the child PNC. Upon reception of the channel time request command, if there is enough capacity in the superframe of the parent PNC, it shall accept the request of the child PNC and send a channel time response command. If no device in either the parent piconet or the child piconet is supporting any real time traffic with a particular value of $X_{MTD(min)}$, then the parent PNC can allocate a single private CTA to the child PNC. However, if either a device in the parent piconet or the child piconet or in both the piconets intends to request channel time for real time traffic with a certain value of $X_{MTD(min)}$, then if the parent PNC allocates a single private CTA to the child PNC, the QoS for the device in the parent piconet and child piconet can be affected as explained. Since the upper limit on the tolerable delay and jitter for voice applications are the strictest, the CTAP of the superframe is partitioned into equal sized slots called Medium Access Slots (MASs).

The concept of dividing the superframe into MASs is defined in [38], however an appropriate size is not specified. We define the size of the MAS to be 8ms so that the QoS for voice applications can be supported easily. If the maximum size of superframe is considered, i.e., 65535 μs, there has to be at least 8 CTAs per superframe to support voice applications. Therefore the value of *Np* becomes 8. Since the QoS requirements of video are more relaxed than voice [30], the CTA rate factor for video traffic can be in factors of 2 per superframe according to the throughput requirements and the available capacity in the superframe. The proposed structure of the superframe when inter-PAN communication is considered is shown in Fig. 4.59.

It can be seen in Fig. 4.59 that there is a Beacon Period (BP) [38] in which the parent and child PNCs send their beacon. The BP can be extended in presence of multiple piconets and more than two beacons can be sent in it. A single CAP is shared between the parent and child piconets for simplicity so that the inter PAN association requests by the devices from either the parent PNC or the child PNC can be sent in it. When the parent PNC receives a request for a private CTA from

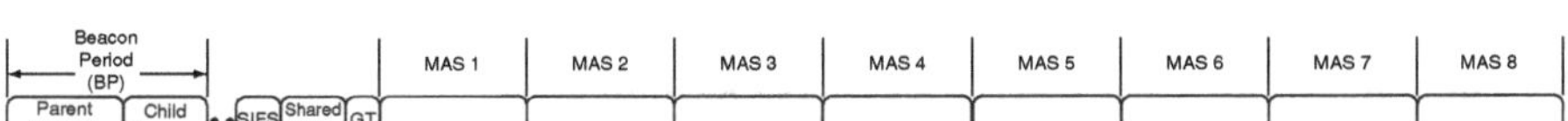

Fig. 4.59 Superframe sharing in inter-PAN communication

the child PNC, it checks the requested CTA duration $\eta_{CTA\text{-}R}$ and compares it with the available time in all of the 8 MAS durations (η_{MAS}). If $\eta_{CTA\text{-}R} < \eta_{(MAS\text{-}A)i}$ (where $\eta_{(MAS\text{-}A)}$ is the available time in a MAS and the index i indicates the MAS number and $1 \leq i \leq 8$), then the parent PNC can accept the channel time request from the child PNC. If there are devices in the child piconet which intend to request time for voice traffic, the parent PNC shall allocate 8 private CTAs to the child PNC spread evenly throughout the 8 MASs in the superframe. In this way, the QoS can be supported for devices in both the parent and child piconets subject to available capacity. For video traffic, the parent or child PNC shall allocate CTAs to requesting devices in factors of 2 depending on the available capacity and throughput requirement specified in the request. The number of child piconets that can be supported depends on the superframe capacity which is discussed in later sections.

Inter-PAN PNC Selection Criteria

When a PNC receives a beacon from another PNC having a different PNID, it includes the PNC capabilities IE in its subsequent beacon frame. The criteria for PNC selection is given in [30] for PNC capable devices. For inter-PAN communication, since capacity is a major issue, three extra parameters are defined and used in the simulation model apart from the ones mentioned in [30]. The three parameters are number of supported child piconets, number of active devices, type of traffic being communicated by the devices and their CTA durations, and the PNID. The PNC which already has dependent child piconets is given preference. If none of the PNCs are already supporting child piconets, then the PNC which has higher number of active devices communicating in its piconet is given preference. If the number of devices is the same, then the PNC with more superframe utilization is given preference. If none of the above is applicable, then the PNC with the higher PNID becomes the parent PNC. Since both of the PNCs perform this comparison, therefore the child PNC sends an association request to the parent PNC. The child PNC also informs its member devices and starts including the parent piconet IE in its beacon. The child piconet calculates the total utilized time in its superframe along with extra time (500 μs more per superframe in the simulation model) that it requires and sends the request to the parent PNC. If the child piconet has devices which are transmitting voice or video traffic, the parent PNC shall allocate 8 private CTAs to the child PNC. If there is no device with voice or video traffic in the child piconet and no device in the child piconet is capable or have any intention to send such traffic in future, then the parent PNC shall allocate a single private CTA to the child PNC. The efficiency of CTA utilization, CTA overhead and number of devices that can be supported are given in the Section 4.4.2.4.

4.4.2.4 Capacity Analysis of HDR WPANs

Different multi-media codecs encode data at different rates with different data rate requirements. The size of the MSDU received from the higher layers to the MAC layer varies as a result. If the size of the MSDU is larger than the largest MPDU size supported by the MAC layer, the MSDU has to be fragmented into smaller MPDUs. To simplify the fragmentation process, the MSDU is divided into equal size fragments (MPDUs). Through the DEV capabilities field in the DEV association IE, each device indicates its preferred fragment size and supported data rates to all the member devices in the piconet. If a specific application requires a data rate of x Mbps, then the MAC layer has to at least support a data rate of ($x+$*Layer Overhead*) Mbps. The layer overhead can be calculated by considering the preamble added at the network layer and MAC layer as shown in Fig. 4.60 indicates the fact that ∂_A should be at least more than ∂_E by an amount equal to *Layer Overhead* in order to support the application.

To determine the efficiency of utilization of a CTA, it is mandatory to consider the particular acknowledgement scheme which is to be used. The acknowledge scheme in use uses a certain IFS duration between successive frames which has an impact on the CTA overhead in Fig. 4.61.

Since voice and video traffic is considered in the capacity analysis carried out, the Delayed Acknowledgement (Del-ACK) scheme is used. When using the Del-ACK, either the SIFS or the MIFS can be used between successive frames. The CTA overhead when the SIFS is used between successive frames is given by

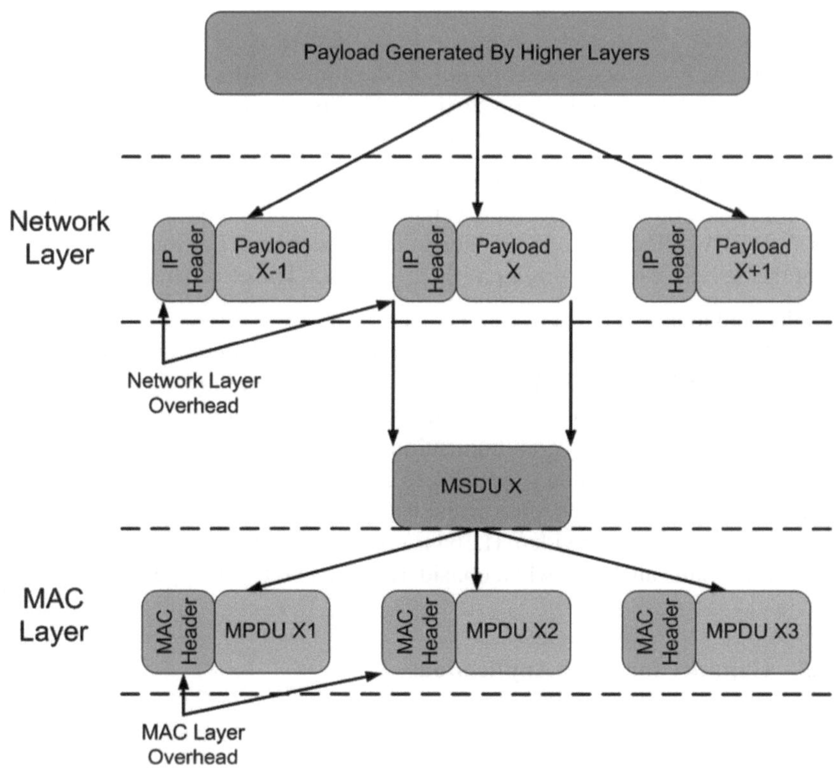

Fig. 4.60 Overhead added at the network and MAC layers

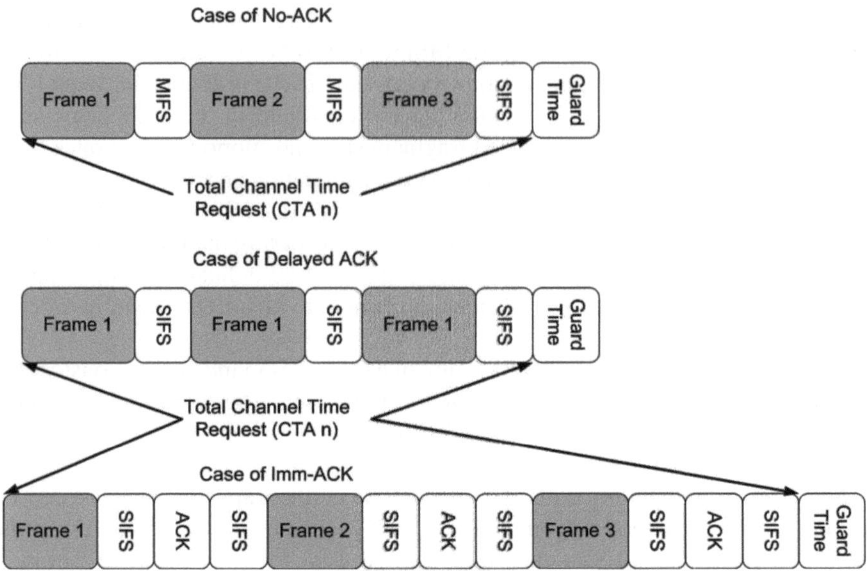

Fig. 4.61 CTA structure in case of different ACK schemes

$\tau_{DACK(SIFS)} = \sum_{n=1}^{xd+b} \eta_{(SIFS)n} + b\,\eta_{DACK} + \eta_{GT}$, where x_d is the number of frames sent in the CTA, η_{SIFS}, η_{DACK} and η_{GT} are the duration of SIFS, time to send the Dly-ACK frame and the guard time, respectively. The parameter b is set to one if there is a Dly-ACK frame in the CTA, otherwise it is set to 0. The total time allocated to each device (η_D) in the superframe can be given by $\eta_D = \frac{\sum_{i=1}^{x}[\eta CTA - \tau_{CTA}]i}{\eta S}$, where η_{CTA} is the duration of a single CTA allocated to the device and x is the total number of CTAs allocated to the device in the superframe. For the actual data rate (∂_A) of the device, we obtain $\eta_S \partial_A = \sum_{i=1}^{x} \eta_{(CTA)i}$. The effective data rate (∂_E) at which the actual payload is delivered is given by $\partial_E = \frac{\sum_{i=1}^{x}[\eta CTA - \tau_{CTA}]i}{\eta S}$, where τ_{CTA} is the CTA overhead for each CTA, which includes the IFS durations, ACKs and the GT. Since the IEEE 802.15.3 MAC is a TDMA MAC, the following equation holds: $\partial_E < \partial_A < \partial_T$. In order to make sure that the applications running on the devices are running smoothly during communication, it should be made sure that $\partial_E \geq \partial_R$. To calculate the capacity of an IEEE 802.15.3 superframe, voice and video traffic is considered since it consumes most of the networks resources. For voice, the G.711 codec is considered and for video, H.264 is considered. The following sections shall focus on the results analyzed when considering the two traffic types.

4.4.2.5 Capacity for Voice Applications

The values for different parameters used for calculating the capacity when using the G.711 codec are given in Table 4.17.

Table 4.17 Parameters considered for voice traffic

S. No.	Parameters	Value
1	Frame duration	20 ms
2	Number of frames/s	50
3	Size of each frame	160 octets
4	Network (IP) layer overhead	40 octets
5	Fragment (MPDU) size considered	256 octets
6	Data rate for G.711	64 kbps

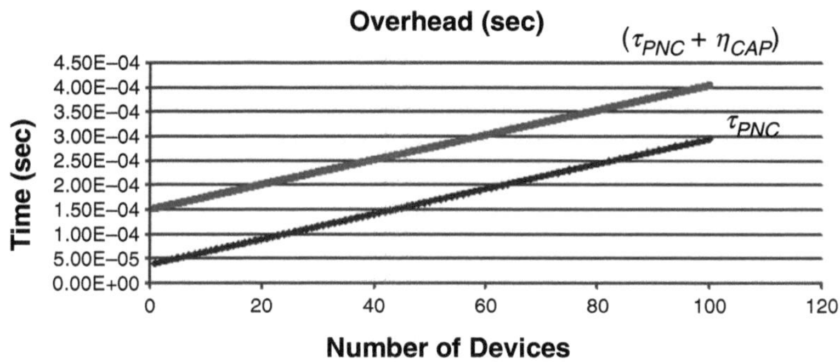

Fig. 4.62 PNC overhead

It should be noted from Table 4.17 that the IP overhead is 40 octets and the number of frames per second is 50. Therefore an additional 16 kbps is required to send the IP layer overhead apart from the 64 kbps data rate which is for voice payload. Since the MSDU size < MPDU size, therefore an MPDU size of 256 octets is chosen to send the MSDU which is 200 octets in size. The MPDU size of 256 octets also includes the Frame Check Sequence (FCS) which is 4 octets in length. While sending the MSDU, an additional overhead of MAC header has to be taken into account also. Therefore the total data rate that needs to be supported is 106.4 kbps. The time required for the device per second can be calculated by ∂_R/∂_T. Since the superframe size considered is 65535 μs, there are 1/65535 μs superframes in 1 s. Therefore the time required per superframe for a device with a required data rate of 106.4 kbps is $\eta_S \partial_R/\partial_T$. The maximum tolerable delay and jitter for voice applications should be <10 ms. Therefore the available channel time i.e. CTAP is divided by 8 to limit the delay and jitter to be less than 10 ms. Figure 4.62 shows us the PNC overhead and the $(\mathfrak{r}_{PNC} + \eta_{CAP})$ sum for 100 devices.

The variation in overhead with increase in number of devices is because of additional information put into the CTA IE by the PNC. The PNC sends the beacon at a base rate of 22 Mbps; therefore, the overhead is the same for devices operating at different transmitter data rates. When a device has been allocated a CTA by the PNC, then depending on the number of frames sent in the CTA and the transmitter data rate, the superframe overhead $(\mathfrak{r}_S)$ increases or decreases. Although η_{MIFS}, η_{SIFS}

and η_{GT} remain the same with the increase or decrease in the transmitted data rate, the time required to send the MPDUs (η_{MPDU}) within a CTA increases or decreases respectively. Since with the increase in transmitted data rates, the ratio of (η_{MPDU}) to $ʈ_{CTA}$ decreases, the overhead increases. This can be seen in Fig. 4.63 where the percentage superframe overhead is plotted against data rates of 22, 33, 44 and 55 Mbps with different number of frames sent in a CTA.

As the number of frames per CTA increase, the ratio of (ηMPDU) to $ʈ_{CTA}$ also increases and hence the overhead decreases.

The superframe overhead apart from the transmitter data rate, also depends on the number of MPDUs sent in the CTA and is calculated as $ʈ_S = ʈ_{PNC} + (ʈ_{CTA} \times N_{CS} \times N_D)$, where $ʈ_S$ is the superframe overhead, N_{CS} denotes the number of CTAs per superframe and N_D denotes the number of supported devices. The CTA overhead $ʈ_{CTA}$ is calculated by $ʈ_{CTA} = \eta_{CTA} - (\eta_{MPDU} \times N_{MC})$, where η_{MPDU} is the time required to send an MPDU and N_{MC} denotes the number of MPDUs per CTA. Increasing ∂_T also has an advantage. The CTA duration (η_{CTA}) decreases with the increase in ∂_T and as a result, the superframe capacity increases. This can be shown in Fig. 4.64 where the superframe capacity is plotted against ∂_T with different number of frames per CTA.

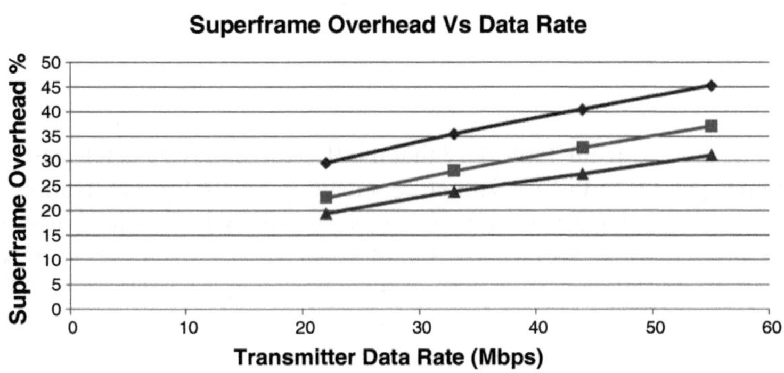

Fig. 4.63 Overhead compared with transmitted data rate

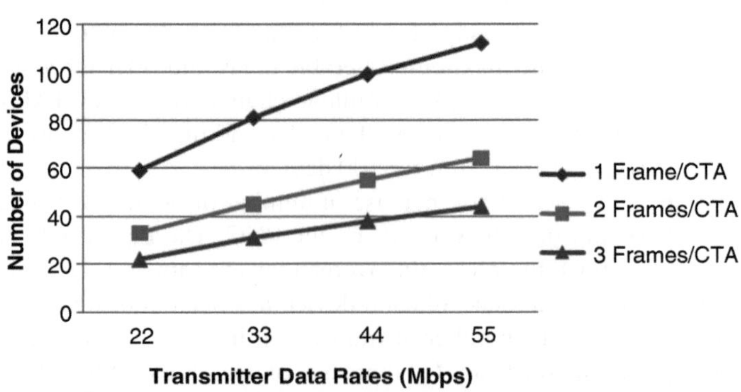

Fig. 4.64 Superframe capacity vs. data rate

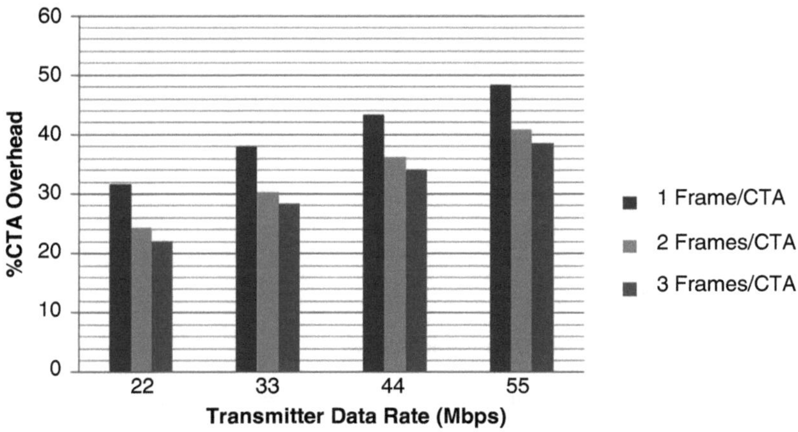

Fig. 4.65 Percentage CTA overhead (MPDU size = 256 octets)

Figure 4.65 shows that the CTA overhead increases with the increase in the transmitter data rate. The reason is that with the increase in ∂_T, the time required to send the MPDU decreases but the IFS remains constant. Furthermore the time required to send the MAC header remains the same as it is always sent at the base rate of 22 Mbps. The CTA overhead decreases when the number of frames per CTA is increased.

Capacity for Video Applications

Since the requirements for video traffic are more resource intensive than voice, the capacity for video traffic is analyzed in order to find an upper limit of data rate. For video traffic, 4 different levels of H.264 are considered for mobile content (3G video), Internet/Standard Definition (SD), High Definition (HD) and Full HD. Each level has different data rate requirements and number of frames sent per second. When sending mobile content at a resolution of 176 by 144 and a frame rate of 24 fps, the data rate required is about 160 kbps. The average size of each frame comes up to 834 octets. If the IP overhead is considered, the frame size becomes 874 octets. The nearest fragment size of 1,024 octets is used to efficiently carry an MSDU size of 874 octets. If the MAC layer overhead is taken into account, ∂_R becomes $\approx$200 kbps. Since the maximum tolerable delay and jitter for video applications should be less than 100 ms, there is more flexibility in assigning super rate CTAs to video applications depending on the required data rate. Figure 4.66 shows us the superframe capacity when considering mobile content with an MPDU size of 1,024 octets.

Different number of frames is sent per CTA and the capacity of superframe is analyzed. ∂_A is also mentioned in Fig. 4.67 where it is shown that by increasing ∂_T, ∂_A only increases by 2%.

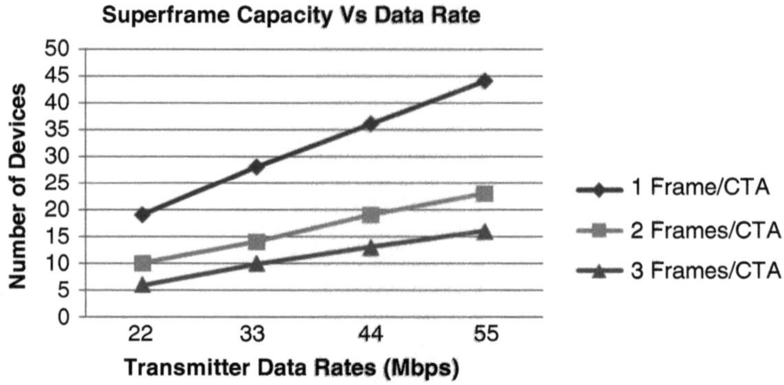

Fig. 4.66 Superframe Capacity against data rate (MPDU size = 1,024 octts)

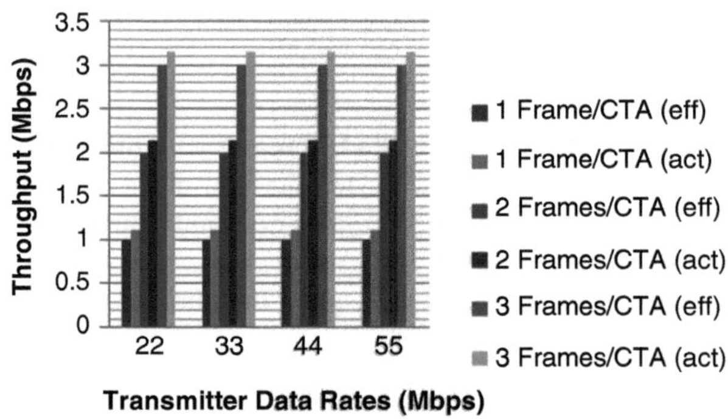

Fig. 4.67 Throughput obtained (MPDU size = 1,024 octets)

However there is a two fold increase in ∂_A by sending more frames in the CTA. Also ∂_A can be increased by increasing or decreasing the number of CTAs in the superframe. ∂_A does not change with the increase in ∂_T due to the TDMA MAC format. Since the number of bits sent per superframe remains the same for a device, ∂_A remains the same. For Internet/Standard Definition (SD), HD and full HD, the data rate requirements are much higher than those for 3G mobile content. The MSDU size > MPDU size and therefore an MPDU size of 2048 is chosen. Data rates considered are 2, 6 and 8 Mbps.

Figure 4.68 shows the capacity of superframe when an MPDU size of 2048 is considered for up to 4 frames/CTA.

It can be seen that for lower values of ∂_T e.g. 22 Mbps, only 5 devices can be supported for 2, 3 and 4 frames per CTA. Figure 4.69 shows that for the same 2, 3 and 4 frames per CTA, ∂_A achieved when ∂_T is 22 Mbps, ∂_A is up to 8 Mbps.

Therefore it can be noted that a practical limit of 8 Mbps can be set for the devices (when the full duration of 65535 μs is used) when the number of devices is low i.e. 5–10 in the piconet. To achieve fairness among higher number of devices, the upper limit should be further dropped.

Figure 4.70 shows that the increase in CTA overhead is relatively less when compared with the use of smaller MPDU sizes.

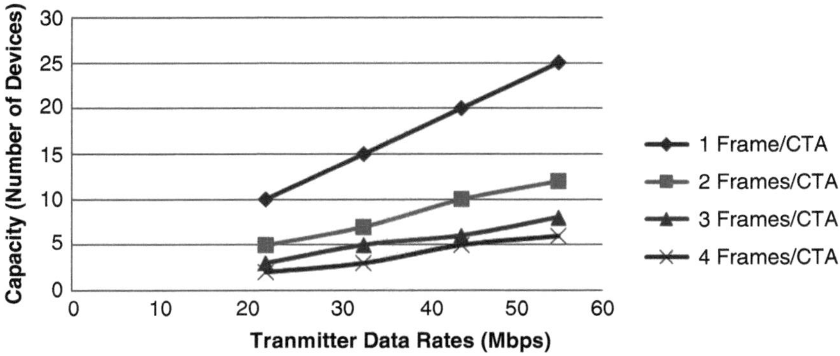

Fig. 4.68 Superframe capacity (MPDU size = 2,048)

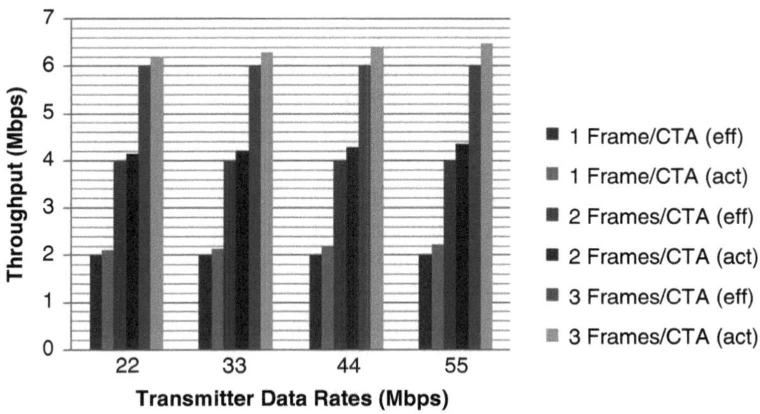

Fig. 4.69 Actual data rate (MPDU size = 2,048 octets)

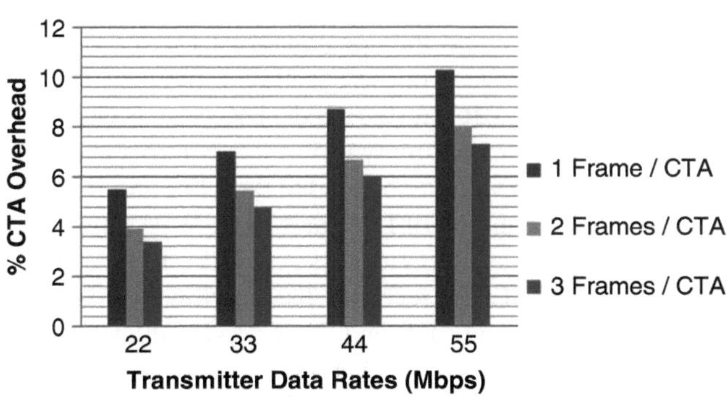

Fig. 4.70 CTA overhead (MPDU size = 2,048 octets)

4.4.3 Multimode Operation

Different transmission schemes and medium access protocols have been proposed in order to comply with the different requirements of HDR and LDR WPANs. These requirements are complementary in terms of data rate, but not in terms of coverage. It is thus an interesting topic to integrate HDR and LDR air interfaces in a dual-mode wireless device. However, because of the unlicensed nature of WPANs, interference issues exist between devices associated with different WPANs. Here, we consider interference issues arising for a MAGNET Beyond HDR air interface being closely located with a MAGNET Beyond LDR air interface. Performance of HDR and LDR WPANs located in close proximity is evaluated and a coexistence mechanism is proposed.

4.4.3.1 Introduction

New generation of WPAN devices require dual-mode, i.e. LDR/HDR air interfaces in order to achieve high spectrum efficiency and being able to span from LDR to HDR (from a few bps to hundred of Mbps) [39]. At the same time also the MACs should be different because of different needs in terms of application requirements, duty cycle and complexity. Typical applications for LDR devices (e.g. sensor networks) can run with low duty cycles (under 1%). The simultaneous use of different and/or uncoordinated wireless networks that overlap (at least partially) in range, time and frequency generates mutual effects of interference which decreases the performance of such networks. The evaluation of the interference effects of collocated LDR and HDR AIs has to be carried out before claiming the need of a coexistence mechanism.

There are two categories of coexistence mechanisms: collaborative coexistence mechanisms where the two interfering networks exchange information and non-collaborative coexistence mechanisms where the exchange of information is not allowed [40]. The possibility of exchanging information is quite easy when the two air interfaces are co-located in the same dual-mode terminal.

The scope of this section is to evaluate the performance of collocated HDR and LDR WPANs and to provide the specifications for the coexistence mechanisms between the two AIs. We propose a collaborative coexistence mechanism between MAGNET Beyond LDR interface and MAGNET Beyond HDR interface, here named Alternating Wireless Activity (AWA). It controls and synchronizes the access to the network of the devices associated to the LDR and HDR WPANs. Since it relies on time division alternation of LDR and HDR WPANs, it totally avoids interference even in the worst case that occurs when the LDR (HDR) interferer is closely located to the HDR (LDR) receiver. Its functionalities are positioned in a common protocol layer above the LDR and HDR MAC sublayers.

Types of Interference

In general terms, interference is any distortion agent to the desired signal. There are two broad categories of interference: additive interference and multiplicative interference. They are discussed in the following paragraphs.

Additive Interference

The additive interference is generated by an undesired signal which is added to the desired signal and it includes: co-channel interference, adjacent channel interference, intersystem intermodulation interference and intersymbol interference.

1. **Co-channel Interference** occurs when the interfering signal has the same carrier frequency of the useful information signal. Co-channel interference can be reduced by using e.g. power control, directional antenna beam pointing control or interference cancellation schemes.
2. **Adjacent Channel Interference** can be categorised into in-band interference and out-of-band interference.

    (a). *In-band interference*: it occurs when the centre of the interfering signal bandwidth falls within the bandwidth of the desired signal.
    (b). *Out-of-band interference*: it occurs when the centre of the interfering signal bandwidth falls outside the bandwidth of the desired signal. This kind of interference can be experienced when transmitters and receivers operate close together in terms of the two main variables that determine their degree of isolation from each other: distance and frequency separation. Out-of band interference may be caused over short to medium distances when there is insufficient isolation. This interference is not directly caused by co-channel emissions, but by having the energy of emissions at other frequencies transferred to co-channel frequencies through a number of special mechanisms. Out of band interference can be reduced with filtering.

3. **Intersystem Intermodulation Interference** occurs when non-linear devices (e.g. a power amplifier) are used simultaneously by a number of carriers, and, hence, intermodulation products are generated which cause distortion of the desired signal. In this case, non-linear system components, especially in analogue signal transmission, generate spurious signals that may play the role of interference by adjacent channels.
4. **Intersymbol Interference** is the interference contribution of other symbols to the symbol under consideration in the demodulation phase. Intersymbol interference is due to a relatively large delay spread in a multipath medium (dispersive channel) or a relatively high transmission bit rate.

Multiplicative Interference

This type of interference is caused by the non ideal characteristics of the propagation environment due to multipath, diffraction and dispersion.

Coexistence Issues

Channel conflicts lead to in-band interference, which is one of the strongest types of interference. However, if a channel conflict does not occur, we can experience out-of-band interference especially when the interferer is closely located to the receiver.

The effects of UWB systems on existing radio systems are essential for the commercialization of UWB technologies. Because of the large spectrum range of UWB systems, they can much likely generate either in-band or out-of-band interference on other systems.

In [41], Tesi at al. pointed out that although impulse radio signals are not Gaussian signals, their interference effects on narrowband systems are equivalent to that of a Gaussian noise. Supporting this result, the bit error rates degradation of a 2 GHz digital wireless transmission system caused by impulse radio and DS-SS UWB on an overlapping narrowband 2 MHz DQPSK system have been experimentally evaluated in [42].

Considering a desired-to-undesired signal power ratio (D/U) of 10 dB, the BER curve shows the floor characteristics which means that the BER improvements were saturated (around a value of $5 \cdot 10^{-2}$) even increasing E_b/N_0.

In a multi mode terminal this effect is even more critical since the two air interfaces are close, and thus it is necessary to approach the problem in an exclusive way. In the next section we propose a novel coexistence mechanism between 802.15.4-based LDR and 802.15.3-based HDR WPANs, that exploits the MAC features of the two standards, but can be properly used by any kind of PHY proposed for the two standards.

4.4.3.2 Coexistence Mechanisms

MAGNET HDR WPAN – MAC

The MAC layer of [30] includes several mechanisms that allow a flexible resource management and support of QoS, which are key requirements for high data rate WPAN devices. An IEEE 802.15.3-based WPAN operates as a centrally controlled ad hoc network called piconet; data exchange in a piconet is performed in a peer-to-peer manner. A piconet consists of a Piconet Coordinator (PNC) and one or more devices (DEVs) that are synchronized with the PNC. Synchronization is required since the MAC superframe is structured in time slots. Two devices in the piconet can communicate directly by either randomly accessing the time slots in the CAP of the superframe or by accessing the channel in some assigned time slots of the superframe during the CTAP.

Figure 4.71 shows the structure of the MAC superframe, which consists of three parts:

- **Beacon**, which is used to set the timing allocations and to communicate management information for the piconet. The beacon consists of the beacon frame, as well as any announce commands sent by the piconet coordinator as an optional beacon extension. Control messages broadcasted by the PNC within the beacon contain information such as timing parameters and assigned time slot for the communication.
- **CAP**, where devices access the channel according to a CSMA/CA mechanism. The CAP is the contention period where the devices and the PNC can send each other frame commands or short asynchronous data.

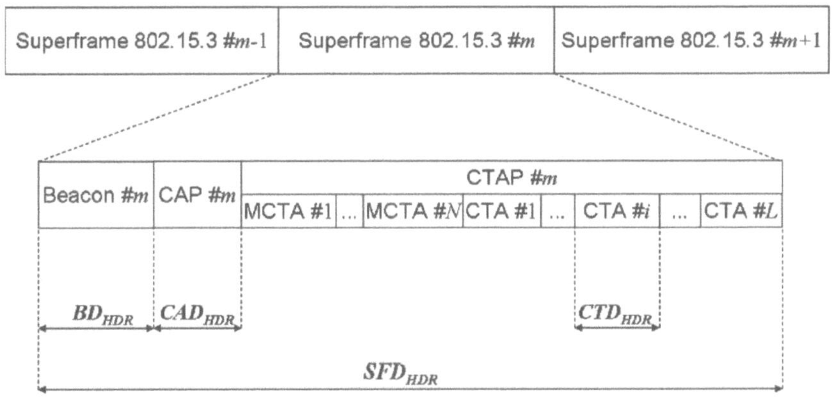

Fig. 4.71 IEEE 802.15.3 MAC Superframe structure

- **CTAP**, where the access to the channel is controlled by the PNC, which assigns CTA time slots for a communication in response to the request message that contains information about the requested bandwidth, delay constraints and other QoS requirements of the communication to be established.

In Figure 4.71, *m* denotes the transmission index.

The following time durations are identified and shown in Fig. 4.71:

- SFD_{HDR} is the total superframe duration which is constrained by: $0 < SFD_{HDR} < 65535 \cdot 10^{-6}$ s.
- CTD_{HDR} is the duration of a CTA.
- BD_{HDR} is the beacon duration.
- CAD_{HDR} is the CAP duration.

There are three methods for communicating data between DEVs associated to a piconet:

- Sending asynchronous data in the CAP (if present);
- Allocating time slots for isochronous streams;
- Allocating time slots for asynchronous streams.

Two types of time slots in a CTAP can be assigned to one communication:

- CTA time slots, used for information data transfer.
- Management CTA (MCTA) time slots, used for exchanging management information to or from the PNC.

CTAs are either dynamic or pseudo-static. The PNC may move dynamic CTAs within the superframe on a superframe by superframe basis, while it should allocate pseudo-static CTA in a superframe with same position and duration as it was in the previous superframe. If the PNC needs to change the duration or location of pseudo-static CTA within the superframe, it shall change the corresponding CTA blocks in the beacon. Pseudo-static CTAs shall be allocated only for isochronous streams.

To increase the coexistence of several piconets in the close vicinity and to extend the coverage area, IEEE 802.15.3 MAC introduced the concepts of child piconet and neighbouring piconet. When a PNC capable device that is a member of an existing piconet wants to form a child piconet, the device shall request a private CTA. A private CTA is pseudo-static CTA where the same DEV is both the source and destination DEV. The PNC of the child piconet is able to communicate with the PNC of the parent piconet. Moreover, the time slots that the child piconet can use for the communication of its devices are assigned by the parent PNC. The same holds for the neighbouring piconet. The main difference with the child piconet is that the there is only a control communication message from the parent PNC and the neighbouring PNC.

No other information exchange is possible among the two piconets.

MAGNET LDR WPAN – MAC

Depending on the application requirements MAGNET LDR WPAN may operate in either a star topology or a peer-to-peer topology. In the star topology, the communication is established between devices and a central controller called PAN coordinator. The peer-to-peer topology also has a PAN controller, however it differs from the star topology since any device may communicate with any other device.

The PAN coordinator bounds its channel time by using a superframe structure (Magnet LDR MAC allows only beacon-enabled PAN). A superframe starts with the transmission of a beacon frame. In a superframe enabled WPAN, the superframe can have an active and an inactive portion; the active portion is divided into 16 equally sized slots. Figure 4.72 shows the general structure of the MAC superframe based on the IEEE 802.15.4 MAC standard [43], which consists of four parts:

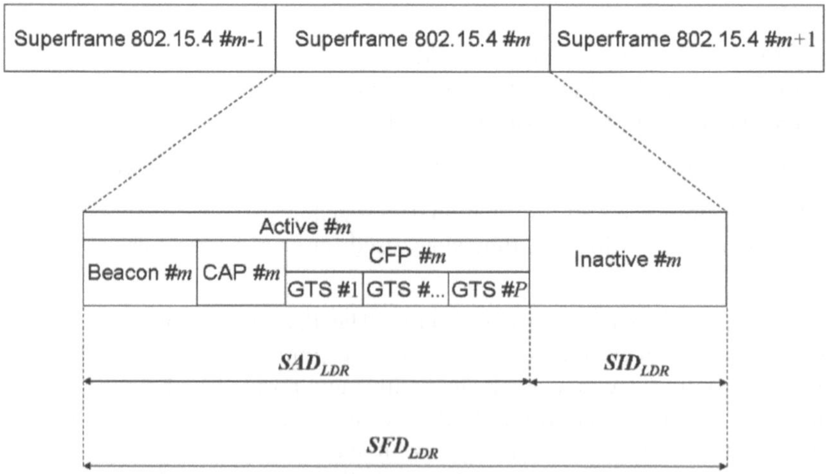

Fig. 4.72 IEEE 802.15.4 MAC Superframe structure

- The beacon frame, which is used to: synchronize the devices associated to the WPAN, identify the WPAN and describe the structure of the superframe. The beacon frame is transmitted in the first slot of each superframe.
- The CAP, where devices may communicate using a slotted CSMA/CA mechanism.
- The Contention Free Period (CFP), where the access to the channel is controlled by the PNC, which assigns Guaranteed Time Slots (GTSs) for that communication in response to the request message. The PAN coordinator can allocate up to seven GTS and a GTS can occupy more than one slot.
- The inactive period, during which devices may enter a low power mode.

It is worth noting that the coexistence mechanism proposed here exploits the capability of this standard to run with low duty cycles. The total duration of the superframe SFD_{LDR}, also called Beacon Interval (BI), is computed as follows (s):

$$\mathrm{SFD_{LDR}} = \mathrm{aBaseSuperframeDuration} \cdot 2^{\mathrm{BO}}/\mathrm{R_s}$$

where $aBaseSuperframeDuration = 16 \cdot 60$ symbols, $R_s = 62.5\,\mathrm{ksymbol/s}$ and the Beacon Order (BO) is constrained by: 0<BO<14; On the other hand, the duration SAD_{LDR} of the active portion of the superframe is computed as follows:

$$SAD_{LDR} = \mathrm{aBaseSuperframeDuration} \cdot \mathrm{2SO}/\mathrm{R_s}$$

where the Superframe Order (SO) is constrained by:$0 < \mathrm{SO} < \mathrm{BO}$. The duration of the inactive portion of the superframe is here denoted as SID_{LDR} and the following equation holds:

$$SFD_{LDR} = SAD_{LDR} + SID_{LDR}$$

The following section briefly introduces interference issues that arise when more than one WPAN is in the same range.

Above several studies have been cited which proved the interference issues generated between different PHY foreseen for WPAN systems: narrowband ISM transmission systems or UWB and narrowband systems. Because of the interference that can be generated between a LDR and a HDR WPAN device, especially when the two devices are located in a common terminal (dual-mode radio interface), at least one coexistence mechanism should be used. Below we propose a novel coexistence mechanism.

AWA Coexistence Algorithm

The coexistence mechanism proposed here is termed Alternating Wireless Activity (AWA). It works by controlling and synchronizing the access to the network of the two air interfaces. Its functionalities are positioned in a common protocol layer over the two MAC sublayers. It makes use of the IEEE 802.15.3 child piconet functionality and the inactive period of the IEEE 802.15.4.

Since no 802.15.3 devices are transmitting during a private CTA allocated to a child piconet, this CTA can be allocated to a 802.15.4 WPAN that will not be interfered by any HDR device. On the other hand, since no 802.15.4 devices are transmitting during the inactive portion of the superframe, this inactive portion shall be synchronized to overlap the entire 802.15.3 superframe except the i-th private CTA that is overlapping in time with the active portion of the 802.15.4 WPAN. In this case the 802.15.3 WPAN will not be interfered by any LDR device.

The synchronization of the 802.15.3 and 802.15.4 superframes is shown in Fig. 4.73.

The i-th CTA of the m-th 802.15.3 superframe is allocated to the active portion of the m-th 802.15.4 superframe. The inactive portion is virtually divided into two parts: the m-th 802.15.3 superframe starts simultaneously to the second part of the inactive portion of the $(m - 1)$-th 802.15.4 superframe, while the m-th 802.15.3 superframe ends simultaneously with the first part of the inactive portion of the m-th 802.15.4 superframe. The synchronization of the two superframe sequences allows to free from interference all LDR and HDR devices associated to the common LDR and HDR PAN controller.

The duration of the first part of the inactive portion of the m-th superframe is denoted as $SID^{m'}_{LDR}$, while the second part of the inactive portion is denoted as $SID^{m''}_{LDR}$ (see Fig. 4.73). Since the LDR and the HDR WPAN controllers are expected to exchange information for network coordination and synchronization of the respective superframes, the AWA mechanism is a collaborative coexistence mechanism.

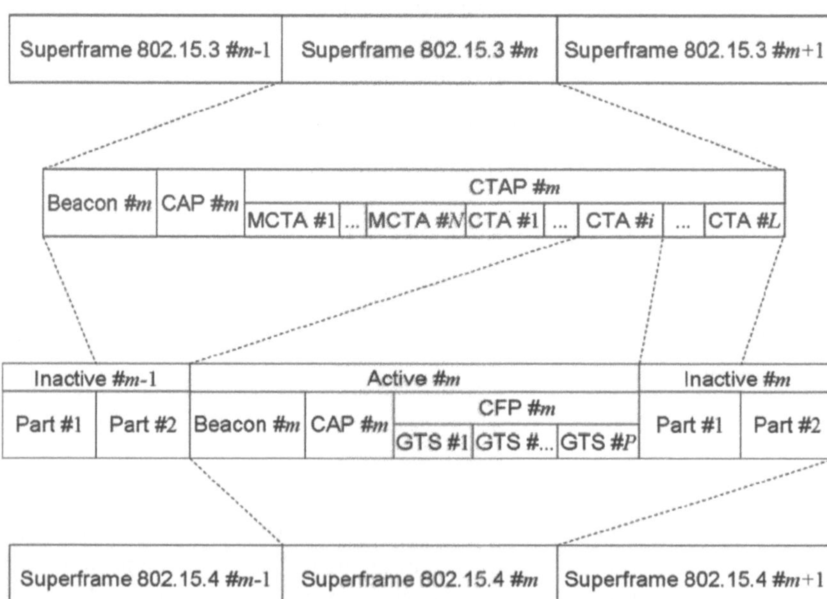

Fig. 4.73 Synchronization of the 802.15.3 and 802.15.4 superframes (AWA)

There are two restrictions to the exploitation of the AWA coexistence algorithm:

- There must be a dual mode HDR/LDR WPAN device within the common coverage area of the 802.15.3 and 802.15.4 WPANs. This device shall be the coordinator of both 802.15.3 and 802.15.4 WPANs.
- The 802.15.4 WPAN must be beacon enabled with an active and an inactive period. This restriction is compulsory according to MAGNET LDR MAC.

From now on it is assumed that the coordination of the HDR WPAN is performed by the 802.15.3-based air interface of the LDR/HDR dual mode device, while the coordination of the LDR WPAN is performed by the 802.15.4-based air interface of the LDR/HDR dual mode device.

The steps of the algorithm are listed in the following:

1. The coordinator of the 802.15.3 WPAN starts a piconet, allocating a private CTA (the i-th CTA) to the 802.15.4 WPAN. If a CTA is allocated to a child piconet, the DestID and SrcID related to this CTA shall both be the DEVID of the DEV that is the child piconets PNC. This means that the 802.15.3 PNC shall use its own ID as DestID and SrcID of the 802.15.4 virtual child piconet. The 802.15.3 PNC should use a pseudo-static private CTA.
2. The coordinator of the 802.15.4 WPAN sets the superframe duration equal to the superframe duration of the 802.15.3 WPAN. Therefore, the superframe periodicity is the same for both 802.15.3 and 802.15.4 WPANs.
3. Under the assumption that all the private CTAs allocated to the 802.15.4 WPAN are pseudo-static, we have $CTD_{HDR}^{(m)} = CTD_{HDR}^{(m+1)}$, where $m = 1, 2 \ldots$ is the transmission index. Furthermore the position of the pseudo-static time slot (i.e. i-th CTA) of the m-th superframe 802.15.3 is equal to the position of the pseudo-static time slot (i.e. i-th CTA) of the $(m+1)$-th superframe 802.15.3.
4. For the synchronization of the two networks, the following equations shall hold:

$$
\begin{aligned}
SFD_{HDR}^{(m)} &= SFD_{LDR}^{(m)} \\
CTD_{HDR}^{(m)} &= SAD_{LDR}^{(m)} \\
SID_{LDR}^{m'} &= SID_{LDR}^{m+1'} \\
SID_{LDR}^{m''} &= SID_{LDR}^{m+1''}
\end{aligned}
$$

Because of the constraint on SFD_{HDR}, which is: $0 < SFD_{HDR} < 65535 \cdot 10^{-6}$ s, and the constraint on SFD_{LDR} where $0 < \mathrm{SO} < \mathrm{BO} < 14$, $SFD_{HDR}^{(m)} = SFD_{LDR}^{(m)}$ can only be satisfied with BO $= 1, 2$.

Furthermore, in order to use the inactive portion, possible values of BO and SO are: (BO, SO) $= (1,0)$, (BO, SO) $= (2,0)$ and (BO, SO) $= (2,1)$ which provide a 802.15.4 duty cycle of 50%, 25% and 50% respectively (see Table 4.18).

In situations where the duty cycle of the HDR network shall be higher than 75% (i.e. the LDR duty of cycle lower than 25%), an improved version of the AWA mechanism should be considered.

Table 4.18 IEEE 802.15.4 timings

@250 kbps		Symbols	Duration (ms)	Size
Beacon order = 1	Beacon interval	1920	30.72	7,680 bit
Superframe order = 0	Superframe duration	960	15.36	3,840 bit
	Time slot	60	0.96	240 bit
	Maximum CFP	520	8.32	8 slot
	Duty cycle	50%		
Beacon order = 2	Beacon interval	3840	61.44	15,360 bit
Superframe order = 0	Superframe duration	960	15.36	3,840 bit
	Time slot	60	0.96	240 bit
	Maximum CFP	520	8.32	8 slot
	Duty cycle	25%		

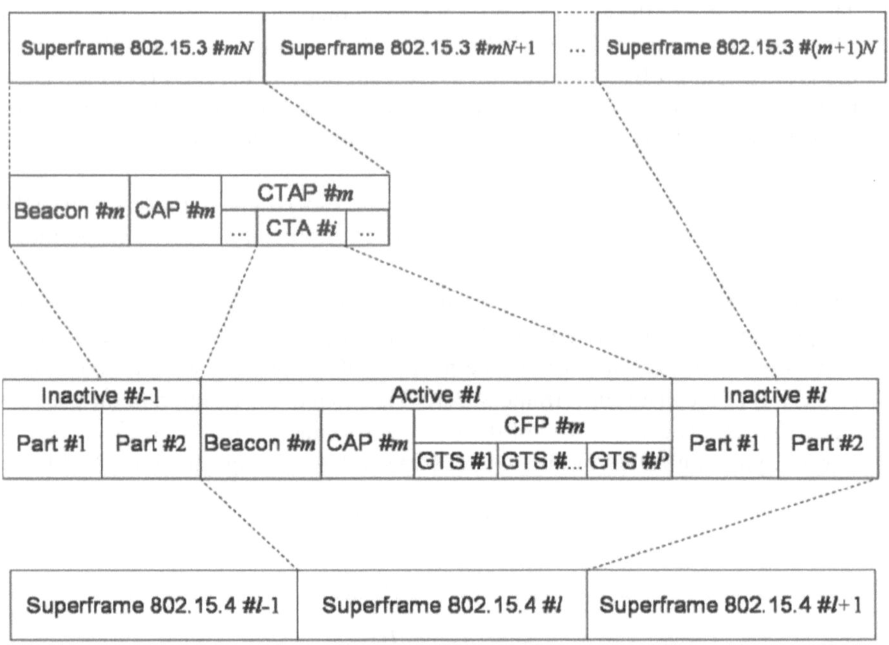

Fig. 4.74 Synchronization of the 802.15.3 and 802.15.4 superframes (IAWA)

Improved AWA Coexistence Mechanism

In the improved version of the AWA coexistence mechanism, the private CTA for LDR is not allocated once per superframe, but it is allocated once per N superframes. In other words, the private CTA is allocated at the superframe no. mN (where $m = 1, 2, \ldots$ is the transmission index), while it will skip the private CTA allocation for the next $N-1$ HDR superframes. The synchronization of the 802.15.3 and 802.15.4 superframes is shown in Fig. 4.74.

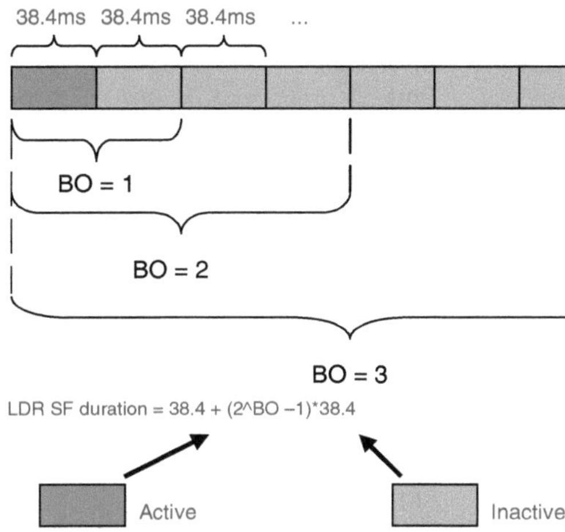

Fig. 4.75 LDR Superframe structure

The improved version of the AWA coexistence mechanism is compulsory when the LDR Beacon Interval is higher than $65535 \cdot 10^{-6}$ s that is the maximum value of SFD_{HDR}, this happens with BO > 2 or when the data rate is lower than 120 kbps.

In the previous basic version of the AWA mechanism, the PNC of the HDR network computes the superframe structure once per time. With this improved version, the PNC shall compute the superframe structure once per N superframes. It is worth noting that, for both AWA basic mechanism and its improved version, no modification to the IEEE 802.15.3 and 802.15.4 MAC standards is required.

An overview of the structure of the Magnet LDR superframe is shown in Fig. 4.75 where it is possible to notice that the minimum SF duration with Magnet LDR air interface is equal to 38.4 ms.

As already mentioned with the improved version of AWA it is possible to use more HDR superframes, this leads to a more flexible solution. An extensive list of the possible combination of LDR BO and number of HDR Superframe is shown in Table 4.19. Due to timing constraints the time duration of the HDR Beacon plus the HDR CAP and the HDR CTA cannot be higher than 27.135 ms (maximum allowed HDR SF duration is 65.53 ms and minimum duration for LDR active part is 38.4 ms) thus only the green selected combinations are allowed.

4.4.3.3 Performance Evaluation and Comparison

The proposed MAGNET physical layer schemes for LDR and HDR WPAN are subject to mutual out of band interference when closely-located devices (e.g. dual-mode devices) are considered. In fact, the wide band structure of the one transmission schemes and the relatively close respective center frequency lead to a situation where the out-of-band emission of the interferer device is added to the wanted signal

Table 4.19 Combination of LDR beacon order and HDR superframe

LDR BO	n° of HDR SF	SF duration (ms)	HDR beacon + CAP (ms)	LDR duty cycle (%)
1	2	76.8	0	100.00
2	1	153.6	115.2	33.33
2	2	153.6	38.4	33.33
2	3	153.6	12.8	33.33
2	4	153.6	0	33.33
3	3	307.2	64	14.29
3	4	307.2	38.4	14.29
3	5	307.2	23.04	14.29
3	6	307.2	12.8	14.29
3	7	307.2	5.48571	14.29
3	8	307.2	0	14.29
4	6	614.4	64	6.67
4	7	614.4	49.3714	6.67
4	8	614.4	38.4	6.67
4	9	614.4	29.8667	6.67
4	10	614.4	23.04	6.67
4	11	614.4	17.4545	6.67
4	12	614.4	12.8	6.67
4	13	614.4	8.86154	6.67
4	14	614.4	5.48571	6.67
4	15	614.4	2.56	6.67
4	16	614.4	0	6.67
5	12	1228.8	64	3.23
5	13	1228.8	56.1231	3.23
5	14	1228.8	49.3714	3.23
5	15	1228.8	43.52	3.23
5	16	1228.8	38.4	3.23
5	17	1228.8	33.8824	3.23
5	18	1228.8	29.8667	3.23
5	19	1228.8	26.2737	3.23
5	20	1228.8	23.04	3.23
5	21	1228.8	20.1143	3.23
5	22	1228.8	17.4545	3.23
5	23	1228.8	15.0261	3.23
5	24	1228.8	12.8	3.23
5	25	1228.8	10.752	3.23
5	26	1228.8	8.86154	3.23
5	27	1228.8	7.11111	3.23
5	28	1228.8	5.48571	3.23
5	29	1228.8	3.97241	3.23
5	30	1228.8	2.56	3.23
5	31	1228.8	1.23871	3.23

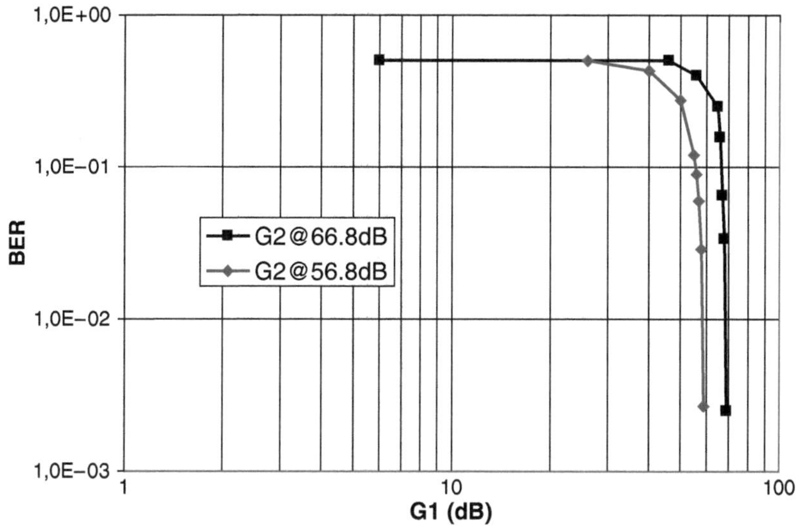

Fig. 4.76 BER vs. HDR path loss G1 (dB)

of the receiving device resulting in a degradation of the link performance. For a fixed path loss between LDR transmitter and LDR Receiver, the path loss between an HDR transmitter and the LDR receiver has been varied. Simulations have been carried out with different values of the receiver-interferer distance and acceptable results have been obtained with path losses higher than 66.8 dB (distances higher than 10 m). Distances below 10 m lead to a complete corrupted reception, thus it is impossible for two devices (HDR and LDR) to coexist in the same room.

In Fig. 4.76, the BER curves are shown for two distances of the LDR link (path loss of 66.8 and 56.8 dB corresponding to 10 and 5 m, respectively).

In this situation, the Packet Error Rate (PER), calculated assuming that the minimum size is 64 octets, is very high, as also shown in Fig. 4.77, thus the IAWA mechanism is almost mandatory.

With the adoption of IAWA the data rate will be lower, but there will be no packet corruption. IAWA allows choosing between different duty cycles according to the beacon order and the number of HDR Superframes considered (cf. Table 4.19). Some examples are shown in Table 4.20.

Considering an LDR with beacon order = 2 and a number of 3 HDR SF there will be a data rate drop from 100 to 33 kbps. A performance comparison with and without IAWA mechanism is shown in Tables 4.21 and 4.22.

4.5 Conclusions

This chapter describe the two PAN-optimized air interfaces selected in MAGNET, one for HDR, a multi-carrier spread spectrum (MC-SS) and an IEEE802.15.3 MAC layer, and one for LDR, a frequency modulation UWB (FM-UWB) with an IEEE

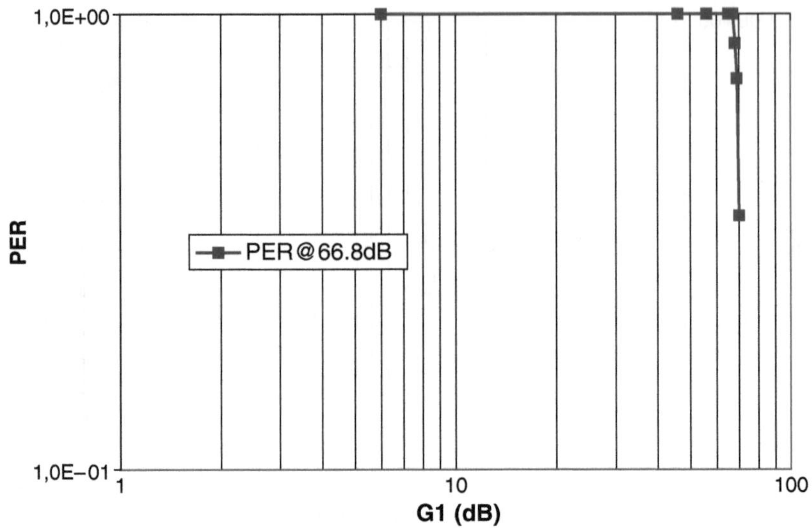

Fig. 4.77 LDR PER vs. HDR path loss G1 (dB)

Table 4.20 Data rate available with IAWA

Beacon order	LDR duty cycle (%)	HDR duty cycle (%)	LDR source data rate (bps)	HDR source data rate (bps)
2	33.33	66.67	33,330	57,749,554
3	14.29	85.71	14,290	74,242,002
4	6.67	93.33	6,670	80,842,446
5	3.23	96.77	3,230	83,822,174

802.15.4 MAC layer. These two air interfaces have been compared with exiting technologies, demonstrating that:

- MC-SS system provides much higher radio coverage than WiMedia system if we consider maximum transmit powers. However, WiMedia provides much higher data rates for very short ranges.
- FM-UWB in the multipath channel with interference is significantly better than Bluetooth. Among the four systems, FM-UWB, ZigBee and WiBree have similar throughput and range. Bluetooth can obtain higher data rate and larger coverage range, but the cost is higher transmission power.

From the perspective of the MAC layers, for HDR MAGNET utilizes a centralized control structure, while WiMedia is fully distributed. Further several scenarios have been simulated to evaluate the performance of LDR-UWB and Bluetooth systems. The results reveal that depending on certain scenarios and conditions the performance of LDR-UWB is better than Bluetooth.

Table 4.21 Performance comparison with G2 equal to 66.8 dB, goodput with IAWA = 33,330

G1 (dB)	LDR source data rate w/o IAWA (bps)	LDR source data rate with IAWA (bps)	HDR source data Rate w/o IAWA (bps)	HDR source data Rate with IAWA (bps)	BER w/o IAWA	BER with IAWA	PER w/o IAWA	PER with IAWA	Goodput w/o IAWA
6	100,000	33,330	86,620,000	57,749,554	0.5	0	1	0	0
26	100,000	33,330	86,620,000	57,749,554	0.5	0	1	0	0
46	100,000	33,330	86,620,000	57,749,554	0.5	0	1	0	0
56	100,000	33,330	86,620,000	57,749,554	0.4	0	1	0	0
65	100,000	33,330	86,620,000	57,749,554	0.25	0	1	0	0
66	100,000	33,330	86,620,000	57,749,554	0.11	0	1	0	0
67	100,000	33,330	86,620,000	57,749,554	0.065	0	1	0	1.11022E-10
68	100,000	33,330	86,620,000	57,749,554	0.025	0	0.999997	0	0.234619

Table 4.22 Performance comparison with G2 equal to 56.8 dB, goodput with IAWA = 33,330

G1 (dB)	LDR source data rate w/o IAWA (bps)	LDR source data rate with IAWA (bps)	HDR source data rate w/o IAWA (bps)	HDR source data rate with IAWA (bps)	BER w/o IAWA	BER with IAWA	PER w/o IAWA	PER with IAWA	Goodput w/o IAWA
6	100,000	33,330	86,620,000	57,749,554	0.5	0	1	0	s0
26	100,000	33,330	86,620,000	57,749,554	0.5	0	1	0	0
46	100,000	33,330	86,620,000	57,749,554	0.4	0	1	0	0
56	100,000	33,330	86,620,000	57,749,554	0.09	0	1	0	0
57	100,000	33,330	86,620,000	57,749,554	0.060402	0	1	0	1.399E-09
58	100,000	33,330	86,620,000	57,749,554	0.029090	0	0.99999972	0	0.027253
59	100,000	33,330	86,620,000	57,749,554	0.002781	0	0.75980142	0	24019.85
66	100,000	33,330	86,620,000	57,749,554	0.000001	0	0.00051186	0	99948.81

Further the chapter describes advanced techniques for increasing the spectrum efficiency and mitigating the effect of interference in the MAGNET HDR wireless communication link, extensions of IEEE 802.15.3 for inter-PAN communication and finally interference issue among LDR and HDR AIs.

References

1. J.F.M. Gerrits, M.H.L. Kouwenhoven, P.R. van der Meer, J.R. Farserotu, J.R. Long, Principles and limitations of ultra wideband FM communications systems. EURASIP J. Appl. Signal Process. (special issue UWB-STATE OF THE ART) **2005**(3), 382–396 (Mar 2005)
2. Y. Zhou, J. Yuan, An 8-Bit 100-MHz CMOS linear interpolation DAC. IEEE J. Solid State Circ. **38**(10), 1758–1761 (Oct 2003)
3. J.F.M. Gerrits, J.R. Farserotu, J.R. Long, Multiple-access interference in FM-UWB communication systems, in *Proceedings of the WPMC2005*, Aalborg, Denmark, 19–22 Sept 2005, pp. 2027–2031
4. T. Messerges, J.I. Curkier, T.A.M. Kevenaar, L. Puhl, R. Struik, E. Callaway, A security design for a general purpose, self-organising, multihop ad hoc wireless network. ACM Workshop on Security of Ad Hoc and Sensor Networks (SASN), TR2003–114, http://www.merl.com (Dec 2004)
5. C.-S. Chang, J.A. Thomas, Effective bandwidth in high-speed digital networks. IEEE J. Select. Areas Commun. **13**(6), 1091–1100 (1995)
6. F.P. Kelly, Notes on effective bandwidths. *Stochastic Networks: Theory and Applications*, vol 9 (Oxford University Press, UK, 1996), pp. 141–168
7. A. Dembo, O. Zeitouni, *Large Deviations Techniques and Applications*, 2nd edn. (Springer-Verlag, New York, 1998)
8. P.W. Glynn, W. Whitt, Logarithmic asymptotics for steady-state tail probabilities in a single-server queue. J. Appl. Prob., **31A**, 131–156 (1994)
9. I.C. Paschalidis, S. Vassilaras, On the estimation of buffer overflow probabilities from measurements. IEEE Trans. Inform. Theory, **47**(1), 178–191 (2001)
10. Q. Liu, S. Zhou, G.B. Giannakis, Cross-layer combining of adaptive modulation and coding with truncated ARQ over wireless links. IEEE Trans. Wireless Comm. **3**(5), 1746–1755 (2004)
11. Q. Liu, S. Zhou, G.B. Giannakis, Queuing with adaptive modulation and coding over wireless links: cross-layer analysis and design. IEEE Trans. Wireless Comm. **4**(3), 1142–1153 (2005)
12. Q. Liu, S. Zhou, G.B. Giannakis, Cross-layer scheduling with predictable QoS guarantees in adaptive wireless networks. IEEE J. Select. Area Commun. **23**(5), 1051–1066 (2005)
13. D. Wu, R. Negi, Effective capacity: a wireless link model for support of quality of service. IEEE Trans. Wireless Commun. **2**(4), 630–643 (July 2003)
14. J. Tang, X. Zhang, Cross-layer-model based adaptive resource allocation for statistical QoS guarantees in mobile wireless networks. QShine'06 The Third International Conference on Quality of Service in Heterogeneous Wired/Wireless Networks, Waterloo, ON, Canada, 7–9 (Aug 2006)
15. J. Tang, X. Zhang, Quality-of-service driven power and rate adaptation over wireless links. IEEE Trans. Wireless Comm. **6**(8) (Aug 2007)
16. J. Tang, X. Zhang, Cross-layer modeling for quality of service guarantees over wireless links. IEEE Trans. Wireless Commun. **6**(12) (Dec 2007)
17. M.S. Alouini, A.J. Goldsmith, Adaptive modulation over Nakagami fading channels. Kluwer J Wireless Commun. **13**(1–2) (2002), 119–143
18. I.C. Paschalidis, Class-specific quality of service guarantees in multimedia communication networks, in *Automatica* (*Special Issue on Control Methods for Communication Networks*), ed. by V. Anantharam, J.Walrand, **35** (1999), 1951–1968

19. D. Bertsimas, I.C. Paschalidis, Probabilistic service level guarantees in make-to-stock manufacturing systems, Operation Res., **49**(1), 119–133 (2001)
20. A. Sendonaris, E. Erkip, B. Aazhang, User cooperation diversity – part I: system description, IEEE Trans. Wireless Comm. **51**, 1927–1938 (Nov 2003)
21. J.N. Laneman, D.N.C. Tse, G.W. Wornell, Cooperative diversity in wireless networks: Efficient protocols and outage behaviour. IEEE Trans. Inform. Theory **50**, 3062–3080 (2004)
22. A. Host-Madsen, Capacity bounds for cooperative diversity. IEEE Trans. Inform. Theory **52**, 1522–1544 (Apr 2006)
23. A. Bletsas, A. Khisti, D.P. Reed, A. Lippman, A simple cooperative diversity method based on network path selection. IEEE J. Select. Areas Commun. **24**, 659–672 (Mar 2006)
24. A. Nosratinia, T.E. Hunter, A. Hedayat, Cooperative communication in wireless networks. IEEE Commun. Mag. **42**, 74–80 (Oct 2004)
25. A. Bletsas, H. Shin, M.Z. Win, Outage optimality of opportunistic amplify-and-forward relaying. IEEE Commun. Lett. **11**(3), 261–263 (Mar 2007)
26. G.M. Kraidy, J.J. Boutros, A.G.I. Fàbregas, Approaching the outage probability of the amplify-and-forward relay fading channel. IEEE Commun. Lett. **11**(10), 808–810 (Oct 2007)
27. IST MAGNET Beyond, Prototype specification for the FM-UWB and MC-SS RA schemes IST 027396, Deliverable D3.2.1 (June 2006)
28. H.L. Van Trees, *Optimum Array Processing* (New York, Wiley, 2002)
29. T. Hunziker, M. Westmeier, D. Dahlhaus, Amplify-and-forward relaying for reducing outages in TDMA-based WPANs operating in unlicensed bands. *Proceedings of the 10th International Symposium on Wireless Personal Multimedia Communications*, Dec 2007, pp. 627–631, Jaipur, India
30. IEEE Standard for Information Technology–Telecommunications and Information Exchange Between Systems–Local and Metropolitan Area Networks-Specific Requirements, IEEE Standard 802.15.3, 2003
31. http://www.ieee802.org/15/pub/2003/Jul03/03268r2P802–15_TG3a-Multi-band-CFP-Document.pdf
32. ftp://ftp.802wirelessworld.com/15/07/15–07–0693–03–003c-compa-phy-proposal.pdf
33. Y.H. Tseng, E.H. Wu, G.H. Chen, Maximum traffic scheduling and capacity analysis for IEEE 802.15.3 high data rate MAC protocol, Proc. Vehicular IEEE Technol. Conf. **3**, 1678–1682 (2003)
34. X. Chen, Y. Xiao, Y. Cai, J. Lu, Z. Zhou, An energy diffserv and application-aware MAC scheduling for VBR streaming video in the IEEE 802.15.3 high-rate wireless personal area networks. Elsevier Comp. Commun. **29**, 3516–3526 (2006)
35. R. Mangharam, M. Demirhan, R. Rajkumar, D. Raychaudhuri, Size matters: Size-based scheduling for MPEG-4 over wireless channels, *Proceedings of the SPIE Conference on Multi-Media Networking and Communications*, 2004, pp. 110–122, Santa Clara, CA
36. L. Vajda, A. Torok, K.J. Youn, J. Sun-Do, Hierarchical superframe formation in 802.15. 3 networks. Proc. IEEE ICC **7**, 4017–4022 (2004)
37. S.H. Rhee, K. Chung, Y. Kim, W. Yoon, K.S. Chang, An application-aware MAC scheme for IEEE 802.15. 3 high-rate WPAN. Proc. WCNC **2**, 1018–1023 (2004)
38. IEEE Draft Recommended Practice to Standard for Information Technology–Telecommunications and Information Exchange Between Systems–Local and Metropolitan Networks-Specific Requirements-Part 15.5: Mesh Enhancements for IEEE 802.15 WPANs, IEEE Draft 15–06–0237–02–0005 (2006)
39. M. De Sanctis, J.F.M. Gerrits, J.P. Vila, Coexistence concept for the implementation of LDR/HDR WPAN multimode devices. Teletronikk Journal (by Telenor), special issue on Personal Networks, 2007, pp. 101–112
40. IEEE, Coexistence of wireless personal area networks with other wireless devices operating in unlicensed frequency bands. IEEE Standard 802.15.2 (August 2003)
41. R. Tesi, M. Condreanu, I. Opperman, Interference effects of UWB transmission in OFDM communication systems, in *Proceedings of the International Workshop on Ultra Wide Band Systems*, Oulu, Finland (June 2003)

42. A. Tomiki, T. Ogawa, A. Fukuda, N. Terada, T. Kobayashi, Evaluation of interference from impulse-radio and direct-sequence-UWB sources to 2-GHz digital radio transmission, in *Proceedings of the IEEE International Symposium on Electromagnetic Compatibility*, Istanbul, Turkey (May 2003)
43. IEEE Std. 802.15.4–2003, Standard for Telecommunications and Information Exchange Between Systems Local Area Metropolitan Area Networks Specific Requirements Wireless Medium Access Control (MAC) and Physical Layer (PHY) Specifications for Low Rate Wireless Personal Area Networks (WPAN)
44. K. Schoo, Y. Wang, H.T. Nguyen, I. Siaud, A.-M. Ulmer-Moll, N. Malhouroux, PHY/MAC Benchmarking of the Target MAGNET FM-UWB and MC-SS Air Interfaces, Deliverable 3.2.2 MAGNET Beyond (June 2007)

Chapter 5
Security in PNs

Hossam Afifi, Dimitris Kyriazanos, Shahab Mirzadeh, Jordi Jaen Pallares, Andreas Pashalidis, Neeli Rashmi Prasad, Antonietta Stango, and Jan Stoter

5.1 Introduction

A PN can provide the opportunity to personalize applications, services and the whole networking environment to the current needs of the user, making the security and privacy the key features in the PNs formation.

In MAGNET Beyond the interest was concentrated on the importance of interactions between multiple PN users with common interests for various services, i.e. on PN Federation (PN-F). Establishing a PN Federation, the user should be able to control which information or services to share with others. For this reason the security is a very important aspect in PN-F as multiple people are involved and takes place at different levels: access to the PN Federation based on membership, secure transport of data within the federation and the access rights to resources and services of the federation.

In this chapter, we present important contributions in the study of security of personal networks. To begin with, a threat analysis methodology tailored for PN and PN-F environments is described and applied to evaluate security. A reflection

H. Afifi (✉)
Groupe des Ecoles des Télécommunications – Institut National des Télécommunications, rue Dunois 69, Paris 75013, France
e-mail: hossam.afifi@it-sudparis.eu

D. Kyriazanos
Institute of Communication and Computer Systems of the National Technical University of Athens, Greece

S. Mirzadeh
The University of Surrey, UK

J.J. Pallares
Fraunhofer Institut FOKUS, Germany

A. Pashalidis
NEC Europe Ltd., Germany

N.R. Prasad and A. Stango
Aalborg University, Denmark

J. Stoter
Twente Institute of Wireless and Mobile Communications, The Netherlands

R. Prasad (ed.), *My Personal Adaptive Global NET (MAGNET)*, Signals and Communication Technology, DOI 10.1007/978-90-481-3437-3_5,

on user perspectives mapped onto security applications, privacy safeguarding and related protocols is encapsulated into the proposed Context Aware Security Manager component. Finally, we present a novel authentication protocol right tailored to specific low power devices needs.

5.2 Security Evaluation: Threat Analysis

Within a PN federation, the privacy of the user needs to be especially protected, and the user needs to be in full control of his/her context information, they should be able to control which information or services to share with others [1].

To identify how potential adversaries exploit system weakness to achieve their goals and, in particular to highlight internal vulnerabilities and drawbacks of mechanisms that could be used against the security infrastructure, a threat analysis is needed.

A general methodology for a complete and general threat analysis in a system has been carried out, and also a new approach to analyze the system has been proposed [2, 3].

5.2.1 Threat Analysis Methodology

The threat analysis process divided in three main phases: threats modelling, assets mapping and building a mitigation plan.

Threat modelling is a method of assessing and documenting the security risk associated with an application that involves also understanding the goals of an adversary in attacking a system based on the assets of interest. This allows to enumerate the threats and also to discover the vulnerabilities. The threat modelling is very useful especially if is done in the earliest stage of the system development and then, as the applications evolve and requirements are better defined, the lists of threats and vulnerabilities can be updated as needed.

Asset mapping involves documenting the tangible and intangible resources of the system and identify the related entry points of the system. The assets value is used as the basis for calculating threat risks and for prioritizing countermeasures, ergo assets need to be prioritized. It is often easier for the analyst to identify system assets via the process of analyzing specific threats. This implies an iterative approach of mapping assets and enumerating threats.

The third phase of the threat analysis is building a mitigation plan, namely selecting from the list of all the proposed countermeasures, the most effective combination. The analysts will decide which of the proposed countermeasures will be included in the actual mitigation plan according to their experience. The result of this analysis is a set of countermeasures that mitigate the threats identified. In Fig. 5.1 the proposed methodology is shown step by step.

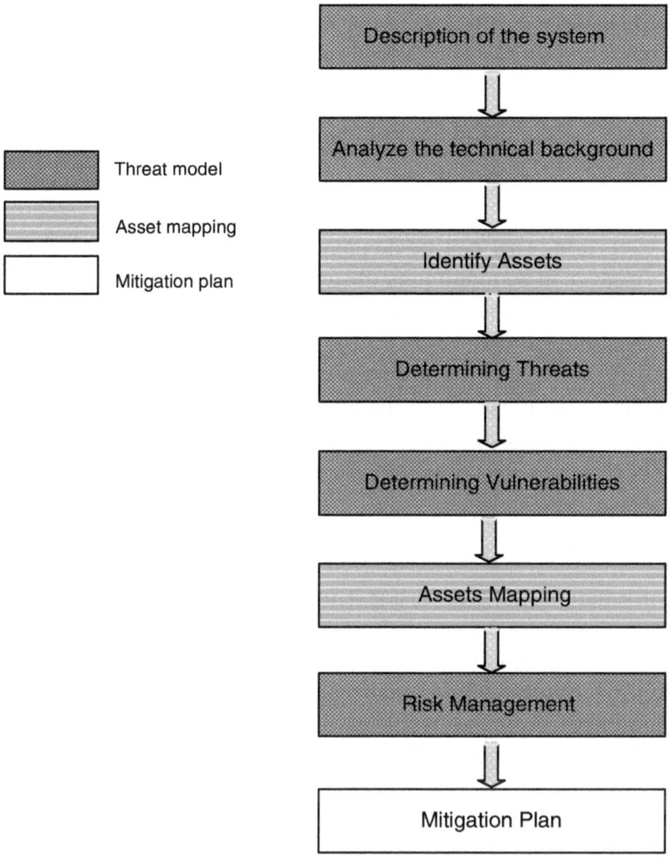

Fig. 5.1 Steps of threat analysis

Step 1. Description of the system: network overview and use cases

To describe a system it is needed to understand every component and its interconnections, defining the scenarios and the use cases. To help the system characterization the Unified Modelling Language (UML) use cases diagram and the related descriptive tables are very useful, as they allow to describe what the system must be able to do, by showing the interaction between the use cases and the actors involved. The use case description is in a table that contains all the information related to the system and the use case, i.e. the goal, the devices and involved technologies, the description of the actor and the stages to realize the use case.

Step 2. Analyze the technical background of the use cases

To analyze the timing sequence of all the devices and actors involved in the use case, the UML sequence diagrams are suitable.

A sequence diagram shows object interactions organized according to their timing sequence. The sequence view describes the system in execution. It can be used to model the behaviour of the system by representing the realization of a use case scenario. It depicts the objects involved in the scenario and the sequence of messages exchanged between the objects.

In this step it is possible to analyze how the technologies are used and who is using them.

Step 3. Identify Assets

In this step everything that can be damaged or violated in the network should be determined. Assets can be tangible or abstract, general or related to a use case.

In general the assets depend on the situations and on the users, but it is possible also to identify some general assets for the system, e.g. the IDs of the owner or the devices itself. By analyzing the use case diagrams and the descriptive tables, it is possible to identify the general and specific assets of each use cases. In this step the assets are enumerated and stored in a table as a record with ID, name and a brief description. In the next step, this table is checked and updated in case some new assets are found.

Step 4. Determining Threats

Using the information gathered so far it is possible to start to identify the threats and the potential threat-sources of the system. A threat-source is defined as any circumstance or event with the potential to cause harm to a system. By analyzing the use cases, technical functionalities and sequence diagrams, it is possible to identify the threats and threat-sources. Afterwards the threats have to be correlated with the assets and with the entry points. The output of this step is the threats profile, a table where every threat is associated with an ID, a name or classification, the source of threat, the assets involved and the entry points associated. Finally the threats must also be analyzed to determine whether the system is susceptible to them.

Step 5. Determining Vulnerabilities

The goal of this step is to develop a list of system vulnerabilities that could be exploited by the potential threat-sources. When all the threats and their scenarios have been described it is possible to see what the threats are exploiting. A table will be filled in with the main real vulnerabilities and the corresponding threats that are exploiting in the specific use cases analyzed in the previous sections. Each vulnerability will be assigned an ID, followed by the description, the name and the corresponding threat.

Another way to evaluate if the system is susceptible to the threats that have been identified is to use the attack trees.

Step 6. Asset Mapping

In this step the list of assets determined in the step 3 is checked to determine if all the assets have been included. It is important also determining the valiance of the assets and the risk that the owner of the assets is willing to accept [4], and based on these values prioritizing them.

To assign a value to the assets is not easy because the value can be personal and the priorities of the people can be different, nevertheless here it has been suggested three different values:

- *High*, assets with this value have to be protected with a high level of security; they are directly linked to the control of the system, with services that require highly secure level, or that have a big financial value.
- *Medium*, assets linked to access to common services, not critical, but still important with an intermediate financial value.
- *Low* value for assets of minor importance.

The values have to be assigned taking in to the account the scenarios and the particular use cases.

Step 7. Risk Management

The risk management helps to balance between what it is acceptable and what it is possible.

From the threat and vulnerability list it is possible to extract the information about which threat pose the highest risk value. The aspects, which have to be taken into account to assess the risk, are the impact, the damage to the assets when the threat would materialize, the size of the vulnerabilities and the likelihood that the threat will attempt to materialize.

Step 8. Mitigation Plan

The last step of the threats analysis is the construction of a mitigation plan that involves the selection of the countermeasures. In this step the threats selected for mitigation must be addressed by one or more countermeasures. To build a mitigation plan it is necessary to identify the countermeasures, i.e. have a list of the countermeasures and a map of the relationship between countermeasures and vulnerabilities, and from this list to select the most effective combination. The decision of which of proposed countermeasures will be included in the actual mitigation plan is taken by the analyst.

The result of the process is a set of proposed countermeasures that would mitigate the threats that were identified. Since the implementation of all the proposed countermeasures is, in most of the cases, impractical due to constraints in budget, time and resources, the goal of a beneficial threats analysis process is to propose the set of the most cost-effective countermeasures against the identified threat.

5.2.2 Threat Analysis of PN-F

The methodology presented above has been applied to the case of PN-F. A scenario with the use cases has been selected and analyzed following step by step the methodology [2].

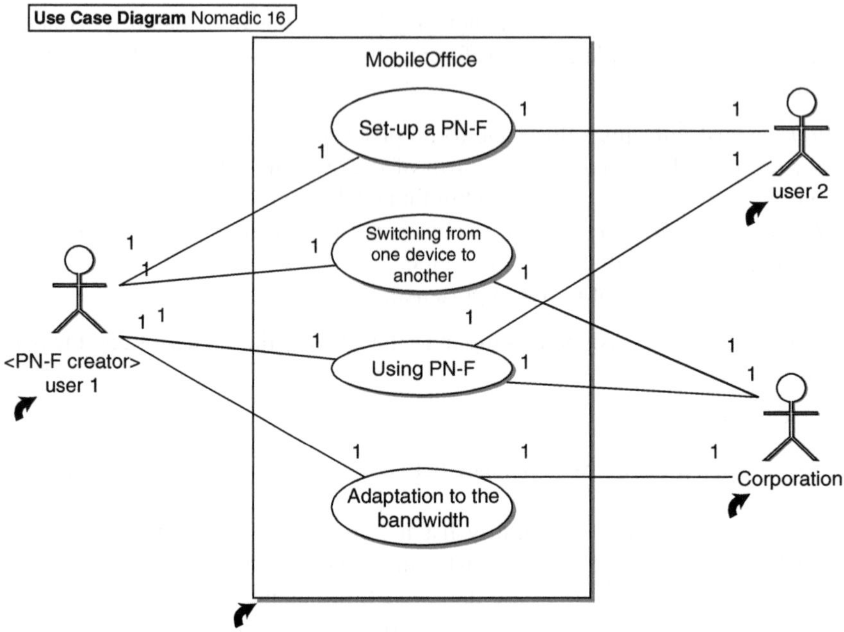

Fig. 5.2 Nomadic@Work 16 Use cases UML Diagram [2]

In the first step the UML use case diagram are used to describe what the system must be able to do and to help in the description of the system in next figure an example (Fig. 5.2).

From the UML use cases diagram and from the tables, like Table 5.1, the description of the system is carried out. In every use case the technologies used, the actors and devices involved, the goals of the systems are identified and also the environment can be deduced. All this information will be utilized in the next steps to obtain the sequence diagrams, which can describe the interactions that will be triggered when a particular use case is executed and in what order those interactions will occur.

In the Fig. 5.3 an example of sequence diagram related to the use case considered above.

Analyzing the use cases diagram and the descriptive tables and the sequence diagrams it is possible to identify, the general and specific assets of the system, in the third step these will be ordered in tables as shown in the following (Tables 5.2 and 5.3).

In the fourth step the possible threats and entry points for every use cases are identified and filled in a table as shown in the following Table 5.4.

In step 5 the vulnerabilities are collected by analyzing the enumerated threats. A table will be filled in with the main real vulnerabilities that are exploiting in the specific use cases analyzed in the previous steps and each vulnerability will be referred with the corresponding threats. An example is in the Table 5.5.

Table 5.1 Set-up a PN-F use case

Use case name	Set-up a PN-F
Goal in context	Collaborative work, service discovery
Preconditions	The actors are carrying portable devices, which incorporate MAGNET Air Interface/WiFi/UMTS capabilities as well as general office accessories. The devices are interconnected with each other
Successful end	A PN-F is created
Failed condition	No federation
Primary actors	Two colleagues (PN-F creator, user)
Secondary Actor	
Trigger	The creator ask for a federation
Main flow	
1	User1 (the creator) ask for the PN-federation by selecting the *PN-federation definition* in the user GUI
2	The user GUI asks for the identifiers of the Personal Networks allowed to participate in the PN-federation
3	User1 gives the PN identifiers of both colleagues
4	User1 opens the PN manager, which recognizes the colleague's device as a foreign node
5	By using the PN manager the creator selects the device of User2 and sends an invitation to it for the PN-federation with the PN manager command: *Invite to PN-Federation*
6	User2 opens the PN manager and accepts the received invitation
7	When the PN manager of User2 has shown a message about a secure PN-federation connection between the personal networks, User2 opens the PN directory manager and selects devices and the directories in the devices, which will be included in the PN-federation. These devices are the PDA and the laptop
8	User2 sends this information to the creator by the PN manager command: *PN-federation participation devices*
9	The creator selects the PDA and laptop by using the PN directory manager. He sends this information to user2 by the PN manager command: *PN-federation participation devices*
Extensions	
7.1	No secure connection for PN-F federation
7.2	No federation

The output of the asset mapping is a final table (Table 5.6) with all the assets and the corresponding value assigned taking in to the account the scenarios and the particular use cases.

From the list of asset and their value, the list of threat and the list of vulnerabilities it is possible to extract the information about which threat pose the highest risk (Table 5.7).

Analyzing the previous table it is possible to notice that threats with a high level of risk, for PN-F, are these related to spoofing and identity theft. This means that it is necessary to pay particular attention to find the countermeasures and a mitigation plan mainly for these threats.

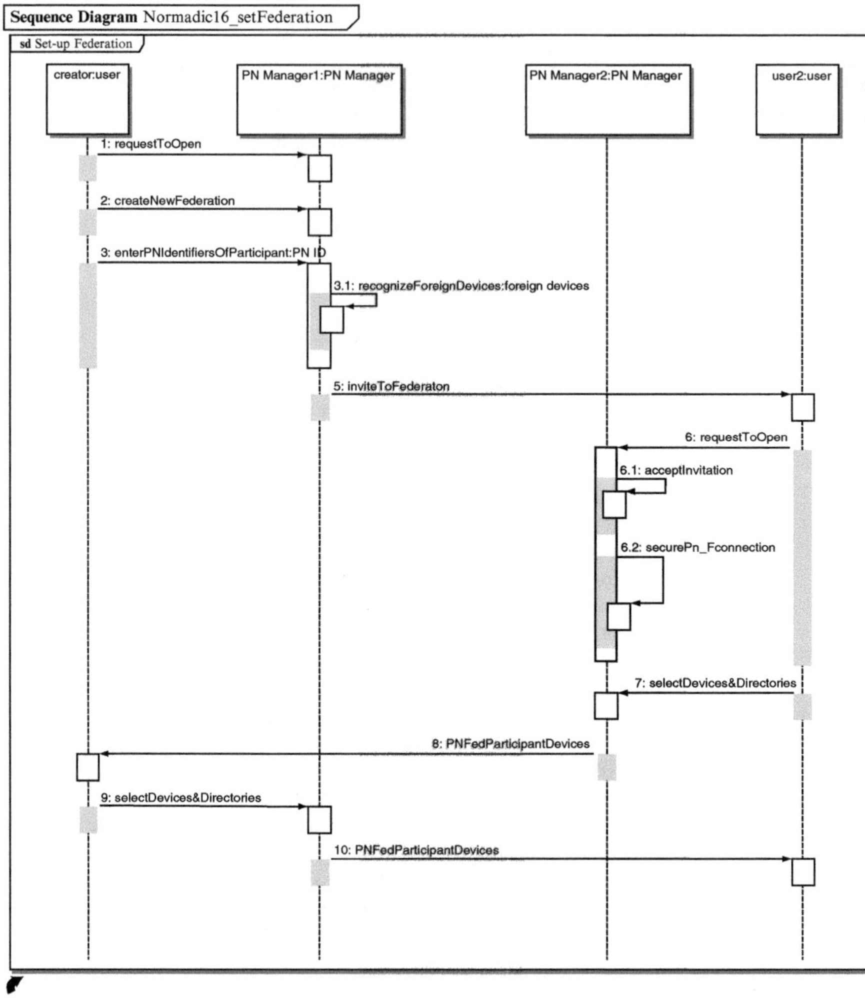

Fig. 5.3 Sequence Diagram of Set-up PN-F use case

Table 5.2 General assets

ID	Name	Description
G.1	IDs	User ID, VID, PNID
G.2	Personal data, profiles	Information about the user, what hi likes....
G.3	Availability/access of the services	The rights to access some services
G.4	Confidentiality of data	Data stored in PN
G.5	Access control	Rules to access data and services
G.6	Reputation	Level of trust that the user put in the other part
G.7	User devices	Physical asset
G.8	Proxy server	Physical asset

Table 5.3 Assets related to Nomadic@Work 16 mobile office

ID	Name	Description
M.1	Personal data stored in devices involved in PN-F	Personal data that the user enter
M.2	Work data stored in devices involved in PN-F	Information related to the work
M.3	User's login data	User credential: password and VID
M.4	Data stored remotely in PN	Information in PC at the corporation
M.5	Contact information	Phone numbers, email addresses
M.6	Access to corporation remote computer	The possibility to access remotely to the PC in corporation
M.7	Video stream	Download a video from the corporation

The last step of the threats analysis is the construction of a mitigation plan that involves the selection of the countermeasures.

The threat with higher risk, spoofing and identity theft can be secured respectively with appropriated mechanisms of authorization and authentication and with a properly encryption of the identity information.

Eavesdropping and disclosure of information can be mitigated providing the access at services and information only at authenticated and authorized users according with privacy regulation, and sending such information only on encrypted channels.

Denial-of-Service (DoS) attacks can be mitigated with appropriate network infrastructures like firewall and intrusion detection system, but also adapting the level of security in the system.

The combination of these countermeasures can be considered in this specific case the mitigation plan.

Further detail about the threat analysis methodology and the case of PN-F can be found in [2] and [3].

5.2.3 Security Evaluation of PN-F Architecture

To assure security and privacy of context information inside the PN-F the Context Aware Security manager (CASM) has been designed, in the next sections it will be described in details and all the features of the CASM will fit with the requirements that came out from the mitigation plan that has been built after the threat analysis done in the previous sections.

The User's Identity together with other sensitive information is properly flagged in order to avoid disclosure. Trust relationship issues are resolved with the Trust agent while different levels of security are also integrated dynamically with the Security Agent.

To mitigate the threats classified with higher risk, appropriated mechanisms of authorization and authentication, with a properly encryption of the identity

Table 5.4 Threats Nomadic@Work 16

ID	Description	Name (classification)	Source	Assets	Entry points
T.N.1	Recognizing the foreign devices to set-up the federation someone can act to be the colleagues if they are not visible each other	Spoofing to accede private information	Human	G1 IDs G3 access services	PN-F Manager (PDA or laptop)
T.N.2	To receive an invitation for federation from not real creator of federation	Eavesdropping on federation members	Human	G2 profiles G6 reputation	PN-F Manager (PDA or laptop)
T.N.3	The channel for federation can seem secure	Identity theft	Human	G1 IDs G2 profiles G5 access control M2 data stored in PN-F device	PN-F Manager (PDA or laptop)
T.N.4	An adversary can ask for streaming from the office computer	Eavesdropping on federation members	Human	G2 personal data G3 access services G4 confidentiality M7 video	PN-F Manager (PDA or laptop)
T.N.5	An adversary can gain the access to the devices involved in the federation	Eavesdropping on federation members DoS	Human	G1 IDs M2 work data G2 profiles G3 access services M1 personal data M5 contact list	PN-F Manager (PDA or laptop)
T.N.6	Someone not authorized can access to data stored in the devices of the conference room	Information disclosure	Human	G4 confidentiality M2 work data	Devices involved with low level of security
T.N.7	Someone can intercept the phone call	Information disclosure	Human	G4 confidentiality M5 contact list	Laptop or PSTN

Table 5.5 Vulnerabilities

ID	Description	Name	Corresponding threats
V1	A user leaves the device without logout and an adversary steals it	An adversary gain access to the PN	TN1 TN2 TN5
V2	The channel encryption is not enough	An adversary intercept communication between PN-F members	TN3 TN7 TD4 TM1
V3	The user gives personal information without check if the recipient is trusted	User too trusting	TN4 TD1
V4	The encryption of the password is insufficient	An adversary decrypt the password	TN4 TN5
V5	The access for a service is denied	An adversary rejects a service access	TM3
V6	Someone poses as service to steal information	An adversary steals private information	TM3 TD2
V7	Someone not authorized access to data	An adversary steals working information	TN6 TM2 TM4
V8	Someone poses as another user	An adversary steals contact information	TD3

information are needed. A functionality of the CASM, as it has been say, is to ensure that all the information and request are authenticated and authorized, further the identity is protected by appropriate encryption. The cryptographic solutions and techniques used to provide an acceptable level of security are discussed also in next sections. However and even in the unfortunate and unlikely event of an identity theft, the Trust Agent can revoke such identities and together with the Privacy Agent all connected information to this identity can be unconnected, denying access to any kind of information.

The Security Agent of the CASM ensures that appropriate network infrastructure exists (such as firewalls and intrusion detection systems) in PN-F areas where we have a DoS. Whenever such a supporting infrastructure is missing the security level is lowered by the CASM, effectively decreasing legitimate requests in such areas, and therefore the probability of a DoS occurring.

5.3 A User Centric Security Perspective

The difference between MAGNET Beyond Security approach and major security proposals in the personal network field comes from the integration of the user centricity in security decisions from the beginning. Contrary to other approaches like VPN (IPSEC), TLS, etc, our protocols were designed after and in concordance to user needs.

Table 5.6 Assets mapping

New ID	Name	Description	Value
A1	User's login data	User credential: passwords, IDs, PNIDs, etc.	High
A2	Banking information	Information about bank account and credit cards	High
A3	User devices		High
A4	Proxy server		High
A5	Access/availability of services	The rights to access services	Medium
A6	Confidentiality of data	Data stored in PN and PN-F	Medium
A7	Access control	Rules to access data and services	Medium
A8	Reputation	Level of trust that the users put in the other part	Medium
A9	Data stored remotely in PN	Data stored in devices not in proximity of the user	Medium
A10	Working data stored in PN-F	Information about work shared in PN-F	Medium
A11	Contact information	Phone numbers, email address, addresses	Medium
A12	Access to remote device	The possibility to access remotely to devices	Medium
A13	Video streaming	Download a video from a remote computer	Medium
A14	List of PN identity known	List of the contact received from a service	Medium
A15	Photos	Private photos or with property rights	Medium
A16	Flash film	Film about a product of the company	Medium
A17	Information shared in PN-F	Information about business and about the member of PN-F	Medium
A18	Projection on public screen	Projection of private or working files	Medium
A19	Personal data, profiles	Personal information that the user enter	Low
A20	Position	Position of the user	Low
A21	Context information	Information about the context around the user	Low
A22	Destination	Destination of the travel	Low
A23	Journey-file	The route information (travel plan or route on a map service)	Low

This has implied the collaboration with groups working on users and social behaviour. Two main contributions are provided in MAGNET Beyond user centric security studies:

- The Context Aware Security Manager (CASM)
- Integration of the Virtual Identity (VID) concept

The first point is detailed hereafter. The VID concept was initially proposed by IST project DAIDALOS and integration studies have been performed in MAGNET Beyond project.

The Context Aware Security Manager (CASM) is part of the MAGNET Secure Contemxt Management Framework. It has been designed by integrating AAA (authentication, authorisation and accounting) functionalities, profile management

Table 5.7 Threats associated with risk

ID	Name (classification)	Assets	Entry points	Vulnerabilities	Risk
T.N.1	Spoofing to accede private information	A1 A5	PN-F Manager (PDA or laptop)	V1	High
T.N.2	Eavesdropping on federation members	A19 A8	PN-F Manager (PDA or laptop)	V2	Medium
T.N.3	Identity theft	A1 A19 A7 A10	PN-F Manager (PDA or laptop)	V2	Medium
T.N.4	Eavesdropping on federation members	A19 A5 A6 A13	PN-F Manager (PDA or laptop)	V4	Medium
T.N.5	Eavesdropping on federation members DoS	A1 A19 A5 A10 A11	PN-F Manager (PDA or laptop)	V4	Medium
T.N.6	Information disclosure	A6 A10	Devices involved with low level of security	V7	Medium
T.N.7	Information disclosure	A6 A11	Laptop or PSTN	V2	Medium
T.D.1	Spoofing Identity theft	A2	Portable device of user	V3	High
T.D.2	Eavesdropping on federation members	A19 A20	Portable device of user	V6	Low
T.D.3	Spoofing Eavesdropping Information disclosure	A19 A11	PN-F Manager (PDA or laptop)	V8	Medium/low
T.D.4	Information disclosure	A16 A10	PN-F Manager (PDA or laptop)	V2	Medium
T.D.1	Spoofing Identity theft	A2	Portable device of user	V3	High
T.M.1	Information disclosure	A19 A10 A11 A20	Mobile phone	V2	Medium/low
T.M.2	Information disclosure	A10	Public screen	V2	Medium
T.M.3	DoS	A7	PN-F Manager (PDA or laptop)	V5	Medium
T.M.4	Eavesdropping	A22 A23	Mobile device	V6	Low

that translates user requirements into rules, a policy engine to verify rules, and it also provides other advanced security options like:

- Identification
- Authorization/Access control/checking clearance according to the policies

- The Privacy Enforcement
- Security and Trust Management

The Security, Privacy and Trust Manager is a logical entity, responsible for defining the PN security, privacy and trust and its decisions are based on (context) information provided for the environment, node location and the requested service.

We assume that this context information is provided to the system. This Manager performs an adaptive secure context-aware management, the goal of which is to ensure context-aware secure interactions. The adaptability and flexibility is ensured with integrating different levels of security, privacy/anonymity and trust. These provided levels of security, privacy and trust are defined as a compromise for the security, privacy and trust policies of the PNs, the user preferences and the device capabilities.

5.3.1 CASM Design

As an x-ray to the CASM box depicted in Fig. 5.4, the model for the CASM block consists of the main building blocks of the security, privacy and trust architecture [5].

We now proceed with providing descriptions for each of the internal blocks, together with corresponding security information models that feed each block.

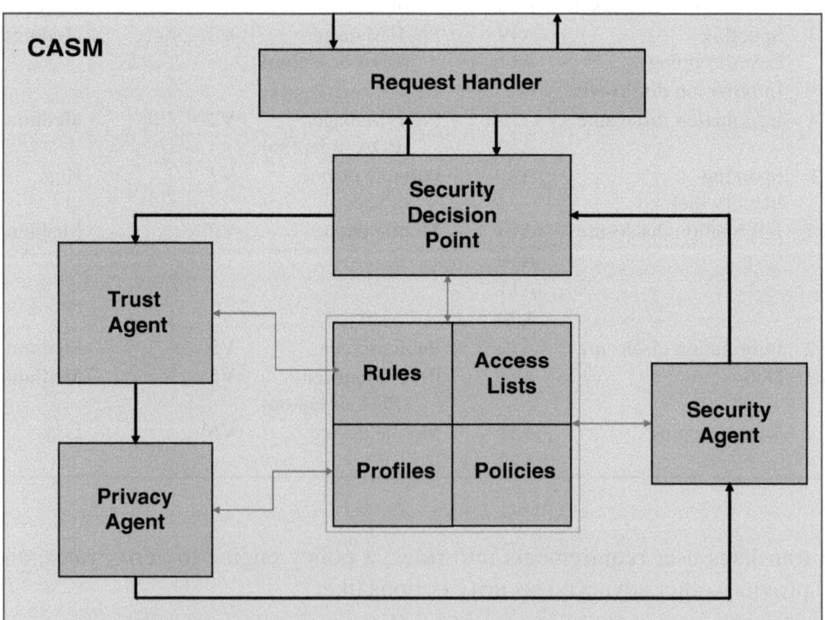

Fig. 5.4 The CASM block for Security, Privacy and Trust for PNs

5.3.1.1 Security Agent

The **Security Agent** assures data confidentiality and data integrity in accordance with the respective profile.

The Security Agent also realizes "data origin authentication" and after that the Trust Agent becomes aware of the results of any authenticity verification.

It should be noted that the role of the Security Agent is linked to the authentication in terms of "data origin authentication" while we also use the term of "authentication" (in the sense of "identity authentication") as a means for the Trust Agent to build trust.

The security mechanisms that are applied depend on the security needs of the communication and the change in the context (for example change in location or transfer of more sensitive user data). The adaptability of the security mechanisms is ensured by the Security Agent, which assigns appropriate Security Level for the communication from three possible levels – low, medium, high.

- *Low* – provides non-privileged services and allows exchange of non-sensitive data
- *Medium* – the communication needs some kind of protection, even if the data exchanged is not necessarily sensitive
- *High* – provides privileged access to service and/or exchange of highly sensitive data. The provision of this security level may compromise the network performance.

The determination of the Security Level is based on the combinations of rules, profiles and context information, as it is shown in Fig. 5.5. The application/service requirements are also taken into account as part of the context, indirectly, via the profile of the application/service.

5.3.1.2 Trust Agent

The trust establishment mechanisms are used to prevent unauthorized or compromised nodes from injecting false data into the network. Trust establishment can be based on identification, roles, behaviour, and plausibility of data. Hence, for some trust establishment mechanisms, authentication and access control may be a prerequisite.

The CASM is also responsible for establishment of trust relationships between communicating parties. This is done by **Trust Agent** (Fig. 5.6), responsible for establishing trust relationships and managing access lists.

The trust levels are defined as follows:

- Unknown – entities that enter PN/P-PAN and request access to some service for the first time (in the joining phase).
- Untrusted – entities that are not allowed to access the P-PAN/PN for any reason even if they have previously been granted access (these mainly involve entities with revoked rights or entities explicitly declared as untrusted by the user.

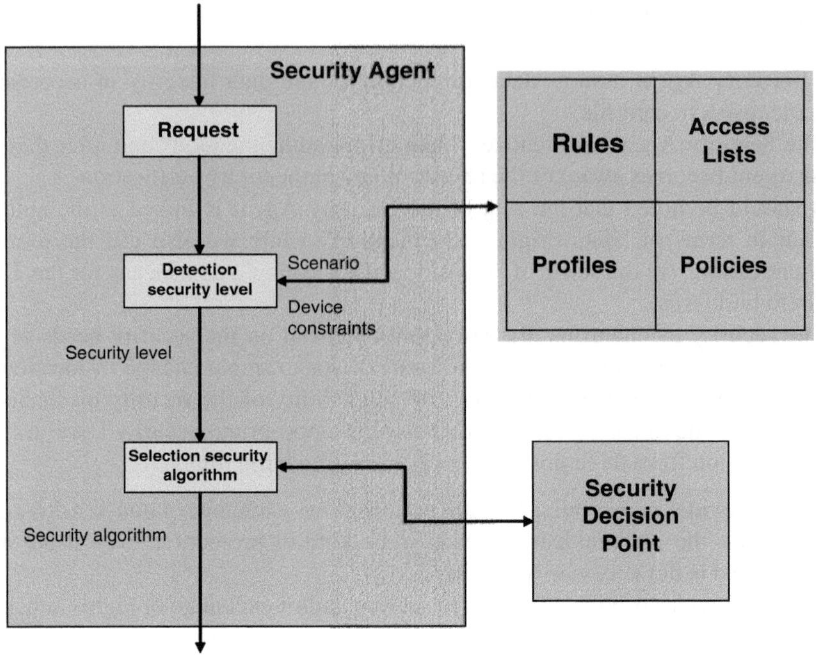

Fig. 5.5 The Security Agent

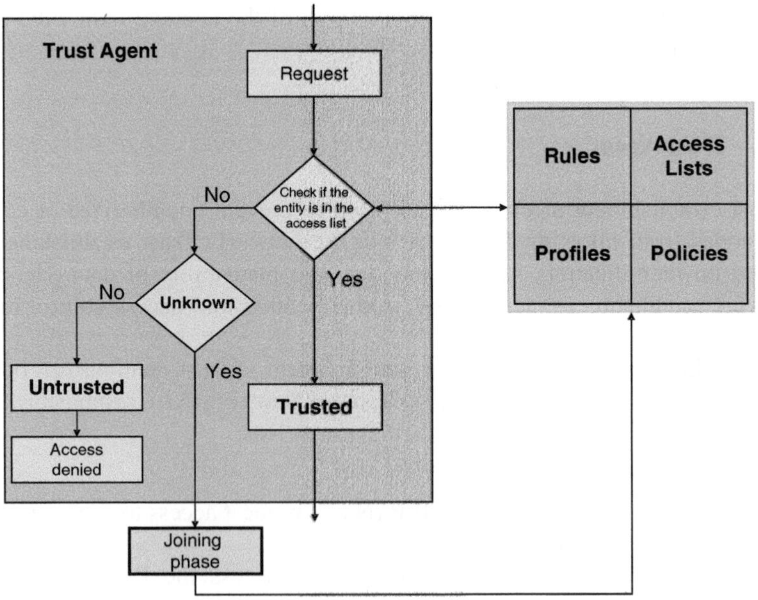

Fig. 5.6 The Trust Agent

- Trusted – entities that have previously established trust relationship and already share a trust key (in the data transfer phase).

5.3.1.3 Privacy Agent

Usually privacy is understood as sender and receiver anonymity. However information for the communication might be deducted from other parameters, like traffic or traffic patterns, size of the messages, time and location, etc. In a nutshell – the privacy can be violated by **tracking a node** for a certain time, **identifying the user** that uses a certain node, **accessing data** against the will of the data owner, **recognising a user** that uses a certain node.

The following aspects of privacy should be considered:

- Maintaining information privacy, i.e. to prevent the disclosure of personal information to attackers by giving away information only to trusted entities referred to as controlled disclosure of information.
- Preserving anonymity of the users for distinct scenarios, i.e. preserving their "state of being not identifiable within a set of subjects". Anonymity affects also location privacy, because as long as a user or a node is anonymous, location privacy is provided.
- Maintaining location privacy of a node, i.e. to deny an attacker the knowledge of a node's current and past location.

The **Privacy Agent** (Fig. 5.7) is responsible for determining if data should be disclosed, and if it should be provided anonymously (for the scenarios which involve users). Privacy level flags indicate how the user wants the data in question to be handled and revealed by the privacy agent. The privacy flags are:

- "always give" – give data without asking for confirmation.
- "check policies and context" – check entity profile for exception list and priority rules from policy module before giving the data. This also includes checking the context, which may or may not lead to exceptions on the access lists. This aspect of the privacy agent enables the context-awareness of our security manager.
- "ask the user" – ask the user before allowing access to the data; if user is unavailable (e.g. offline) this automatically results into automatically assuming a negative response, as this is the safest to assume from a security point of view.
- "never give" – never disclose the private data.

5.3.1.4 Security Decision Point

The main aim of CASM is to provide clearance according to policies, and this is done by the Security Decision Point. A Policy is in general a rule or rules that consist of a set of defined values, value-sets or value ranges for parameters that exist inside the security profiles. The Security Decision Point consists of set of rules

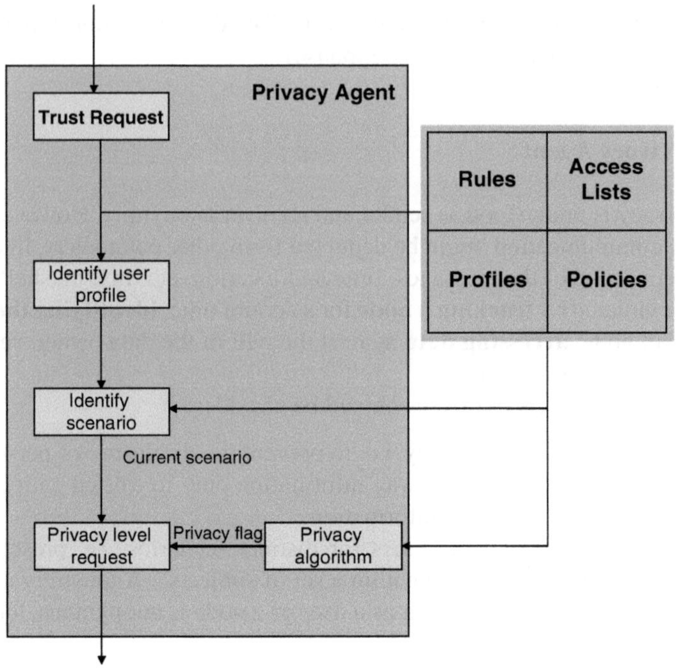

Fig. 5.7 The privacy Agent

Table 5.8 User social roles and user sensitive information to be protected

Profiles considered	Source_Role	Destination_Role	Location context Role	User sensitive data – general categories
User	Patient	Doctor	Home	User identity
Node (device)	Spouse	Family member	Office	Medical history
Situation	Family member	Boss	Hospital	Medical data
Service (application)	Employee	Friend	Car	Social security number
	Friend	Spouse	Public	Address
	User	Child	Unknown	Telephone
		Unknown		Location

and security parameters, responsible for taking policy decisions. By properly applying these policies towards authenticated requests, access control is realized. In the Security Decision Point, the profiles of node, service, application, user, situations are considered. Default profiles exist too. From them, the profiles of the user can be modified, updated and deleted. The profiles, the user roles, the locations and the user sensitive data, which is privacy protected, are presented in Table 5.8.

The "Current Source Role" is assigned to the user with a probability-based approach. Although the rules are predefined, there is always the choice for the user to change the rules for disclosing information, as well as the default security level.

The profile description holds as much of the necessary data as possible for the authorized users, services and applications that allow the system to decide in alliance with the rules and priorities in the Rules Module. In addition to the required data, each Profile Description has optional data ensuring adaptation to new context configurations. In this proposal four main profiles are considered – user, device/node, application/service, and situation (reflecting the context).

Based on their identity and role, users are categorised with User's Profiles. Users' profiles also contain data descriptions for area of interests, preferences, organisations, etc. The rightful users dispose of different template models in the setting up stage. Any update to the profile is done through the devices able to handle the update or the control unit. The templates are made with data abstractions, which the user maps to his profile. Each data abstraction is assigned type and a flag for privacy purpose.

1. User information: private and personal such as first & given names, user identity, person description, category, preference and exception lists, address and job description, etc…
2. Security attributes: policies (security, privacy and trust); the description also includes data like service secure levels lists, password, encryption keys, etc…
3. Access information: preferences and priorities, etc…
4. Node/Device profiles
5. Node/Device information: ID number, type, manufacture information, processing power, memory, battery life
6. Security attributes: encryption key, etc…

The *Situation Profile* describes the parameters of the smart context (home, office, and hospital) in terms of location description, time and date, person presence, physical parameters, devices and sensors list, communication mediums, etc…

It is the location's context, as well as the users and their presence, the devices' and the time's (date, time, season, working day vs. holiday…) contexts that are essentially the main driving force to set up security requirements in all data connections and communication (such as authentication, confidentiality, privacy…). These security requirements concern the users, the devices' linking them and the services' being accessed. The smart home (city house, residence, and country house), smart car and smart office (administrative office, doctor office) have all completely different security requirements at different time and depending on the person's presence, the context changes completely.

The data that is part of the *Service Profile* description is divided into main descriptions like Service name, type, version number, service ID number, description, and local policy information.

5.3.2 Security Profiling and Associations to the User Profile

CASM is a profile-driven module. The Profiles provide structured information about all PN and PN-F elements and conceptual entities (Users, Clearances, Roles, PNs,

Federations, Services, Nodes-Devices, SMN-Devices (devices and services also known as Resources) along with related security policies). As far as the service authorization is concerned, these policies state what rights users have for service access according to devices they use. By properly applying these policies towards authenticated requests, access control is realized. In the extended architecture, Users, Groups, Roles and PN Federations are new or modified entities in the extended architecture that have to be profiled.

User profiles hold all user attributes that – among other – determine the access rights for users:

- User information: the identity, organization, role, group membership, areas of interest, UI related information and preferences of the user.
- The pair-wise long term keys shared with other paired devices/nodes that are referenced by their unique device IDs and their group.
- Services: Information around service subscriptions, charging rates, credit limits and usage.
- Trust level: the profile holds the trust level required for each resource. Trust level is determined from the position of the user inside a well-defined trust framework e.g. hierarchy of certificates.
- PN Federation identifier, credentials, timestamps and related information.
- Clearance associations, which will grant the corresponding User appropriate rights. This also includes appropriate credentials for PN Federation status: Creator, simple or privileged member.

Clearance/Security role profiles contain:

- Role identifier, role name and UI related information.
- Credentials and timestamp certifying the Role creator and subsequently the credibility of the profile.
- The information here must be compliant according to specifications written in Chap. 4.

PN – Federation profiles:

- Federation identifier, name and UI related information.
- Information, Credential and timestamp for the creator.
- Reference to the formation policies, namely to the policies set by creator, and define the scope, goals and duration of the Federation.
- Reference to the sign-up policies, namely to the policies managed by the PN-F administrators, and describe rules that provide decisions for potential new members.
- Reference to revocation/kick-out policies, namely to the policies managed by the PN-F administrators, and describe rules that provide decisions for cancelling memberships to the Federation prematurely.
- As described in this section, all the subsequent protocols are instantiated according to the CASM decisions. We explain hereafter two major security protocols in MAGNET Beyond: the pairing protocol and the federation security part.

5.3.3 *Integrating the VID Concept*

In this section we briefly list all CASM design considerations that were implemented in order to make VID support feasible.

To begin with, CASM information modelling in the Security Profile was adapted to support the VID concept. Any PN Entity may be linked to more than one identity. A main one always exists, along with an indefinite number of alias identities. Each Entity is connected to a set of policies, security attributes and access rules. These three elements identity-entity-security attributes may or may not be linked with each other, effectively creating linkable or unlinkable identities. Namely, a user may create many instances of a PN asset structure. One instance may be connected with a virtual identity and a specific set of security attributes while another instance may connect him with his professional identity and a different, perhaps more privileged, set of rights. These two instances may or may not be linkable, according to user preference. Unlinkability is ensured as long as these instances don't include common identifiers such as IP and MAC addresses.

Given the Information Structure considerations, subsequent instances of any Entity objects are also VID compliant, leading to VID compliant application logic for CASM.

5.4 PN Key Management

With lack of permanent access to a common trusted third party in PN environments and also user unwillingness to delegate her trust to a centralized entity outside her personal territory, classical network security mechanisms based on the conventional public key infrastructure (PKI) and certificate authorities (CA) cannot be directly applied to the PN. In MAGNET phase 1, a key agreement protocol based on an authenticated Diffie-Hellman (DH) protocol, named PN Formation Protocol (PFP), was developed, which fulfils the security needs of small networks [6]. In MAGNET phase 2, we introduced Certified PN Formation Protocol (CPFP) as a new key agreement protocol based on a personal public key infrastructure (Personal PKI) [7] and Elliptic Curve Cryptography (ECC), which is scalable to larger PNs and provides an enhanced level of authentication and non-repudiation with ease of key revocation and key update.

CPFP is based on a personal public key infrastructure (Personal PKI) in which instead of global certificates issued by a trusted third party, the local certificates issued by the PN certificate authority (PNCA) will be applied. CPFP has two different stages. In the first stage, all PN devices get imprinted with the PNCA i.e., establish the PNCA signature public key as the PN root key and get a certificate on their long term Diffie-Hellman public key. In the second stage, PN nodes use their certificates to authenticate each other and establish pairwise keys based on the Elliptic Curve Menezes-Qu-Vanstone (ECMQV) [8] protocol.

The ECMQV is the elliptic curve variant of MQV [9] key establishment protocol which is incorporated in the public key standard IEEE P1363 and is based on two sets of long term (static) and ephemeral (dynamic) Elliptic Curve Diffie-Hellman (ECDH) public and private keys. As a prerequisite in ECMQV, peers should a priori possess authenticated copies of each other's long term public keys which will be done through the issued certificates within the first stage of CPFP.

5.4.1 CPFP Stage 1: Initializing and Imprinting with PNCA

PN security depends on the security of the imprinting procedure which is subject to the following assumptions:

- The user is in full control of the imprinting procedure and determines when and how new devices get imprinted with PNCA and taken as members of her PN.
- The personal devices share two different communication interfaces with PNCA including Proximity Authenticated Channel (PAC) and usual (insecure) wireless communication channel.

A proximity authenticated channel is a communication interface between two devices, which is authenticated by physical means of user. Typically, proximity authentication is performed by touching the device, or by reading from or entering to a device's interface. We distinguish between two types of PAC channels, private and public PAC channels, with respect to the level of security the PAC channel can provide. A private PAC channel provides authenticity, integrity and confidentiality, while a public PAC channels provide authenticity and integrity only.

A typical example of a private PAC channel is realized by a user, who reads an alphanumeric string from the display of one device and then enters it to the other device using the keypad. Clearly such a channel sets some limits to the length of the string that can be transferred from one device to another, e.g., typically 32–40 bits, which is feasible to be transferred using devices' user interfaces by the user. Typical realizations of public PAC are RFID tags, Infrared communication, and public displays on the devices (such as an overhead display over a cashier, printer, or network access point [10]). If the PAC is public, the protocol requires that at least 160 bits of information can be transferred over it.

The user starts CPFP by choosing one device with keypad and display as the PN certificate authority (PNCA) and imprints all PN devices with it. The PNCA initializes itself with the generation of a pair of public and private ECDSA (Elliptic Curve Digital Signature Algorithm) signature keys and other PN components initialize themselves with the generation of their long term ECDH public and private keys. The parameters are based on a fixed elliptic curve with standardized coefficients e.g. P-192 recommended by NIST.

PNCA and PN components exchange their public keys (signature and long term) over the insecure wireless channel and the user authenticates the procedure with help of the complementary PAC channel. The outcome of this stage is that the PNCA issues certificates for the long term public key of each paired component which can be at the same time verified by all PN components. Based on the used PAC, there are two different procedures for this stage of the protocol:

5.4.1.1 Imprinting Over Private PAC

In this version of the protocol (Fig. 5.8), after the public keys are exchanged over the insecure wireless channel, the PNCA generates a key K which is suitable to be used in a Message Authentication Code (MAC) function which is shared by all PN components. Using this key K, the PNCA computes a MAC of the exchanged public keys. Both the key K and the MAC value should be feasible to be transferred by the user interfaces of the devices over the private PAC (at most 8 digits). This means that the MAC value should be truncated to 4 digits. One possible way of doing it is to take the 32 least significant bits of it, turn it to an integer, and then take the 4 least significant digits of it.

There are different scenarios, depending on the types of available interfaces. For example, if the PNCA has a display and the PN device has a keypad, then the key K and the truncated MAC are displayed by the PNCA to the user who enters them into the pairing device. The pairing device uses the received key value K to compute the truncated MAC value on the exchanged public keys (received over the insecure wireless channel) on its own. In a second step, it compares the result with the entered information and shows an accepted or rejected signal (peeps or blinks a light) to the user who updates the PNCA.

As the key K is chosen randomly each time and the private PAC provides confidentiality, an attacker gets no knowledge on the key K or on the MAC from the protocol runs. Hence, the only possible attacks are to block the messages over the

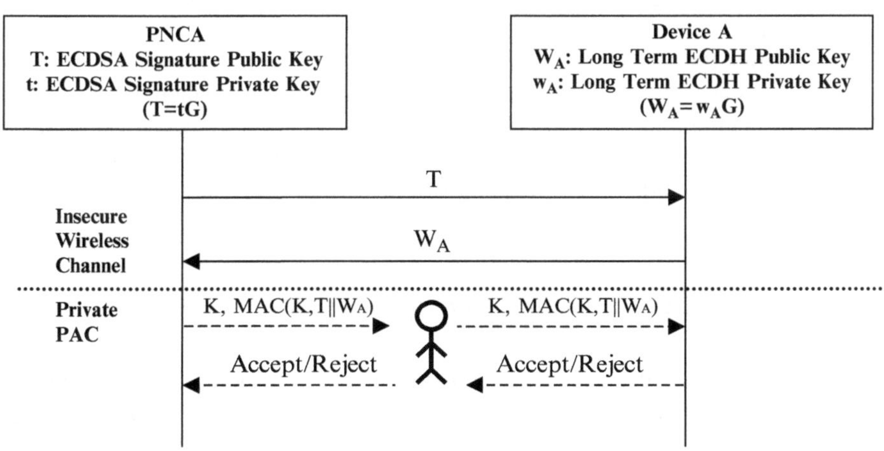

Fig. 5.8 Imprinting over Private PAC

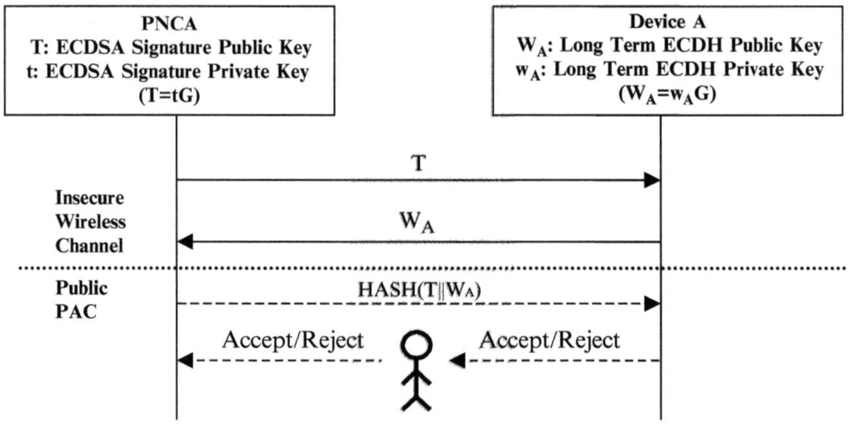

Fig. 5.9 Imprinting over Public PAC

Private PAC to prevent that the imprinting stage from finishing or to replace the key of the PN device with its own key for impersonation and to hope that the MAC value remains valid by coincidence. Assuming a message size of 8 digits, the probability for success is less than 2^{-16}.

5.4.1.2 Imprinting Over Public PAC

In this version of the protocol (Fig. 5.9), after exchanging the signature and the long term public keys over the insecure wireless channel, the PNCA generates a hash of the exchanged public keys and sends it to the pairing device over the public PAC. The pairing device calculates the hash of the exchanged public keys, compares the result with the received data over the public PAC, and shows an accepted or rejected signal to the user who updates the PNCA.

As the public PAC provides integrity and authenticity, an attacker can either again block the messages to prevent the completion of the imprinting stage or replace the PN key W_A with another key to achieve impersonation. The replacement remains only undetected if the hash value would be the same. However, if the hash function is collision resistant, this is possible only with a negligible probability.

5.4.2 CPFP Stage 1: Getting Certificates from PNCA

The use of digital certificates is an established method to generate trusted identities in network communications. A certificate provides a binding between identity information and a public key; a key pair can subsequently be used for key exchange to set up secured communications as well as for digital signatures to validate transactions. In CPFP, certificates are used to bind the user friendly identities of PN components to their long term ECDH public keys. This ensures that once the certificates are issued by the PNCA and while they are not revoked or expired, the identities and their long term ECDH public keys are trustable by all PN components.

The PN components' identities are locally chosen in our key management system and can be any unique name in the PN environment. Because of the dynamic and heterogeneous nature of the PN and also because of the distribution of PN nodes in different clusters (fixed or mobile), MAC address or IP address (main candidate for homogeneous static network) can not be used as identities in the considered scenario. On the other hand, in CPFP all PN devices get certificates on their long term ECDH public keys and rarely change them, so a hash value of these long term ECDH public keys is a good candidate for a PN identity in MAGNET. To make the recognition of different components as easy as possible for the user, she will choose a user friendly name (UFN), including the PN name and/or the owner name, for each component during the imprinting and use these UFNs as their identities.

RSA, DSA and ECDSA are three standard algorithms that are usually used for digital signatures [11]. The use of ECC-based signatures with digital certificates provides both size and performance advantages. ECC-based signatures on a certificate are smaller and faster to create; and the public key that the certificate holds is smaller too.

The process of issuing certificates by the PNCA is as follows:

- After receiving the authenticated copy of the device's long term public key (during the imprinting procedure), the PNCA asks the user for extra information which should be included in the certificates like a user friendly name (UFN) and a validity period. Based on the received information and on the device's long term public key, the PNCA constructs a message m.
- The PNCA selects an ephemeral random secret private key k from the interval $[1, n-1]$ which has an inverse modulo n.
- Then, it computes $R = kG$ with G being the generator of the used elliptic curve and converts its x-coordinate to an integer x1.
- Next, it computes $r = x1 \bmod n$. If $r = 0$, it goes back to step 2.
- Otherwise, it computes $e = h(m)$ with h being a hash function.
- Then, it calculated $s = k - 1(e + tr) \bmod n$, where t is its ECDSA signature private key. If $s = 0$, it goes back to step 2.
- Finally, it outputs the message m with its signature (r, s) as the issued certificate for the paired device.

Each PN component is equipped with the PNCA's public key during the imprinting procedure. Given a certificate m and a signature (r, s), a PN component verifies its validity by performing the following procedure:

- Verify if r and s are from the interval $[1, n-1]$. If they are not, stop and reject the signature
- Compute $e = h(m)$
- Compute $w = s\text{-}1 \bmod n$
- Compute $u1 = ew \bmod n$ and $u2 = rw \bmod n$
- Compute $R = u1G + u2T$, if $R = \infty$ reject signature

- Convert x-coordinate of R to integer x1 and calculate v = x1 mod n
- The device accepts the signature if v = r

Observe that the algorithms described above are the established ECDSA algorithm (e.g., see [12]). It is believed to be secure according to the current state of knowledge if the parameters are appropriately selected.

5.4.3 CPFP Stage 2: Using ECMQV to Drive Shared Keys

In the last stage of CPFP, the Elliptic Curve Menezes-Qu-Vanstone (ECMQV) [8] key agreement protocol is used to establish a shared secret key between PN components which have already imprinted and have got PNCA' certificates on their long term public keys. The PNCA itself participates in this stage to establish shared pairwise keys with other PN components, with issuing a self signed certificate on its long term ECDH public key.

While based on ECDH, ECMQV offers attributes – such as key-compromise impersonation resilience and unknown key-share resilience – that are not found with ECDH. ECMQV has many desirable performance attributes, including the fact that the dominant computational steps are not expensive while the protocol also has low communication overhead, is role-symmetric, non-interactive and does not use encryption or time-stamping. This makes it ideal in the development of security protocols and systems that require efficient and authenticated key agreement protocol and was chosen as a one of the three recommended key management protocols in NSA Suite B cryptographic primitives to be used to protect classified and unclassified sensitive information. For example, ECMQV is proposed for securing US Federal government communications up to the TOP SECRET classification (for more information, see [13]).

We are using a three-pass version of ECMQV [14] with the protocol messages shown by Fig. 5.10.

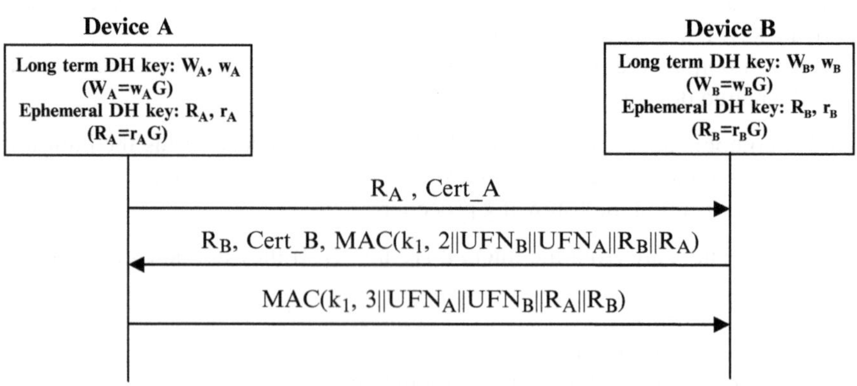

Fig. 5.10 PFP Stage 2 – Using ECMQV to derive shared keys

In this protocol:

- A generates its ephemeral (dynamic) public and private keys (rA, RA) and sends its ephemeral public key (RA) along with its long term public key certificate (Cert_A) to B.
- Upon receipt the first message, B does the following:
  - Performs an embedded public key validation of R_A to verify it possesses certain arithmetic properties.
  - Generates its ephemeral public and private keys (r_B, R_B).
  - Computes an implicit signature "$s_B = (r_B + \check{R}_B w_B) \bmod n$" and a shared key "$K = hs_B(R_A + \check{R}_A W_A)$" and verifies that $K \neq \infty$ ($\check{R}_B$ and $\check{R}_A$ are the first "$L = [((\log_2 n) + 1)/2]$" bits of the first component of the point R_B and R_A).
  - Using shared key derivation function (KDF), B derives k_1 and k_2 from the x-coordinate of the shared key K.
  - Compute $MAC(k_1, 2||UFN_B||UFN_A||R_B||R_A)$ and send the result along with its ephemeral public key R_B and its long term public key certificate Cert_B to A.
- With receiving the second message, A does the following:
  - Perform an embedded public key validation of R_B to verify it possesses certain arithmetic properties.
  - Compute an implicit signature "$s_A = (r_A + \check{R}_A w_A) \bmod n$" and a shared key "$K = hs_A(R_B + \check{R}_B W_B)$" and verify that $K \neq \infty$.
  - Using shared key derivation function (KDF), derive k_1 and k_2 from the x-coordinate of the shared key K.
  - Compute $MAC(k_1, 2||UFN_B||UFN_A||R_B||R_A)$ and verify it based on the received message 2.
  - Compute $MAC(k_1, 3||UFN_A||UFN_B||R_A||R_B)$ and send the result to B.
  - B computes $MAC(k_1, 3||UFN_A||UFN_B||R_A||R_B)$ and verifies it based on the message 3. The session key is k_2.

5.4.4 Key Revocation Mechanism

Like a certificate authority in a normal PKI, the PNCA is not only in charge of inviting nodes into the PN but also to revoke them in the case of need. Since the user is the centre of the PN architecture, only the user herself should be able to decide whether a node has to be revoked or not. In practice, we envision the following procedure from a user's point of view to revoke one node.

Whenever the user logs into one PNCA device, he can choose to have a list of the currently valid PN members displayed. Given the list of current nodes, a user can select one or several devices and choose the REVOKE option to revoke these nodes.

When the revocation procedure is initiated, the actually used PNCA updates the Certificate Revocation List (CRL). The CRL is a file which contains all necessary information on the nodes that need to be revoked. This information include at least:

- PN's node identifier
- A time stamp and/or a CRL version number
- Serial number identification of the revoked certificate
- A code implying the reason of revocation

PNCA keeps a record of revoked certificates in a CRL up to their expiry date (each certificate has a specified expiry date). Each CRL has either a version number and/or a time stamp. With every revocation procedure, the PNCA updates the CRL and changes its version number and/or refreshes the time stamp. The CRL is signed with the private key of the PNCA (SK_{PNCA}) to ensure the non-repudiation, integrity, and message authenticity. As the revocation list is signed, each node can check its validity with the public key of the PNCA (PK_{PNCA}) obtained at imprinting time.

The new CRL is either distributed whenever a new revocation has happened and/or periodically (even if nothing has changed except the version number/time stamp). Nodes keep a record of the CRL locally, update it with revocation messages, and check its version with other communicating peers. If a node does not have the latest version of the CRL (or if it is overdue), it will update it.

It goes without saying that the CRL has only its value if it is ensured that every node has at any point in time the actual version. If two nodes exchange data, both must be sure that the other one has not been revoked since the last time they communicated. Thus, we envision that each node checks the current version of the CRL (either stored locally or retrieved from appropriate places) before a new communication starts (or at least in regular time intervals). This requires the following functionalities in the context of the CRL:

- The updated CRL can be distributed within the whole PN in a reliable way.
- The current version of the CRL can be provided upon request.

This can be realized in several ways. One possibility is to make use of the existing upper layer facilities in MAGNET, e.g., by using the Secure Context Management Framework (SCMF) [15] which is able to distribute and provide data on demand or adding an extra service to a MAGNET PN, a kind of 'revocation list service' which is discoverable through the MAGNET Service Management Platform (MSMP) [15]. The other approach is going for an ad-hoc CRL distribution scheme, where PN nodes ask each other for the latest version of the CRL and in case of difference, both nodes update to the latest CRL version.

5.4.5 PNCA Resilience

The fact that the PNCA plays a central role in the PN's key management brings the problem of resilience. If the PNCA is broken or out of reach, the basic operations as inviting new devices and revoking keys should not be abandoned.

In the currently discussed approach, we use the fact that in principle the difference between the PNCA and an ordinary PN node is that the PNCA knows the secret key SK_{PNCA} that is mandatory for the operations mentioned above. This means that PNCA is rather a functionality than a certain device and if other devices share its knowledge of SK_{PNCA}, they can take over its functionality if necessary. Therefore, we propose to store SK_{PNCA} on different devices on several, strategically well-chosen locations. Each of these devices can act as the PNCA in the case of need, e.g., if the previous PNCA is unreachable or broken. As a device acting as PNCA has in principle full control over the PN as it can invite or revoke devices, it is of utmost importance that SK_{PNCA} is stored only in encrypted form to prevent an attacker to take over control of the PN if she steals a PNCA. If the value SK_{PNCA} is only protected by a key chosen by the user, e.g., derived from a password, this requirement is unfortunately most probably not fulfilled. Observations have showed that humans rather tend to choose insecure password which would compromise the security of the whole PN. A possible countermeasure could be to force the user to choose a strong password by refusing weak ones. Alternatively, one could imagine that the encryption is additionally protected by a piece of hardware. As it is already common practice for mobile phones, one could require the usage of a smart card together with a password to decrypt SK_{PNCA}.

Observe that decrypting SK_{PNCA} is only necessary once in the beginning of an epoch to turn a device into the PNCA. As long as the same device keeps this functionality, no user interaction is required in this point. After this epoch, the unencrypted SK_{PNCA} need to be erased from the memory so that only the encryption of SK_{PNCA} remains. At this time, the device looses its "superior knowledge" and becomes an ordinary node again.

Of course, knowing SK_{PNCA} is only half of the battle. It is likewise required that the PNCA has an actual list of PN members and revoked nodes. Therefore, the "old" PNCA and the "new" PNCA have to synchronize their lists to provide full functionality. If synchronization cannot be handled by the PN itself, one could think of storing this information on a portable medium like an SD card, possibly encrypted as well. Thus, at the end of one epoch, when a device looses its PNCA role, it stores the current data on the medium. The device which becomes the next PNCA should have access to this medium to restore the actual data.

5.5 PN-F Key Management

Based on how the cooperation between the devices in different PNs is realized, we can distinguish between two general type of infrastructure and ad hoc based federations. While in ad hoc based PN federation, our trust establishment is based on direct users' involvement, in infrastructure based federations our solutions involve high level of trust relationship with a central entity which acts as the trusted third party (TTP) for all the PN participants. In this regard, we will define our solution for key establishment in infrastructure and ad hoc based federations and clarify how the authentication and access control process can be done in each case. In description of the protocols, we are using the Table 5.9 notations.

Table 5.9 Notations

Symbol	Meaning
E(k, m)	Symmetric encryption of data m with key k
$S_X(m)$	Signature of data m using X's private key (assumed that the signature scheme does not provide message recovery, e.g. RSA signature by hashing input)
$P_X(m)$	Public key encryption of data m using X's public key
$Cert_X$	Certificate binding X's identity to its public key (suitable for both encryption and signature verification)
PK_X	X's Public Key
SK_X	X's Private Key
X	X's Identity
MAC(k, X)	Keyed hash of X with key k
HASH(X)	Hash function of X
r_X	Fresh random number generated by X
||	Concatenation
- - - - - - - - -	Secure channel
–.–.–.–.	Insecure (wireless) channel
———	Proximity Authenticated Channel (PAC)

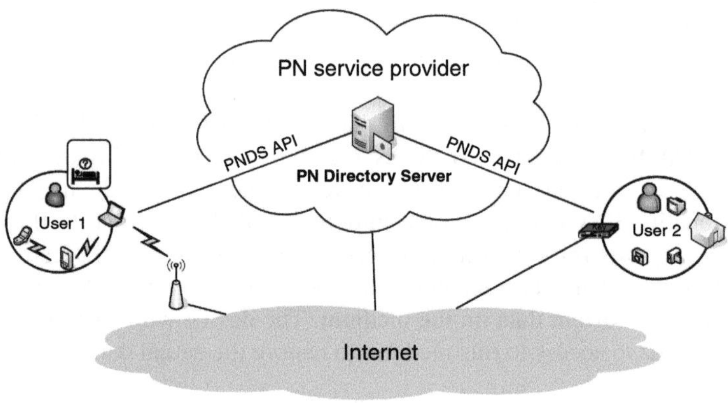

Fig. 5.11 High level PNDS view [15]

5.5.1 *PN-F Key Management in Infrastructure Based PN Federations*

MAGNET Beyond project has studied the PN federation architecture in details and has specified a central entity named PN directory service (PNDS) in infrastructure based PN federations. PNDS embraces a business model where users register their PN to a PN service provider and establish their federations via it (Fig. 5.11). PNDS plays a central role in infrastructure based federations with acting as a directory service for federations.

Based on how the PN-F creator or PN-F participants publish their federation(s) or announce their willingness to participate in federations, there is two mode of publish based and invitation based PN federation. In publish mode, the creators uses the PNDS to advert their PN-F and candidate participants use it to look-up for the adverted PN-Fs. In the Invitation based, in contrast, the candidate PN-F participants announce their willingness of participation to PNDS and PN-F creators browse and inquire it to find and invite the interested participants to their PN federations [16].

Our key management solution in infrastructure based federations is based on existence of a high level trust relationship with the PNCA as the common trusted third party (TTP) for all the PN-F participants. Without loosing any generality, the proposed solutions can be also easily applied to more general case of hierarchical certificate authorities (CAs) of a common public key infrastructure (PKI). In our solution, all the PN-F participants, before participating in any federation, authenticate themselves with the PNDS and get certificates binding their identities to public keys suitable for both encryption and signature verification.

From the security point of view, both cases of publish and invitation mode impose the same requirements on trust establishment and key management solutions and so we exemplify our solution in publish mode. As a result of publish stage, PN-F candidate participant knows PN-F creator and can send it a PN-F joining request. PN-F candidate participant and PN-F creator authenticate each other based on their PNDS certificates and establish a secure channel, over which the participant sends its PN-F Participation Profile, mainly consisting on the resources that it makes available to this federation. The Creator then checks whether the Candidate fulfils the federation policies and if this is the case, the private part of the PN-F Profile, consisting on the complete list of PN-F members and a group key is securely forwarded to the new PN-F member. The secure channel should ensure that no adversary can get the group key or replay previously captured messages.

The shared group key allows a light authentication and trust establishment between the PN-F participants, but it does not provide an individual member authentication by its own. If PN-F application mandates the individual member authentication, in extra to group key, the PNDS certificate and PN-F profile (including the members list), can be used to provide mutual authentication. The PN-F creator can also optionally issues PN-F certificates for all the PN-F participants, which can be used in mutual authentication and security association. In this case, the PN-F members have a certificate issued by PN-F creator as the trusted CA of their common PN-F and use it to prove their membership and establish security association with other PN-F participants.

Figure 5.12 depicts a typical example of authentication and security establishment in publish mode of infrastructure based PN federation, the details is as follow:

- PN-F creator and participant authenticate with PNDS and get certificates on their public keys. It can be done through the normal and complicated PKI methods or other simple and more usable ways (it has been shown by secure channel in figure). In the current implementation of PNDS in project, PN-F creator and participants authenticate the PNDS based on its certificate (which is preloaded in

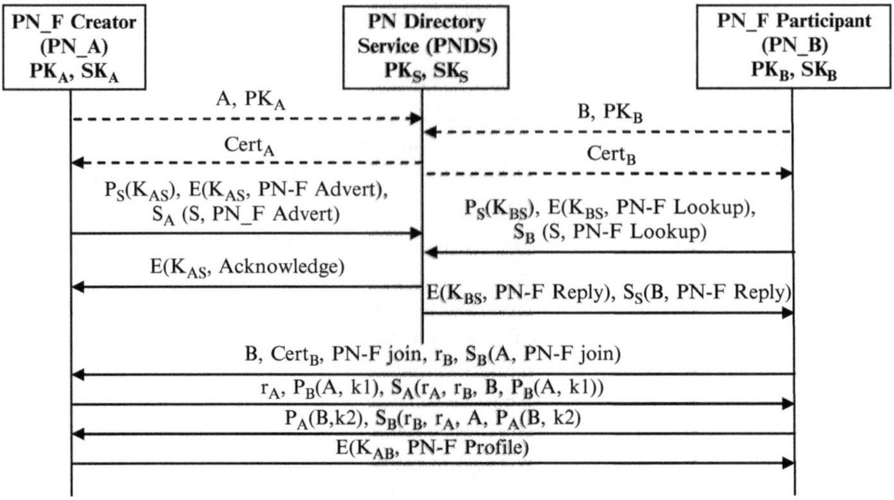

Fig. 5.12 Infrastructure based PN federation

their terminals), and the PNDS authenticates them through a kind of two factor authentication based on their User ID and Password. The User ID is set by user and includes a mobile phone number which is used to receive the generated Password by PNDS in SMS (Short Message Service) format [17].

- Creator uses the PNDS public key to establish a symmetric session key with it and publishes its PN-F in secure way (encrypted by established symmetric key and signed by its private key). In publishing the PN-F, creator can limit its visibility to authorized participants by specifying the credentials that participants should have.
- Candidate participant uses the PNDS public key to establish a symmetric session key with it and enquires the PNDS for the available federations. The PNDS controls the discoverable federations based on the participants' credentials and replies with primary information, including PN-F creator names and their point of contacts and certificates, for the visible registered federations. All the communications are encrypted by established symmetric key and signed by the private keys.
- PN-F creator and participant authenticate each others based on their PNDS certificates and proceed to establish a secure connection between them over the main wireless link. To this end, they can use any established public-key-based key exchange protocol which requires them to prove possession of a particular private key such as Secure Socket Layer (SSL) or Transport Layer Security (TLS). As an example, figure shows how they can establish a pair-wise symmetric key through a three-pass challenge-response protocol as follow [18]:
- Candidate participant sends the PN-F creator a signed PN-F joining request, its certificate and a fresh random number rB.

- PN-F creator verifies the authenticity of participant' certificate; extracts its public key and verifies its signature on joining request. Creator then generates a fresh random number rA and a symmetric key k1; encrypts k1 with participant' public key and send the result along with rA and its signature on rA, rB, participant identity and encrypted k1 to the participant.
- Participant decrypts key k1 and verifies creator' signature. Participant then generates symmetric key k2; encrypts k2 with creator public key and sends the result along with its signature on rA, rB, creator identity and encrypted k2 to the creator. Participant uses a known key derivation function (KDF) on both key k1 and k2 to drive the shared symmetric key KAB.
- PN-F creator decrypts key k2, verifies participant' signature and uses the same KDF on both key k1 and k2 to drive the shared symmetric key K_{AB}.
- Creator sends the PN-F profile, encrypted by the shared symmetric key, to participant. PN-F profile includes an updated list of current joined members and a group key (PN-F key).

5.5.2 PN-F Key Management in Ad Hoc Based Federations

In contrary to infrastructure based federations, the dynamic nature of ad hoc based federations does not guarantee that a trusted third party will always be available for the trust establishment and authentication between the PN-F members. In the other hand, key pre-distribution schemes are not also generally applicable, since all the participants within the PN-F may not be known a priori. Based on these facts, our key management solution in ad hoc based federations is based on direct users' involvement and using extra proximity authenticated channel (PAC) in authenticating the exchanged keys.

Similar to infrastructure based federations, there is also two kind of publish and invitation modes in ad hoc based federations. While in invitation mode, the PN-F creator sends its invitation with the public part of PN-F profile to the known participants (known from their adverts for participation or known by neighbor discovery mechanisms), in publish mode, PN-F creators publish their federation(s) by broadcasting their adverts including public part of PN-F profile. The key management and trust establishment in both cases is similar and we concentrate our discussion on invitation based federations which is depicted by Fig. 5.13 and includes the following stages:

- Creator invites the candidate participants by sending them signed invitation messages including the public part of PN-F profile and its public key.
- Candidate participants study the received PN-F profile, verify the signature and send back a signed Join_Request with their public keys (if they are interested in invited federation).
- PN-F Creator and participant authenticate the exchanged public keys over the private or public PAC as follow:

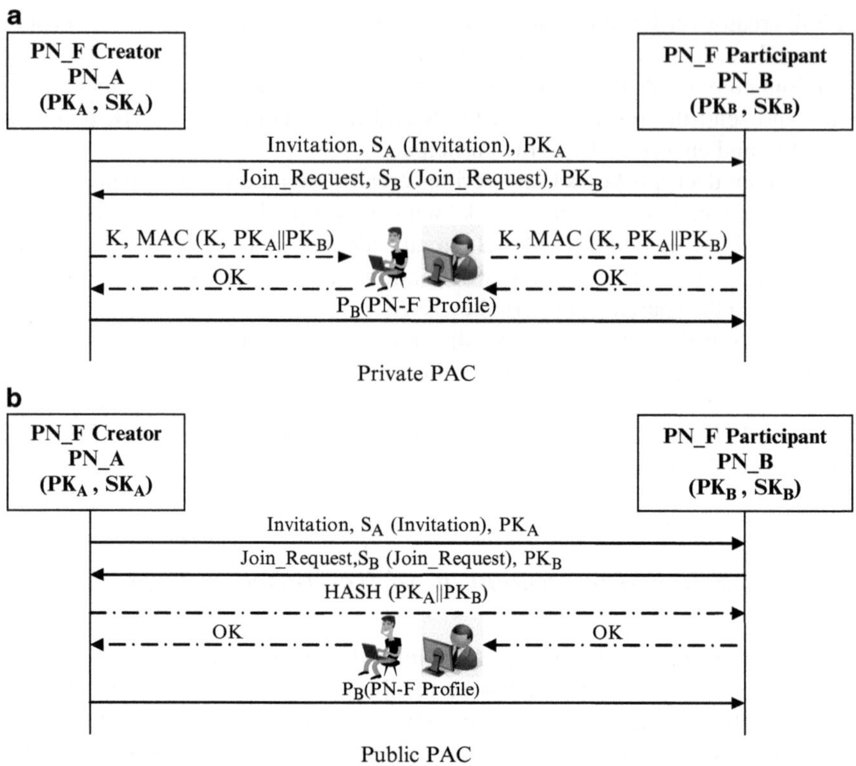

Fig. 5.13 Ad hoc based PN federation

- Private PAC (Fig. 5.13a): Creator generates a key K and computes a keyed hash on both public keys using the key K, and shows the key and MAC result in truncated form to its user. Candidate participant's user enters the result in its federation manager which uses the key to compute the similar keyed hash and verifies authenticity of the received public keys and updates the creator about the outcome.
- Public PAC (Fig. 5.13b): Creator generates a hash of both public keys and sends it to participant over the public PAC. The participant calculates the similar hash, compares the result and updates the creator about the outcome.

- After authentication, creator sends an encrypted copy of the PN-F profile to each participant which includes the PN-F key or PN-F certificate issued by creator. PN-F participants use the PN-F key as the group key or the PN-F certificates for authentication and secure communication in PN-F.

5.5.3 Security Association Between the PN-F Members

The security association between the PN-F members can be established using shared PN-F key, PN-F certificates and PNDS certificate plus creator-signed PN-F member list. In the first case, all the PN-F members share PN-F key as the group key and use it to prove their PN-F membership. In the second case, all PN-F members have a PN-F certificate on their public key, issued by creator as the PN-F common root certificate authority (CA) and use their private keys to prove they are owners of such that certificates. In the last one, PN-F participants use their PNDS certificate and creator-signed member list (part of PN-F profile) as a proof for their PN-F membership. In this section we investigate these mechanisms in establishing security association between the PN-F members and discuss pros and cons of each method.

5.5.3.1 PN-F Key Based Security Association

In this method, PN-F creator sends all the PN-F participants a shared PN-F key as a part of PN_F profile which will be used in authentication and security association establishment between the PN-F participants. PN-F participants authenticate each others by showing their knowledge of shared PN-F key and use the PN-F key for secure communication within the PN-F. Figure 5.14 shows a typical challenge-response protocol that can be used in PN-F participants' authentication as follow:

- Participant B generates a random number rB and send it to participant A.
- Participant A generates a random number rA and sends back a shared PN-F key encrypted version of both random number and participant B identity to participant B.
- Participant B decrypts the packet, verifies its random number and sends back a shared PN-F key encrypted version of both random number to participant A.
- Participant A decrypts the packet and verifies its random number.

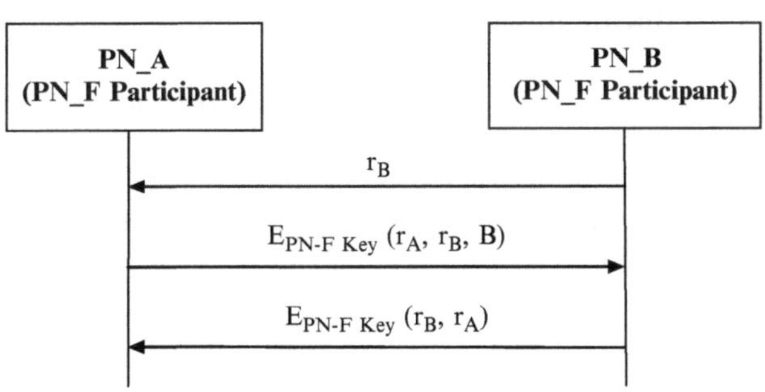

Fig. 5.14 PN-F key based security association

The shared group key does not need any asymmetric cryptography or storage for different key or certificates and allows a light authentication and trust establishment between the PN-F participants, but it does not provide an individual member authentication. When there is a need to expel a participant, a new group key should be used; in this case, the creator shall update the group key of all remaining members and remove the revoked PN from the member list in the PN-F database.

5.5.3.2 PN-F Certificate Based Security Association

In this solution, PN-F creator acts as a certificate authority (CA) for its PN-F and issues PN-F certificate for all its participants. It means that every PN-F member get a PN-F certificate on their authenticated public key (authenticated via proximity authenticated channels in ad hoc based federation or via PNCA certificate in infrastructure based federation) which is valid for that PN-F and include PN-F′ ID, member's identity, issue date, validity and also shows whether the member can invite new member to that federation or no (creator's right delegation).

The PN-F certificates are used by PN-F participants as proof of membership in PN-F by proving that certificates stem from the same root and participants poses the respective private key. PN-F members also use the certificate in authentication and security association with each other by using any established public-key-based key exchange protocol such as Secure Socket Layer (SSL) or Transport Layer Security (TLS).

PN-F certificate allows individual authentication in price of higher processing. It shall be also possible to revoke a PN from the federation. In case of PN-F certificates, a certification revocation list (CRL) will be used, which will contain all revoked certificates that are still within their validity period. Members will make sure that they are aware of the revoked certificates. The PN-F profile shall define how this CRL will be made known to the members. An obvious option is to retrieve the CRL at the federation manager of the PN-F creator.

5.5.3.3 PNDS Certificate Based Security Association

In infrastructure based federations, when each member has a PNCA certificate, PNCA certificates and creator-signed PN-F member list which is part of PN-F profile can be used to prove PN-F membership. In this case, PN-F participants' mutual authentication and security association establishment is based on PNCA certificate and will be done through the standard public-key-based key exchange protocol such as SSL or TLS.

5.5.4 Security Evaluation of PN-F Key Management Protocols

Our PN-F key management solution in infrastructure based federation uses the established cryptographic algorithms which according to the current state of knowledge are secured. In ad hoc based federations, our protocols based on using the PAC in authenticating the exchanged public keys, are such designed to be usable (e.g. just entering 4 digits which is easy even for non-technical people) and secure against any man-in-the-middle (MITM) attack.

In private PAC scenario, as the key K is chosen randomly each time and the private PAC provides confidentiality, an attacker gets no knowledge on the key K or on the MAC from the protocol runs. Hence, the only possible attacks are to block the messages over the Private PAC to prevent that the imprinting stage from finishing or to replace PN device' key with its own key for impersonation and to hope that the MAC value remains valid by coincidence. Assuming a message size of 8 digits, the probability of success in that case is less than 2^{-16} which is an accepted target for our protocol.

In public PAC scenario, as the public PAC provides integrity and authenticity, an attacker can either again block the messages to prevent the completion of the imprinting stage or replace either of public keys with another key to achieve impersonation. The replacement remains only undetected if the hash value would be the same. However, as our hash function is collision resistant and its output is at least 160 bits, this is possible only with a negligible probability.

5.6 Conclusions

Security has to be considered as a service to users and hence has to be tailored in a top down approach. After all, the users themselves have been blamed to be the "weakest links" in the security chain [18].

Towards this direction, MAGNET and MAGNET Beyond projects have provided some new concepts:

- A user-centric approach for threat analysis methodology and security evaluation
- A security manager safeguarding transparently the user's privacy and enforcing user security preferences across personal overlay networks

Morover, innovative authentication protocols and key management techniques have been provided to meet the specific needs of personal networking.

We hope that these contributions will help in improving the global usability of security in the future.

References

1. S. Gritzalis, T. Karygiannis, C. Skianis (eds.), *Security and Privacy in Mobile and Wireless Networking* (Troubador Publishing, UK, Mar 2009)
2. A. Stango, N. Prasad, J.J. Pallares, Analysis, Verification and Evaluation, IST MAGNET deliverable D4.4.2, June 2008
3. A. Stango, D.M. Kyriazanos, N. Prasad, A threat analysis methodology for security evaluation and enhancement planning. Securware, Athens, June 2009
4. N.R. Prasad, Threat model framework and methodology for personal network. Communication Systems Software and Middleware, COMSWARE 2007
5. D.M. Kyriazanos et al., Specification of user profile, identity and role management for PNs and integration to the PN platform. IST MAGNET B deliverable D4.3.2, Mar 2007
6. S. Mirzadeh et al., Final version of the Network-Level Security Architecture Specification. IST MAGNET deliverable D4.3.2, Feb 2005
7. C.J. Mitchell, R. Schaffelhofer. The personal PKI, in *Security for Mobility*, ed. by C.J. Mitchell (IEE, London, UK, 2004), Chapter 3, pp. 35–61
8. L. Law, A. Menezes, M. Qu, J. Solinas, S. Vanstone, An efficient protocol for authenticated key agreement. Designs Codes Cryptogr. **28**(2), 119–134 (2003)
9. http://en.wikipedia.org/wiki/MQV
10. D. Balfanz, D.K. Smetters, P. Stewart, H. Chi Wong. Talking to strangers: Authentication in ad-hoc wireless networks. Technical Report, Xerox Palo Alto Research Center, Palo Alto, 2002
11. NIST FIPS PUB 186–2, DIGITAL SIGNATURE STANDARD (DSS), Jan 2000
12. D. Johnson, A. Menezes, S. Vanstone, The elliptic curve digital signature algorithm (ECDSA), Int. J. Inf. Secur. **1**(1), 36–63 (2001) – Springer
13. Fact Sheet NSA Suite B Cryptography, http://www.nsa.gov/ia/industry/crypto_suite_b.cfm?MenuID=10.2.7
14. D. Hankerson, A. Menezes, S. Vanstone, *Guide to Elliptic Curve Cryptography* (Springer-Verlag, New York, 2004)
15. M. Jacobsson et al., Specification of PN networking and security components. IST-MAGNET Beyond deliverable D2.3.1, Dec 2006
16. J. Hoebeke, G. Holderbeke, I. Moerman, M. Jacobsson, V. Prasad, C. Wangi, I. Niemegeers, S. Heemstra de Groot, Personal network federations, in *Proceedings of the IST Mobile Summit 2006*, Myconos, Greece, June 2006
17. M. Alutoin, S. Lehtonen, K. Ahola, J. Paananen, Personal network directory service, Telektronikk **103**(1), 85–92 (2007)
18. A.J. Menezes, P.C.V. Oorschot, S.A. Vanstone, *Handbook of Applied Cryptography* (CRC Press, Boca Raton, FL, 1996), ISBN: 0-8493-8523-7

Chapter 6
Link Level Prototypes

Dominique Noguet, Gerrit van Veenendaal, Jan Mikkelsen, Lionel Biard, Marco Detratti, Balamuralidhar P., Deepak Dasalukunte, John Gerrits, Manuel Lobeira, Jaouhar Ayadi, Tian Tong, Marc Laugeois, Yunzhi Dong, Yi Zhao, and Hamid Bonakdar

6.1 Introduction

Chapter 4 described the design and selection of the short range communication air interfaces tailored to Personal Networks application. These air interfaces consist of a low data rate (LDR) FM-UWB system and a high data rate (HDR) MC-SS system. The present chapter focuses on the hardware and software design and implementation that was carried out to prove the aforementioned concepts and to assess the performance of these air interfaces, taking into account all the impairments coming from a real hardware implementation, as well as the impact of real usage conditions.

The LDR FM-UWB system has been derived into two platforms operating at 4 and 7.25 GHz respectively. Specific chipsets have been designed and implemented in order to show the low power potential the FM-UWB. These chips are described as well as other main features of the LDR system such as low power channel coding and IEEE 802.15.4 compliant MAC implementation.

The High Data Rate MC-SS system operates in the 5.2 GHz ISM (Industrial, Scientific and Medical) band and is implemented on top of a state of the art off-the-shelf chipset. As explained in Chap. 4, the MC-SS system is a very versatile

D. Noguet (✉), L. Biard, and M. Laugeois
Commissariat à l'Energie Atomique, rue des Martyrs 17, Grenoble Cedex 9 38054, France
e-mail: dominique.noguet@cea.fr

G. van Veenendaal, Y. Dong, Y. Zhao, and H. Bonakdar
NXP Semiconductors Netherlands B.V, The Netherlands

J. Mikkelsen and T. Tong
Aalborg University, Denmark

M. Detratti and M. Lobeira
Advanced Communications Research and Development S.A

P. Balamuralidhar
Tata Consultancy Service, India

D. Dasalukunte
Lund University, Sweden

J. Gerrits and J. Ayadi
Centre Suisse d'Electronique et de Microtechnique – Recherche et Development SA, Switzerland

R. Prasad (ed.), *My Personal Adaptive Global NET (MAGNET)*, Signals and Communication Technology, DOI 10.1007/978-90-481-3437-3_6,

air interface, providing modulation and coding flexibility. In the present chapter, a specific emphasis is drawn on the features that have been designed to achieve flexibility in hardware.

6.2 Low Data Rate FM-UWB Prototype

FM-Ultra WideBand (FM-UWB) radio is an analogue implementation of a spread-spectrum system that targets short range (i.e., 1–15 m) applications requiring bit-rates up to 100 kbps. It offers the advantages of a simple and robust modulation scheme and a low complexity circuit implementation. Thanks to this low complexity, FM-UWB radio represents a low-power and low-cost solution for robust UWB communications. Besides, considering its flat power spectral density and steep spectral roll-off, the FM-UWB scheme optimizes exploitation of the available spectral mask approved for UWB transmission [1]. This section presents the implementation and performance analysis of the FM-UWB system which enable to precisely quantify the previous features.

6.2.1 General Architecture

Figure 6.1 presents the FM-UWB inner transceiver architecture overview. In the transmitter (upper branch), a triangular frequency-shift keyed (FSK) sub-carrier signal generated by Direct Digital Synthesis (DDS) modulates an RF VCO thereby yielding a constant-envelope FM-UWB signal. In the lower branch of the picture, the FM-UWB receiver comprises a preamplifier (hereafter also referred to as Low Noise Amplifier, LNA), a wideband FM demodulator, a direct-conversion low-frequency sub-carrier filtering, an amplifier and an FSK demodulator circuitry.

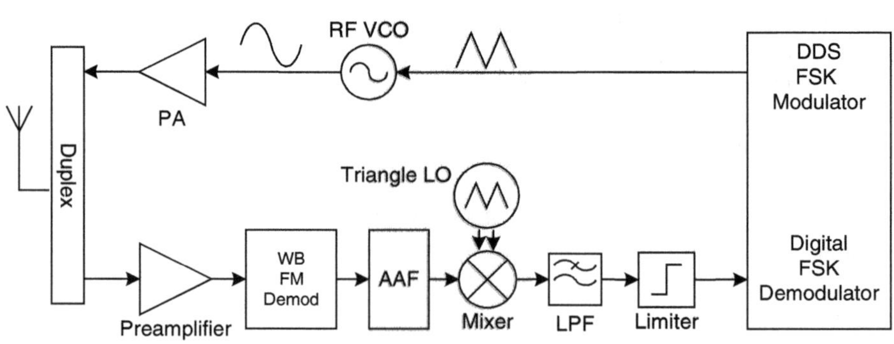

Fig. 6.1 FM-UWB radio transceiver architecture

The modulation mechanism in FM-UWB (Fig. 6.2) is double FM: narrowband FSK ($\beta_{SUB} \approx 1$) followed by high modulation index ($\beta_{RF} > 100$) analogue FM with a deviation $\Delta f > 250$ MHz. The approximate bandwidth B_{SUB} of the FSK subcarrier signal m(t) modulated by lowpass filtered digital data at bit rate R [bit/s] equals

$$B_{SUB} = R(\beta_{SUB} + 1)$$

The bandwidth of the FM-UWB signal V(t) with modulation frequency f_{SUB} is well approximated by Carson's rule

$$B_{RF} \approx 2(\beta_{RF} + 1)f_{SUB} = 2(\Delta f + f_{SUB})$$

Figure 6.3 shows the data d(t), the subcarrier m(t) and the UWB V(t) signals in the time domain for a data transition at t = 0 and subcarrier frequency of 1 MHz; the centre frequency of the UWB signal V(t) was chosen to be 10 MHz for the sake of clarity.

Since the transmitter uses double FM, the receiver needs to perform two FM demodulations; one at RF and another one at the subcarrier frequencies.

Fig. 6.2 UWB transmitter block diagram

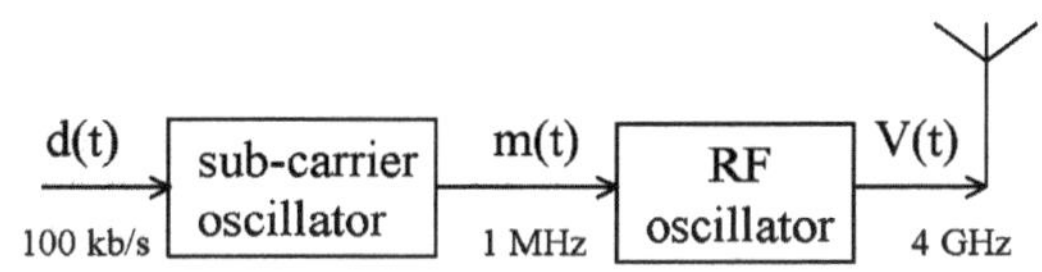

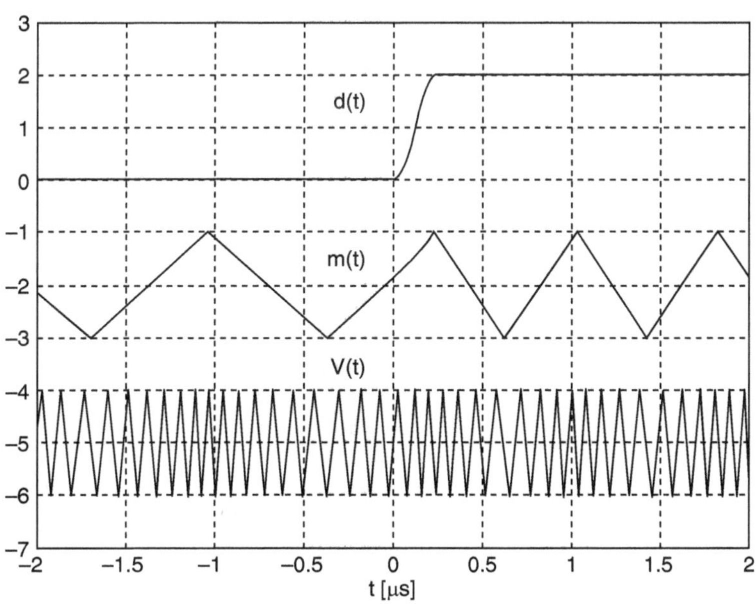

Fig. 6.3 Time domain view of data d(t), subcarrier m(t) and UWB signal V(t)

In the simplest and most low power receiver architecture, the FM-UWB signal is demodulated without frequency translation. No local oscillator and no carrier synchronisation are required which leads to a low complexity implementation of the receiver. A more detailed description can be found in Chapter 4.

6.2.2 Key Specifications

The first band that was targeted by UWB systems was the below 5 GHz band, hereafter referred to as the low band. Table 6.1 presents specifications of the low-band FM-UWB air interface.

Coexistence with legacy systems has been a major concern for UWB systems. As an example, the UWB systems have to guarantee operation without interfering with WIMAX operated at 3.5 GHz. To enable unlicensed usage in the low band, regulators have demanded the UWB systems to implement Low Duty Cycle (LDC) or Detect And Avoid (DAA) schemes.

As the regulatory policy evolved, new opportunities for UWB system emerged. More precisely, it became clear that the band between 7.25 and 8 MHz was well suited for UWB operation worldwide as it is not required to implement Low Duty Cycle (LDC) or Detect And Avoid (DAA) schemes. Hence this band, hereafter referred to as high band is more suited to low power implementation.

Moving to the high band translates the specification of the transceiver. The FM-UWB specifications for the high band specification are shown in Table 6.2.

As illustrated in Table 6.3, four 500 MHz wide channels, spaced 512 MHz apart have been defined for the high band FM-UWB system, with channel H3 being the privileged option.

The low band and high band system specifications mainly defer from their RF specifications. Baseband processing and Medium Access Control (MAC) are common to both systems. As far as the MAC is concerned, an IEEE 802.15 compliant MAC was considered. Thus, the rest of this section is organised as follows: first

Table 6.1 FM-UWB low band system specifications

RF centre frequency	3.35–4.75 GHz
RF bandwidth	500 MHz to 2 GHz
RF output power	>14 dBm
Antenna gain	0 dBi
Sub-carrier frequency	1–2 MHz
Sub-carrier modulation	FSK, $\beta = 1$
Raw bit rate	$\leq$125 kbps
Receiver sensitivity	<80 dBm
TX, RX switching time	$\leq$10 μs
Latency (at PHY level)	<1 ms@ 100 kbps
RX synchronization time	<50 bits
Current consumption RX	<7 mA
Current consumption TX	<4 mA

Table 6.2 FM-UWB high band specifications

RF centre frequency	6.4–8.7 GHz
RF bandwidth	500 MHz
RF output power	Greater than −14 dBm
Subcarrier frequency	1–2 MHz
Subcarrier modulation	FSK, $\beta = 1$
Raw bit rate	≤125 kbps
Receiver sensitivity	Less than −83 dBm
TX, RX switching time	≤10 μs
Latency (at PHY level)	<1 ms@ 100 kbps
RX synchronisation time	<50 bits
Power consumption RX	<12 mW
Power consumption TX	<4 mW
Transmitter output power	Greater than −6 dBm

Table 6.3 High band channel centre frequencies

Channel number	Centre frequency (MHz)
H1	6,464
H2	6,976
H3	7,488
H4	8,000

we address RF chipset targeted to the low band, then to the high band; secondly we discuss the parts of the system that are common to both systems, namely the base-band and the MAC. Finally, the system prototypes and performance are discussed.

6.2.3 *Low Power RF Chipsets for Low Band FM UWB*

6.2.3.1 RF Low Band Transmitter

As this was explained in the previous section, the FM-UWB communication system exploits a double FM modulation where a narrowband FSK part generates a triangular sub-carrier signal followed by wideband analogue FM (see Fig. 6.2).

The generation of the sub-carrier signal d(t) can be advantageously done in the digital domain using DDS techniques. The wideband FM modulation is implemented as open loop modulation on the RF VCO as shown in Fig. 6.4. The output amplifier (OA) drives the antenna and also isolates the antenna from the RF VCO.

The frequency synthesizer is only powered up during calibration of the VCO or when the VCO changes for a different carrier frequency. Calibration data (typically the RF VCO tuning curve) is stored in a digital memory and next used to generate the appropriate bias voltage. In this way the RF centre frequency (F_C) and the FM-UWB signal bandwidth are well controlled avoiding, for instance, out-of-band operation. Due to the low duty cycle of its use, the frequency synthesizer does not represent a critical block from a power consumption point of view. However an IC implementation of the complete PLL is clearly advantageous for reducing both the

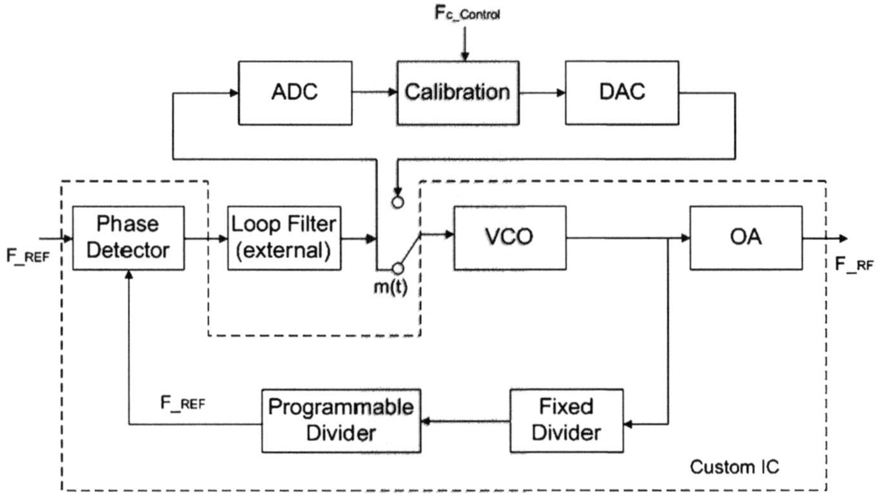

Fig. 6.4 Block diagram of the RF signal Generation

overall component count and also for reasons of size and cost. The RF VCO and the output amplifier blocks are always turned on during transmission and are specific to the FM-UWB implementation. Due to the ultra wide band, relaxed phase noise and output power requirements, these blocks are not available as low-power and low cost off-the-shelf components, so a custom IC implementation in appropriate IC technologies was required. Given the low power and low cost requirements, CMOS technology is probably the best choice for this block. It also offers the possibility of integrating analogue and digital functions into the same chip and could become advantageous for the integration of the complete system on a chip.

The most critical part of the transmitter chain is the RF VCO. However, the non-coherent detection scheme of the UWB-FM architecture allows for relaxed phase noise requirements [1], so tuning range and modulation bandwidth are more important features. Several wide-band CMOS VCOs have been reported in literature [2–6], but due to limited tuning range or too high power consumption, these circuits are not fully suited for the intended application. A linear tuning characteristic is also required by the FM modulation scheme and this indirectly sets a lower limit in the current dissipation as more current is required to shift the VCO frequency tuning saturation to higher frequencies. A solution which can effectively address all the specific requirements in the low band is a third order ring oscillator. The VCO circuit was implemented using differential delay cells because of their greater immunity to supply noise. Variable delay is achieved by simultaneously controlling the biasing of the transistors acting as tail current source and PMOS loads of the basic differential cell of the ring oscillator. A linear dependence of the oscillation frequency is achieved by simultaneously controlling the load resistance and the tail current through proper biasing. In particular, the tail device is biased in the triode region resulting in a linear dependence of the frequency with the tail bias [7]. The differential outputs of the VCO core circuit are connected to two single-ended inverting buffers. One is used as OA and one to drive the PLL prescaler of the integer PLL with external passive loop filter, which has been implemented around the VCO (Fig. 6.5). With a reference frequency F_{ref} of 250 kHz, the centre frequency of the UWB signal will have a resolution of 64 MHz.

The PLL is first tuned to a desired channel frequency within the 3.1–5 GHz band and, in a second step; modulation is introduced into VCO control voltage to generate the UWB signal. At this stage, the loop needs to be opened. This is done by introducing a power-down feature in the prescaler buffer that sets its output to ground, limiting the power consumption when the loop is left open.

The circuit has been implemented in a 0.18 μm 1P6M MMC/RF 1.8/3.3 V low cost RFCMOS technology. The layout of the complete IC is shown in Fig. 6.6. It occupies less than 1 × 1.5 mm including pads, ESD circuitry and by-pass and decoupling capacitors.

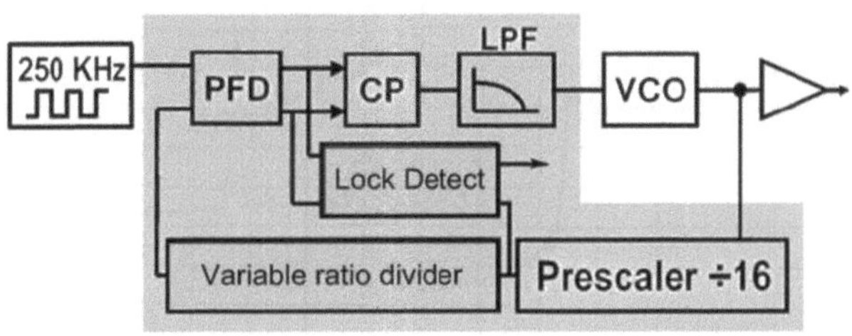

Fig. 6.5 PLL block Diagram

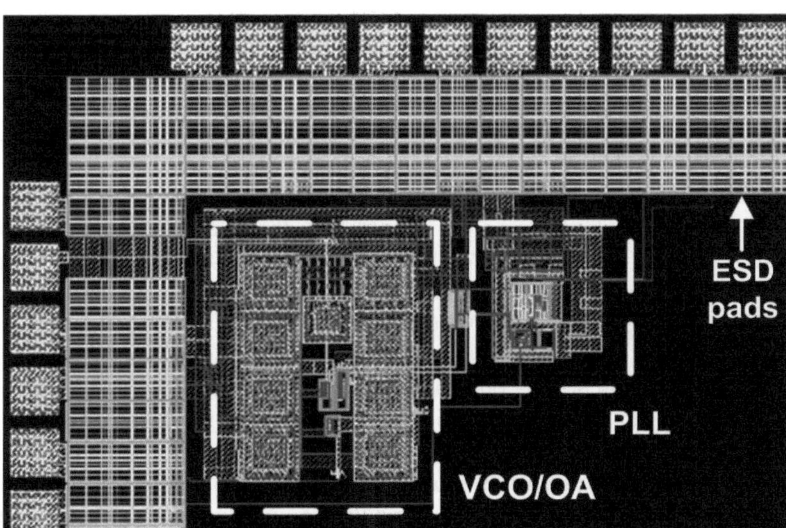

Fig. 6.6 Layout of the complete Transmitter IC

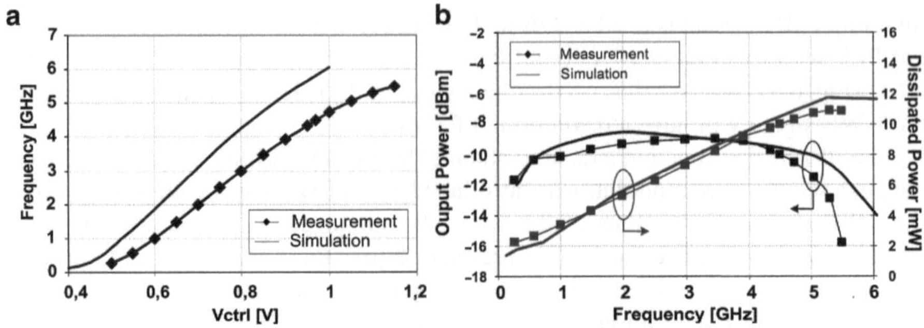

Fig. 6.7 VCO tuning range (**a**), output power and DC power consumption (with OA) (**b**)

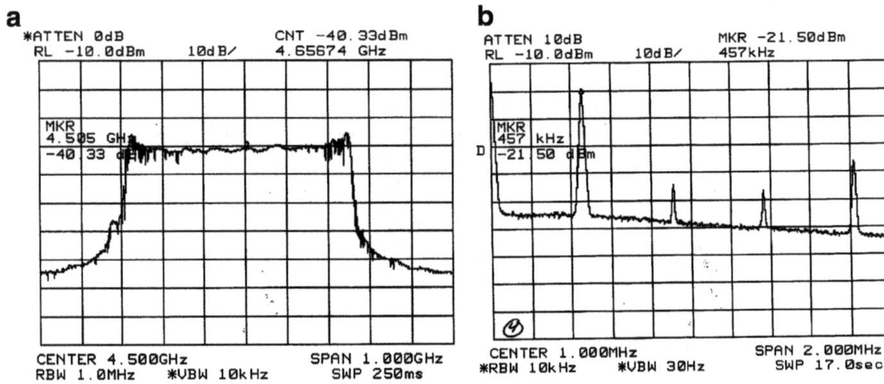

Fig. 6.8 Modulated Spectrum at 4.5 GHz with f_{sub} 457 kHz (**a**), FM demodulated signal (IEEE International Workshop on Radio-Frequency Integration Technology (**b**)

In Fig. 6.7a, the measured VCO frequency tuning characteristic is compared to simulation. The results confirm an extremely linear tuning range from 0.5 to more than 5 GHz under very low control voltage span. The output power variation is shown in Fig. 6.7b, where it can be noticed how the operating band, limited by the buffers, ranges from about 0.5–5.0 GHz. The power dissipation of the complete circuit is also shown in Fig. 6.7b. It is seen that the VCO, including buffers, dissipates from 2.5 to 10.5 mW at 1.8 V supply depending on the operation frequency. The maximum core dissipation is measured at the highest frequency, and represents only 50% of the overall power consumption. The VCO phase noise results are better than −75 dBc/Hz at 1 MHz offset from the carrier across the 3.1–5 GHz band and it can be considered adequate for the target application.

In order to test the transmitter performance, the VCO frequency was first set within the 3–5 GHz band, and then the modulation introduced to the V_{ctrl} pin using a triangular subcarrier signal with $f_{sub} = 457$ kHz. The VCO output was demodulated by a custom FM-UWB demodulator made from commercial off-the-shelf components. The modulated VCO output spectrum is shown in Fig. 6.8a. The flat output spectrum is in agreement with the VCO highly linear frequency tuning curve. A measured output of the FM demodulator is shown in Fig. 6.8b, where it can be seen that the subcarrier has been correctly recovered.

Compared with other published ring oscillator configurations, the proposed VCO demonstrates outstanding performance in terms of power consumption, linearity, and tuning range. This combination of properties makes the circuit able to fulfil the requirements of FM-UWB applications.

6.2.3.2 RF Low Band Receiver

The initial approach was therefore a truly wideband receiver, where a single receiver band covering the entire 3.1–4.9 GHz band was attempted. The LB RX system implementation compromises a wideband LNA and a wideband FM demodulator as illustrated in Fig. 6.9.

Low Noise Amplifier

Apart from the bandwidth, the gain is the most important parameter for the LNA. Due to elevated noise figure of the wideband FM demodulator the noise performance of the receiver is not determined only by the LNA noise figure, but rather by its gain in combination with the demodulator noise performance. The LNA gain lowers the noise contribution of the wideband FM demodulator on the overall receiver noise figure. To achieve a receiver sensitivity of −83 dBm an LNA gain of 30 dB is required.

To determine the best option to achieve 30 dB wideband gain LNA, the four classical options, consisting of the resistive termination, 1/gm termination, resistive feedback and inductive source degeneration were evaluated. Out of these four topologies more than one is interesting from a wideband perspective. However, in the end the inductive source degeneration topology was preferred. This structure offers the best noise performance and at the same time features a potentially wideband input match. Also, the gain of the topology is not limited by input requirements per say. The LNA is implemented as a multi-stage, fully differential topology where a T-type input matching network is used to achieve the required wideband operation.

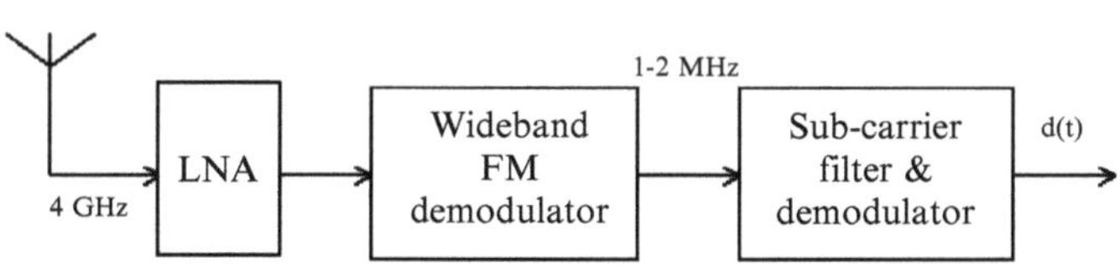

Fig. 6.9 FM-UWB receiver structure

Wideband FM Demodulator

The key FM-UWB specific block in the receiver is the wideband FM demodulator. For this block a delay-line based architecture, as illustrated in Fig. 6.10, was chosen.

The delay-line demodulator operates by converting an FM modulation into PM, which then is followed by a phase detector here implemented as a multiplier. With this architecture, a delay τ is added to input signal s(t), giving the delayed signal $s(t-\tau)$ with odd 90° phase shift at RF carrier frequency f_C. The original s(t) is then multiplied by the delayed signal $s(t-\tau)$ in order to produce the demodulated output [1]. The demodulator output response is ideally given by Eq. (6.1), where K is the multiplication gain, A is the amplitude of input signal and KD is the delay gain (i.e., phase shifter and gain block in Fig. 6.10).

$$V_{DEMOD}(f) = KK_D A^2 \sin(2\pi\tau(f - f_c)) \tag{6.1}$$

It can be shown that the sensitivity and dynamic range (DR) of the demodulator, which is defined by the ratio of maximum RF input power to the minimum power required for a given SNR at the demodulator output, is given by the following equations

$$P_{Sensitivity} \approx \frac{1}{\pi\tau B_{RF} R}\sqrt{\frac{SNR_{SUB} B_{SUB} SN_{Multiplier}}{AV_{AMP}{}^4 K_D{}^2 K^2}} \tag{6.2}$$

$$DR = \frac{P_{RFinMax}}{P_{Sensitivity}} \approx \frac{3V_T}{K_D}\frac{2\pi\tau\, B_{RF} K}{\sqrt{SNR_{SUB} B_{SUB} SN_{Noise}}} \tag{6.3}$$

Here, SNR_{SUB} is the baseband SNR required for FSK demodulation, B_{SUB} is the baseband channel bandwidth and $SN_{Multiplier}$ is the noise spectral density at the demodulator output in V^2/Hz. Noise contributed by the preamplifier is assumed to be much smaller than $SN_{Multiplier}$ and is therefore neglected in Eqs. (6.2) and (6.3). In practice, the noise figure of the receiver is dominated by the demodulator stage.

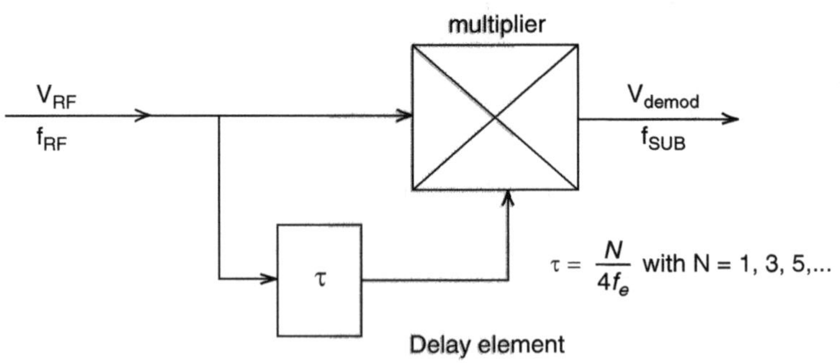

Fig. 6.10 Structure of a delay-line based FM demodulator

It can be seen from Eq. (6.2) that increasing the voltage gain of the preamplifier or gain in the delay path, increasing the delay time τ, and minimizing the noise contributed by the multiplier are effective methods of improving the receiver sensitivity. However, as seen from Eq. (6.3), the gain in the delay path reduces the overall dynamic range, so the DR is optimized by minimizing noise arising from the multiplier and increasing the delay time. Also, once the sensitivity of demodulator is improved, the noise from preamplifier cannot be neglected, see in Eq. (6.2).

The multiplier circuit is a key component in the demodulator. Often multiplier designs are based on Gilbert cell like implementations. Indeed, the double-balanced Gilbert cell provides suppression of both RF and LO signals, as well as any common-mode interference. However, the Gilbert cell may not be the best-suited candidate for ultra wide band applications as parasitic effects easily limit the circuit's performance. After a careful study and analysis of the different implementation options a transconductance-based multiplier is found to be the best candidate. A fully differential configuration is adopted for the whole demodulator, to mitigate common mode interference and harmonics as much as possible.

Turning to the implementation of the actual delay-line and the options available, there is a fundamental choice to be made between an active and a passive implementation. One fundamental difference between the two is that an active delay-line always shows a better insertion loss performance than a passive topology. However, considering the required delay time (62.5 ps) and the focus on power consumption, a passive delay-line could be a better choice. In addition, a passive delay-line induces less delay imbalance than an active delay-line implementation. This is very important since any imbalance in the delay will induce undesired effects such as a higher DC-offset and a reduced demodulation gain. Considering these performance tradeoffs, a passive approach was chosen. An additional phase imbalance correction circuitry was added to improve the performance of the delay line based demodulator. A schematic of the full demodulator is provided in Fig. 6.11.

For the final prototype implementation the designed wideband LNA is combined with the wideband FM demodulator using a system-in-package (SiP) approach. A picture showing the test board as well as the SiP based implementation of the low band receiver is shown in Fig. 6.12.

A summary of the performance achieved for the LB receiver is provided in Table 6.4.

6.2.4 Low Power RF Chipsets for High Band FM UWB

The blocks requiring specific IC implementation for the high band FM-UWB system are the same as the low band. However, due to higher operating frequency, the specifications are changed (see Table 6.2).

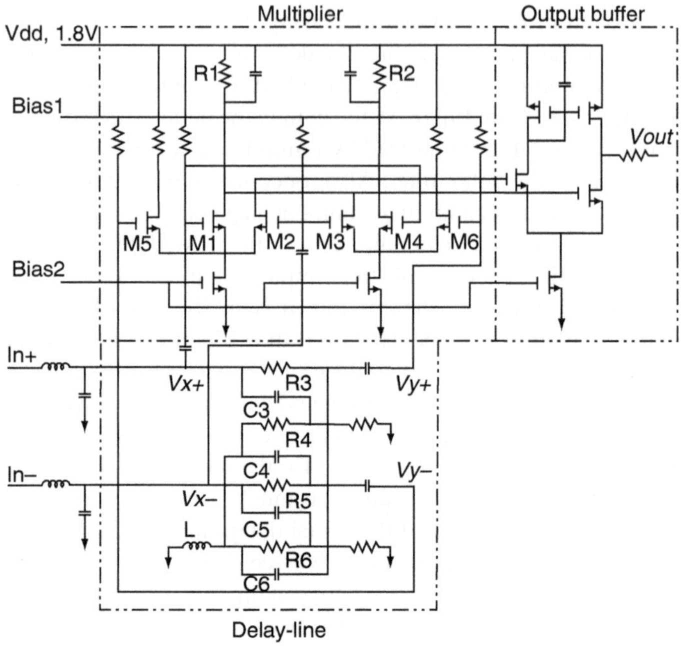

Fig. 6.11 Schematic of the combined FM demodulator

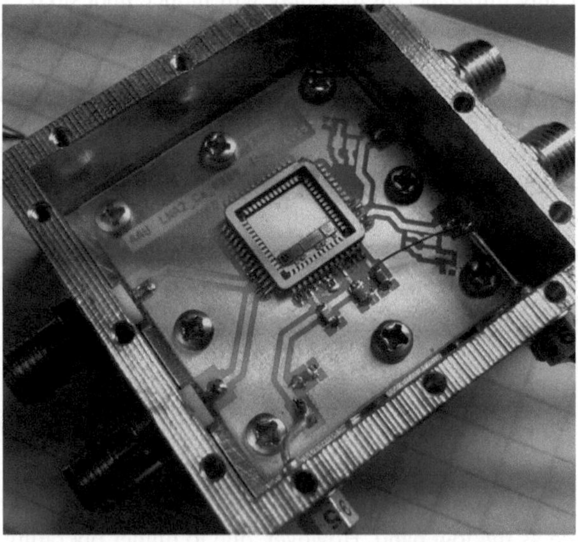

Fig. 6.12 Photo of the SiP based test board for the LB receiver prototype

Table 6.4 Summary of the measured LB receiver performance

	LNA	WBDM	Combined RX
Process	UMC 0.18 μm CMOS	UMC 0.18 μm CMOS	UMC 0.18 μm CMOS
Power dissipation (incl. buffers) (mW)	10	18.7 (Core: 5.4 mW)	28.7
Power supply (V)	1.8/1.7	1.8	1.8
Gain	20 dB	–	–
−10 dB bandwidth (GHz)	1.5	>1.5	1.5
Sensitivity	–	−30 dBm/500 MHz	−50 dBm/500 MHz
Area (including pads) (mm^2)	2.3	1.26	3.6

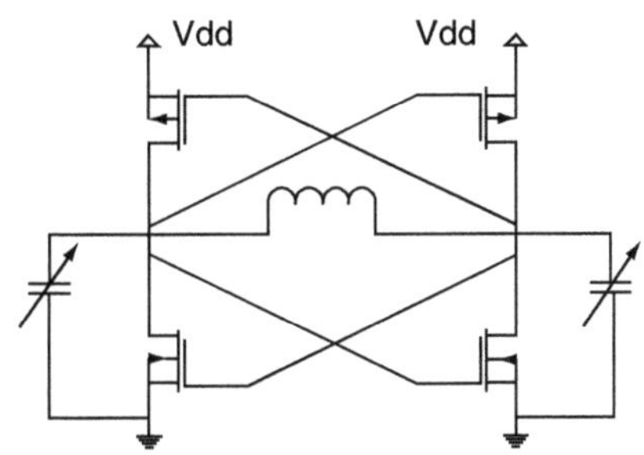

Fig. 6.13 High band VCO architecture

6.2.4.1 RF High Band Transmitter

The most critical part of the transmitter chain is yet the RF VCO. The tuning range required is less than 30% of the carrier frequency, so significantly lower than the one required for the low band system (∼50%). However, in the case of the high band a simple ring oscillator solution, as used for the low band, would not be appropriate due to its very high phase noise at high frequencies. For this reason, and to maintain reduced power consumption, a LC-type oscillator was selected. One of the main challenges of wideband LC VCO design consists of expanding an intrinsically narrow tuning range without significantly degrading noise performance or incurring in excessive tuning sensitivity. In recent years, band-switching techniques have been used extensively and have proved to be a successful way to increase tuning range. In the first design iteration such a solution was tested, but the excessive load of the VCO core (which draws very low current and is thus capable of supporting only low capacitive loads) by the multiple switches together with the poor noise model of the transistor used as a switch, led to the dismissal of this solution for the final implementation. Instead, the VCO topology and layout was made simpler to avoid any kind of parasitic effect which could degrade its performance. A complementary cross-coupled differential structure was used to achieve higher transconductance for a given current [8].

A double cross-connected pMOS and nMOS differential pairs provide the negative resistance (Fig. 6.13). The bias points of the transistors were selected in order to maximize the output voltage swing, thus reducing phase noise while maintaining low power consumption. Tail current source was omitted in order to simplify the design (an additional tail bias circuit would have been required) and avoid additional voltage headroom on the VCO core transistors. To achieve a wide tuning range and cover the required frequency bands, the transistor's sizes, as well as the resonant LC tank were carefully sized finding the optimum value for the inductance and capacitance of the varactor in the tank circuit, to achieve maximum tank quality factor under all process parameter variations.

The symmetrical differential outputs of the VCO core are connected to two properly sized single-ended inverting buffers. One is used as output amplifier and one to drive the PLL prescaler of the conventional integer PLL with external loop filter, which was implemented around the VCO to fix the centre frequency of the selected UWB channel. With a 64 MHz UWB signal resolution ($f_{ref} = 250$ kHz), four channels (H1–H4) which are multiple of this frequency and spaced 512 MHz (to avoid overlapping between adjacent bands) have been fixed as the PLL transmitter design target. The circuits which have been integrated are a modulus-16 prescaler, a variable ratio digital divider, a phase frequency detector (PFD) a charge pump (CP) and a double lock-detect circuit. This ensures that multiple lock signals are available at the output depending on the required precision. To disable the PLL once the VCO is locked to the desired frequency, a power down digital control signal is connected to the prescaler, limiting the power consumption when the loop is left open and the VCO is being modulated. This VCO design with OA and the PLL have been fabricated into a single IC with multiple test points and separate supply voltages for the various subcircuits in a commercial low-cost 0.13 μm 1P6M MMC/RF 1.8/3.3V RFCMOS technology. A microphotograph of the complete IC is shown in Fig. 6.14.

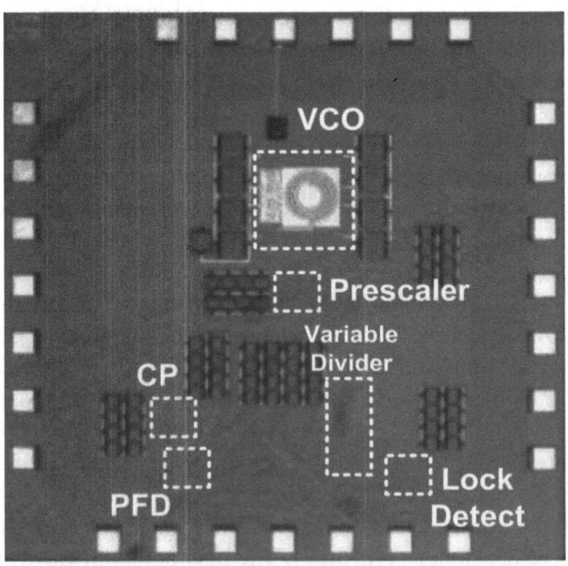

Fig. 6.14 Microphotograph of the complete Transmitter IC. Size: 1.5 × 1.5 mm

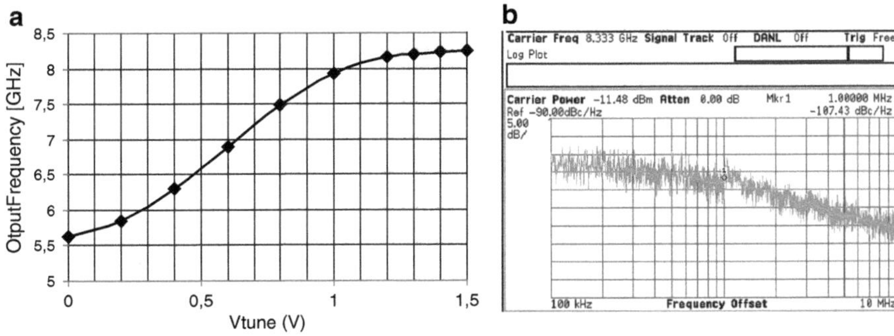

Fig. 6.15 VCO tuning range **(a)**, and phase noise **(b)**

Table 6.5 High band PLL locking conditions

Locking frequency (GHz)	VCO lock voltage (mV)	Overall current (VCO, OA and PLL) (mA)
6.464	450	4.23
6.976	615	4.2
7.488	790	4.1
8.000	102	4.1

Figure 6.15a shows how the measured VCO tuning characteristic can cover the specified frequency range. The problem is that this would imply a control voltage higher than 1.2 V, which could be provided by the charge pump. A solution was found by decreasing the VCO supply voltage down to 1.0 V without sacrificing performances and also reducing the power consumption. Phase noise performance better than −100dBc@1MHz offset can be easily obtained even at the highest operating frequency (Fig. 6.15b). The complete circuit with the OA capable of providing an output power in excess of −5 dBm, dissipates less than 5 mW at 1.2 V supply.

By optimizing VCO, PLL and OA voltages it was possible to cover the four bands with the transmitter IC delivering the proper lock detect signals as described in Table 6.5.

Compared with other published ring oscillator configurations, the proposed VCO demonstrates outstanding performance in terms of power consumption, linearity, and tuning range. This combination of properties makes the circuit able to fulfill the high band requirements of LDR FM-UWB system.

6.2.4.2 RF High Band Receiver

The receiver structure for the high band [9] gathers the same building blocks as the one of the low band which was presented in Fig. 6.9. In the case of the high band a gain stage was included after the phase shifter, leading to the front end structure presented in Fig. 6.16. In order to satisfy the requirements for sensitivity, and the low-power/low-cost constraints of the intended applications, a fixed time delay (FTD) demodulator is selected as for the low band.

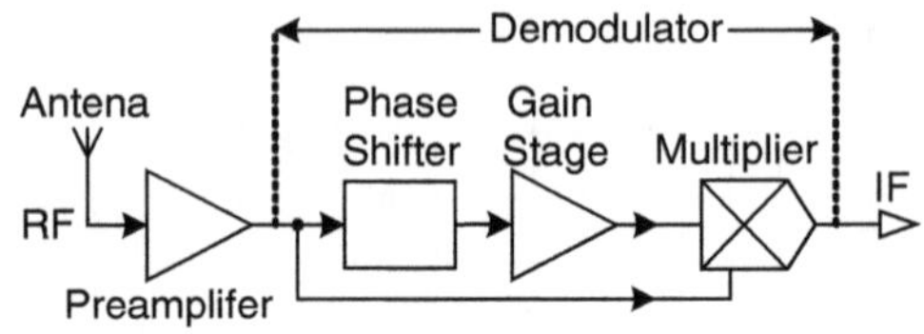

Fig. 6.16 Front-end with fixed time delay demodulator

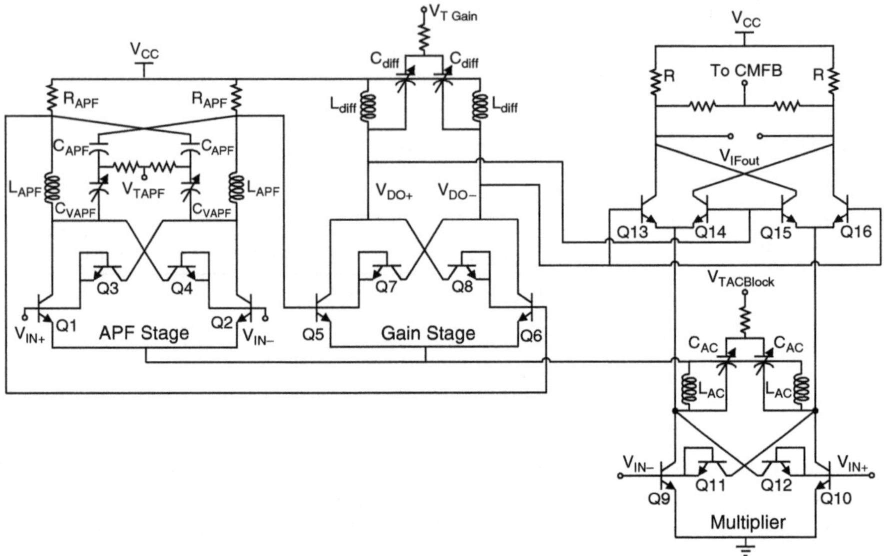

Fig. 6.17 Schematic of the FM-UWB demodulator

The QUBiC4X 0.25 μm SiGe:C BiCMOS technology from NXP Semiconductors [10] is used to implement the receiver front-end circuits. Bias current reuse is applied to both the preamplifier and FM demodulator blocks in order to minimize power consumption from a fixed supply voltage.

In the case of the high band, a balanced Gilbert architecture was used for the multiplier. Indeed, the use of bipolar transistors which have less parasitic capacitance than the CMOS transistors makes this architecture a good option. Besides, the bandwidth of the high band receiver is reduced compared to the low band which further limits the impact of parasitics. Simulations of a balanced bipolar Gilbert multiplier in the technology used for implementation of the receiver show that the maximum input signal voltage is approximately three times the BJT thermal voltage (i.e., $3V_T$ or ~75 mV at 25°C). Input greater than $3V_T$ drive the multiplier into the large-signal regime, which affects proper demodulation of subcarriers in a multi-access FM-UWB system [1].

The demodulator implementation is shown in Fig. 6.17. The phase shifter – or All Pass Filter (APF) – and gain stages drive a Gilbert multiplier consisting of input pair Q_9–Q_{10} (neutralized by diodes Q_{11}–Q_{12}) loaded by an LC tank (L_{AC} and C_{AC})

and transistor quad Q_{13}–Q_{16}. Current used to bias the APF and gain stages is fed via L_{AC} into Q_9–Q_{10} in order to boost its transconductance and the overall gain of the demodulator, K. This enables independent biasing of the input pair from the quad, which is used to optimize the multiplier's noise performance. Non-linear noise simulations predict that noise produced by the switching quad accounts for 93% of the total noise at the demodulator output, and that collector shot noise (i.e., $2qI_{Bias}$) is the main source of noise in these devices. Therefore, as the bias current in the switching quad decreases, the output noise (i.e., $SN_{Multiplier}$) also decreases. Noise present at the quad inputs is rejected at the IF outputs by the balanced, cross-coupled circuit topology, whereas noise originating from the switching quad remains.

The emitter area of each transistor in the switching quad is chosen based on a trade-off between output noise, gain-bandwidth performance and the extrinsic base and emitter resistances, which affect the gain. In the final design, a voltage gain of 20 with a mid-band group delay in the delay path of 500 ps is realized. Bias currents for the APF, gain and multiplier quad are set at 1.2, 1.7 and 0.32 mA, respectively. The sum of these bias currents flow into the input transconductor of the multiplier.

The preamplifier, also referred to as LNA, in a radio receiver front-end suppresses the impact of noise contributed by other components (e.g., mixer and baseband circuitry) thereby increasing the sensitivity. However, in the FM-UWB receiver design, emphasis is placed on voltage gain (~30 dB) and minimizing power consumption from the 1.8 V supply. In addition, a single-ended input and differential output are required in order to minimize implementation costs on the antenna side and interface to the demodulator, respectively. The overall sensitivity of the FM-UWB receiver is set by the demodulator, so noise figure on the order of 5 dB can be tolerated.

A simple active circuit balun (see Fig. 6.18) consisting of parallel common-base (Q_1) and common-emitter (Q_2 ac grounded via C_P) amplifiers is adopted [11], because the relatively low input impedance forced by the common-base stage simplifies interfacing to a 50 Ohm antenna. An input return loss of 10 dB (including packaging parasitics) is expected.

The Miller effect seen at the input of Q_2 is compensated by current feedback via diode-connected transistor Q_6. The input transistors drive a tuneable resonant tank load formed by inductor L_1 and varactor C_{T1}. A second, differential common-emitter stage (Q_3, Q_4 driving tuneable LC tank L_2 and varactor C_{T2}) increases the overall gain of the preamplifier to 30 dB when loaded by the demodulator input impedance of 1.4 kOhms. Bias current for the second stage is shared with the input stage as the tail current of pair Q_3, Q_4 flows through L_1. The neutrodyne formed by Q_3, Q_4 and diode-connected transistors Q_5 and Q_6 suppresses the Miller effect in the second stage, further increasing the overall gain. An automatic gain control (AGC) function giving a gain range of 25 dB is realized by tuning the load impedance of both stages via MOSFETs M_4–M_7.

The bias current for Q_1 and Q_2 (I_{Bias}) is generated by a PTAT bias block (not shown in Fig. 6.18) which is then mirrored by NMOS transistors M_1–M_3. A total 2 mA current excluding emitter followers is consumed under 1.8 V supply which fits the 4 mW design specification. The pi-topology network formed by the input

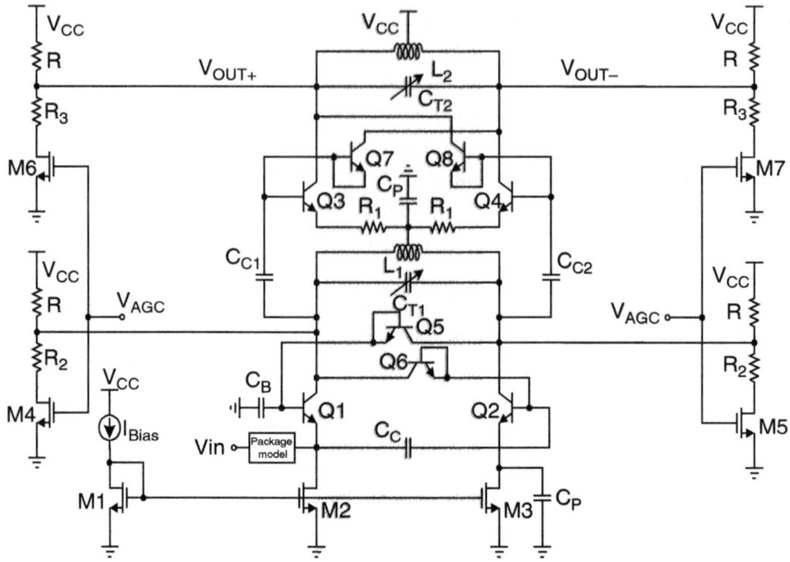

Fig. 6.18 High band preamplifier schematics

bondpad parasitic and ESD protection diode capacitance and bondwire and package lead inductance is absorbed in the RF input impedance matching network.

The die photograph of the high band receiver front-end (preamplifier and demodulator) is shown in Fig. 6.19. The preamplifier occupies an active area of 0.4 mm^2 while the demodulator occupies 0.4 mm^2 of the 1.62 mm^2 die.

A stand-alone version of the preamplifier has been fully-characterized for S-parameters (Fig. 6.20), noise figure (NF) and linearity from on-wafer probing. Emitter follower buffers are added to the stand-alone preamplifier to interface standard 50 Ohm measurement equipment. However, these buffers degrade the amplifier linearity slightly (−16 dBm P_{IIP3}, measured from a two-tone test) and are not used in the final receiver test chip. The measured input return loss is between 7 and 8 dB over the 7.2–7.7 GHz frequency range which is higher than intended because packaging parasitics are not included. The single-ended input to differential output power gain (S_{21}) is 21–22.5 dB when the amplifier is biased at 2 mA (excluding emitter followers) from a 1.8 V supply. Excellent isolation is seen as the measured S_{12} is better than −50 dB up to 10 GHz.

The measured noise figure at different gain and bandwidth settings is shown in Fig. 6.21. Here Vctrl and Vctrl_inter are tuning voltages for varactor CT1 and CT2 in Fig. 6.18. Varying the tuning tanks to set the gain and bandwidth of the preamplifier has little effect on the noise figure. The average noise figure of 5.5 dB measured across the 7–8 GHz band is expected to be improved by 1 dB when the amplifier input is impedance matched. The noise figure increases from 5.5 to 9 dB when the AGC voltage is used to tune the gain/bandwidth of the preamplifier. Tuning the resonant tanks via varactors C_{T1} and C_{T2} to set the tank gain/bandwidth has much less effect on the noise figure.

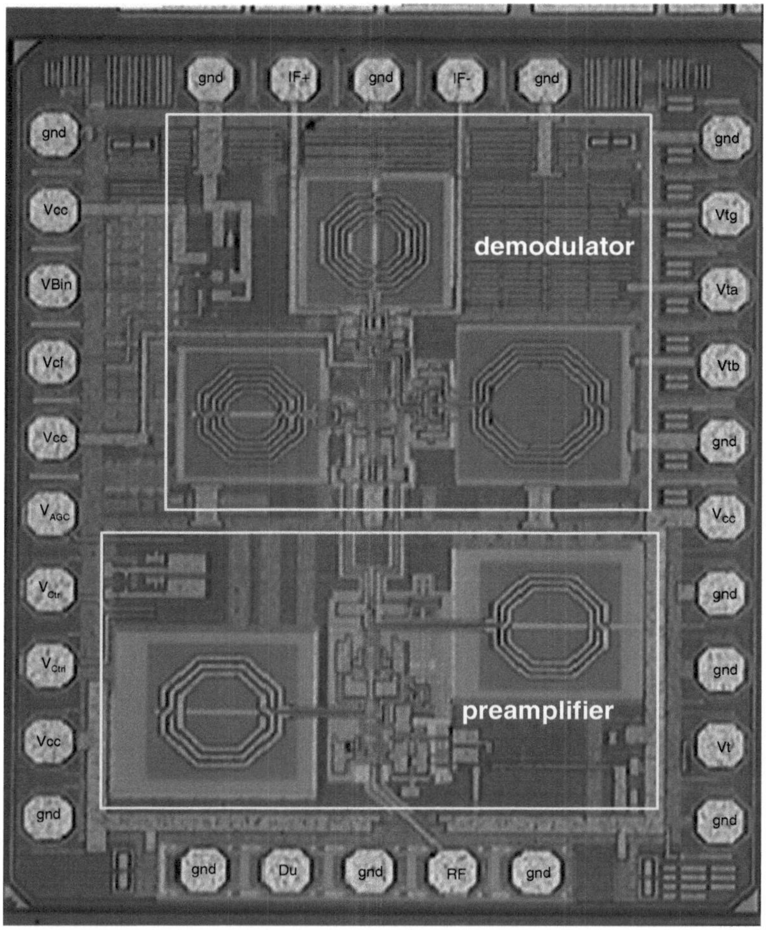

Fig. 6.19 High band front end receiver die photograph

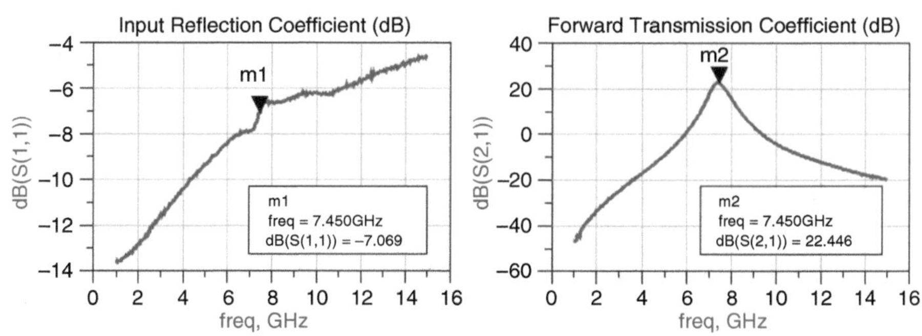

Fig. 6.20 High band preamplifier measured S_{11} and S_{21}

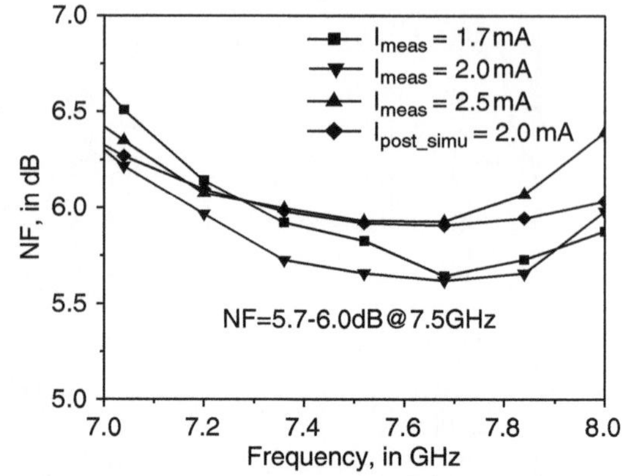

Fig. 6.21 High band preamplifier measured NF

Fig. 6.22 High band demodulator **(a)** and complete front-end **(b)** test circuits

Evaluation of the demodulator (Fig. 6.22) and receiver performance focused on the measurement of sensitivity. It is assumed that an SNR of 14 dB is sufficient for FSK demodulation with a bit-error-rate less than 10^{-6}, and that the RF input power corresponding to this SNR at the demodulator output is defined as the sensitivity. The results are summarized in Table 6.6.

6.2.5 Baseband Processing and Channel Coding

6.2.5.1 Receiver Sub Carrier Processing

After the RF section, the FSK modulated subcarrier signals (between 1 and 2 MHz) are processed by the Sub Carrier Processing (SCP) circuitry. The SCP is composed of the blocks lying between the wideband demodulator and the digital FSK de-

Table 6.6 Summary of measured high band front end results

Parameter	Target spec	Measurements	
	FM-UWB demodulator		
Sensitivity (dBm)	−55	−66.8	−61.8
Power consumption (mW)	6	5.8	2.8
Power supply (V)	1.8	1.8	1.2
Active chip area (mm^2)	–	0.4	0.4
	Receiver front−end		
Sensitivity (dBM)	−85	−86.8	−84.3
Power consumption (mW)	10	9.1	6
Power supply (V)	1.8	1.8	1.5
Active chip area (mm^2)	–	0.88	0.88

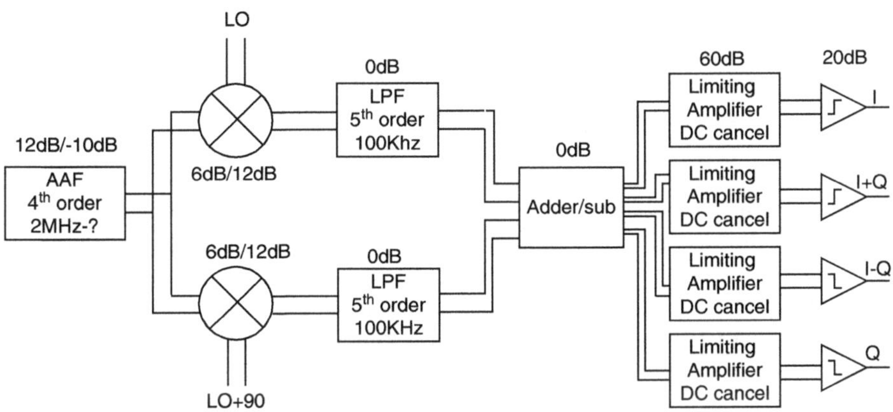

Fig. 6.23 Block diagram of SCP

modulator in Fig. 6.1, takes care of the selection and amplification of the wanted subcarrier signal. As shown in Fig. 6.23, the subcarrier signal is filtered by the Anti-Aliasing Filter (AAF), downconverted to baseband, filtered by lowpass filters (LPF) and finally amplified to a full-swing digital CMOS signal (0–1.8 V).

For correct demodulation of the low modulation index FSK signals, not only the usual quadrature I and Q (0° and 90°) signals, but also additional signals I + Q (45°) and I − Q (−45°) are generated to increase the number of zero crossings available for FSK demodulation [12]. These signals are generated prior to the hard-limiting process.

The AAF is the first block of the SCP. The dynamic range requirement is the most stringent parameter of this block. The AAF is also the block that impacts the most the overall noise performance. Finally a compromise noise versus power consumption was chosen. The performance of the AAF is shown in Table 6.7.

The mixer used for downconverting the FSK signals to baseband is a simple Gilbert cell. In order to avoid downconversion of signals at odd harmonics of the LO frequency, a triangular wave is used as LO signal. This suppresses the third harmonic mixing by 30 dB at the acceptable cost of 2.5 dB less gain. Noise performance of

Table 6.7 AAF performance

Parameter	Measured
Output Noise @ 1 MHz	50 nV
Output Noise @ peak	70 nV
IIP3 @ 1 MHz	129 mV/1.17 V
Gain	12.15 dB/−6.9 dB
BW range	1.53–2.72 MHz

Table 6.8 Mixer performance

Parameter	Measurement
Gain (low)	7.48 dB
Gain (high)	12.59 dB
IIP3	74 mV_{RMS}
LO leakage	−14 dB/−18 dB

the mixer is good enough for this application and it has minimal contribution on the overall noise budget of the SCP chip. Global performance of the mixer is given in Table 6.8.

For the low pass filters that follow the mixer, the gm-C architecture was preferred over the active-RC topology, for it requires smaller required chip area and despite the better performance of active-RC filters at a given power consumption. In filters, capacitors determine the chip area. The capacitors in the gm-C filter are placed differentially and this drastically reduces the required silicon area.

A fifth order Chebyshev filter with 1 dB of ripple was chosen, since it provides sufficient attenuation for the adjacent subcarrier channels. Two LPFs are used in the SCP chip for I and Q channels. Each one of them consumes 600 μA current at 1.8 V supply voltage. Since the filter cut-off frequency was specified for a wide frequency range (50–300 kHz), both digital and analogue tuning methods are combined to correctly set the filter bandwidth. According to the configuration, the bandwidth can be selected between 46, 97 and 289 kHz.

After filtering out the unwanted signals by the LPF, the subcarrier signal needs to be transformed into a digital full-swing CMOS signal, since the digital FSK demodulator uses only the zero-crossings of the signal. A cascade of a limiting amplifier and a comparator provides the necessary gain. The gain of the limiting amplifier is 60 dB and the comparator around its equilibrium position provides an additional 50–60 dB gain. Due to this high gain, any offset at the input – even a few μV – will set the limiting and the comparator into overdrive, thereby masking the subcarrier signal. Therefore, the DC offset needs to be suppressed by the limiting amplifier. Besides, the limiting amplifier needs to have a high pass characteristic. Then, an architecture that was most tolerant against input offset was chosen. The limiting amplifier consists of three identical cascaded stages.

Figure 6.24 shows the four phases of the hard limited signals: I, Q, I+Q and I−Q. The oscilloscope time base is running at 2 μs/div. The minimum input signal for which the output signal showed no unwanted artifacts (spikes, undesired transitions) equals −82 dBm.

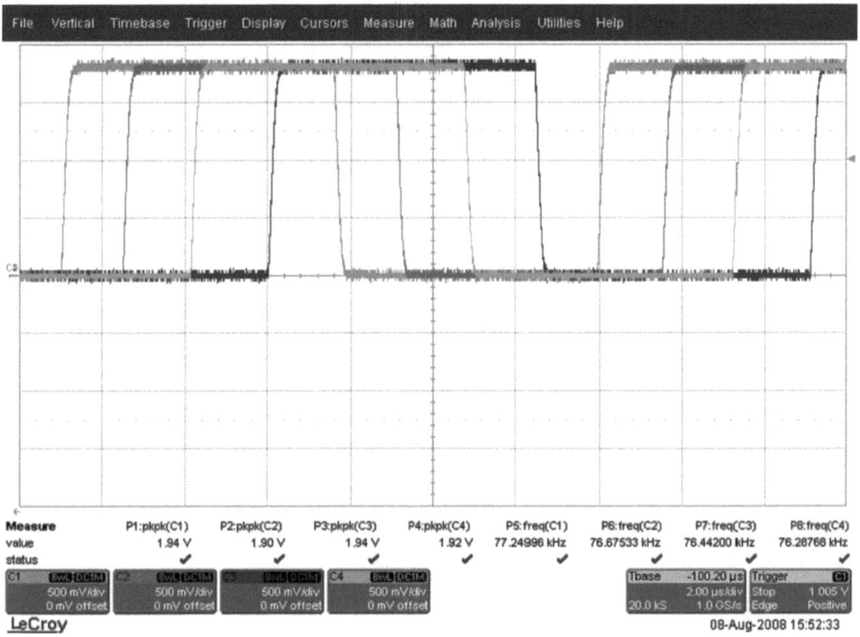

Fig. 6.24 SCP measured output signal

6.2.5.2 FSK Demodulator

The purpose of the design and implementation of the FSK demodulator and combiner block is to process the I and Q signals in the receive path coming from the subcarrier processing in order to recover the bit stream. Its block diagram is shown in Fig. 6.25.

Rising edge of I (Q stable)	$D \leq \text{not}(Q)$
Falling edge of I (Q stable)	$D \leq Q$
Rising edge of Q (I stable)	$D \leq I$
Falling edge of Q (I stable)	$D \leq \text{not}(I)$
Any other combination	$D \leq D$

The FSK demodulator block receives the I and Q hard decision signals and performs the demodulation according to the following rules:
This means that in case of positive frequency (I leads Q) the demodulator output is HIGH. Else, in case of negative frequency (Q leads I), the demodulator output is LOW. In order to implement this design with digital synchronous components, the asynchronous I and Q signals coming from the low pass filter need to be converted to the synchronous clocked domain. This is done by using an oversampling frequency clock of 2 MHz to sample the input signals and process them to decide

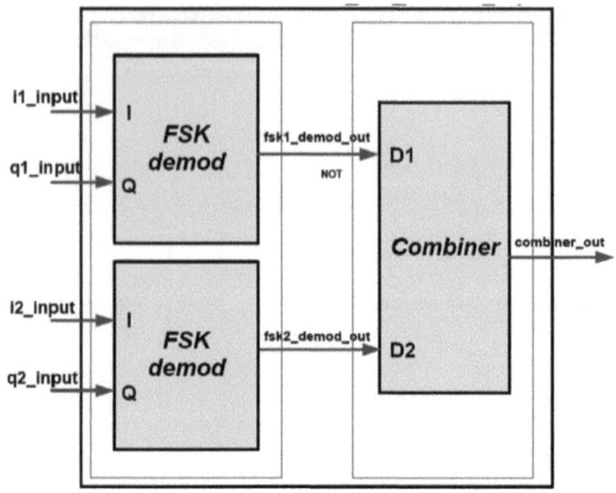

Fig. 6.25 FSK demodulator overview

for the demodulator output value. The 2 MHz correspond to an oversampling rate of 16 of the I and Q signals running at 125 kHz. The sampling error thus is kept below 500 ns.

The implementation is done by synchronizing initially I and Q signals to the digital clock domain and then by applying the edge detection mechanism that will be used as a trigger for the output decision. In case of a simultaneous transition of I and Q signals (conflict in the edge detection and decision), the demodulator outputs its previous value.

In order to eliminate the glitches that may occur in the combiner's output under increased noisy environments (SNR $= \sim$6 dB), a digital low-pass filter (LPF) was inserted at the combiner output. The mathematical equations of the LPF are shown below:

$$h(n) = \begin{cases} 1, & 0 \le n \le (N-1) \\ 0, & others \end{cases}$$

$$H\left(e^{j\omega}\right) = \frac{\mathrm{Sin}\left(N\frac{\omega}{2}\right)}{\mathrm{Sin}\left(\frac{\omega}{2}\right)} e^{-j\omega\frac{N-1}{2}} \Rightarrow$$
$$\left|H\left(e^{j\omega}\right)\right| = \frac{\mathrm{Sin}\left(N\frac{\omega}{2}\right)}{\mathrm{Sin}\left(\frac{\omega}{2}\right)} \tag{6.4}$$

The addition of the LPF filter in the combiner's output eliminates the glitches in noisy conditions (SNR $=$ 6 dB). This is shown in Fig. 6.26.

6.2.5.3 Channel Coding

The development of low power communication systems leads to strong requirements on the power consumption of the transceivers. For such systems, the

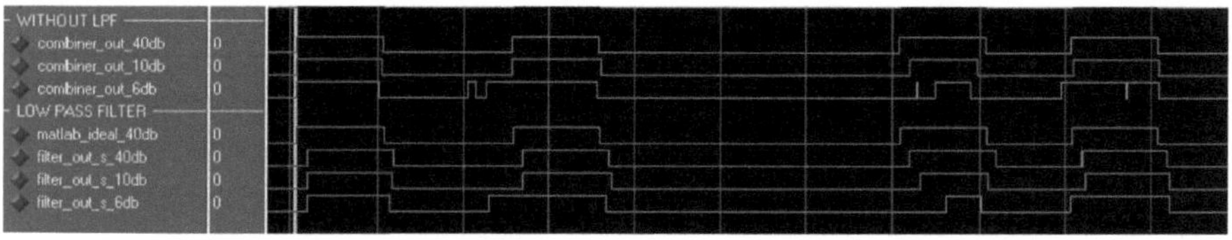

Fig. 6.26 FSK demodulator and combiner with output LPF filter

traditional way of selecting a FEC scheme only based on its performance shall be reconsidered. Indeed, its correcting capabilities shall be seen more as a way to decrease the transmit power than as a way to improve communication robustness. However, a high error correction capability usually comes at the cost of a significant power consumption overhead caused by the additional encoding/decoding computations. In this context, the use of the most advanced FEC schemes specifically designed to reach very high performances is not the most appropriate for the FM-UWB system. The challenge in choosing the FEC scheme for a low power communication system is to find the best trade-off between its performance and power consumption. Different types of codes, including soft-decoding methods, have been compared in [13]. However, several code parameters selected are not targeted for low power consumption. On the contrary, the current paper focuses only on one type of codes, and aims at choosing an optimal set of code parameters to meet low power constraints. RS codes have been selected in this study because they accommodate well with hardware complexity restrictions of LDR WPAN transceivers.

Reed Solomon (RS) codes (see e.g. [14]) are defined by the set of three parameters (n,k,t), k and n being the number of symbols respectively before and after encoding, and $t = (n - k)/2$ the number of symbols which can be corrected among n. The code rate is denoted by $R = k/n$. Symbols take their values in a Galois Field $GF(2^m)$, and are thus represented with m bits. The n parameter is bounded by 2^m. A lower value for n specifies a shortened RS code. From the different existing decoding methods, frequency domain algorithms traditionally show the lowest computational complexity [15]. The complexity figures of RS coder/decoder, obtained after a first implementation analysis, are displayed in Table 6.9. They are expressed in terms of Galois Field (GF) operations. GFadd represents a Galois Field addition. GFmulαi corresponds to the multiplication by a specific Galois Field element α^i, whereas GFmul is the multiplication of two unspecified Galois Field elements. GFinv provides the inverse of an element. And finally, register storage and memory storage of a Galois Field element are differentiated by GFreg and GFmem, because of their significantly different power consumption. For the Key Equation Solving, which is the core of the decoder, several algorithms can be used [14], Berlekamp-Massey Algorithm (BMA), Extended Euclidean Algorithm (EEA) or Peterson-Gorenstein-Zierler (PGZ) algorithm. Considering a classical implementation, it can be seen from Table 6.9 that EEA [16] requires the least computations, except for low values of t ($t \leq 3$) where PGZ algorithm [17] performs better.

Table 6.9 Complexity of the RS coders/decoders

	GFinv	GFmul	GFmulαi	GFadd	GFreg	GFmem
Encoder			n.(2.t)	n.(2.t)	n.(2.t)	
Syndrome calculation			n.(2.t)	n.(2.t)	n.(2.t)	
BMA	2.t − 1	(2.t − 1). (2.t + 1) + t^2		(2.t − 1). (2.t) + t^2	(2.t − 1). (5.t − 1) + t	
EEA	T	t.(4.t)		t.(4.t)	t.(6.t + 1)	
PGZ						
For t = 1	1	1		2		
For t = 2	1	9		4	4	
For t = 3	1	27		6	15	
Chien Search			n.(2.t − 1)	n.(2.t − 1)	n.(2.t − 1)	
Forney Algorithm	T	T				
Error correction				t		
Delay line						k
Total (with EEA)	2.t	t.(4.t + 1)	n.(6.t − 1)	n.(6.t − 1) + t.(4.t + 1)	n.(6.t − 1) + t.(6.t + 1)	k

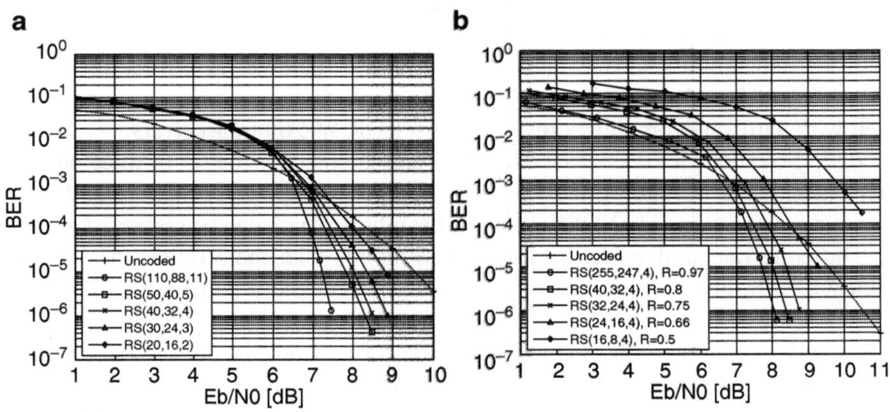

Fig. 6.27 Comparison of RS codes over GF(2^8) with R = 0.8 (left) and over GF(2^8) with t = 4 (right)

By comparing shortened RS codes with the same code rate R and same m, it can be noticed that the Computational Complexity per Information Bit (CCIB) grows like *O(t)*. Therefore, it seems more interesting to choose a low-t RS code, implying a low value for n and k. Besides, a similar comparison is shown in Fig. 6.27 (left) from a performance point of view. The BER vs. Eb/N0 curves have been obtained by simulation for several codes over GF(2^8) with the same rate R = 0.8. A very simple transmission scheme is used, including an AWGN channel and a BPSK modulation. Same rate implies that the error correction capability per information symbol t/k is constant. But even if this ratio is the same for every code, it can be noted that for low BER, the larger the block size k (or n), the better the performance. As a result, for a given code rate, an optimal (n,k) pair should be determined in order to achieve a good trade-off between the computational complexity and the performance of the code.

In the same way, RS codes can be compared by fixing the error correction capability t, and choosing different code rates. In Fig. 6.27 (right), performance of codes with $t = 4$ and different code rates is illustrated. As known, a high code rate is preferable in order to reduce the shift of the curve. However, concerning the slope of the curve, two antagonist effects can be distinguished. With a fixed t, a higher code rate corresponds to a larger block size k, but also results in a smaller t/k ratio. The former aspect, like before, has a positive impact on performances, while a smaller t/k decreases the slope of the curve. With $t = 4$, it can be seen that the former aspect is preponderant, strengthening the choice of a high code rate.

To sum it up, some general trends can be drawn. On one hand, computational complexity is reduced by choosing a low t and by increasing R. On the other hand, for performance optimisation, a high code rate R is also preferable to reduce the shift of the curve, while a high parameter t will improve its slope. However this last point has not a very significant impact, and thus, a quite low value should be chosen for t, in order to improve computational complexity. Considering this, RS(255,249,3) and RS(255,247,4) over GF(2^8) would be good choices, for instance.

Furthermore, when integrating the RS code in a communication system, the data to be transmitted rarely fit exactly in a multiple of RS data blocks. Some padding has to be performed on the last data block and the consumption increases uselessly. That is why, when the frame size is not very large, a fixed shortened RS code is better suited than a large code, even if "on-the-fly" shortening is performed. For example, RS(40,32,4) which shows only a 0.3 dB poorest coding gain could be more attractive than the RS(255,247,4) previously mentioned.

At last, the influence of the GF size is analysed. Regarding the CCIB, only GFmul and GFmulαi, which are $O(m^2)$, are impacted by the GF size. As they do not represent the major part of the CCIB, increasing m will not have a strong influence on consumption increase.

From a performance point of view, Fig. 6.28 shows the BER achieved with a shortened RS(40,32,4) code over different GF. It appears that the smallest GF code performs slightly better. Indeed, when the number of errors considered in one data block of a large GF code exceeds its error correction capability, a smaller GF code might still be able to correct them, as the errors might be dispatched over several of its shorter input data blocks.

Consequently, depending on the requirements, a small Galois Field could be satisfying, as it improves both computational cost and performance. However, an excessively small GF size would limit the data block size and thus the code rate, which is not in agreement with the previous recommendations.

As a case study, the RS(40,32,4) code over GF(2^8) is selected to quantify the worth of some low-power implementation improvements. Detailed results can be found in [18]. The impact of some simple design improvement is emphasized with

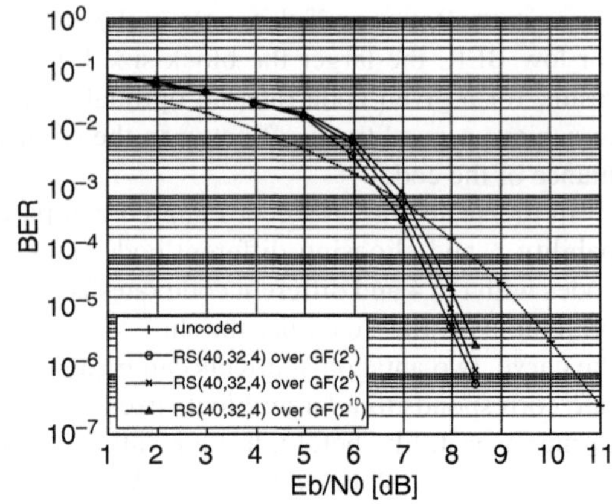

Fig. 6.28 Comparison of RS codes over different Galois Fields

the optimisation of the delay line which allowed for 78% memory power savings. Besides, further advanced considerations like switch-off strategies and Composite Galois Field approach resulted in up to 18% extra logic power reduction.

6.2.6 *MAC Layer and Connectivity*

The MAC layer used in the FM-UWB prototype is a subset of the IEEE 802.15.4 standard MAC. An overview of the functionality and the underlying primitives is discussed in this section.

6.2.6.1 MAC Functionality

In this section a brief introduction to MAC functionality is provided. A more detailed description of FM-UWB MAC layer functionality can be found in Chap. 4.

Starting a Piconet

An LDR piconet is started by a coordinator device. It selects a suitable channel for operation by passive scanning of the available channels and then starts advertising for association by devices in the vicinity by sending out beacons that contain the signatures of the piconet (i.e. only the so-called "beacon-enabled configuration" in the terminology of IEEE 802.15.4 is considered). Devices that can listen to these network beacons send association requests to the coordinator. The coordinator then accepts or rejects the association based on the available resources.

Channel Access

An LDR piconet uses a slotted CSMA-CA channel access mechanism, where the backoff slots are aligned with the start of the beacon transmission. Each time a device wishes to transmit data frames during the Contention Access period (CAP), it locates the boundary of the next backoff slot and then waits for a random number of backoff slots. If the channel is busy, following this random backoff, the device waits for another random number of backoff slots before trying to access the channel again. If the channel is idle, the device can begin transmitting on the next available backoff slot boundary. Acknowledgment and beacon frames are sent without using a CSMA-CA mechanism.

Channel Time Management

The LDR MAC uses a superframe structure for channel time management. The superframe is bounded by beacon frames, which are sent by the coordinator, and is divided into 16 equally sized slots. The beacon frame is transmitted in the first slot of each superframe. The beacons are used to synchronize the attached devices and to identify the LDR WPAN. Any device wishing to communicate during the CAP between two beacons shall compete with other devices using a slotted CSMA-CA mechanism. All transactions shall be completed by the time of the next network beacon.

Data Transfer

Three types of data transfer transactions exist. The first one is the data transfer to a coordinator in which a device transmits the data. The second transaction is the data transfer from a coordinator in which the device receives the data. The third transaction is the data transfer between two peer devices. In star topology only the two first of these transactions are used, because data may be exchanged only between the coordinator and a device. In a peer-to-peer topology data may be exchanged between any pair of devices on the network; consequently all three transactions may be used in this topology.

Acknowledgement and Retransmission

In order to detect bit errors, a frame check sequence mechanism, employing a 16-bit cyclic redundancy check (CRC), is used to protect every frame. A successful reception and validation of a data frame can be optionally confirmed with an acknowledgment. If the receiving device is unable to handle the received data frame for any reason, the message is not acknowledged. If the sender does not receive an acknowledgment after some time, it assumes that the transmission was unsuccessful and retries the frame transmission. When the acknowledgment is not used, the sender assumes the transmission was successful, meaning that the frame is lost in case of unsuccessful transmission.

Security Service

The higher layers determine when security is to be used at the MAC layer and provide all keying materials necessary to provide the security service. The security mechanisms are symmetric key based. Secured modes include Access Control, Data Encryption, Frame Integrity and Sequential Freshness.

6.2.6.2 MAC Implementation Architecture

The network layer and applications implemented on the host platform. Figure 6.29 depicts the high level architecture of MAC with its interfaces, highlighting the HW/SW partitioning of the MAC primitives. The upper layers consist of a network

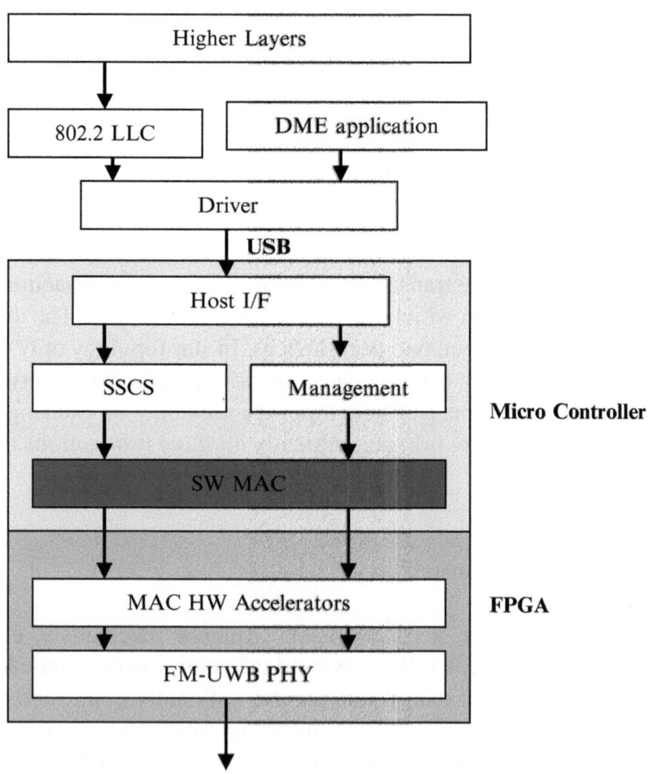

Fig. 6.29 MAC HW/SW architecture

layer, which provides network configuration, manipulation, and message routing, and an application layer, which provides the intended function of the device. An IEEE 802.2 Type 1 logical link control (LLC) can accesses the MAC sublayer through the service specific convergence sublayer (SSCS).

There are two types of frames exchanged with host, namely data frames (LLC) and management frames (Control path). The host interface encapsulates these two type messages at the sending side and separates them at the receiving side.

Primary consideration for the hardware software partitioning is that the time critical and compute intensive functionalities are implemented in hardware and targeted to an FPGA. The control and management functionality is targeted in software running on a microcontroler. Hardware acceleration of many of the lower MAC functions includes ciphering, CRC checksum, timers, auto-acknowledgement with retries and CSMA/CA. This reduced the microcontroller overhead allowing operation with low-end processors and minimizes the system power consumption.

The LDR MAC sublayer provides two services: the MAC data service and the MAC management service interfacing to the MAC sublayer management entity (MLME) service access point (SAP) (known as MLME-SAP). The MAC data service enables the transmission and reception of MAC protocol data units (MPDUs) across the PHY data service. The features of the MAC sublayer are beacon management, channel access, frame validation, acknowledged frame delivery, association, and disassociation. In addition, the MAC sublayer provides hooks for implementing application specific security mechanisms. A set of messaging primitives are implemented for providing following baseline MAC functionalities:

- PAN startup
- PAN Discovery
- Device Synchronization with a coordinator
- Joining the PAN
- Data Transmission
- Leaving the PAN

The standard IEEE 802.15.4 frame formats have been used for messaging. These messages are classified into four categories namely Request, Response, Indication, and Confirm. Respective message primitives are:

- Start (Request, Confirm)
- Associate/Disassociate (Request, Indication, Confirm)
- Get/Set Parameters (Request, Response)
- Synchronisation (Request)
- Polling (Request, Confirm)
- Data (Request, Indication, Confirm)

Two types of devices are implemented namely PNC capable, and End devices. PNC capable devices support beaconing, association request handling, superframe management and data transmission. End devices have the capability to synchronise with network, send association request and data transmission.

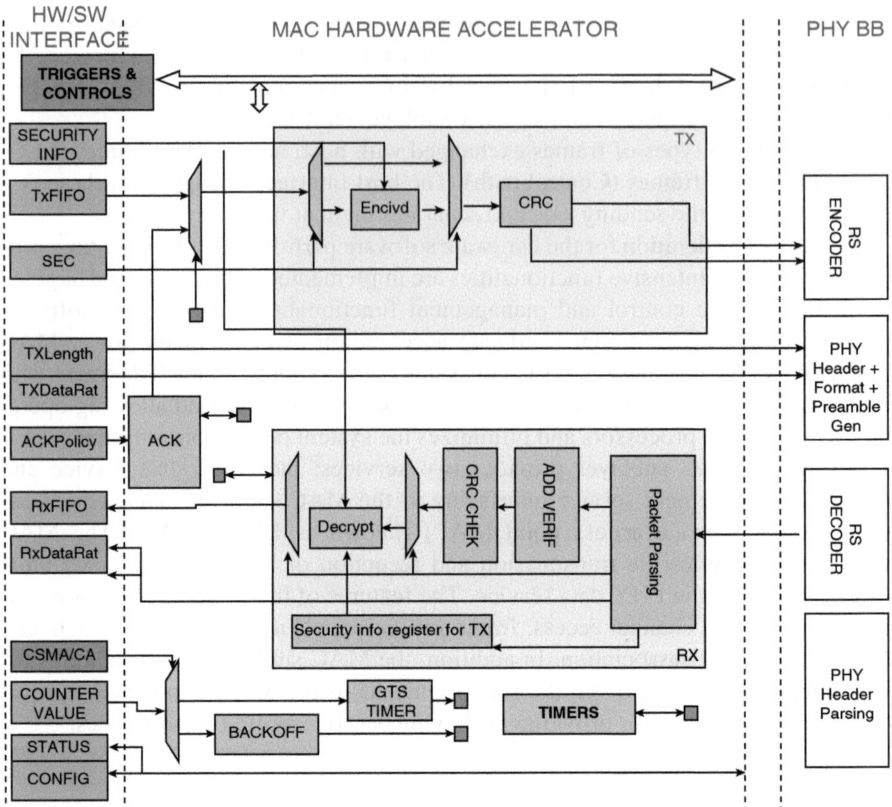

Fig. 6.30 MAC HW/SW interface

The physical interface to access HW blocks implemented in FPGA is shown Fig. 6.30. This interface consists of FIFO for data path connectivity, configuration registers for setting parameters of various hardware blocks, status registers for monitoring the events and interrupts for registering time critical events in real-time.

6.2.7 LDR Hardware Prototype

The architecture of the LDR prototype is illustrated in Fig. 6.31 and includes five main parts:

- Software MAC primitives (SW-MAC) implemented onto a 8051 micro-controller on the digital board
- Hardware MAC (HW-MAC) primitives implemented onto an Altera Cyclone II FPGA on the digital board

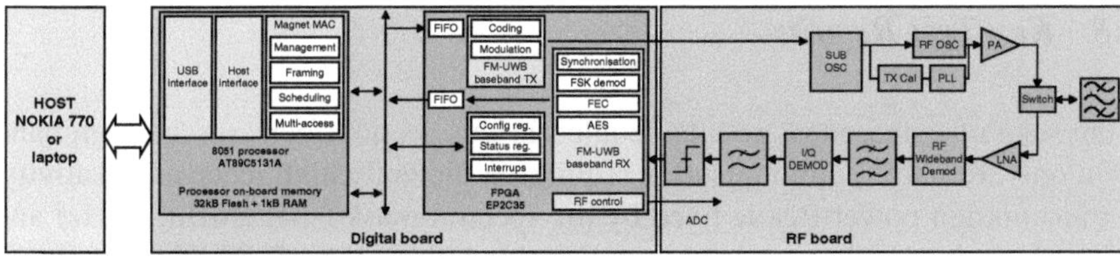

Fig. 6.31 LDR prototype architecture

Fig. 6.32 LDR low band prototype

Fig. 6.33 LDR high band prototype

- Digital PHY blocks including FEC implemented onto the same FPGA on the digital board
- Analogue PHY blocks implemented on the RF board
- RF blocks implemented on MAGNET ICs on the RF boards

The prototypes provide host connectivity via a USB interface. This USB link conveys frames to the LDR prototype which implements the MAC and PHY building blocks. The hardware prototypes include all the blocks described in the previous sections. Two versions have been implemented and tested: one targeting the low band and the other one the high band. They differ by their RF sections only. Figures 6.32 and 6.33 show the low band and high band prototypes respectively.

6.2.8 Key Test Results

For the sake of conciseness results detailed in this section focus on the high band system only. Reliable communication requires sufficiently high receiver sensitivity. The transmission power P_{TX} is fixed by the spectral mask (−41.3 dBm/MHz) and bandwidth of the UWB signal. For a RF bandwidth $B_{RF} = 500$ MHz, maximum transmission power $P_{TX} = -14.3$ dBm.

The LDR system targets short range indoor communication under line of sight (LOS) conditions. Figure 6.34 shows the received power for operation at 7.5 GHz as function of the distance for a path loss exponent n = 2 and antenna gain of 0 dBi. Measurements of commercially available small UWB antennas show that antenna gain values of 0 dBi can be realistically considered.

It can be seen that a receiver sensitivity around −85 dBm is required. As a first test BER measurements were made using a wired setup as shown in Fig. 6.35. The left board is in TX mode, the right board is in RX mode. A variable attenuator is connected between the transmitter and receiver. A splitter is used to monitor the transmitted signal on a spectrum analyzer and a power meter. A logic analyzer is also connected to monitor the various digital control signals on t he boards. A BER test set based upon the RJ-013 Bit Error Rate Test IC from RAD Electronics is also connected to the boards.

Figure 6.36 shows the spectrum of the transmitter output signal as observed on the spectrum analyser.

BER testing was performed using a 511-bit long PRN sequence at a data rate of 50 kbps. The results are provided for two receivers (Fig. 6.37), one of which has an optimized receiver input. It is worthwhile to see that sensitivity can be increase up to 4 dB after optimization of the receiver input matching circuit.

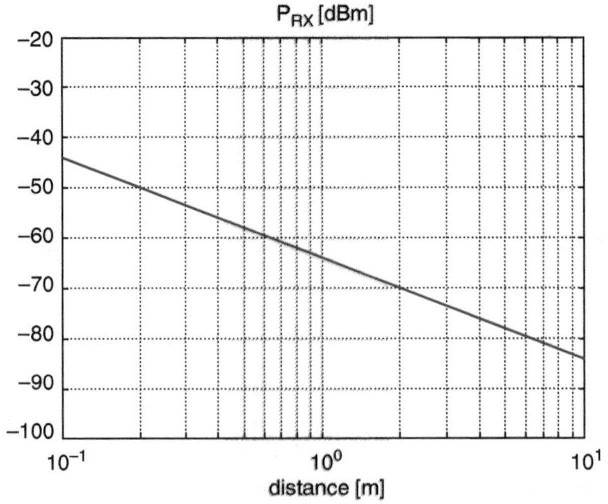

Fig. 6.34 Received power as a function of distance at 7.5 GHz

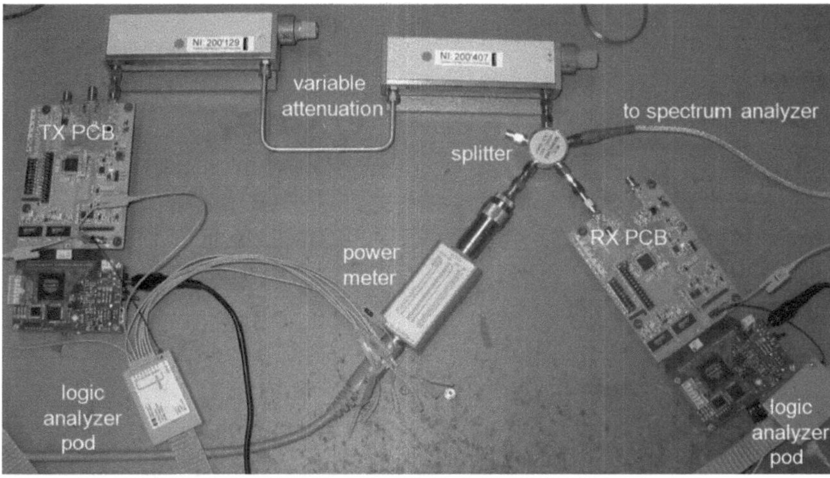

Fig. 6.35 Wired setup for BER measurements

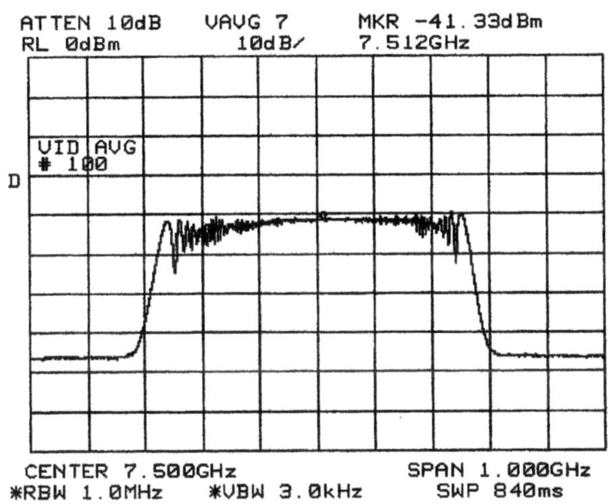

Fig. 6.36 Spectrum of the transmitter output signal

From these results it can be concluded that the FM-UWB receiver sensitivity equals −85 dBm for a BER $= 1 \times 10^{-3}$, whereas at −83 dBm the BER $= 1 \times 10^{-6}$. Results have shown to be reproducible over four prototypes.

The FM-UWB system was also measured with narrowband interference. FM-UWB signal power was −80 dBm and the Continuous Wave (CW) interferer was adjusted to check with interference level could be tolerated. It was observed that the tolerance to the CW interferer depends on its frequency. From these measurements, it was concluded that the FM-UWB radio successfully coped with this 15 dB stronger in the worst case.

Finally, the main LDR prototype figures are captured in Table 6.10.

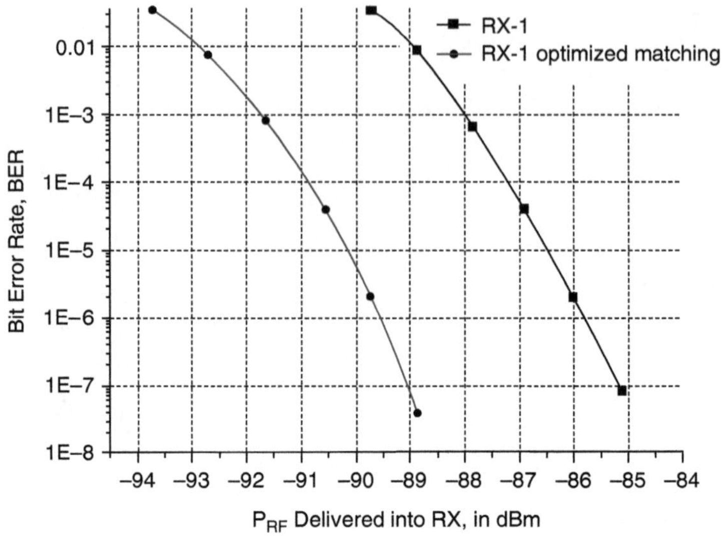

Fig. 6.37 High band receiver BER performance

Table 6.10 Comparison of initial specifications and obtained results

Parameter	Obtained
RF center frequency	6.5–8.0 GHz (Tx);7.5 GHz (Rx)
RF bandwidth	500 MHz
RF output power	−7 dBm
Subcarrier frequency	1–2 MHz
Subcarrier modulation	FSK, $\beta = 1$
Raw bit rate	62.5 kbps (125 kbps manchester enc.)
Receiver sensitivity	−85 dBm
TX, RX switching time	100 μs startup, 100 μs switch off
Latency (at PHY level)	150 μs
RX synchronisation time	8 bits
Current consumption RX	7.3 mA × 1.8 V = 13.5 mW (@max sensitivity)
Current consumption TX	3.8 mA × 1.2 V = 5 mW

6.3 High Data Rate MC-SS Prototype

The air interface presented in this section targets data rates up to 130 Mbps at reasonable implementation cost and power consumption. It is a mixture of multi-carrier OFDM based technique together with spreading, referred to as Multi-Carrier Spread Spectrum (MC-SS). This approach exploits the advantages of OFDM systems, namely a potentially low complexity equalizer and robustness against frequency selective channels (e.g. [19]) that is strengthened by code spreading. The use of Time Division Multiplex Access (TDMA) prevents the system from inter code interference experienced in Code Division Multiple Access (CDMA) approaches when many asynchronous users are sharing the same band at the same time. Considering the degrees of freedom brought by the use of frequency, time and codes, the MC-SS air interface exhibits a very high degree of flexibility from which link adaptation techniques can benefit [20]. The hardware design and implementation of this MC-SS air interface operating in the 5.2 GHz ISM band is detailed in this section. Its MAC layer is compliant with the IEEE 802.15.3 standard [21].

6.3.1 General Architecture and Key Specifications

The MC-SS High Data Rate (HDR) air interface has been optimised for Wireless Personal area Networks (WPAN) in the PN context [22]. Unlike for cellular and wireless LAN systems with which the MC-SS may be compared, peer-to-peer communications (especially from data traffic point of view) will happen in such a context. In this case, simultaneous communication between different users will lead to high interference for which CDMA would require high complexity multi-user detectors which is not compliant with the low complexity requirement of the M-HDR system. Therefore a TDMA scheme was chosen whereas the spreading codes are only used for additional diversity and flexibility. Moreover, the TDMA scheme which is considered has the advantage of being compliant with the IEEE 802.15.3 standard. An overview of the baseband PHY operations is illustrated by the block diagram of Fig. 6.38.

The HDR air interface is based on a coded OFDM modulation using convolutional coder. In the inner modem, data are spread over the subcarriers by the Spreading and Multi-Code block. This function aims at a better exploitation of channel diversity, thus yielding to more robustness. Preambule information is then appended in the time domain (after OFDM modulation) to build the PHY frame structure described in Fig. 6.39.

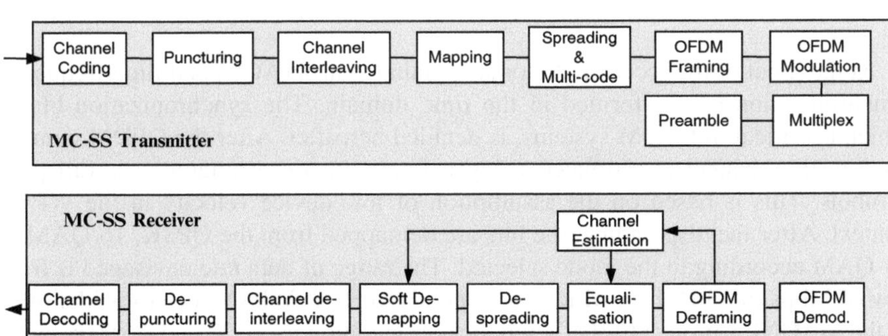

Fig. 6.38 MC-SS PHY functional diagram

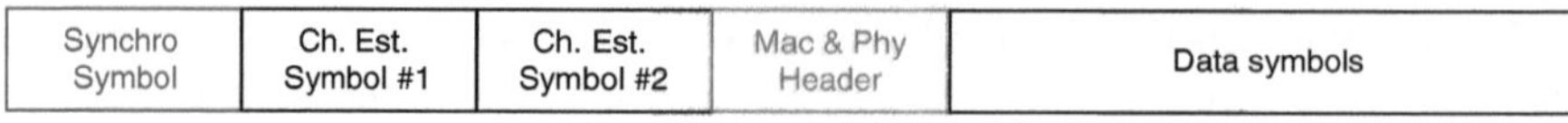

Fig. 6.39 HDR PHY frame format

Table 6.11 HDR air interface main parameters

	40 MHz	20 MHz	
Carrier frequency	5.20	5.20	GHz
Sampling frequency	40	20	MHz
FFT size	256	128	
Total subcarriers	256	128	
Subcarriers for guard band	45	23	
Subcarriers for pilot	19	9	
Subcarriers for data	192	96	
Percentage of guard band	17.578	17.969	%
Subcarrier spacing	156.250	156.250	kHz
Occupied signal bandwidth	33.13	16.56	MHz
Number of time samples per data symbol	256	128	Samples
Samples for guard interval	10	5	Samples
Samples for total OFDM burst	266	133	Samples
Maximum delay spread	0.213	0.213	μs
Sample duration in time	0.025	0.050	μs
Length of data interval in time	6.40	6.40	μs
Length of guard interval in time	0.250	0.250	μs
Length of total OFDM interval in time	6.65	6.65	S
Percentage of guard interval	3.91	3.91	%
Channel coding	Convolution code		
Generator polynomial	G1 = 133, g2 = 171		
Tail	6		Bits
Spreading factor	8	8	Chips
Maximum velocity	3	3	km/h
Maximum Doppler spread	14.4	14.4	Hz
Coherence time = $9/(16\pi f_D)$	12.4	12.4	ms

At the input of the receiver, Automatic Gain Control (AGC) and time/frequency synchronization are performed in the time domain. The synchronization block, which is critical in OFDM systems, is detailed hereafter. After the OFDM demodulation, the channel is estimated using a Least Square estimator over full pilot symbols. This is based on the assumption of low device velocity in the WPAN context. After the dispreading, the bits are demapped from the QPSK, 16-QAM or 64-QAM according to the mode selected. The range of data rate envisaged is from few of Mbps to 130 Mbps, which corresponds to HDR-WPAN scenarios identified in the MAGNET project [22]. Two modes of operation using 20 and 40 MHz bandwidth handling up to 65 and 130Mbps respectively are considered for additional flexibility.

The maximal spectral efficiency of 3.5 bits.s^{-1} Hz^{-1} is achieved using the 64-QAM. The specifications of the HDR air interface are presented in Table 6.11.

6.3.2 *RF Front End for the MC-SS System*

For the HDR platform, several receiver front end architectures have been considered, among which two have emerged as possible candidates. On one hand a zero Intermediate Frequency (IF), and on the other hand a modified Weaver [23] which achieves a rejection of the image frequency generated by the down conversion of a heterodyne receiver.

In the Weaver architecture, the signal is first mixed with the quadrature phases of the local oscillator, to be then low pass filtered (Fig. 6.40, in which IF = $RF_1 - LO = LO - RF_2$, being RF_1 the desired signal and RF_2 the image frequency that would lead to the same IF after the synthesis).

One drawback of this architecture is that it introduces the problem of a secondary image, if the second mixer translates the spectrum to a non-zero frequency. With the frequency plan considered for the HDR system, this effect may cause UMTS image frequency to interferer with desired signal.

The performance of the modified Weaver architecture in terms of rejection depends on the phase and gain mismatch between the two reception paths. For a 1–5° phase mismatch or 0.2–0.6 dB gain mismatch, such architecture achieves 30–40 dB rejection.

The parameters of the second approach, the zero IF based architecture, are specified in Fig. 6.41. The global noise factor is similar to the one of the Weaver architecture.

The zero IF receiver does not suffer from image interference thanks to the direct conversion nature of this architecture, but potential interference may come from the IEEE 802.11a systems operating in the same ISM band. Therefore, rejection filtering concerns fall on this WLAN system. The filtering contribution is shared

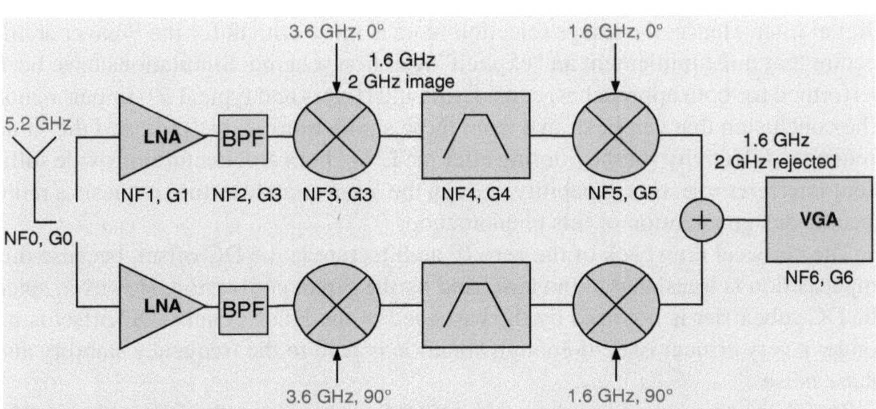

$$NFtotal = NF0 + \frac{NF1-1}{G0} + \frac{NF2-1}{G0.G1} + \frac{NF3-1}{G0.G1.G2} + \frac{NF4-1}{G0.G1.G2.G3} + \frac{NF5-1}{G0.G1.G2.G3.G4} + \frac{NF5-1}{G0.G1.G2.G3.G4.G5}$$

Fig. 6.40 Weaver RF architecture

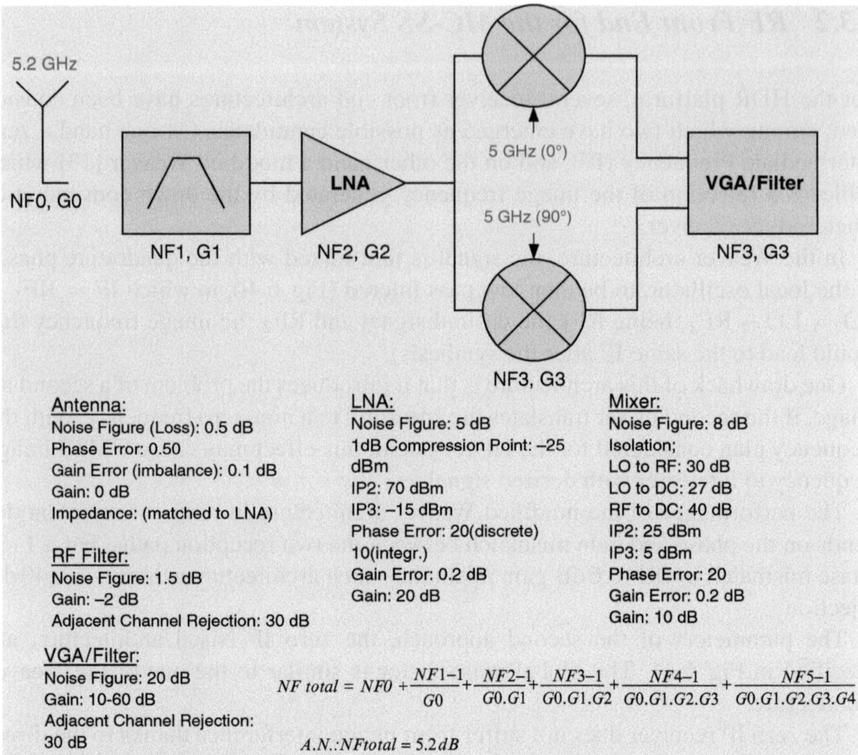

Fig. 6.41 Zero-IF RF architecture

between the Radio Frequency (RF) filter, the Analogue Base Band filter and the Digital filter. Hence, the image rejection issue is more critical for the Weaver architecture that must implement an "explicit" rejection scheme. Simulations have been performed for both approaches, considering interferers and typical RF impairments. The conclusion that can be drawn from these simulations is that provided the same frequency selectivity for the filtering after the LNA, both architectures provide sufficient interferer rejection capability, though the Weaver architecture requests a more specific design attention of this phenomenon.

The classical drawback of the zero IF architecture is the DC offset, because this imperfection is translated to the baseband by the direct conversion. However, since the DC subcarrier is not used by the baseband of the HDR system, DC offset is no longer a very critical issue if enough attention is paid to the frequency stability and phase noise.

The phase noise is imposed on each OFDM subcarrier by the RF synthesis. The phase noise is generated by the RF frequency synthesis of Phase Locked Loop (PLL) and mixed with the RF signal, thus affecting down-converted baseband signal by a random phase shift in the time domain (before FFT). The influence of the phase noise in the OFDM signal appears in two different ways in the frequency domain as reported in [24]:

- A Common Phase Error (complex value) is multiplied to all subcarriers. This error comes from close to carrier phase noise. This error can be tracked and removed by equalization.
- Due to further to carrier phase noise, subcarriers are mixed together at FFT process such a way Inter Carrier Interference appears as hardly-removable extra noise in the signal.

This leads to the need for a trade-off between signal processing extra-computation (Common Phase Error tracking) and requirements on the PLL and crystal choices.

Innovative design works [25–28] presented different techniques that provided improved reliability and yield of CMOS RF transceivers, what has made, after the proper evolution in the research areas, CMOS process a real player in the cost-effective radio market. Single chip solution offers as well several advantages such as reduction in manufacturing and packaging costs due to the elimination of the routing between different integrated circuits, leading to a Printed Circuit Board (PCB) multilayer complexity reduction. Smaller areas and diminished consumption (simplification of internal interfaces between blocks) jointly with shorter factory test times and higher test yields are other benefits of the single chip designs. For these reasons the zero IF approach was preferred and the MAXIM MAX2829 chip was used as the heart of the RF part of the design. Besides, the included PLL bandwidth and the chosen crystal reference made negligible the extra distortion caused by the phase noise effects.

6.3.3 Baseband Processing and Channel Coding

Like any OFDM system, the HDR air interface presented herein is sensitive to synchronization error and a particular attention has been made to handle robust synchronization at the receiver. Another specific concern for real-time digital design of the HDR air interface is the clock management. Finally, hardware implementation errors (quantization noise, operator bias etc.) impact on processing precision has to be quantified to properly adapt the datapath dynamic range. Implementation loss induced by the baseband processing is scarcely addressed in the literature. In this section the error introduced by the digital baseband processing is quantified and its impact is given in terms of equivalent Aditive White Gaussian Noise (AWGN) signal on the ideal signal.

6.3.3.1 Synchronization

The synchronization aims at referencing in time the FFT vector for OFDM demodulation and at estimating the Carrier Frequency Offset (CFO) in the time domain (pre-FFT). CFO corresponds to the TX/RX oscillator frequency shift. Correcting the CFO is of paramount importance for OFDM systems which are very sensitive to this impairment [29]. Synchronization is processed on the fly and runs continuously once the AGC is locked. It seeks a specific synchronization pattern contained in each frame header [21]. The synchronization process is ruled by a Finite State Machine (FSM) whose state is updated every received sample. It synchronizes the data flow according to the strongest path of the channel which is used as time reference.

The time synchronization is performed as follows. First, the autocorrelation of the received signal is computed. The periodic nature of the synchronization pattern enables the autocorrelation to show a typical flat region when the synchronization symbol is received [30]. When the flat region is detected, the synchronization sample is coarsely indexed. To refine the position of this synchronization point, a more restricted window is considered and the cross-correlation of the input signal with the known synchronization pattern is analyzed throughout this window. In fact, the window is active when the autocorrelation signal is higher than the threshold over more than a predetermined time. This time is related to the synchronization pattern duration. Peaks appear on the cross-correlation profile as soon as the known pattern is completely received. As previously, a criterion to detect those peaks is defined. When the last cross-correlation peak is received, the system can be synchronized accurately.

In order to determine the best threshold value, the synchronization Probability of False Alarm (pfa) or MisDetection (pmd) are analyzed. The pfa and pmd as a function of the autocorrelation threshold are given in Fig. 6.42 for an AWGN channel. The threshold is represented as a percentage of the maximum value of the autocorrelation.

The pfa is defined as the probability of finding a synchronization sample while no synchronization symbol was transmitted. Obviously, it decreases when the

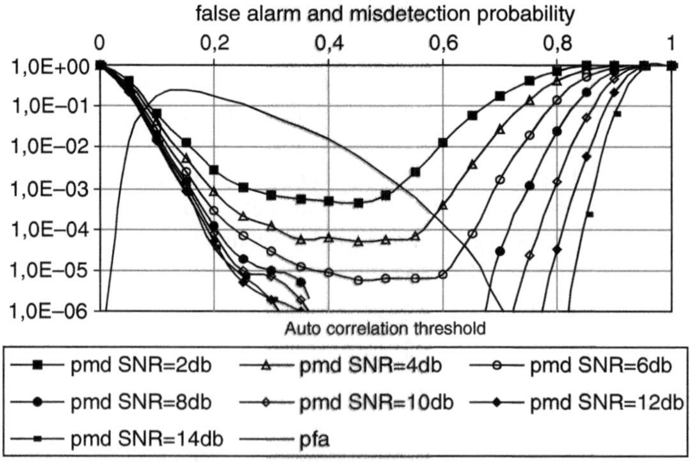

Fig. 6.42 False alarm and misdetection probability

autocorrelation threshold increases. The pfa does not depend on the Signal to Noise Ratio (SNR). This is due to the fact that the flat region never appears when no synchronization pattern is sent, whatever the noise level.

The pmd is defined as the probability of missing the synchronization point despite the transmission of a synchronization symbol. For the lowest thresholds, the misdetection is mainly due to bad flat region localisation, or as for the false alarm, to the absence of autocorrelation flat region falling edge. For high thresholds, misdetection is also high but mainly due to non detection of the flat region. Between these two threshold regions, a minimum is obtained around 0.5.

A pfa vs. pmd trade-off value can be obtained for each SNR as the crossing point of the misdetection and false alarm curves. For instance, the SNR = 8 dB provides pfa $< 10^{-5}$ and pmf $< 10^{-5}$ choosing the threshold equal to 68%. When higher SNR are targeted, increasing the threshold will reduce the false alarm probability. For 10 dB, setting a 70% threshold brings about pfa $< 10^{-6}$ and pmd $< 10^{-6}$ [29].

6.3.3.2 Clock Management for Flexible Design

Bringing flexibility to the baseband in terms of data rate increases the complexity of clock management. For the sake of clarity, the focus of this section is on the 40 MHz system but can be transposed to the 20 MHz case easily. The convolutional encoder is fed with data at sampling frequency f. The coder produces two parallel bits which are serialized before being punctured. Let N be the number of bits per symbol, D the serial output data rate of the convolutional encoder, R the global code rate, P the puncturing rate and f the processing frequency if only one frequency were used in the design. Since each OFDM symbol of 266 samples carries 192 data, the serial bit rate at the output of the coder is $\mathrm{D} = 192 \times 40 \times \mathrm{N}/266 \approx 29 \times \mathrm{N}$. At the output of the puncturing, the data are at the frequency f_s. Table 6.12 recaps the frequency to use at the coder module according to the MAGNET modulation scheme implying different clock frequency. Considering all the configurations of Table 6.12, a very flexible clock management needs to be implemented. This can be advantageously achieved through the use XILINX Virtex 4 tunable DLL feature that enables to dynamically change the clock frequencies in the design. Hence, a single design can handle all configurations by simply configuring the mode register without the need to reload the FPGA. This is a must to support highly dynamic change in the

Table 6.12 Modulation and coding configurations

						Nb bit/OFDM symbol	
Modulation	N	R	D	P	F	Coder input	Coder output
QPSK	2	1/2	58	1	58	192	384
QPSK	2	3/4	58	2/3	87	288	384
16 QAM	4	1/2	116	1	116	384	768
16 QAM	4	3/4	116	2/3	174	576	768
64 QAM	6	2/3	174	3/4	232	768	1,152
64 QAM	6	3/4	174	2/3	261	864	1,152

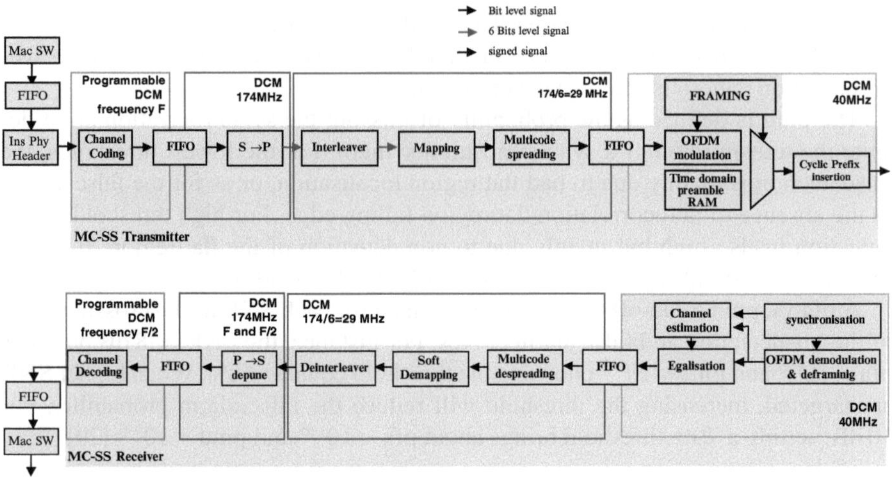

Fig. 6.43 M-HDR baseband clock management

modulation and coding, as for instance to support adaptive modulation and coding schemes or when several communications use different coding/modulation schemes in TDMA mode.

The interleaver is using a parallel architecture which width is determined by the one of a symbol, although the interleaver intrinsically processes bits. This parallel approach was chosen due to frequency requirements for real-time operation. A serial implementation would indeed have had to sustain 174 MHz operation rate in the worst case. In the case of a parallel implementation, the operating frequency is determined by $D/N = 29\,MHz$. As a consequence, the parallelisation which is usually performed before the mapper in OFDM modems, sources here the input of the interleaver. In order to simplify the clock management, the serial to parallel converter always works at the highest frequency and the data validation signal duty cycle is adjusted according to the modulation. This choice leads to a very small part of the design working at high frequency. This part does not need to be changed according to the modulation. The mapper and the spreader that follow, process at the modulation symbol rate, namely 29 MHz. Then, pilots are inserted increasing the rate up to 40 MHz for the OFDM modulation. Figure 6.43 shows the resulting clock domains.

6.3.4 MAC Layer and Connectivity

The generic MAC architecture for a device capable of supporting HDR air interface has been developed with the functional partitioning between the Host and the Network Interface Card (NIC). The MAC primitives implemented are, from a functional point of view, very similar to the ones described in Sect. 6.2.6.1. The NIC implements the HDR air interface prototype which consists of MAC and PHY layer with an appropriate interface to the host platform. USB is chosen as the default

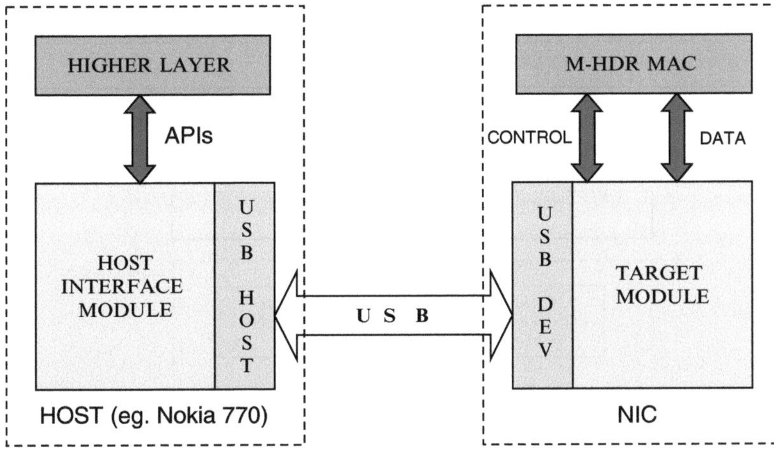

Fig. 6.44 HDR MAC Implementation Architecture

physical interface to the host. The network layer and the applications are implemented on the host platform (NOKIA 770 PDA). Figure 6.44 depicts a high level MAC architecture for the HDR air interface.

From the implementation point of view the three following modules were implemented.

- The HDR MAC module contains the implementation for the core MAC functionalities, e.g. beacon transmission for piconet formation, channel scanning for piconet discovery, synchronization with other devices, association/disassociation requests to join and leave piconet, and asynchronous/isochronous data transmission. On the data path this module exchanges Logical Link Control (LLC) frames with the host while the control path is used to exchange various management commands, e.g. set or fetch configuration parameters. In order to achieve the required real-time performance the MAC is partitioned into hardware (HW-MAC) and software (SW-MAC). The time critical and computation intensive blocks like CRC generation and ciphering are implemented in hardware as part of HW-MAC. In the following sub-sections we elaborate on both the software and hardware parts of the MAC implementation.
- The Target Module of Fig. 6.44 acts as an interpreter for the messages it receives from the Host over the USB link. It translates these commands into IEEE 802.15.3 format and forwards them to the HDR MAC module for further processing.
- The Host Interface Module implements the Application Programmable Interfaces (APIs) which are used by the higher layers to access various MAC functionalities.

To facilitate message exchange between the Host and the NIC, a frame format has been specified. As shown in Fig. 6.45, it contains a frame identifier field which uniquely identifies the type of the frame, a payload size field of two bytes which

0	1	3
FRAME IDENTIFIER	PAYLOAD_SIZE	PAYLOAD

Fig. 6.45 Frame format for Message Exchange between the host and HDR NIC

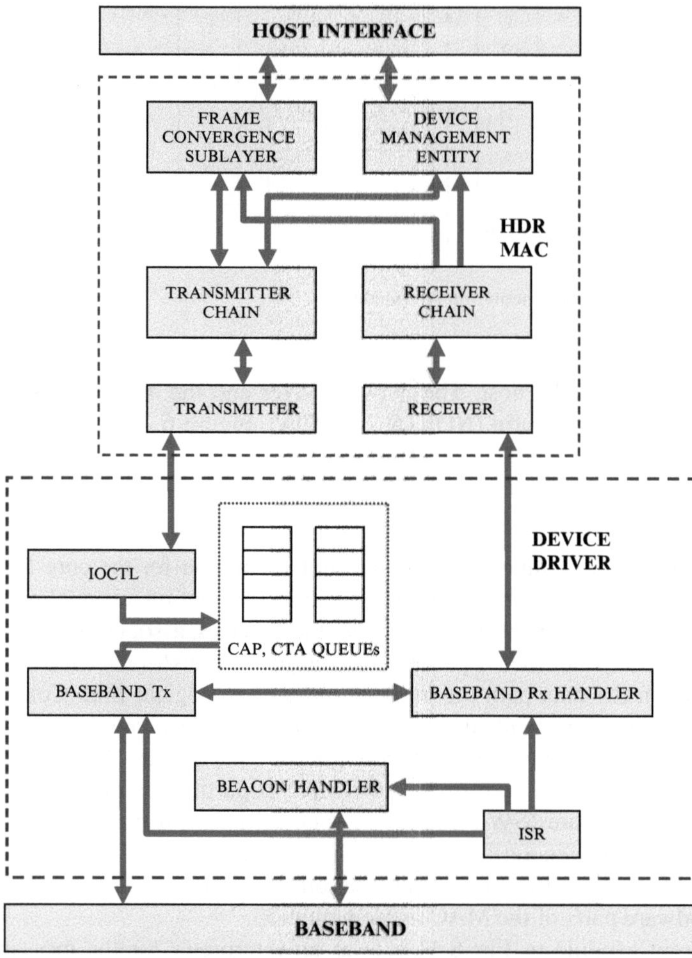

Fig. 6.46 Architecture of the IEEE 802.15.3 MAC implementation

provides the length of the attached payload. The payload field consists of the parameters specified with the command and can be a maximum of 2,048 bytes.

As mentioned earlier the HDR MAC is derived from IEEE 802.15.3 and the implementation architecture is shown in Fig. 6.46.

6.3.5 SW-MAC Design

Non real-time critical primitives of the MAC are implemented in software (SW-MAC) running on an embedded General Purpose Processor (GPP).

The TX-frame processing block is mainly responsible for the formation of the data and command frames to be transmitted. Upon receiving the data/command request from the Frame Convergence SubLayer (FCSL) or the Device Management Entity (DME), the transmitter chain validates the request e.g. the sourceid, dstid, data length and stream index parameters. The MAC frame prepared by attaching the MAC header and the payload are then sent to the transmitter for the transmission over the air.

The Transmitter Block puts these frames into appropriate device driver queues for transmission. The device driver implements two transmission queues – one for transmissions during the Contention Access Period (CAP) and the other for transmissions during the allocated channel time (CTA).

The RX-frame processing module is responsible for receiving the frames from the baseband and forwarding them to the FCSL or DME. The receiver upon receiving frames from the baseband verifies the frame for the command or the data. The command frames are forwarded to the DME block and the data frames are passed to the FCSL block.

The Receiver Block coordinates the packet reception between the Receiver chain and the baseband device driver.

From an implementation point of view each of these blocks are implemented as a separate thread. These threads communicate with each other using the Linux message queues as the Inter-Process Communication (IPC) mechanism. The synchronization between the threads is achieved by the use of semaphores. The Linux system calls are implemented as a thin Operating System Abstraction Layer (OSAL). The OSAL implements the generic wrapper functions over the OS dependent system calls.

The multi-threaded program implementation of the HDR SW-MAC is illustrated in Fig. 6.47. The module is activated by a call to the main function which in turn invokes the initMAC() function. The initMac() function initialises the framework

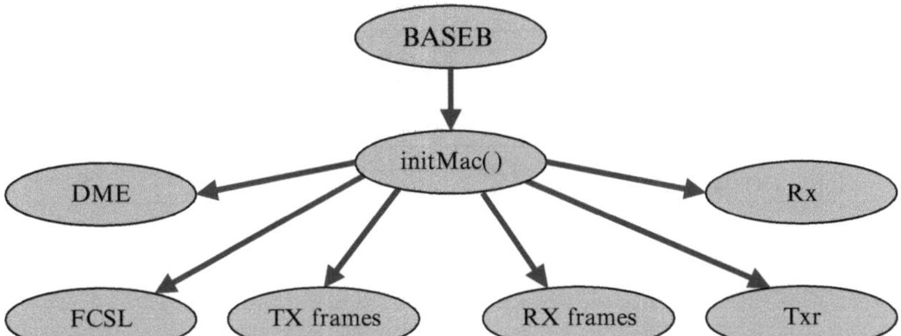

Fig. 6.47 A Multi-threaded Implementation

by creating the threads for each of the DME, FCSL, Transmitter Chain, Receiver Chain, Transmitter and the Receiver block. The associated message queues, registers, memory pool, PIB (PAN Information Base) parameters are also initialised.

6.3.6 HW-MAC Primitives

The hardware MAC (HW MAC) is present at the interface between the PHY layer and the SW-MAC layer. It inherits some terminal functions of the MAC layer to achieve improved real-time performance as compared to that performed when in software. The HW-MAC handles all the data processing in order to provide the PHY layer with the required format of the packet to be transmitted and the received packet from PHY BB to the SW MAC. The HW-MAC consists of several blocks like the hardware 128 bit Advanced Encryption Standard (AES) [31, 32], unit which benefits from the implementation described in Fig. 6.48, CRC generation/verification units and register address space along with an address decoder. The top-level finite state machine is the intelligence behind the working of HW MAC. It schedules and synchronizes the data flow between SW MAC and PHY BB depending on type of configuration defined by the SW-MAC.

The block diagram of the HW MAC depicting the flow of data between SW MAC and PHY BB is shown in Fig. 6.48. The presence of HW-MAC makes the SW-MAC perceive the PHY layer as any other peripheral. This is because the HW-MAC provides the SW MAC an interface similar to a memory. Various configurations and status registers including the data and header FIFOs are mapped onto an address space to which the SW-MAC can write. If the SW MAC requires transmitting a data packet over the air, it writes the configuration in the registers and the data to be transmitted into the FIFOs. The HW MAC delivers it to the PHY layer according

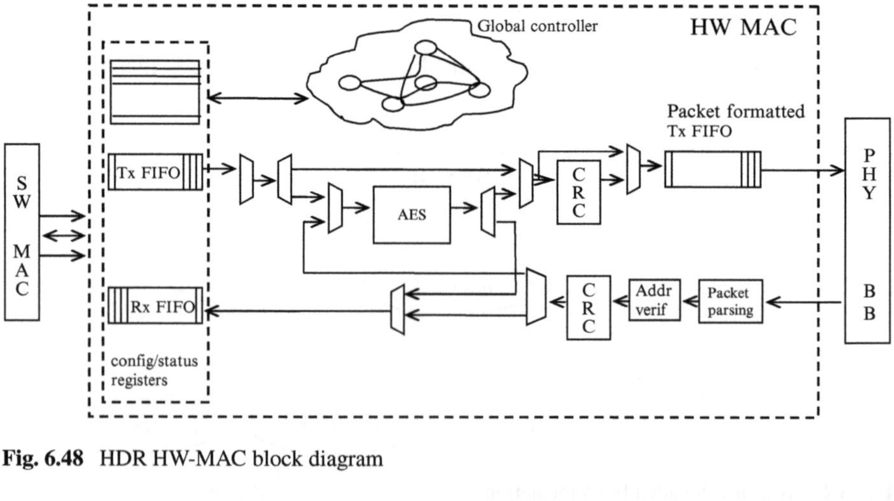

Fig. 6.48 HDR HW-MAC block diagram

to the configuration set by the SW-MAC. Conversely, when a packet is received from the PHY layer and if it is intended for a device in receive mode, the HW-MAC verifies the packet for its integrity and interrupts the SW-MAC to inform about the received packet. Besides these scheduling functions, the HW-MAC also implements primitives to speed up the computation of data. This concerns encryption, CRC generation/verification and other minor functions like packet parsing, packet formatting, timers, etc.

6.3.7 HDR Hardware Prototype

The HDR prototype consists of a set of boards that embed the components needed for the implementation of MAC and PHY layers. Namely, an RF board implementing TX and RX RF functions (from/up to the converters up to/from the antenna) and a digital board implementing digital PHY and MAC functionalities. The latter also includes some host bridging features in order to plug the HDR prototype to a host device. An overview of the HDR prototype is illustrated in Fig. 6.49.

As mentioned above, the HDR RF subsystem (or board) is based on an off the shelf component from MAXIM (MAX2829). The MAX2829 is designed for dual-band 802.11 a/g applications covering especially world-band frequencies of 4.9–5.875 GHz. The IC includes all circuitry required to implement the RF transceiver functions, providing a fully integrated receive path, transmit path, VCO, frequency synthesizer, and baseband/control interface. Only the power amplifier, RF switches, RF bandpass filters (BPF), RF baluns, and a small number of passive components are needed to form the complete RF front-end solution.

The digital board houses the programmable chips that implement baseband PHY and MAC functions. For SW MAC primitives an ARM9 has been selected (AT91RM9200). The SW MAC primitives run on top of a Linux OS. For the HW-MAC and PHY primitives, a Xilinx Virtex 4 has been chosen due to hardware resource available and flexible clock management capability (XC4VSX55–10). Complexity analysis that led to the selection of this chipset is provided in Table 6.13. The NIC is used by its host as a USB device.

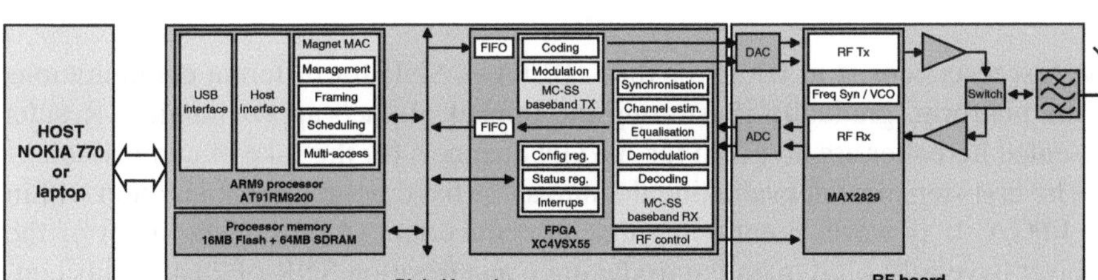

Fig. 6.49 HDR platform block diagram

Table 6.13 HDR digital complexity analysis

PHY/MAC HW (FPGA)	Logic (slices)	Multipliers (18 × 18)	Block RAM (18 kb)	Clock domains (DCM)
Required for TX baseband	3,800	18	30	4
Required for RX baseband	10,200	118	49	4
Total required	14,000	136	79	4

MAC SW (processor)	Computational requirement	ROM requirement	RAM requirement
Requires for MAC	∼180–200 MIPS	8	64

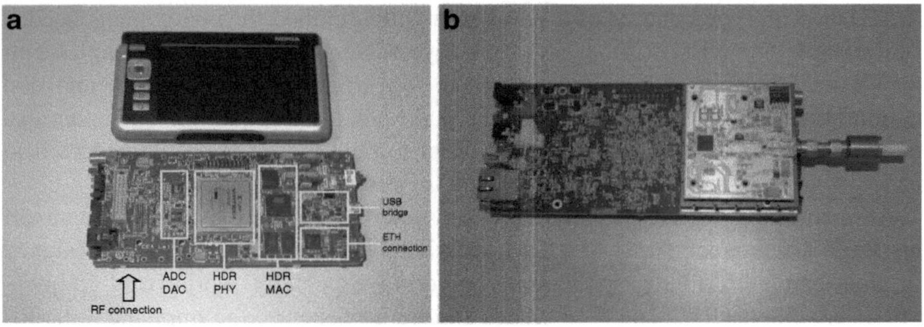

Fig. 6.50 HDR prototype–digital side **(a)**, RF side **(b)**

Figure 6.50 shows the HDR prototype. On the left (a) the digital side of the board is shown with the main components identified. In terms of form factor, it can be seen that the prototype has approximately the size of the host PDA, a NOKIA770. On the right (b) the other side of the board is shown with the RF daughter module than can be seen clearly in Fig. 6.50.

6.3.8 HDR Key Test Results

The first tests consist in Bit Error Rate (BER) vs SNR for different configurations of the platform, gradually illustrating the impact of each approximation. Results presented hereafter are all given for AWGN channels for the sake of comparison.

The first step aims at evaluating the impact of fixed point implementation within the FPGA. It is worth mentioning that the converters (ADC) at the input of the receiver have a 12 bit dynamic introducing a quantization SNR of 72 dB. This conversion noise is negligible within the SNR range addressed by the receiver. In order to see the impact of fixed point computation, the BER vs SNR performance of the prototype was compared with the floating point simulation model. In both cases

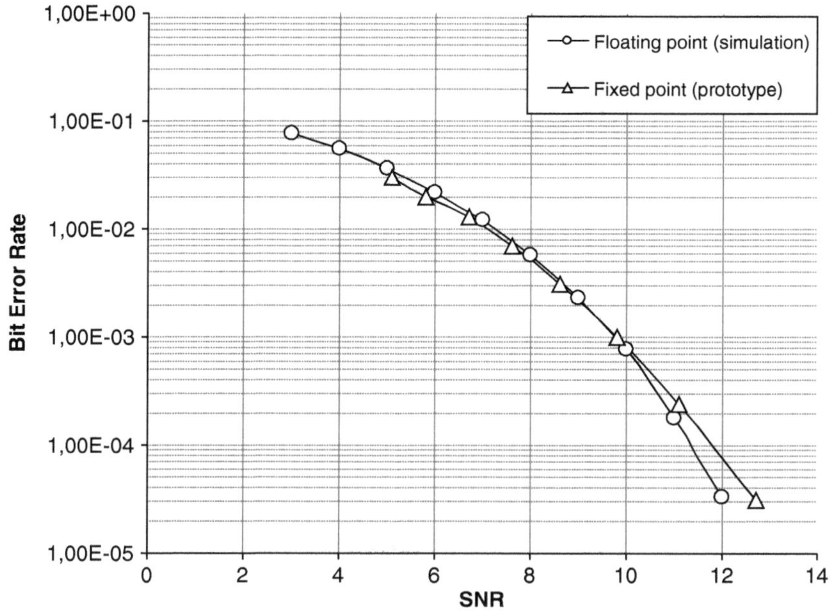

Fig. 6.51 Impact of fixed point computation for non-coded QPSK configuration

measurements are performed under perfect channel estimation and without CFO estimation (both TX and RX share the same clock reference and CFO estimator is disabled). The results are shown in Fig. 6.51. It can be noted that no significant degradation is introduced by the fixed point representation.

Results provided hereafter are coming from prototype measurements. Figure 6.52 shows the additional impact of CFO estimation and channel estimation on the HDR baseband performance. The previous curve obtained with perfect estimation is given as a reference. This reference curve is similar to the one of Fig. 6.51.

It can be observed that the major degradation is brought by the channel estimator linked to the zero-forcing equalizer. However, the 3 dB shift is rather due to the kind of estimator chosen rather than its implementation, since floating point simulations provide similar results. It can be noticed that the CFO estimation and correction has little impact on the overall performance.

Finally, Fig. 6.53 shows the impact of the RF on the performance of the QPSK 1/2 system. The baseband performance curve obtained previously is provided as a reference.

From this result it can be concluded that the RF section does not introduce significant degradation on the phase modulation. This was confirmed by analysing phase noise performance of the RF section alone.

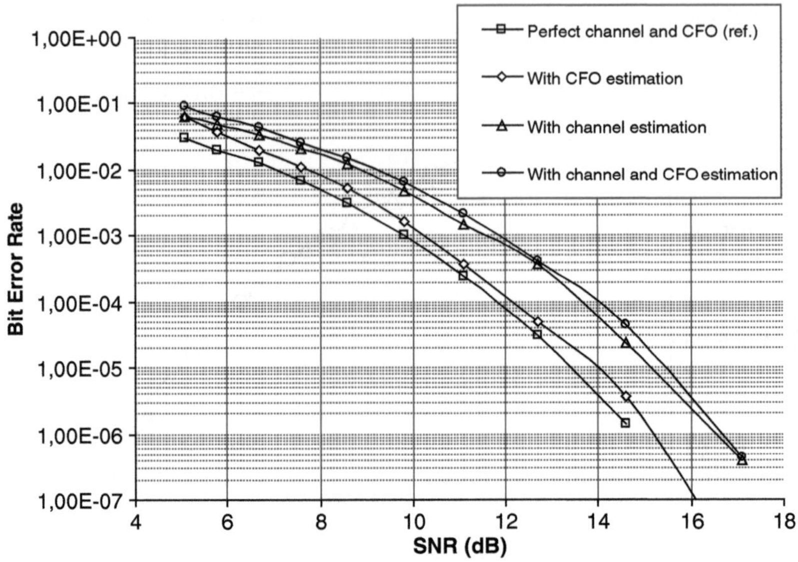

Fig. 6.52 Impact of CFO and channel estimation for non-coded QPSK configuration

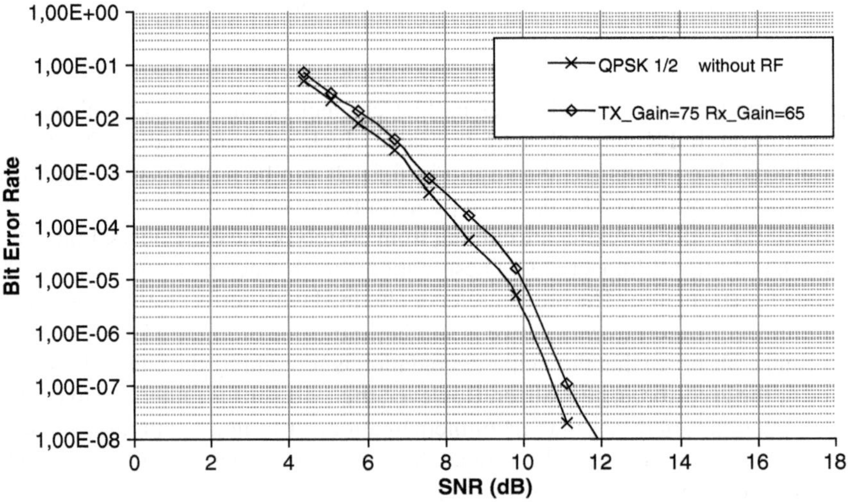

Fig. 6.53 Digital baseband vs system including RF performance for QPSK $^1/_2$ configuration

6.4 Conclusions

This chapter presented the design implementation and test of three prototypes dedicated to WPAN applications.

The LDR prototype, which relies on FM-UWB modulation shows very high sensitivity (−86.6 dB) and low power consumption figures. This high sensitivity enables communication range of 10–15 m even when very low radiated UWB power is considered. Two versions of the FM-UWB system have been implemented covering low and high UWB band operation. They are built around specific chipsets using advanced RF IC design implemented in CMOS (low band) and BiCMOS (high band) technologies.

The HDR prototype has also proved performance very close to the simulations. The focus of the HDR design emphasizes the baseband section which needed to be flexible. Using a versatile clock management system, this flexibility could be achieved. A specific attention has also been put on RF impairment correction such as CFO. These correction schemes have proved to be efficient based on the measurement performed.

These implementations pave the way towards further integration that could target System On Chip design implementation. Such system on chip would make sense from a technical perspective. The FM-UWB prototype is used as a proof of concept demonstrator towards the IEEE 802.15.6 standardization group, in which the FM-UWB technology is considered as a candidate.

References

1. J.F.M. Gerrits, J.R. Farserotu, J.R. Long, Principles and limitations of ultra-wideband FM communication systems. EURASIP J. Appl. Signal Process. **3**, 382–396 (2005)
2. T. Tong, Z. Wenhua, J. Mikkelsen, T. Larsen, A 0.18 um CMOS low power ring VCO with 1 GHz tuning range for 3–5 GHz FM-UWB applications. IEEE 10th International Conference on Communication Systems, Singapore, Oct 2006, pp. 1–5
3. C.-C. Wei, H.-C. Chiu, W.-S. Feng, An ultra-wideband CMOS VCO with 3–5 GHz tuning range. IEEE International Workshop on Radio-Frequency Integration Technology, Singapore, Nov 2005, pp. 87–90
4. W. Tu, J. Yeh, H. Tsai, C. Wang, A 1.8V 2.5–5.2 GHz CMOS dual input two stage ring VCO. IEEE Asia-Pacific Conference on Advanced System Integrated Circuits, Fukuoka, Japan, Aug 2004, pp. 134–137
5. T. Rui, M. Berroth, The design of 5 GHz voltage controlled ring oscillator using source capacitively coupled current amplifier. IEEE Radio Frequency Integrated Circuits Symposium (RFIC), Philadelphia, Pennsylvania, Jun 2003, pp. 623–626
6. A. Rezayee, K. Martin, A coupled two-stage ring oscillator. IEEE Midwest Symp. Circ. Syst. (MWSCAS) **2**, 878–881, 2001
7. A. Georgiadis, M. Detratti, A linear, low power, wideband CMOS VCO for FM-UWB applications, Wiley Microwave and Optical Technology Letters, **50**(7), 1955–1958, July 2008
8. D. Ham, A. Hajimiri, Concepts and methods in optimization of integrated LC VCOs. IEEE J. Solid-State Circ. **33**, 179–194 (Feb 1998)
9. Y. Dong, Y. Zhao, G. van Veenendaal, J. Long, J. Gerrits, A 9mW high band fm-UWB receiver front-end. IEEE ESSCIRC'08, Edinburgh, UK, Sept 2008
10. P. Deixler, A. Rodriguez, W. De Boer, H. Sun, et al., QUBIC4X: An fT/fmax= 130/140GHz SiGe:C-BiCMOS Manufacturing Technology with Elite Passives for Emerging Microwave Applications, IEEE Bipolar/BiCMOS Circuits and Technology Meeting, Sept 2004
11. B. Nauta, Single-to-differential converter. US Patent 5,404,054 4 Apr 1995
12. S. Samadian, R. Hayashi, A.A. Abidi, Demodulators for a zero-IF bluetooth recevier. IEEE J. Solid-State Circ. **38**(8), 1393–1396 (Aug 2003)
13. C. Desset, Selection of channel coding for low-power wireless systems. Vehicular Tech. Conf. **3**, Jeju, Korea, 1920–1924 (Apr 2003)

14. S.B. Wicker, *Error Control Systems for Digital Communication and Storage* (Prentice Hall, Englewood Cliffs, NJ, 1995)
15. S. Choomchuay, B. Arambepola, Time domain algorithms and architectures for Reed-Solomon decoding. IEE Proc. Commun. Speech Vision **140**(3), 189–196 (Jun 1993)
16. H. Lee, M.-L. Yu, L. Song, VLSI design of Reed-Solomon decoder architectures, *IEEE International Symposium on Circuits and Systems* (ISCAS), Geneva, Switzerland, **5**, 705–708 (May 2000)
17. S.-F. Wang, H.-Y. Hsu, A.-Y. Wu, A very low-cost multi-mode Reed-Solomon decoder based on Peterson-Gorenstein-Zierler algorithm. *IEEE Workshop Signal. Processing Systems*, Antwerp, Belgium, Sept 2001, pp. 37–48
18. L. Biard, D. Noguet, Reed-Solomon codes for low power applications. *Journal of Communication (JCM)* (Academy publisher, Grenoble, France, 2008)
19. R. Prasad, *OFDM for Wireless Communication Systems* (Artech House, Boston, 2004)
20. K. Schoo, F. Bauer, K. Strohmenger, Adaptive modulation and coding in a PAN optimized air interface considering computation complexity. IST Mobile Summit, Myconos, Greece, June 2006
21. IEEE 802.15.3 Standard. Part 15.3: Wireless Medium Access Control (MAC) and Physical Layer (PHY) Specifications for High Rate Wireless Personal Area Networks (WPANs)
22. R. Prasad, K. Skouby, Personal network (PN) applications. Wireless Pers. Commun. **33**(3–4), 227–242 (2005)
23. T.E. Dodgson, E. Lee, P. Gardner D. Noguet, Reconfigurability in its application to platforms for Private-Personal Area Networks and Personal Networks. 15th Wireless World Research Forum, Dec 2005
24. L. Maret, C. Dehos, M. Bouvier Des Noes, D. Morche, J. Barletta, Sensitivity of a MC-CDMA beyond 3G system to RF impairments. 14th IST Mobile and Wireless Communications Summit 2005, Dresden (Germany)
25. R.L. Hovald, The communications performance of single-carrier and multi-carrier quadrature amplitude modulation in RF carrier phase noise. Ph.D. thesis, Drexel University, Dec 1997
26. C.W.S. Tim, E.R. Fledderus, P.F.M. Smulders, Performance impact of IQ mismatch in direct-conversion MIMO OFDM transceivers. Proceedings of the IEEE Radio Wireless Symposium 2007, Long Beach CA, Jan 2007, pp. 329–332
27. K. Vavelidis, et al., A dual band 5.15–5.35 GHz, 2.4–2.5 GHz 0.18 um CMOS Transceiver for 802.11a/b/g Wireless LAN. IEEE J. Solid State Circuits **39**(7), 1180–1184 (July 2004)
28. K. Ming-Dou, H. Yuan-Wen, On-chip ESD protection strategies for RF circuits in CMOS technology. 8th International Conference on Solid-State and Integrated Circuit Technology, ICSICT '06, Shanghai, China, October 2006
29. M. Laugeois, D. Noguet, N. Cassiau, Robust timing synchronization for OFDM based transmission. Wireless Personal and Multimedia Communication (WPMC), Jaipur, India, 2007
30. T.M. Schmidl, D.C. Cox, Robust frequency and timing synchronization for OFDM. IEEE Trans. Commun. **45**, 1613–1621 (Dec 1997)
31. H. Li, J. Li, A high performance sub-pipelined architecture for AES. Proceedings of ICCD 2005, San Jose, CA, USA, pp. 491–496
32. H. Li, Z. Friggstad, An efficient architecture for AES mix columns operation. IEEE Int. Symp. Circuits Syst. **5**, 4637–4640 (May 2005)

Chapter 7
PN Platforms

Juha Zidbeck, Luis Sánchez, Kimmo Ahola, Mikko Alutoin, Martin Bauer, Sandford Bessler, Marc Girod Genet, Jeroen Hoebeke, Jorge Lanza, Ingrid Moerman, Rasmus L. Olsen, Jordi Jaen Pallares, and Joachim Zeiss

7.1 Introduction

The development of new research paradigms is usually not supported by a proof-of-concept that helps to showcase the potential impact of the research concept behind. Personal Networking is an emerging concept which combines pervasive computing and strong user focus. The idea is that the user's personal devices organize themselves in a secure and private personal network transparently of their geographical location or the access technologies used. The user expects the network to be always ready for supporting her/his necessities without requiring too much involvement on the user's side. Additionally, the PN must be ready to share the services it provides to the user with other users that have been authorised in order to allow the collaboration between the PNs' users. The PN Federation concept is presented as a secure cooperation between a subset of devices belonging to different PNs for the purpose of achieving a common goal or service by establishing an alliance. This chapter presents the highlights of the implementation of a full-blown

J. Zidbeck (✉), K. Ahola, and M. Alutoin
Technical Research Centre of Finland, P.O. Box 1000, FIN-02044 VTT, Finland
e-mail: Juha.Zidbeck@vtt.fi

L. Sánchez and J. Lanza
Universidad de Cantabria, Spain

M. Bauer
NEC Europe Ltd., Germany

S. Bessler and J. Zeiss
Forschungszentrum Telekommunikation Wien Betriebs-GmbH, Austria

M.G. Genet
Groupe des Ecoles des Télécommunications – Institut National des Télécommunications, France

J. Hoebeke and I. Moerman
Interuniversitair Micro-Elektronica Centrum vzw, Belgium

R.L. Olsen
Aalborg University, Denmark

J.J. Pallares
Fraunhofer Institut FOKUS, Germany

R. Prasad (ed.), *My Personal Adaptive Global NET (MAGNET)*, Signals and Communication Technology, DOI 10.1007/978-90-481-3437-3_7,

Personal Networking system carried out and the set up of a pan-European testbed where the system can be subject of functionality and performance tests as well as be used to demonstrate the potentiality of Personal Networking concept.

Take the concept of pervasive computing and combine it with strong user focus and you get Personal Networks (PN) [1, 2]. PN is a collection of one's most private devices referred to as personal nodes. The PN consist of devices sharing a common trust relationship. Security and privacy are the fundamental properties of the PN, as well as its ability to self-organize and adapt to mobility and changing network environments.

The IST project MAGNET vision is that PNs will support the users' professional and private activities, without being obtrusive and while safeguarding privacy and security [3]. A PN can operate on top of any number of networks that exist for subscriber services or are composed in an ad hoc manner for this particular purpose. These networks are dynamic and diverse in composition, configuration and connectivity depending on time, place, preference and context, as well as resources available and required, and they function in cooperation with all the needed and preferred partners.

As shown in Fig. 7.1, the PN consists of clusters of personal nodes. One cluster is special, so-called Private Personal Area Network (P-PAN), because it is located around the user. The clusters are connected with each other via an interconnecting structure, which is likely to be infrastructure based.

In order to protect the privacy of the user and the integrity of the PN, security measures are used to encrypt the user's data when it is sent outside of the device, i.e. using a wireless medium or the infrastructure. The user can reach all of his or her devices using a variety of underlying networking technologies, which are invisible to the user. The user only sees the services that are available in the PN and on foreign nodes that have been made available.

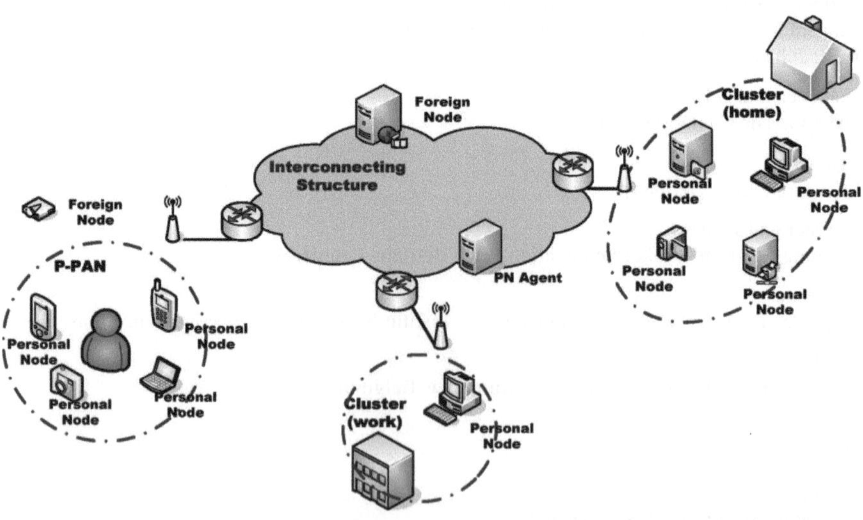

Fig. 7.1 Personal network

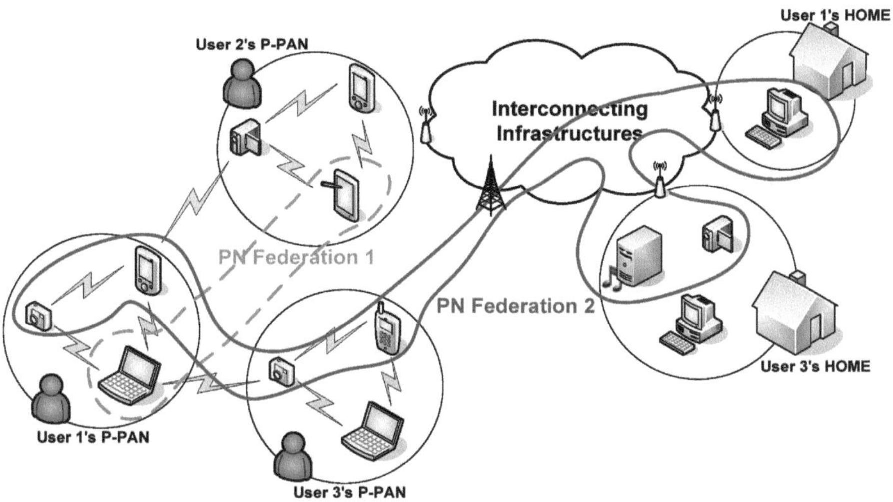

Fig. 7.2 Personal network federation

Nonetheless, personal communications cannot be restricted to the services provided by the devices the user owns, but the possibility to interact with other user's PN has to be enabled in order to support the user in his/her private and professional activities. The concept of PN Federations (PN-F) (Fig. 7.2) is even a more challenging one since the relations between users have to be managed and the security has to be reinforced in order to not open security holes while allowing authorized users to cooperate with you. PN-F is a secure cooperation between a subset of devices belonging to different PNs for the purpose of achieving a common goal or service by establishing an alliance. It can be established through interconnecting infrastructures (namely infrastructure case) or by direct communication between PN nodes (namely ad hoc case).

The evaluation of this concept cannot be fully tackled just by means of simulations or theoretical analyses. Instead, there is a clear need for a real system that based on the requirements imposed implements the required functionalities so that both functional, performance and usability evaluations can be carried out. Additionally, the implementation has to be done taking into account the scenarios in which the system is going to be used. Thus, it is necessary to assure that the system can be run over real portable devices like PDAs and laptops. Finally, the Personal Networking concept has a global footprint that imposes remote operation. In this sense, it is not enough to test the system on reduced laboratory setups but there is the need for extending the range of the tests and embracing multiple sites located at remote places and connected using the current interconnecting infrastructures.

This chapter will present the highlights of the Personal Networking system implementation and the different components that compose it. Additionally, it will describe the main aspects of the pan-European testbed that have been settled between a group of research laboratories in order to assess the system functionality and performance as well as to help on the system integration and in the future to perform usability tests with real users.

7.2 Implementation of the PN and PN-F Concept

For the PN and PN-F concept, conceptual solutions have been proposed and evaluated [4, 5], and the most promising solutions have been selected and implemented on x86 and ARM architectures for a Linux-based platform. In the following subsections we will first briefly summarize the implementation of the PN concept, as reported earlier in [6], followed by a more elaborated discussion on how the different components have been extended to support the PN-F concept. The Bird's Eye View in Fig. 7.3 highlights the different components that have been implemented and integrated into the PN and PN-F platform.

7.2.1 System Overview

Table 7.1 briefly introduces the main building blocks before we describe their implementation detail in the following sections.

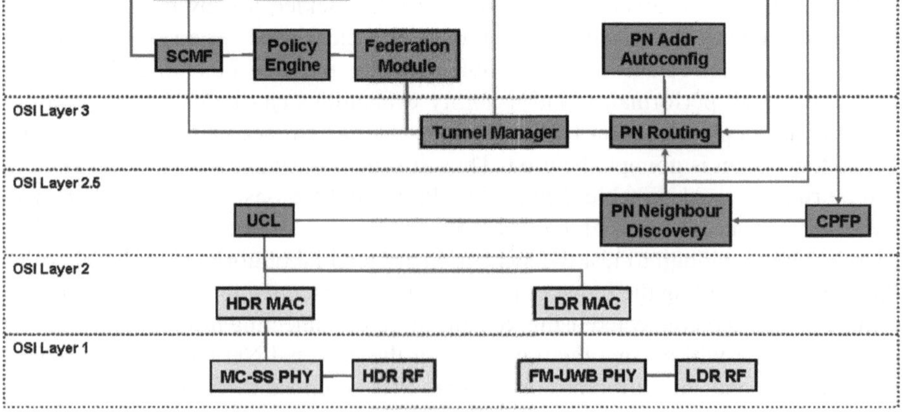

Fig. 7.3 Bird's eye view highlighting PN and PN-F system

Table 7.1 Integrated components on the PN/PN-F system overview

Component name	Main functionalities
Trust Establishment module	This component is the responsible for the Certified PAN Formation Protocol (CPFP), so called imprinting procedure. Every node that wants to be included into the PN needs to establish a long-term trust relationship with the rest of nodes
Neighbour Discovery and Authentication module	This module is in charge of the discovery and authentication of surrounding personal nodes. The first step for the cluster formation is to know which the nodes in the surroundings are. Additionally, it is necessary to authenticate them based on the pair-wise long-term trust relationship they share after the imprinting
Universal Convergence Layer	This component has a twofold objective. On the one hand, it hides the possible heterogeneity in terms of multiple air-interfaces on the same node to the upper layers, so that it supports nodes with multiple interfaces. On the other hand, it provides encryption and filtering of the traffic, so it assures a secure and private connectivity level.
PN/PN-F Routing module	This module is the responsible for routing all the PN internal traffic. Irrespective of whether it is a route within the cluster or it is a packet destined to a node in a remote cluster or even to Internet, they are routed on this module
Dynamic Tunnel Establishment module	This module is in charge of the tunnels negotiation, establishment and adaptation. Each of the PN clusters is interconnected across the Internet via secure tunnels that are established between the different clusters' Gateways. These tunnels are setup automatically and adapted if the cluster mobility requires such a thing
PN Agent framework	The PN Agent Framework is responsible for registering and maintaining up to date information concerning major PN components, mainly in terms of availability, location and contact points. The key components that are registered within this PN Agent are the Cluster attachment points (i.e. Cluster gateways and Edge nodes if any are used), the MAGNET Service Management Protocol (MSMP) SMNs (Service Management Nodes) and the CMNs (Context Management Nodes) on the Secure Context Management Framework (SCMF)
PN/PN-F Directory Service	PN Directory enables PN Federations. It acts as a trusted third party by providing X.509 certificates to MAGNET users. In addition the PN Directory can be used to store and publish PN federation profiles which contain information about PN federations and members of these federations; who created it and who maintains it (i.e. has the ability to add or remove members or edit member attributes). It is also designed to make it possible for people to use aliases instead of their real names.
Federation Manager	It manages the participation of the PN in PN-Fs and the resulting PN-F profiles and participation profiles that result from the establishment of the federations
MAGNET Service Management Platform	Service discovery, service provisioning, control, use and adaptation (i.e. context awareness) features are supported by the MSMP

(continued)

Table 7.1 (continued)

Component name	Main functionalities
Secure Context Management Framework	It constitutes a distributed agent framework which is dedicated to gather, process and distribute various types of information, commonly known as context information. The framework manifest itself in a so-called Context Agent carrying out the required functionality, and provides context sensitive applications, services and other networking components, easy access to context information
MAGNET Air-Interfaces driver	This module provides the interface between the software based platform and the hardware network interfaces developed[a]

[a]See Chapter 6.

7.2.1.1 Personal Network Implementation

The basic approach taken to realize the PN concept was to implement the PN as a secure and self-organising overlay network consisting of all nodes that belong to the PN. This overlay network has its own private IP addressing space, creating a confined and private network in which personal nodes (PN nodes) can freely communicate with each other and on top of which a service discovery platform and PN applications can be deployed.

A basic requirement to realize this overlay network, is the ability to discriminate between personal nodes and non-personal nodes (i.e. foreign nodes). This discrimination is stored as a property of the corresponding trust relationship. This bilateral secure association between the PN nodes is negotiated using the Certified PN Formation Protocol (CPFP). The CPFP protocol is based on asymmetric cryptography and uses the novel Elliptic Curve Cryptography algorithms to generate the secret keys. The concept behind is that each new node must be introduced to the PN by the user during a procedure called *imprinting*. After a successful imprinting, the new personal device receives a valid PN certificate and is ready to establish secure associations with any other personal node based on PN certificates of each other. While the first step (i.e. the introduction to the PN through the imprinting) has to be monitored by the user, the subsequent secure associations that the node establishes with each of the other personal nodes are done automatically and transparently to the user.

Next, physically neighbouring PN nodes can authenticate each other and establish short-term link-level security associations based on the long-term pair-wise keys exchanged during the imprinting. Direct secure communication is then possible at the link level. In order to be able to have IP communication, an address configuration protocol with duplicate address detection allows PN nodes to automatically generate a unique PN IP address from the private IP addressing space assigned to the PN. After the establishment of a secure link, ad hoc routing information is exchanged. The result of the above procedures is a secure and self-organising cluster in which PN nodes can communicate over one or multiple hops.

In order to realize full PN connectivity, clusters at different geographical locations need to be interconnected through PN Gateway Nodes that have access to the Internet. To secure inter-cluster connectivity GW nodes will use CPFP over the insecure channel to derive a secure key which will be used to set up an IPsec tunnel between the clusters. A new PN entity called the PN Agent was designed and implemented for maintaining up to date the information of all the PN cluster attachment points. This PN Agent provides name registration/deregistration/discovery, publish subscribe and name resolution functions at PN and PN Federation level. During the PN formation process, the PN Gateway Nodes register themselves to the PN Agent (mainly in terms of attachment point to the Internet – public/private IP addresses and ports) and get, as registration response, the location information of the Cluster Gateway Nodes of all the remote PN Clusters. This remote PN Gateway information will be maintained up to date by the PN Agent through binding updates. The PN Gateway information in the PN Agent is used to dynamically establish and maintain tunnels between the PN Gateway Nodes. Finally, after the exchange of routing information over these tunnels, full inter-cluster connectivity within the PN IP addressing space is possible, allowing secure communication between every pair of PN nodes.

Additional mechanisms have been implemented to improve communication. A universal convergence layer manages all network interfaces and hides the heterogeneity of the underlying interfaces to the routing layer and PN IP addressing space. Extensions have been implemented to be able to take into account NAT boxes. Next to unicast functionality, cluster-wide and PN-wide broadcasting functionality is also supported. Also, the combination of mechanisms to deal with dynamics (such as cluster splits and merges) and private PN addressing allows applications to maintain connectivity despite mobility. Finally, a PN Manager GUI presents the user an interface to use, manage and monitor the implemented software. This tool gathers the management and control of the system through its GUI. User interacts with the system, triggers service discovery, etc through this GUI.

On top of this network overlay, a service discovery and management platform (called MAGNET Service Management Platform, MSMP) and a Secure Context Management Framework (SCMF) are implemented. The MSMP offers the user viewing, managing and secure access to all PN resources and services. Its structure follows a twofold approach centralized at the PN cluster level and distributed P2P structure at the PN level (i.e. between the PN clusters). A Service Management Node (SMN) is elected for each PN cluster. The SMN discovers and manages services within its cluster and interacts with other clusters' SMNs in a peer-to-peer fashion via a service overlay. This SMN is also responsible for discovering and advertising remote services within the cluster. The user can achieve this in a very flexible manner, through a GUI, by performing SD queries based on any combination of service/name attributes (e.g. a device name, a service name, a service type and wildcards to get all available matching services); selecting nodes presented on the GUI as icons attached to friendly names; and finally triggering the service invocation and control.

The Secure Context Management Framework [7] provides access to all context and user profile information within a PN. The SCMF consists of context agents running on all PN nodes. Applications can access all information through the context agent running on their local node. The internal structure of the SCMF follows the structure of the PN. For each cluster a Context Management Node (CMN) is elected. The CMN keeps index information regarding what information can be accessed from each node in the cluster. On the PN level, the CMNs in the different clusters interact with each other on a peer-to-peer basis. For accessing context information, the Context Access Language (CALA) is used. CALA provides a synchronous query/response as well as an asynchronous subscribe/notify interaction style. The modelling of information is entity-based with an underlying entity type hierarchy. The entity type defines what kind of attributes an entity can have. Access to information can be based on entity id/attribute or entity type/attribute combinations. A scoping concept makes access to information more efficient by limiting the nodes that need to be queried, e.g., only the local node or the cluster. Context agents access local context information through retrievers that provide a uniform interface to context sources. Examples for context sources are sensors, the networking stack, and the operating system. User profile information is stored in a storage component within a context agent.

7.2.1.2 Personal Network Federation Implementation

In order to realize the PN-F concept, a mechanism to define new PN-Fs and to add PNs to this PN-F is needed, resulting in a PN-F creation and participation protocol. The PN-F Creator generates a PN-F Profile containing the main details of the PN-F (i.e. identification, means to proceed with the participation protocol and policies that rule the federation) and stores it in the SCMF. The PN-F Profile is made public and candidates (i.e. other PNs) go on a dialogue with the creator to see whether they are allowed to enter the PN-F or not.

In order to proceed with the next step in the PN-F participation phase, the PN-F Creator and potential PN-F members (i.e. other PNs) need to be able to authenticate each other and to establish a security association that can be used to secure all ensuing communication. A new PN component, called Personal Network Directory Service [8], is also introduced as the identity provider (i.e. trusted third party entity). The PNDS, operated by a service provider, acts as a Certificate Authority (CA) providing X.509 certificates which associate a public key with a particular user. The PNDS authenticates users via GSM's Short Message Service (SMS). The PNDS certificates are leveraged by CPFP to establish bilateral trust relationships between the PNs that are afterwards enforced each time the two PNs communicate under the auspices of any federation.

After this authentication and security association step, the PN-F member can actually join the PN-F. A PN-F participation profile, which lists the services that the new PN-F member will make available within the PN-F, is created and stored in the SCMF. At this stage, each member knows in which PN-Fs she/he participates, which other PNs are currently members of the PN-F and, optionally, what services are made available by these members. This information can in any case be retrieved through a PN-F wide service discovery mechanism since the MSMP implementation has been extended to support also this feature.

The concept of a network overlay has been selected to realize secure PN-F communication. In order to separate the internal PN communication from any PN-F communication, every PN-F will also have its own PN-F addressing space (defined in the PN-F profile) and every involved node will obtain a unique PN-F IP address within this addressing space. PN-F overlays will be established in a similar way to the PN. Neighbouring clusters of different PNs are discovered through the use of beacons. When establishing secure associations with a PN, a pair-wise (one key for each pair of PNs) primary master key is securely exchanged (using the PNDS certificates through CPFP). This key is then used for deriving link level session keys which enables to secure the link between nodes of different PNs. Using this secure link, PN-F routing information can be exchanged, forming a PN-F cluster. For interconnecting clusters of PNs at different locations, the PN Agents of the respective PNs are used. All clusters location information can be retrieved by contacting the PN Agents of the other PN-F members. Tunnels are then established using the primary master key as basis and routing information is exchanged, creating full end-to-end secure PN-F connectivity.

The service discovery framework is extended to allow PN-F service discovery and use. Higher-level SMNs, called PN-F Agents, are introduced. The PN-F Agent implements all PN SMN functionalities but is exclusively devoted to store and to discover PN-F resources and services at PN-F level. One PN-F Agent per federation is activated within a PN. The PN-F Agents of each participant interact in a peer-to-peer manner via a PN-F service overlay to provide PN-F wide service discovery according to PN-F participation profiles. Service related functions provided by the GUI are extended to the PN-F case.

In the PN-F case, the SCMF also provides access to context information from the members of the PN-F [9]. The SCMF of each PN has a dedicated Context Management Gateway (CMG) which interacts with each other, exchanging context information, while enforcing the privacy policies of the user.

7.2.2 *Trust Establishment Module*

7.2.2.1 PN Device Initialization

The trust establishment module provides a set of functionalities that allow nodes to offer privacy and confidentiality to all communication between PN nodes as well as the ability to identify and authenticate other PN nodes. These functionalities are based on the Certified PN Formation Protocol (CPFP) thoroughly described in Chapter 5. What is important for our implementation is to retain the concept that CPFP allows each PN to act as a Certificate Authority (CA) instance capable of distributing X.509 certificates securely to each of the PN participants. The remaining security procedures (such as key generation) remain tightly related to these certificates as we will see later on. This section characterizes implementation aspects involved in the device initialization, imprinting protocol and transitive imprinting, key generation for remote tunnels, and node revocation. The final subsections provide a brief description of the access network interfaces supported by CPFP as well as a reference to the cryptographic primitives used during the implementation. The implementation has been written in C with aid of some bash shell scripts on a Linux platform. Regarding the libraries used for cryptography the choice has been OpenSSL, and to implement the database functionalities SQLite, a SQL database with a small footprint well suited for mobile devices.

The generic steps to be done to initialize MAGNET devices are to personalize the device with a user defined device name, generate the cryptographic functions to be used, and prepare the local database to store the data. We have to make a difference between the procedure to initialize a normal device with the initialization procedure of the PN Certificate Authority (PNCA). The PNCA is the PN entity required to act as a Certificate Authority and therefore special procedures need to be undertaken. Any PN device provided with keyboard and display is capable of becoming the PNCA. It is therefore not necessary that the PNCA is available permanently. Thus the PNCA configuration is to be stored separately from the regular PN node configuration. For example it can be stored in a smartcard that provides password protection.

Node Initialisation

The configuration program takes as parameters the filesystem path where the data will be stored and the new device's user friendly name. It creates a pair of public and private long term elliptic curve keys from the SECG standardized group secp256k1. Despite its 256 bit length, this Koblitz elliptic curve offers good performance and a strength compared to a 3072 bit RSA key. The keys are used to generate a X.509 certificate request that includes user personalized device information. The next step is to create the MAGNET node ID which is a hash of the public long term key. To achieve this, first we need to convert the two coordinates of the EC public key to a BIGNUM with help of the OpenSSL library and then calculate a SHA-1 hash of it. Regarding the database initialization, three tables are initialized: *personal_nodes*, *federated_nodes* and *friend_nodes*. These tables collect information about other PN members such as its MAGNET ID, friendly device name, bilateral encryption keys, imprint date and blacklisted flag information. The database table name defines the type of trusted relationship between the MAGNET nodes.

PNCA Initialisation

The PNCA initialization program receives the following parameters: first the installation path and the path to the previously received PNDS X.509v3 certificate that contains in a certificate extension the Personal Network ID and the PN name provisioned by the PNDS. Later, a set of public and private Elliptic Curve keys from the type secp256k1 and a self signed certificate that will be used as root certificate by the PNCA are initialized. Afterwards, an empty Certificate Revocation List is generated and the database tables are initialized. The PNCA has the same tables as the normal node plus another table with the Personal Network ID and the SHA-256 hash of the PNCA user key. This value is checked prior to grant access to the PNCA functionality: to authenticate the user, he is asked to enter the PNCA password, then a hash of the user input is compared to this value. This is necessary since a failure to keep undisclosed the PNCA cryptographic material can jeopardize the complete PN security. Please refer to Chapter 5 for details.

7.2.2.2 Imprinting

This section covers the imprinting of PN nodes and the imprinting in the PN-F case. The PN node imprinting with the CPFP protocol is always started by the PNCA after the user selects in the user GUI a node in the PNCA vicinity so that a Proximity Authenticated Channel (PAC) can be used. The PAC is used to avoid man-in-the-middle attacks while exchanging the PN node credentials over the insecure channel. First, the PNCA connects to the PN node that provides a daemon that waits for imprinting requests. After the credential exchange the user acknowledges the procedure after the PAC authentication. A screen showing two numbers that identify the procedure will be shown in *both* the PNCA and the imprinted node as in Fig. 7.4.

If the numbers appearing in both imprint dialog screens are the same, then the user can be sure that there's no man in the middle in the connection. After the PAC authentication, the PNCA issues a X.509v3 certificate for the new personal node:

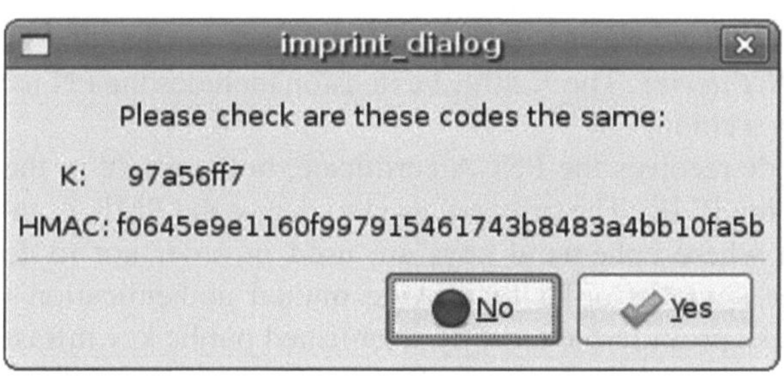

Fig. 7.4 PAC authentication dialog

Certificate:
  Data:
    Version: 3 (0x2)
    Serial Number: 2 (0x2)
    Signature Algorithm: ecdsa-with-SHA1
    Issuer: *C = No, ST = CA, L = testing, O = FP6 MAGNET Beyond, OU = MAGNET WP6, CN = MAGNET Beyond PNCA*
    Validity
      Not Before: Oct 4 03:51:03 2008 GMT
      Not After: Oct 4 03:51:03 2009 GMT
    Subject: *C = PN, ST = MAGNET, L = Mobile, O = MAGNET Beyond, OU = Testing, CN = My first MAGNET device*
    Subject Public Key Info:
      Public Key Algorithm: id-ecPublicKey
      EC Public Key:
        pub:
          04:42:4a:6c:8d:e2:7b:86:5a:e0:62:73:c9:13:dc:
          84:ad:03:3b:41:a4:46:0c:99:f4:4b:5f:18:f5:80:
          8a:21:20:fe:6b:bc:72:5f:7a:8f:8b:ee:70:43:30:
          91:35:0d:31:31:07:7b:57:9a:4b:26:dd:7a:02:ab:
          cf:4b:8f:40:71
        ASN1 OID: secp256k1
    X509v3 extensions:
      X509v3 Subject Alternative Name:
        URI:*+358123456478:5b8f3fe0612e11dc9287001921a6909f00000000*
  Signature Algorithm: ecdsa-with-SHA1
    30:45:02:20:06:cb:5d:69:2c:77:f0:f8:8b:8f:b6:32:a6:5c:
    a3:06:e3:fe:a3:bf:cb:71:1d:0b:fa:2c:5b:06:7e:e2:1c:c0:
    02:21:00:9f:73:e5:c9:1a:b2:f6:76:06:df:f9:1b:85:37:72:
    81:7a:5e:31:46:de:74:e4:da:2a:b3:d8:e6:ac:52:2b:94
—–BEGIN CERTIFICATE—–
—–END CERTIFICATE—–

The above certificate was issued by the MAGNET Beyond PNCA for the device *My first MAGNET device*. The X.509v3 extension includes the PN name and the PN ID separated by a colon.

After the node receives the PNCA certificate, both engage in the generation of the permanent key PMK. The protocol used to derive the PMK is an Elliptic Curve Diffie-Hellman where ephemeral keys are used in order not to compromise the long term node keys. In order to provide mutual authentication while deriving the PMK, we take profit of the recently established public key infrastructure so that the ephemeral keys are signed with the PNCA certificates using ECDSA.

After successful key generation, the new Personal Node information is stored in the local database and the UCL is signaled with help of an ioctl procedure to acknowledge the new node as a Personal Node in the PN so that secure communications between the nodes can be established.

In the PN-F imprinting, the procedure is quite similar: but using the PNDS certificates instead of the PNCA certificates. The PAC authentication does not take place since the imprinting PNs can both check the validity of the PNDS certificates before continuing. In case the PN-F imprinting has to be done with self-signed certificates, additional security measures would be required to ensure the security of the procedure.

7.2.2.3 Transitive Imprinting

One of the design principles of Personal Networks is to be as user friendly as possible. Following this idea, all tedious user tasks need to be automated. After imprinting we have established bilateral trusted relationships between the PNCA and each one of the PN nodes. All remaining trusted relationships between the PN nodes will be set automatically with aid of the transitive imprinting and the fact that all PN nodes are equipped with a valid certificate that was issued by the PNCA.

PN nodes advertise themselves. The neighbour discovery module is used to process this information. Whenever a new node is detected that is part of the same PN but does not have a common PMK, the transitive imprinting module will be triggered and a transitive imprinting request will be sent to the transitive imprinting daemon at the recently discovered node. After this, both nodes will exchange their PNCA certificates and after a successful certificate validation, they will proceed to derive a bilateral PMK with signed ECDH as in the imprinting procedure described above.

7.2.2.4 Tunnel Key Generation

As it will be described later in the section on Dynamic Tunneling Framework, dynamic IPsec tunnels are used to encrypt communication between the PN clusters. In order to derive the keys that will be used in the tunnel setup, the tunnel keys are generated using a slightly modified version of the transitive imprinting module.

Once two routers start the IP communication to set up a tunnel, both derive a key using the signed ECDH as in the PMK derivation. The new bilateral key is passed to the click router via an internal socket. This freshly generated key is then used as a 3DES key in the establishment of the IPsec tunnels.

7.2.2.5 PN Node Revocation

Sometimes during the PN life cycle, a PN node will need to be excluded from the PN due to loss or theft or any other reason. In this case, the public key infrastructure built in the PN will revoke the certificate issued by the PNCA. Later an updated Certificate Revocation List will be distributed among the PN nodes.

First a user selects the device to be revoked in the PNCA. In this example, the revoked device will be the one with the certificate serial number 2 that corresponds to the device we imprinted as an example. The Certificate Revocation List is updated as shown below:

```
Certificate Revocation List (CRL):
    Version 2 (0x1)
    Signature Algorithm: ecdsa-with-SHA1
    Issuer: /C = No/ST = CA/L = testing/O = FP6 MAGNET Beyond/OU =
MAGNET WP6/CN = MAGNET Beyond PNCA
    Last Update: Dec 11 15:59:00 2008 GMT
    Next Update: Jan 10 15:59:00 2009 GMT
    CRL extensions:
      X509v3 CRL Number:
          1
Revoked Certificates:
  Serial Number: 02
    Revocation Date: Dec 11 15:59:00 2008 GMT
  Signature Algorithm: ecdsa-with-SHA1
    30:46:02:21:00:91:5b:f4:61:fb:b7:a2:61:be:00:02:e9:49:
    d7:a2:ba:47:df:c0:8d:e7:d3:98:e2:0e:30:b2:c6:4e:f8:ef:
    63:02:21:00:ad:82:47:3a:98:e7:65:7e:ac:cd:27:df:14:50:
    d4:28:76:1a:38:6e:bd:41:ff:af:71:15:2f:08:06:29:3b:eb
-----BEGIN X509 CRL-----
-----END X509 CRL-----
```

Afterwards, the revoked node is marked as blacklisted in the local database and the UCL is notified about the new status of the node. In order to be effective, the CRL still needs to be distributed to all PN nodes. Updated CRLs arrive to a daemon application provided by the PN nodes that checks the signature of the CRL against the PNCA certificate and then updates accordingly the local information: database and UCL.

7.2.2.6 Interfaces Supported

The CPFP protocol can be used over different network interfaces. In the case of Bluetooth it uses a native L2CAP socket interface for the protocol communication. The MAGNET Beyond imprinting service can be advertised with the Bluetooth Service Discovery Protocol. In order to do this, the service needs to be registered in the local Bluetooth SDP database with a provided application. The result is shown below:

```
~# sdptool browse local

Service Name: MAGNET Beyond
Service Description: Imprinting service
```

Service Provider: Bluetooth Access Point
Service RecHandle: 0x10004
Service Class ID List:
  "MAGNET Beyond" (0x1310)
Protocol Descriptor List:
  "L2CAP" (0x0100)
  PSM: 4369
  uint8: 0x1

Another interface available is WiFi. Over WiFi the imprinting can be done natively or forming first an ad-hoc network and using the IP interfaces available. The socket programming for the IPv4 and IPv6 interfaces has been family agnostic, meaning that the same code is used by both versions of the IP protocol. This is achieved using the *struct addrinfo* instead of the traditional *struct sockaddr_in*. For more information and examples check the *getaddrinfo* manual page.

7.2.2.7 Cryptographic Primitives

This section provides an overview of the cryptographic primitives used in the implementation of the PN security procedures. All of these have been implemented with the OpenSSL libraries version 0.9.8g.

Security procedures often require the calculation of hash values. The algorithms of choice have been Secure Hash Algorithm 1 for 160 bit hash values and the SHA-256 from the same family but with a length of 256 bits.

During the PAC authentication a Hash Message Authentication Code (HMAC) is used to verify the process. In Fig. 7.4 the authentication code used is tagged with the K. The hash algorithm used in the HMAC calculation is the SHA-1.For the symmetrical encryption, the Advanced Encryption Standard (AES) is used in Cipher Block Chaining mode of operation with a key length of 128 bit. Symmetrical encryption is used during imprinting to encrypt data within protocol frames and later to encrypt all node communications.

The remaining cryptographic primitives also implemented with help of OpenSSL are the related to Elliptic Curve Cryptography, the signature algorithm used is ECDSA and the PMK derivation protocol is the signed Elliptic Curve Diffie-Hellman protocol. The type and version of the certificates used in the PN are the X.509v3.

7.2.3 Neighbour Discovery and Authentication Module

As already introduced, a PN is a collection of one's most private devices, referred to as personal devices/nodes, that form a virtual network where collocated personal devices organize themselves in clusters which are in turn interconnected over the Internet. Physically neighbouring PN nodes must authenticate each other and establish short-term link-level security associations, based on the long-term pair-wise keys exchanged during the imprinting, as the first step towards secure cluster self-organization. Direct secure communication is possible in an ad-hoc manner after the two nodes have correctly authenticated each other, exchanged the link-level session key and exchanged each others personal network address.

The neighbour discovery and authentication module is in charge of the aforementioned tasks. In this section we will present this module implementation details and the main insights needed to understand the module operation and its role in the overall PN/PN-F framework.

7.2.3.1 SW Architecture and Implementation Details

The neighbour discovery module is implemented as a Linux kernel module. Besides that some user space daemons are also used in order to ease the interaction with other applications. It is mainly divided in three components, as shown in Fig. 7.5, that interplay to accomplish the tasks the module has been put in charge of.

Neighbour Database

The information that is retrieved by the Neighbour Discovery module is stored in an internal database and made available to the rest of the system through several interfaces.

As it is shown in Fig. 7.6 the structure of the database starts from the PN to which the neighbouring nodes belong. This first table contains a list of nodes. Each entry on this list contains the information of a node, basically its identifier and the Primary Master Key (PMK – the one exchanged during imprinting procedure). Besides, a list of the IP and MAC addresses attached to this node is also included on each node record. For each IP address, the type of address (i.e. IPv4 or IPv6), its value and its category (PERSONAL or PUBLIC) is stored. Finally, for the MAC addresses, not only its value and the network interface to which this entry corresponds are stored but the session and broadcast keys are stored on each MAC record. It is important to note that, although a node is univocally identified by a node identifier

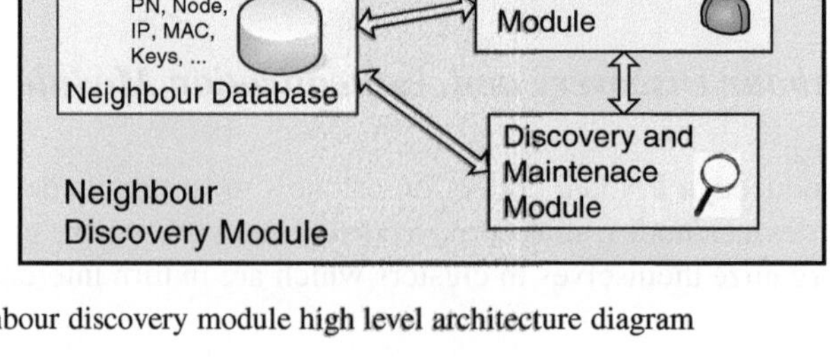

Fig. 7.5 Neighbour discovery module high level architecture diagram

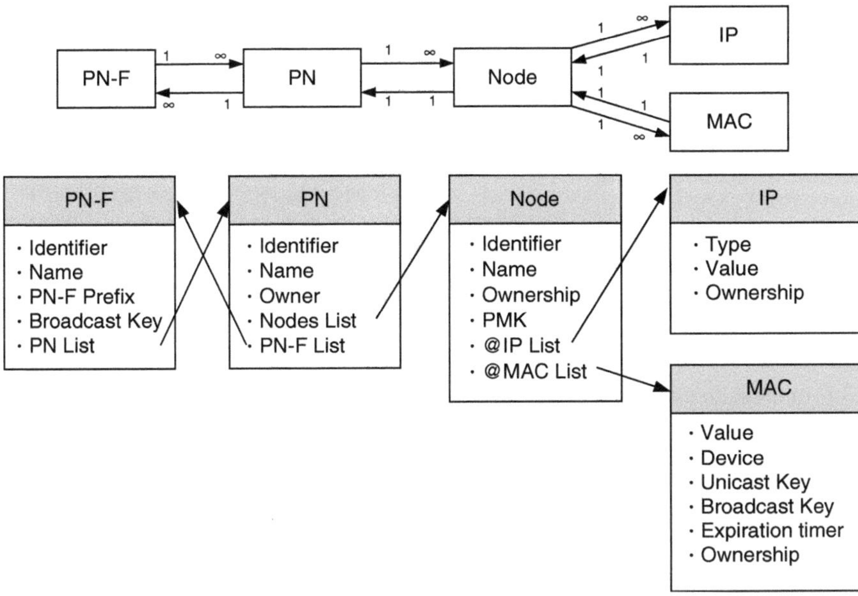

Fig. 7.6 Neighbour discovery module data base

provided within the beacon payload, a different authentication process for each of the air interfaces, through which it is possible to communicate with the neighbour, is performed. Thus, there will be different keys with the same node if they belong to more than one radio domain.

Additionally, the neighbour discovery module also holds the information pertaining to PN-Fs to which the PN belongs and those that are publicly advertised in the node radio domain. In this sense, the information stored refers to the PN-F basic parameters such as name, identifier and list of members. Besides that, as a node enters a PN-F, it is assigned with an IP address of the new overlay it is part of, the node record is also updated to include the new addresses, described as FEDERATED.

Figure 7.5 also presents the model of the possible relations between entities in the Neighbour Database. Hence, a node can only belong to one PN, this is, a node record can only be present in the node's list of one PN. Oppositely, a PN can have any number of node's records in its nodes list. Similar situation appears on the relation between IP addresses and nodes. While a node can have many IP addresses, an IP address can only belong to one node. For the MAC addresses, the relation is the same. A MAC address record can only be associated with one node while this node can be reached through multiple network interfaces, thus the MAC addresses list for that node will contain multiple MAC address records. When it comes to PN-Fs the relation is mutual since a PN-F has multiple members (i.e. multiple PNs associated) and a PN can be member of several federations. Each table is organized as an independent hash list that contains the necessary pointers to link each entry to the associated entries on the other lists. The use of hash lists helps to fetch information when necessary.

```
struct ndisc_fdb_info {
    struct hlist_head pnf[NDISC_PNF_HASH_SIZE]; //ndisc_fdb_pnf_entry
    struct hlist_head pn[NDISC_PN_HASH_SIZE]; //ndisc_fdb_pn_entry
    struct hlist_head node[NDISC_NODE_HASH_SIZE]; //ndisc_fdb_node_
      entry
    struct hlist_head ip[NDISC_IP_HASH_SIZE]; //ndisc_fdb_ip_entry
    struct hlist_head mac[NDISC_MAC_HASH_SIZE]; //ndisc_fdb_mac_entry
};
```

From these *heads* the lists are expanded with the different *entries*.

```
struct ndisc_fdb_pnf_entry{
    struct hlist_node pnf_hlist; //for handling global PN-F list (base on hash)
    //General information
    u8 id[PNF_ID_LENGTH]; //Unique PN-F identifier
    u8 name[PNF_NAME_LENGTH]; //User friendly descriptive name
    union {
        u32 ipv4;
        struct in6_addr ipv6;
        unsigned char ip[0];
    } u;
    u8 bcast_key[PNF_BKEY_LENGTH];
    u8 membership;
};
struct ndisc_fdb_pn_entry {
    struct hlist_node pn_hlist; //for handling global pn list (base on hash)
    struct hlist_head pnf_hlist; //for handling pnf_hlist from pn
    struct hlist_head node_hlist; //for handling node_hlist from pn
    //General information
    u8 id[PN_ID_LENGTH]; //Unique PN identifier
    u8 name[PN_NAME_LENGTH]; //User friendly descriptive name
    //Imprinting and security
    u8 ownership;
    u8 owner[PN_NAME_LENGTH];
    u8 imprinted;
    u8 pmk[PMK_LENGTH];
};
struct ndisc_fdb_node_entry {
    struct hlist_node node_hlist; //for handling global node list (base on hash)
    struct hlist_head ip_hlist; //for handling ip_hlist from node
    struct hlist_head mac_hlist; //for handling mac_hlist from node
    struct hlist_node hlist; //for handling node_hlist from pn
```

```
    struct ndisc_fdb_pn_entry *pn; //pointer to pn
    //General information
    u8 id[NODE_ID_LENGTH]; //Unique node identifier
    u8 name[NODE_NAME_LENGTH]; //User friendly descriptive name
    //Imprinting and security
    u8 ownership;
    u8 imprinted;
    u8 pmk[PMK_LENGTH];
};
struct ndisc_fdb_mac_entry {
    struct hlist_node mac_hlist; //for handling global mac list (base on hash)
    struct hlist_node hlist; //for handling mac_hlist from node
    struct ndisc_fdb_node_entry *node; //pointer to node
    //General information
    struct net_device *dev; //In device
    unsigned char addr[ETH_ALEN]; //device MAC (kept for virtual deleting)
    //Imprinting and security
    u8 ownership; //Personal, foreign
    struct ucl_key ukey; //Unicast key
    struct ucl_key bkey; //Bcast key
    unsigned long key_exp_time; //Session key expiration time
};
struct ndisc_fdb_ip_entry {
    struct hlist_node ip_hlist; //for handling global ip list (base on hash)
    struct hlist_node hlist; //for handling ip_hlist from node
    struct ndisc_fdb_node_entry *node; //pointer to node
    //General information
    int type; //IPv4 or IPv6
    union {
        u32 ipv4;
        struct in6_addr ipv6;
        unsigned char ip[0];
    } u;
    //Imprinting and security
    u8 ownership;
};
```

Each time a new neighbour is detected the corresponding entries are created and indexed such that all the information can be retrieved at any moment in time. The idea behind indexing all the information pertaining to one node is that this way it is possible to check all the report from one node given a known parameter. For example, given the reception of a packet we can check from the MAC address if this is a personal node and which key to use for decrypting the frame.

Discovery and Maintenance

The discovery and maintenance mechanisms are based on periodic beaconing as it has been introduced in Chapter 3. Specific beacon packets are defined by using a pre-defined Ethernet Type field value on the MAC header. The same Ethernet Type value is used for the beacon packets and for the configuration exchange ones. Hence, the reception function (*ndisc_packet_handler*) has to split the handling of the two kinds of packets.

```
static struct packet_type ndisc_packet_type = {
    .type = ETH_P_NDISC,
    .func = ndisc_packet_handler,
};
int ndisc_packet_handler(struct sk_buff *skb, struct net_device *dev,
        struct packet_type *pt, struct net_device *orig_dev)
{
    struct ndisc_hdr *hdr;
    hdr = (struct ndisc_hdr *)skb->data;
    switch (hdr->type) {
      case NDISC_BEACON:
          ndisc_beacon_handler(skb, dev, pt);
          break;
      case NDISC_CONF:
          ndisc_conf_handler(skb, dev, pt);
          break;
      default:
          printk("Unknown packet");
    }
    kfree_skb(skb);
    return 0;
}
```

The beacon handler function parses the information contained in the beacon packet and triggers the corresponding actions. It creates the node and associated MAC and IP entries on the database if this is the first beacon received. If the entries were already created by previous beacon, just the timeouts detecting the node disappearance are re-started. For beacons received from the same node but through different network devices the corresponding entries are created. Upon successful insertion in the database, the authentication procedure is started.

Authentication

When a beacon is received from a personal node the authentication mechanism is triggered in order to certify this status.

As already pointed out the authentication procedure is embedded on EAP protocol. Similarly to the handler defined for the beacon packets a function handling all the packets exchanged during the authentication is declared.

```
static struct packet_type eap_packet_type = {
    . type = __constant_htons(ETH_P_EAP),
    . func = eap_recv,
};
```

It is important to note that when the node is switched on, it already has some information pertaining to other personal nodes as resulting from the imprinting procedure. This information is loaded into the neighbour database prior to the initialization of the node interfaces. The information loaded consists of the node identifier and the shared secret. When the authentication procedure is launched, the PMK is fetched from the database and the LMSK is derived from it. The EAP exchange can then start. In order to increase security periodically link session keys are renegotiated.

7.2.4 Universal Convergence Layer

The first main objective of the UCL is to hide the complexity of the available air interfaces and to offer a unique interface to the upper layers. This module will handle this task by discovering and managing the different network resources (set them up, acquire statistics for feeding cross-layer optimization techniques, etc...).

UCL aims at masquerading multihoming by aggregating the different network interfaces (one per access technology the node is equipped with) on a single interface. Management of the heterogeneity of wireless interfaces is recommended to be implemented in a kernel space level. Although management programs run on user space in order to provide easy access to the list of devices and their main features, it is required to enable some Linux kernel modules which will be the supplier of the information.

In this section we will present the module implementation details and the main insights needed to understand the module operation and its role in the overall PN/PN-F system. It will be focused on the implementation of the UCL framework without developing further the insights of each of the UCL building blocks that has been already addressed on Section 3.3.2.

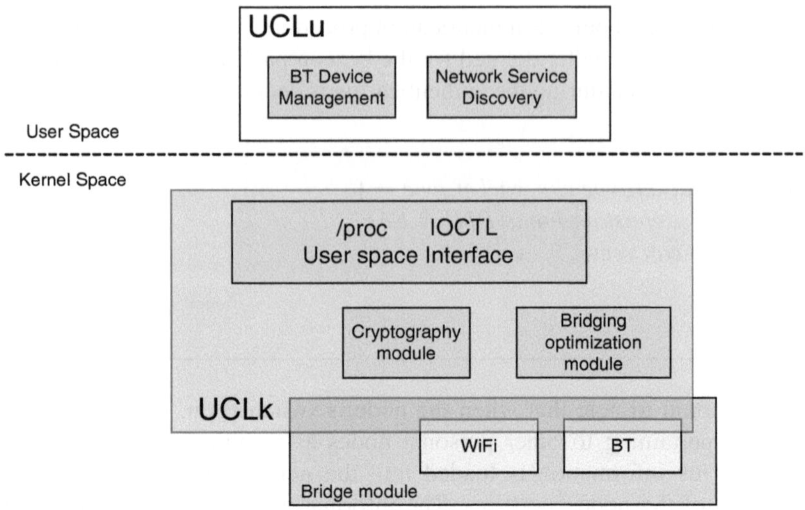

Fig. 7.7 UCL low level architecture specification

7.2.4.1 SW Architecture and Implementation Details

As it can be seen in Fig. 7.7 the UCL is divided in two main parts. One of them is sited on the user space while the other is implemented on the kernel space. While the first one is used during the node start-up for creating the framework and discovering the network interfaces to be managed through the UCL, the kernel space one is the one in charge of all the operations that deal with inbound and outbound traffic flows.

UCLk

The UCL kernel module provides the framework over which the other components operate. It creates the network device and sets it up generating a virtual interface that provides the necessary functions to operate in a pseudo-bridge manner.

```
static struct net_device *new_ucl_dev(const char *name)
{
    ...
    dev = alloc_netdev(sizeof(struct net_bridge), name, ucl_dev_setup);
    ...
    return dev;
}

void ucl_dev_setup(struct net_device *dev)
```

```
{
    memset(dev->dev_addr, 0, ETH_ALEN);

    ether_setup(dev);

    dev->do_ioctl = ucl_dev_ioctl;
    dev->get_stats = ucl_dev_get_stats;
    dev->hard_start_xmit = ucl_dev_xmit;
    dev->open = ucl_dev_open;
    dev->set_multicast_list = ucl_dev_set_multicast_list;
    dev->change_mtu = ucl_change_mtu;
    dev->mtu = 1496;
    dev->destructor = free_netdev;
    SET_MODULE_OWNER(dev);
    dev->stop = ucl_dev_stop;
    dev->tx_queue_len = 0;
    dev->set_mac_address = NULL;
    dev->priv_flags = IFF_EBRIDGE;
}
```

Upon introduction of the network interfaces with which the node is equipped, new ports are added to the UCL structure.

When a packet is received by any of the controlled network interfaces, it is passed to the UCL framework created by this kernel module. Similarly, when a packet is to be transmitted, the UCL transmission function is the anchor point for the network layer.

Additionally, it defines other functions that will be used for interfacing with the UCL such as *ioctl* framework.

Bridge Module and UCLu

By definition, a bridge is a device that separates two or more network segments within one logical network (e.g. a single IP-subnet). The UCL implementation is built around the Linux bridge implementation. On its roots, this implementation enables managing several interfaces on the same machine as a unique virtual interface with a single network identifier. This is theoretically all what is required from the UCL, but the addressing and other control functionalities of Linux bridging are mainly founded on the IEEE 802.1d protocol which is not the required behaviour for the UCL. Thus, we only take the gathering capacity of the Linux bridging framework but trim all the other control characteristics.

Additionally, the UCL is meant to automatically load all the interfaces of a multihomed device into its virtual domain so that they are all managed from it. To fulfil this requirement the user space part of the UCL (UCLu) was implemented. In our implementation it looks for all the wired, WiFi and Bluetooth network interfaces and adds it to the UCL bridging framework.

```
int UCL::serviceStart()
{
    PRINTF("\nUCL Initializing ...\n");
    if (initBridge() != 0) {
        return -1;
    }
    if (m_bRunWired) {
        PRINTF("Adding Wired interface ...\n");
        addNetworkInterface(m_pcWiredInterfaceName);
    }
    if (m_bRunWiFi) {
        PRINTF("Seaching WiFi interfaces ...\n");
        m_bExistsWiFi = (getWiFiInterfaces(&m_mapWiFiIf) == 0) ? true :
false;
        if (m_bExistsWiFi) {
          mapWiFiLLCT_t::iterator mapIter;
          for (mapIter = m_mapWiFiIf.begin();
            mapIter != m_mapWiFiIf.end();
            mapIter++) {

               WiFiLLCT *wifi = mapIter->second;
               addNetworkInterface(mapIter->first);
               wifi->serviceStart();

          }
        } else {
          PRINTF("Not found ...\n");
        }
    }
    if (m_bRunBT) {
        PRINTF("Seaching Bluetooth interfaces ...\n");
        std::set<int> setBT;
        m_bExistsBT = (getBTDevices(&setBT) == 0) ? true : false;
        if (m_bExistsBT) {
          m_BTLLCT = new BTLLCT(this);
          m_BTLLCT->serviceStart();
        } else {
          PRINTF("Not found ...\n");
        }
    }
```

```
    PRINTF("UCL service start ended\n");
    PRINTF("UCL Initiliazing ...OK\n");
    return 0;
}
```

For the Bluetooth case a special mechanism is implemented due to the implicit characteristics of this technology. BNEP profile is used for the Bluetooth devices. This profile creates a point-to-point link between the piconet coordinator and the slave. Thus, when a node is switched on it might be the case that it is the first Bluetooth device on the radio domain and as such will have to act as piconet coordinator. Other nodes with Bluetooth devices entering in the area will be slaves in this piconet and establish the link with the coordinator. The node acting as the coordinator will have as many Bluetooth interfaces as slaves the piconet. Besides, when the coordinator switches off or leaves the coverage area, a new device has to take the responsibility and reorganize the Bluetooth piconet. All these tasks are performed also in the UCLu.

Bridging Optimization Module and Cryptography Module

At the UCL the packet transmission procedure follows the path described in Fig. 3.8. Basically, these two modules implement the functionalities specified in Section 3.3.2.2.

The bridging optimization module operation is the first to be called on the downstream flow. As a result of this module the outbound hardware interface is selected. This information is used for both applying the corresponding security keys on the cryptography module and to appropriately compose the MAC layer header.

```
static void ucl_deliver(const struct net_bridge_port *to, struct sk_buff *skb)
{
    struct ndisc_fdb_mac_entry *fdb_mac;
    ...
    skb->dev = to->dev;
    // UCL Path optimization
    fdb_mac = ucl_path_opt(& skb);
    ...
    //UCL sign and encryption
    ucl_apply_security(& skb, fdb_mac);
    ..
    //Fill MAC source address field with proper address
    memcpy(eth_hdr(skb)->h_source, skb->dev->dev_addr, ETH_ALEN);
    dev_queue_xmit(skb);
}
```

User Space Interface

In order to interact with the rest of the system, special purpose interfaces have to be deployed. It has to be taken into account that opposite to the rest of the system components, the UCL is implemented on the kernel space.

The two ways implemented for interplaying with the UCL are based on *ioctl* commands and on the */proc* filesystem.

Most devices can perform operations beyond simple data transfers; user space must often be able to make special requests or inform about certain parameters to the kernel module handling the device. These operations are usually supported via the *ioctl* method, which implements the system call by the same name.

The */proc* filesystem is a virtual filesystem that permits communication between the Linux kernel and user space. In the */proc* filesystem, virtual files can be read from or written to as a means of communicating with entities in the kernel, but unlike regular files, the content of these virtual files is dynamically created.

It is important to note that although the neighbour discovery module and the UCL are independent components, they are tightly intertwined and some of the interfaces of the first one, described in the previous section, are actually implemented in the UCL module. These are mainly the ones implemented through *ioctl* commands.

7.2.5 PN Agent Framework

7.2.5.1 SW Architecture and Implementation Details

For carrying out the PN Agent framework and functionalities already described in Chapter 3, we choose to base our implementation on a scalable and wide scale name resolution system named INS/Twine and introduced by MIT [20]. This naming system is designed in a P2P fashion and also provides extra functionalities for resources publishing and wide-area discovery. INS/Twine comprises 3 main entities: a name repository called Name Tree, a name resolver peer called Intentional Name Resolver (INR), and finally a name resolver client called ins application. The P2P system that handles the communication among the INR, for name resolution and resource discovery purposes, is a cord Distributed Hash Table (DHT) framework. INS/Twine is also provided a proprietary name description format, called intentional name, which is similar to XML and stores attribute-value pairs hierarchically organized in a tree structure. This description is sufficiently flexible and extensible for implementing intelligent (e.g. semantic-based) name or resource description storage and name-based query for name to address resolution and resource discovery. Therefore, PN Agent framework implemented for the PN platform is INS/Twine-based and is described as follows:

- The PN Agent Server (or PN Agent) functionalities are handled by the INS/Twine INR peer

- The PN Agent client functionalities are handled by an ins application. This ins application has also been extended with an XML RPC interface that provides a standardized communication interface between any PN nodes or components and the ins application, i.e. the PN Agent client
- The communication interface between the PN Agent and the PN Agent client is handled by the INS/Twine system. In that way, PN Agent name publishing and notification messages become ins announcements, while name discovery messages become name queries
- The PN Agent overlay network is now handled by the INS/Twine INR overlay network based on the cord DHT system
- And finally the name descriptions of all the PN components that are registered within the PN Agent, like e.g. – Cluster Gateways, Service Management Nodes, Context Management Nodes -, are given in an intentional name format

Even if the P2P distribution of the PN Agent functionalities among PN Clusters was implemented (using the functionalities provided by INR overlay network) and tested, we have finally decided to activate the PN Agent functionalities/component in only one dedicated PN node, which renders this PN Agent centralized. This final deployment choice was only due to the limited number of PCs/Laptops that was available for the PN Platform. The deployed architecture of the INS/Twine-based PN Agent framework is depicted in Fig. 7.8. This figure also depicts the layer architecture of the PN Agent, the INR, and the PN Agent client, the in application. The

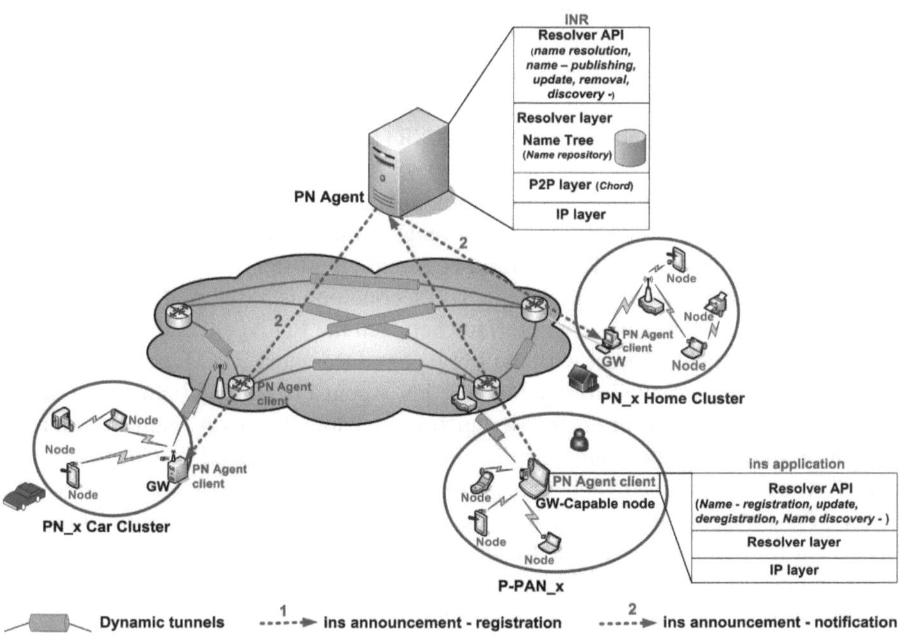

Fig. 7.8 Implemented PN agent for the PN platform

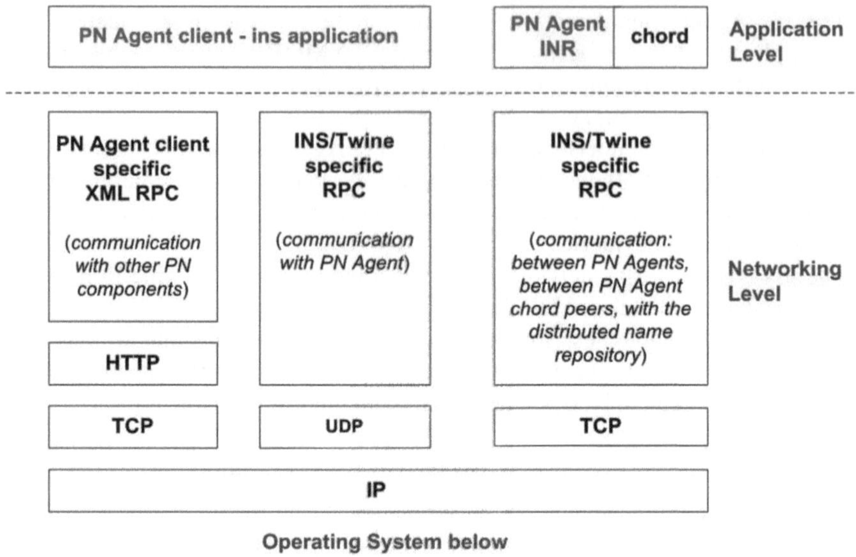

Fig. 7.9 Protocol stack of the INS/Twine-based PN Agent framework

PN Agent client is obviously not part of a P2P overlay, contrary to its attachment PN Agent, and does not implement the P2P layer, as shown in Fig. 7.8. Furthermore, it does not implement the name resolution functionality and the Name Tree since those functionalities are only managed at the PN Agent Level. Finally, both PN Agent and PN Agent client are implemented on top of an IP layer. Figure 7.9 summarizes the protocol stack that was carried out for the communication between the PN Agent client (ins application) and the PN Agent (INR) and for the communication among PN Agents.

Figure 7.9 shows that the PN Agent implemented for the PN platform is provided with two interfaces. The first one corresponds to the communication interface between the PN Agent (an INR) and the PN Agent client (an ins application) and is internal to the INS/Twine environment. The second one is based on XML RPC and is design for providing a generic interface between the PN Agent and any PN components that wants to interact with the PN Agent. Dedicated RPC function calls are specified and implemented for each of the PN Agent/PN Agent client functionalities, namely name registration and deregistration, name discovery, name to address resolution, name update, name notification.As already depicted in Chapter 3, all the PN Gateways register their description within the PN Agent. In addition, PN SMNs, PN-F Agents and CMNs also register and update their descriptions into the PN Agent for being discoverable by any PN component that needs to interact with them. In the CMN case, the information contained in the description name will even be used during the CMN networking establishment procedure.

Table 7.2 Description name registered to the PN Agent

PN component	Description name
GW	[**User_Id**= +33160764578][**Cluster_Id**=INT] [**NodeType**=GWER][**Annoucer_Id**=Gw] [**Nat_Id**=157.159.229.252:8000] [**Type_Id**=Cluster_+33160764578_Subscriber]
Edge node	[**device**=router] [type=edge] [**service**=VPNgateway[**protocol**=IPSec]] [**domain**=INT]
CMN	[**NodeType**=CMN] [**PNId**=5b8f3fe0612e11] [**UserId**= +33160764578] [**NodeId**=e53a6f05866741] [**NodeName**=batekelt1] [**NodeIP@**=10.2.137.187] [**NodePort**=5060]
SMN	[**PNId**=5b8f3fe0612e11] [**Cluster_Id**=Office] [**NodeType**=SMN] [**URL**=http://10.2.137.230:25005/] [**NodeIP@**=10.2.137.230]
PN-F agent	[**PNFId**=EDF48C2A531B11] [**PNId**=5b8f3fe0612e11] [**NodeType**=PNF-Agent] [**URL**=http://10.2.137.231:8091/][**NodeIP@**=10. 2.137.231]

PN Name and Resource Description

In the INS/Twine-based PN Agent framework, the description format is of the type 'intentional name' and all the descriptions of PN nodes/components have to be provided using this PN Agent intentional name format. They are forwarded by the PN Agent clients and the PN Agent using 'ins announcement' messages. Table 7.2 gives instances of the normalized description name format for each of the PN components actually registered within the PN Agent repository.

7.2.6 Dynamic Tunneling Framework

7.2.6.1 SW Architecture

In order to interconnect the different clusters belonging to the same PN/PN-F, these clusters need to know each other's current location, where location means the public IP address of the cluster gateways or, if an Edge Router is used to connect to the Interconnecting Structure, the IP of the Edge Router. To this end, the PN Agent concept has been introduced. The PN Agent stores all information related to the cluster locations for a specific PN/PN-F (see Section 7.2.5.1). Clusters register with their PN Agent to inform the PN Agent of their location and in turn they will be updated about the location of all other clusters of the same PN/PN-F. As such, every Gateway Node will have at all times all information required in order to establish IPSec tunnels to remote clusters. The Dynamic Tunneling Framework is then responsible for the negotiation of, management of and forwarding over these tunnels and will be discussed in this section.

The Dynamic Tunneling Framework is responsible for establishing tunnels and storing all information related to these PN/PN-F tunnels. The module consists of three components: a TunnelNegotiation component, a TunnelManager component and a Tunneling component. The TunnelNegotiation component is responsible for establishing the tunnels. The information needed to establish these tunnels (IP addresses of the tunnel endpoints, PN/PN-F prefix, the tunnel type (i.e. between which entities the tunnel is established), the tunnel maintenance type and the NAT information in case the requesting end point is behind a NAT) is provided by the PN Agent Client (see Section 7.2.5.1) and passed to the module responsible for setting up new tunnels. This information is then, together with the negotiated keys, stored in the TunnelManager. Next, the Tunneling component will use this information to encrypt/encapsulate and decrypt/decapsulate packets sent to or coming from a tunnel using IPSec ESP in tunnel mode [14] or IPSec over UDP [15] in case a NAT [16] box must be bypassed. Finally, when cluster deregistration is triggered explicitly, the action to remove tunnels is also passed from the PN Agent Client to the module responsible for managing the tunnels.

7.2.6.2 Implementation Details

TunnelNegotation and TunnelManagement

In order to start the establishment of a tunnel, the following information is needed:

- The IP address of the start point, i.e. the IP address of the node that wants to establish the tunnel
- The IP address of the end point, i.e. the IP address of the node with which a tunnel wants to be established
- The tunnel type (between which entities – GW, ER or Agent – the tunnel is established)
- The tunnel maintenance type (reactive or proactive)
- The PN/PN-F prefix of the PN/PN-F for which the tunnel is being established
- If the node that triggers the tunnel is behind a NAT, the public IP of the NAT is also provided (currently, if both endpoints are behind NAT, traffic is tunneled via the PN Agent)

Next, if the tunnel does not exist yet or the establishment is not already ongoing, the element will start the tunnel negotiation. The above information is securely exchanged together with dynamically generated keys (see Section 7.2.2.4) that will subsequently be used for the 3DES [13] IPSec encryption. When the negotiation has completed all information is stored in the TunnelManager in a number of tables as shown in Fig. 7.10. Other elements that require information about tunnels can than request this information with the TunnelManager.

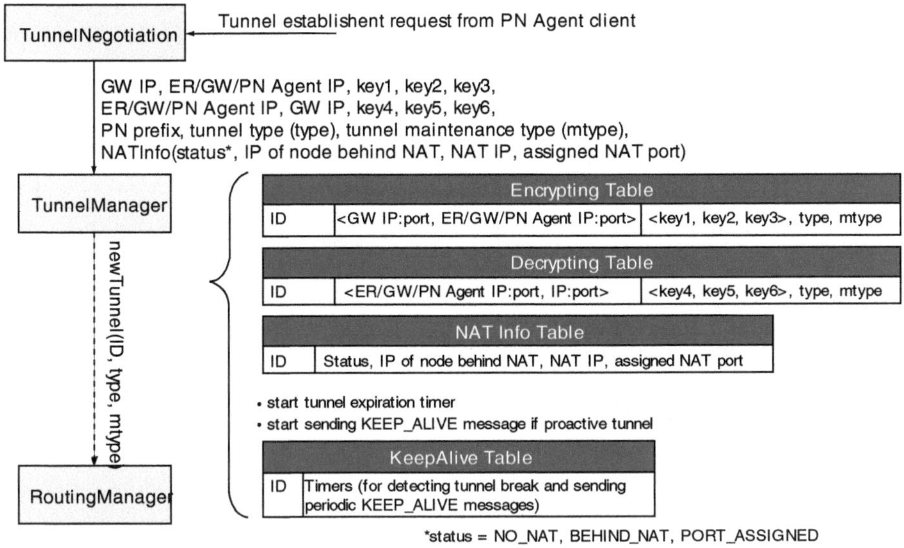

Fig. 7.10 Tunnel establishment and storage of tunnel information

TunnelForwarding

Encryption/Encapsulation

Each dynamically established tunnel receives a unique ID. This ID will be used for encrypting and encapsulating packets that need to be sent over a specific tunnel. Decryption and decapsulation of packets coming from a tunnel is based on the IP addresses of the tunnel end points and, in case a NAT needs to be bypassed and IPSec over UDP is used, also the ports. All PN/PN-F data packets destined for remote clusters, all inter-cluster routing control packets and tunnel keep-alive messages will travel over these tunnels. Before these packets can be sent over a tunnel, they need to be encrypted and encapsulated first. This is done by the PNTunnel3DES and PNTunnelEncap elements that have been developed in Click Router [10]. When the PNTunnel3DES element receives a packet that has its tunnel annotation set to the ID of the tunnel its need to be send over, the following actions will take place. Based on the tunnel annotation, the encryption keys are retrieved from the encryption table stored by the TunnelManager. Using these keys the packet is encrypted using 3DES. The TunnelManager is informed that traffic is sent over this tunnel (in order to restart the expiration timer of a reactive tunnel). Next, the packet is forwarded to the PNTunnelEncap element that retrieves the tunnel end points from the encryption table stored by the TunnelManager and checks whether a NAT box must be bypassed or not. If so, the ports to be used are also retrieved from the encryption table. Finally, it encapsulates the packet using normal IPSec (if no NAT needs to be bypassed) or IPSec over UDP (if a NAT needs to be bypassed) (Fig. 7.11).

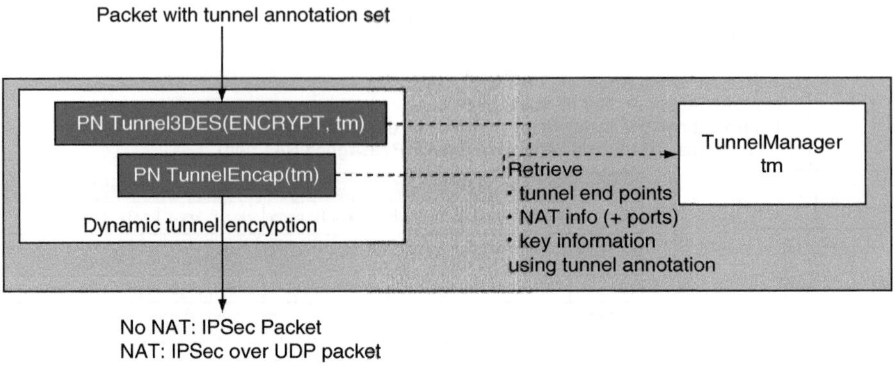

Fig. 7.11 Encryption and encapsulation

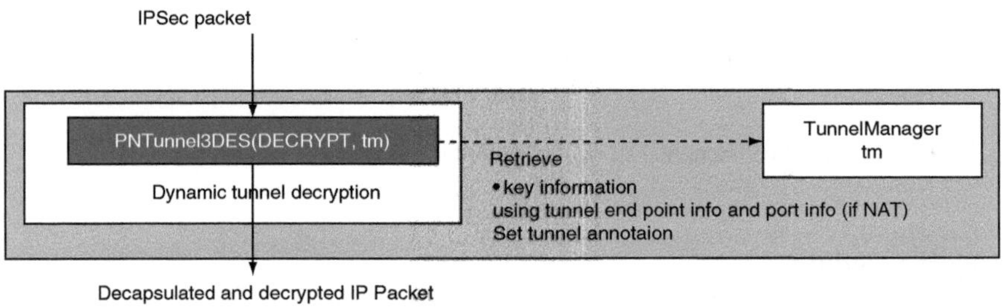

Fig. 7.12 Decryption and decapsulation

Decryption/Decapsulation

When an IPSec packet coming from a tunnel arrives in a Gateway Node or Edge Router, the packet is given to another PNTunnel3DES element, responsible for decryption. The element will use the source and destination IP address of the outer IP header and, if IPSec over UDP is used, also the UDP ports, to retrieve the tunnel ID from the decryption table stored in the TunnelManager. Next, it will set the tunnel annotation of the packet to this ID. This annotation can later be used to know from which tunnel the packet arrived. Then, the packet is decapsulated and decrypted using the 3DES keys retrieved from the decryption table stored in the TunnelManager. Also, the TunnelManager is informed that traffic has been received over this tunnel (in order to restart the expiration timer of a reactive tunnel). Finally, the packet can further be processed and forwarded (Fig. 7.12).

Tunnel Tear Down

Upon deregistration of a Gateway Node, tunnels are torn down in the Gateway Node This is triggered by the PN Agent client in these nodes, which will inform the TunnelManager. In addition, a reactively established tunnel can be torn down when its expiration timer expires due to the absence of traffic traveling over the tunnel. Finally, over the dynamically established proactive tunnels, keep-alive messages are being exchanged in order to detect a tunnel break and the absence of those can also lead to the tearing down of the tunnel. In all three cases, the TunnelManager will need to tear down the tunnel.

When the TunnelManager element in a Gateway Node needs to tear down a tunnel, the corresponding information for that tunnel is removed from the encryption, decryption, NAT info and keep-alive tables. In addition the RoutingManager is informed about the tunnel tear down (see later) and will take appropriate actions. Further, when the tunnel tear down is caused by a deregistration of the Gateway Node, the RoutingManager is informed by the PN Agent client that this node is no longer a PN Gateway Node and will take appropriate actions.

7.2.7 PN/PN-F Routing Framework

7.2.7.1 SW Architecture

The PN/PN-F routing framework implements ad hoc routing techniques capable of establishing paths between any two nodes in a dynamically established overlay (PN overlay or PN-F overlay), where all nodes in the overlay have received an PN/PN-F IP address from a separate private addressing space. The protocol operates in a hierarchical fashion, thereby separating intra-cluster and inter-cluster routing. The intra-cluster and inter-cluster routing protocol can be either proactive or reactive depending on the user preferences. The desired strategy is stored in the PN/PN-F Profile and this information is assumed to have been communicated to all PN/PN-F Members. The following combinations are allowed for intra-cluster and inter-cluster routing: proactive-proactive, proactive-reactive, reactive-reactive. In addition support for cluster-wide and PN/PN-F-wide broadcasting through blind flooding and mechanisms for gateway selection are provided. Also, new PN-F overlays can be added and removed dynamically and a RoutingManager takes care of all interfaces with the other non-routing related components and between different routing modules. Further, components for this protocol are provided both for Personal Nodes and Edge Routers (in case Edge Routers are used), where the Edge Routers only provide the inter-cluster routing part of the protocol. In addition, when no direct tunnels can be established between Gateway Nodes, for example when both Gateway Nodes are behind symmetric NAT, tunneling can be done via the PN Agent. As such, the PN Agent will also have all components needed to provide inter-cluster routing, similar to the Edge Router. This has resulted in a hierarchical, profile-based multi-mode routing framework with components for Gateway Nodes, Edge Routers and PN Agents. In the remainder of this section, we will discuss in more detail the implementation and operation of this routing framework.

7.2.7.2 Implementation Details

The routing framework has been implemented in Click Router [10]. Click is a software architecture for building flexible and configurable routers, but can be used for implementing any network level packet processing functionality. Briefly summarized, a Click router is assembled from packet processing modules called elements. Individual elements implement simple packet processing functions like packet classification, queuing, scheduling, interfacing with network devices. Complete configurations are built by connecting elements into a graph; packets flow along the graph's edges.

Figure 7.13 provides an overview of this framework implemented. As already stated, the framework clearly separates intra-Cluster routing and inter-Cluster routing capabilities. For both intra-Cluster and inter-Cluster routing, a reactive and proactive version has been implemented. Finally, the framework clearly separates forwarding and control, which will now be discussed in more detail.

Control Flow

RoutingManager. The RoutingManager element in every PN/PN-F Member takes care of all interfaces with the other (non-routing related) components. More precisely the RoutingManager will be informed by the Neighboring module (see Section 7.2.3) about any new links and link breaks and by the TunnelManager module (see Section 7.2.6.2) about dynamically established tunnels, tunnel breaks and the status of the node's Gateway functionality. Also, it can also ask the TunnelManager to reactively establish tunnels when needed. Next, the RoutingManager is

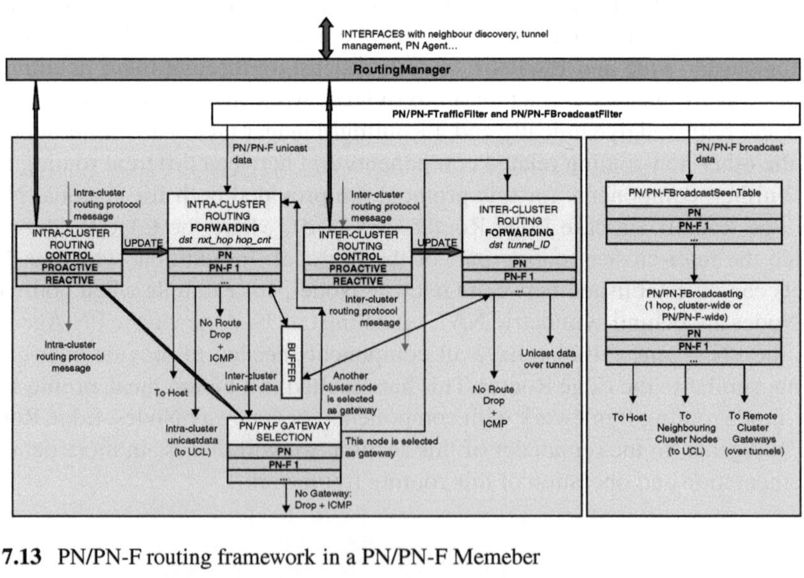

Fig. 7.13 PN/PN-F routing framework in a PN/PN-F Memeber

informed about the selected intra-Cluster and inter-Cluster routing type and ensures compatibility between the intra-Cluster and inter-Cluster routing protocol. Depending on the selected type of intra-Cluster routing protocol and inter-Cluster routing protocol, the RoutingManager takes care of all communication between both control components and provides them with the required information. Finally, the intra-Cluster and inter-Cluster routing protocol components will also process incoming routing protocol messages and, if needed, generate such messages (the same holds for the inter-Cluster routing protocol component in the Edge Router). Every routing component (e.g. reactive intra-Cluster, proactive intra-Cluster...) has its own type, next to a common routing header, allowing classification of incoming routing protocol messages.

Depending on the selected type of intra-cluster and inter-cluster routing strategy, the control flow will be different. Hereafter, the control flow for all allowed combinations is described.

Intra-cluster proactive routing + inter-cluster proactive routing. Proactive routing information (including if the node is a PN/PN-F Gateway Node or not) is propagated within the cluster. This information exchange is triggered upon the detection of new links and link breaks. All this routing information is stored in an intra-cluster forwarding table. The gateway information is stored in a gateway selection table. The implementation of the intra-cluster proactive routing protocol is a highly modified version of the Wireless Routing Protocol [11], a proactive ad hoc distance vector protocol. As such every node within the cluster will have a route to every other node within the same cluster. The content of the intra-cluster forwarding tables and any changes thereof are propagated to remote clusters (i.e. over the existing tunnels with the remote clusters) and will update the inter-cluster routing tables. To this end, a proprietary proactive ad hoc inter-cluster routing protocol has been developed. This protocol does not make use of next hop information, but will make use of tunnel identifiers (Fig. 7.14). Also, only changes in the composition of clusters are communicated to the Gateway Nodes in other clusters, resulting in a much lower overhead.

Intra-cluster proactive routing + inter-cluster reactive routing. Again proactive routing information is propagated within the cluster. Only when an inter-cluster route is needed, a route request is sent to the remote clusters. When a route request is received, the intra-cluster routing table is checked in order to see if the node to which a route is requested resides in the cluster (in case an Edge Router is used, the Edge Router still has an overview of all nodes in the local cluster in its inter-cluster forwarding table, i.e. proactive until the local Edge Router). If so, a route reply is sent and the route is established and stored in the inter-cluster forwarding table (Fig. 7.15). Further, reactive routes to nodes are removed when these nodes leave the cluster. Otherwise, they are removed when the corresponding tunnels are torn down.

Intra-cluster reactive routing + inter-cluster reactive routing. When a route is needed a route request is sent. This request can be propagated within the local cluster only (and if this fails within all clusters) or immediately within all clusters. The

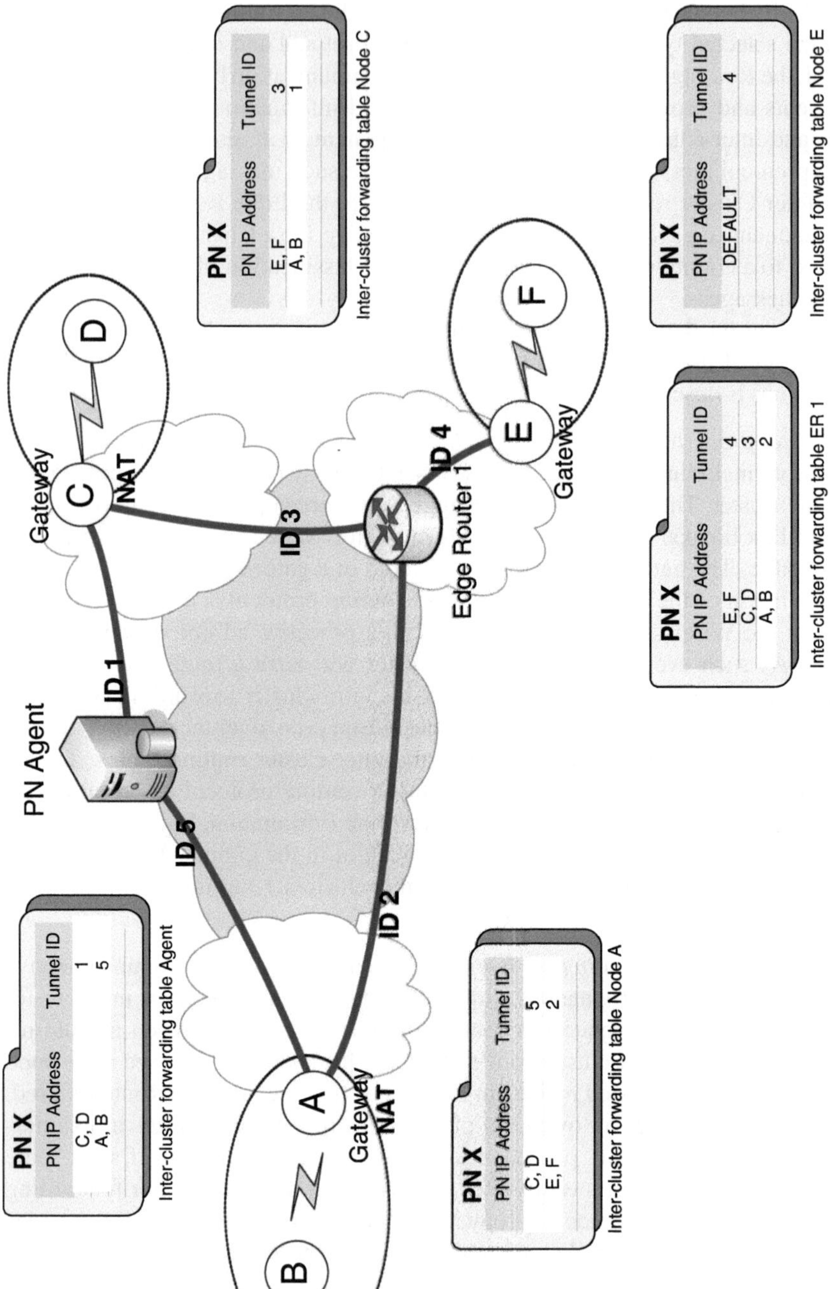

Fig. 7.14 Proactive inter-cluster routing – content of routing tables

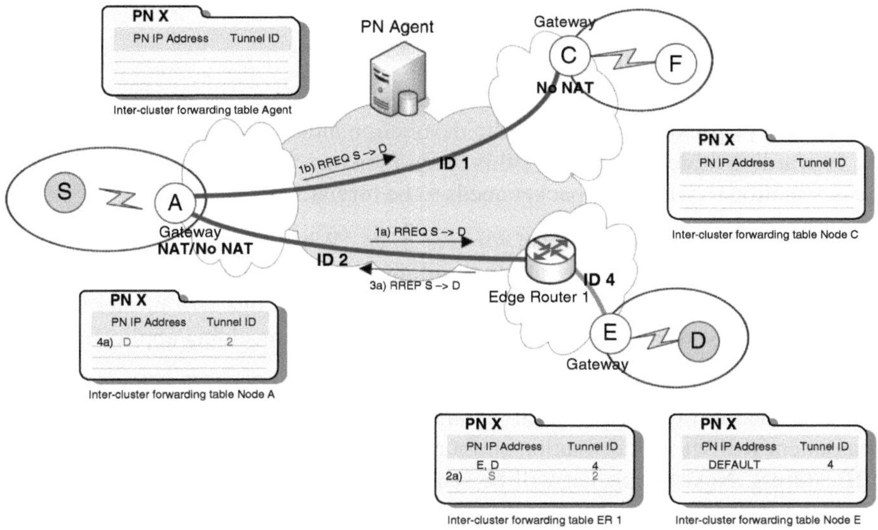

Fig. 7.15 Reactive inter-cluster routing – route establishment between node S and D

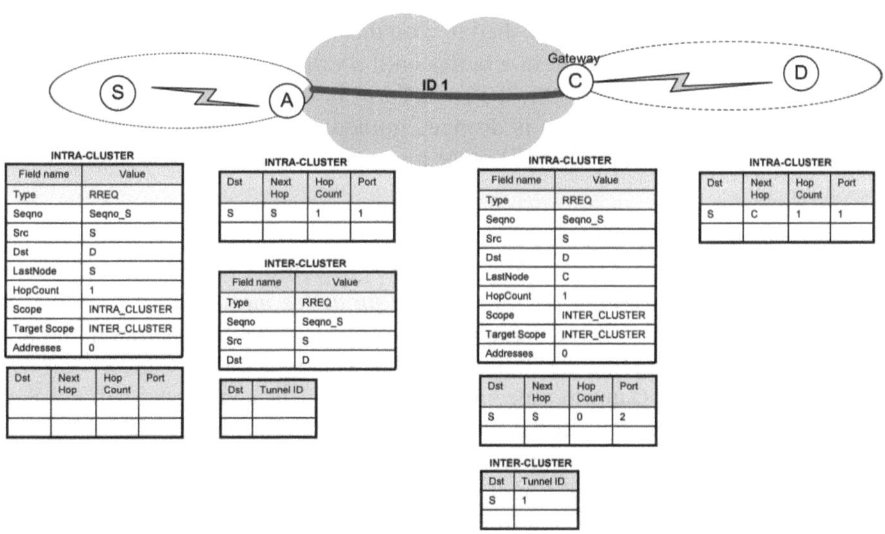

Fig. 7.16 Reactive intra/inter-cluster routing – routing request relaying and routing table updating

destination node will send back a route reply, thereby establishing a bidirectional path between the source and destination (consisting of an inter-cluster part when both nodes are located in different clusters). To this end route request, route replies and route errors are relayed from the intra-cluster level to the inter-cluster level and vice versa (Fig. 7.16). The intra-cluster reactive ad hoc protocol has been based on the Ad hoc On-Demand Distance Vector Routing protocol [12].

Forwarding Flow for Unicast Traffic

PN/PN-F Node (Non-gateway Node)

Forwarding of packets is based on the destination annotation set in the packet, i.e. the forwarding tables will retrieve this annotation and will consider this address as the destination to which the packet needs to be forwarded.

Incoming unicast packet destined for this node. When the packet arrives in the intra-Cluster forwarding table, the forwarding component will see that the packet is destined for this node and will send it to its output port that will further relay the packet to the host itself, where it can be delivered to the receiving application.

Incoming unicast packet destined for another node in this Cluster. When the packet arrives in the intra-Cluster forwarding table and a valid route exists, the forwarding component will set the destination annotation of the packet to the next hop PN/PN-F IP address. Next the packet will be sent to the corresponding output port where it will be further relayed to the UCL. The UCL will take care of encryption and creation of the MAC header (based on the destination annotation) and will send the packet to the correct interface. If no valid route exists, two possibilities exist depending on the selected intra-Cluster routing type. In case of reactive intra-Cluster routing, a new route can be established in case the node is the sender of the packet. Meanwhile, the packet is stored in a buffer until a valid route has been obtained. If it is not possible to obtain a route, the packet will be dropped. In case of proactive intra-Cluster routing, the packet is dropped immediately (since the table contains all valid entries) and an ICMP [17] error message is sent to the originator of the message.

Incoming unicast packet destined for a node in a remote Cluster. When the packet arrives in the intra-Cluster forwarding table, two possibilities exist depending on the selected intra-Cluster routing type. *In case of proactive intra-Cluster routing*, the forwarding component will see that there is no entry for that destination in the forwarding table (meaning that the destination node is not inside the Cluster) and will send the packet to the Gateway selection component. This component will check if there is a Gateway in the Cluster. If there is no Gateway, the packet will be dropped and an ICMP error message will be sent. If there is a Gateway, the annotation of the packet is set to the selected Gateway and the packet is resent to the intra-Cluster forwarding table, which will now relay the packet further to this Gateway Node. *In case of reactive intra-Cluster routing*, a route to the destination will be set up when the node is the sender of the packet, meanwhile buffering the packet, and an entry will be created in the forwarding table. This entry will allow forwarding the packet to the next hop on the path to an appropriate Gateway Node, i.e. one that has obtained an inter-Cluster route to the destination node. If it is not possible to find a route, the packet will be dropped and an ICMP error message is sent to the originator of the message.

PN/PN-F Gateway Node + Edge Router

Incoming unicast packet destined for this node. This will trigger the same behavior as in the case of a non-Gateway Node.

Incoming unicast packet destined for another node in this Cluster. This will trigger the same behavior as in the case of a non-Gateway Node.

Incoming unicast packet destined for a node in a remote Cluster. When the packet arrives in the intra-Cluster forwarding table a number of possibilities exist depending on the selected intra-Cluster and inter-Cluster routing type and the fact whether or not an Edge Router is used. In the case of proactive intra-Cluster and *proactive inter-Cluster routing*, the forwarding component will see that there is no entry for that destination in the forwarding table and will send the packet to the Gateway selection component. This component will select the most appropriate Gateway. If another Gateway than this node is selected, the annotation of the packet is set to this selected Gateway and the packet is resent to the intra-Cluster forwarding table, which will now relay the packet further to this Gateway Node. If this node is selected as the preferred Gateway Node, the packet is sent to the inter-Cluster forwarding table. Inter-Cluster forwarding uses tunnel IDs as the forwarding mechanism instead of next hop addresses. If an Edge Router is used, this table will only contain a default entry with the ID of the proactive tunnel to the Edge Router. All routing and forwarding functionality is outsourced to the Edge Router. The tunnel annotation of the packet will be set to this ID and the packet will be encapsulated and sent to the Edge Router. When the packet arrives in the Edge Router, it is further forwarded in the same way based on the content of the inter-Cluster routing table in the Edge Router. If no Edge Router is used, the forwarding table will look up the destination address in the table in order to retrieve the tunnel ID of the tunnel the packet needs to be sent to. Next, the tunnel annotation of the packet is set to this ID. Based on this annotation, the packet will be encapsulated and sent to the correct tunnel. If no valid entry can be found, the packet is dropped and an ICMP error message is sent to the originator of the message. *In the case of proactive intra-Cluster and reactive inter-Cluster routing*, the packet will also arrive in the inter-Cluster forwarding table in case this node is selected as the Gateway Node that needs to be used. If a valid route exists, the tunnel annotation of the packet will be set to the ID of the tunnel over which the packet needs to be sent and the packet will be encapsulated. If no valid route exists, the reactive inter-Cluster routing protocol will try to establish a route. Meanwhile the packet is stored in the buffer until the route has been set up. Once a valid route has been obtained, the tunnel annotation of the packet will be set to the ID of the tunnel over which the packet needs to be sent and the packet will be encapsulated. If no valid route can be found, the packet is dropped and an ICMP error message is sent to the originator of the message. The described actions will not differ when an Edge Router is used (the Edge Router will only help in reactively establishing the inter-Cluster route and forwarding of the packets). *In the case of reactive intra-Cluster and reactive inter-Cluster routing*, the packet will be forwarded using the reactively established intra-Cluster path and will also arrive in the inter-Cluster forwarding table of the Gateway Node. Next, similar actions will take place as in the previous case.

PN Agent

The PN Agent inter-cluster forwarding capabilities are only used when PN/PN-F Gateway Nodes are not capable of directly establishing a tunnel (e.g. in case they are both behind symmetric NAT). In that case, both Gateway Nodes will establish a tunnel to the PN Agent and forwarding between these two nodes will take place over these tunnels in the same way as described before. All routing functionality is the same as in the Edge Router.

Forwarding Flow for Broadcast Traffic

Next to this unicast functionality, also 1-hop broadcast functionality is foreseen as this is required by the intra-Cluster routing protocol. However, primitives to inform all nodes within the Cluster or to inform all nodes within the entire PN/PN-F can prove valuable for higher layer protocols and applications. For example, if a PN/PN-F -wide broadcast primitive is available, this could be used by the UPnP framework (which normally operates only within a single LAN, i.e. broadcast domain, by using multicasting) in order to deploy UPnP services that are visible within the entire PN/PN-F and hidden for the outside world. Therefore, besides one-hop broadcast functionality, also Cluster-wide and PN/PN-F -wide broadcasting functionality is provided using blind flooding. To this end, within the PN/PN-F private addressing space, both a cluster-wide and PN/PN-F wide broadcasting IP address are defined. When a PN/PN-F Member sends a message to the Cluster-wide broadcast address that has been defined, the message is propagated to all its neighboring nodes. Each neighboring node will send a copy of the message to the higher layers. In addition, each node will check if it has already seen the broadcast or not. If not, the broadcast is again propagated. This process continues until all nodes within the Cluster have received the Cluster-wide broadcast. In order to stop the flooding (by checking if a node has already seen the broadcast), a BroadcastSeenTable element has been implemented in Click Router. When receiving a Cluster-wide or PN/PN-F -wide broadcast, the element checks if it has already seen the broadcast before. If so, the broadcast is dropped. If not, a copy is temporarily stored in a table and the packet is further propagated. Comparison of the packets is done by comparing the entire content of the stored and incoming packet except for the variable fields such as the IP time-to-live value and checksum.

When a PN/PN-F Member sends a message to the PN/PN-F -wide broadcast address, the same actions will take place. In addition, when a PN/PN-F Gateway Node receives such a broadcast, the broadcast is propagated to its Edge Router (if an Edge Router is used), to all remote Edge Routers and PN/PN-F Gateway Nodes (if no Edge Router is used) or to the PN Agent (if a tunnel with the PN Agent has been established). In the Edge Routers and PN Agent, a BroadcastSeenTable element is present that has the same functionality as discussed before. When an Edge Router receives a PN-wide broadcast message from its PN/PN-F Gateway Node for the first time, the message will be propagated to all remote Edge Routers that have Clusters of the same PN/PN-F connected to them and to all remote PN/PN-F Gateway Nodes. Upon reception of the broadcast by remote Edge Routers, they will propagate it to their PN/PN-F Gateway Nodes. Upon reception of the broadcast by the remote PN/PN-F Gateway Nodes, the broadcast will be further propagated within their Cluster. Upon reception of a PN/PN-F -wide broadcast message by the PN Agent for the first time, the broadcast is propagated over all other tunnels the Agent has with Gateway Nodes that are part of the same PN/PN-F.

Remarks

Handling Multiple Overlays

Once a node is part of 1 PN, it and can be part of multiple PN-Fs. Therefore, the routing framework needs to be capable of dealing with multiple PN/PN-F overlays. Concerning the control flow, all routing protocol packets will contain in their header a field of which the value has been set to the PN/PN-F prefix in order to indicate to which PN/PN-F the control messages belong to. Concerning the forwarding flow, all forwarding elements (intra-Cluster forwarding table, inter-Cluster forwarding table and Gateway selection table) will contain separate data structures for every PN/PN-F the node is Member of. These data structures are indexed using the PN/PN-F prefix, which can be derived from the IP addresses in the IP header of the PN/PN-F unicast data packets. These data structures are created dynamically upon starting the software and reading in the different PN/PN-F Profiles. In addition, when adding a new PN/PN-F when the software is running, the data structures needed for the new PN/PN-F are also added dynamically.

Overlay Versus Internet Traffic

All PN/PN-F traffic is handled by the routing framework implemented in Click Router. Next to this, normal Internet communication can still take place, but this traffic is managed by the normal Linux routing stack.

7.2.8 PN/PN-F Directory Service

Personal Network/Personal Network Federations Directory Service (PNDS) is a web service whose basic role is to provide an authentication service. Within a single PN the PNDS is not mandatory, since the intra-PN authentication is based on the long-term pair-wise secrets between personal devices. However, when thinking about real deployments where people may authenticate each other's PNs, the PNDS is clearly needed. It embraces a business model where users register themselves to a PN service provider in the very beginning of the PN life-cycle. The PNDS then acts as a trusted third party by acting as a certificate store. The PN service provider can also host other types of servers, in order to provide much more services to the PN users.

PNDS Certificates

As previously mentioned, the PNDS service provider is a trusted third party. It stores public keys and provides *PN certificates* for those public keys. A PN certificate binds together a public key and a PN name so that others may authenticate the user, provided that they trust the certificate issuer and its ability to authenticate the user to whom it has signed the PN certificate. Therefore, in order to deliver the certificate signing service, the PNDS must authenticate the user credibly. Otherwise anyone could take over a well-established PN name and thereby steal one's digital identity. To this end, the user is required to create a PNDS account as shown in Fig. 7.17.

The above data are sent via PNDS *Register new user account* method call and the PNDS account is created in the database. A random password is associated with the account so that the subsequent method calls from the user can be authenticated. This PNDS password is sent to the user via GSM's Short Message Service (SMS) as illustrated in Fig. 7.18. Use of SMS ensures fairly reliable user authentication.

All the PNDS method calls are encrypted using Secure Sockets Layer (SSL), so that the PNDS password is never sent in clear-text through the Internet. The PNDS service provider itself can also be authenticated via SSL, if the PNDS client terminals have its root certificate. Except for the *Register new user account* method, all other PNDS interfaces require the user's GSM Number and the PNDS password as arguments. Therefore the user needs to log in, before using the actual PNDS API and

Fig. 7.17 Creating a PNDS account

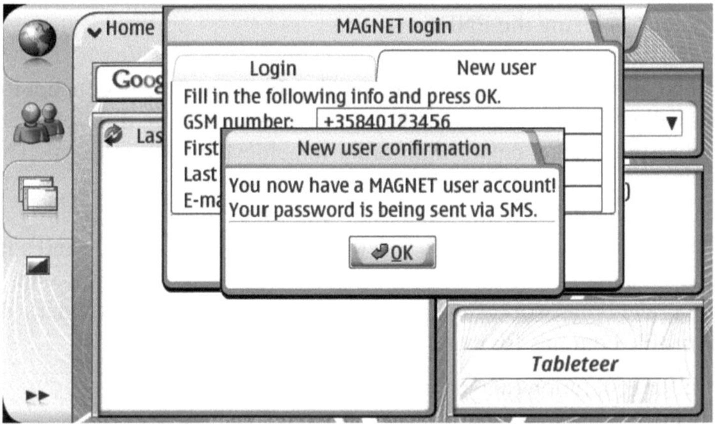

Fig. 7.18 The user's PNDS password is sent via SMS

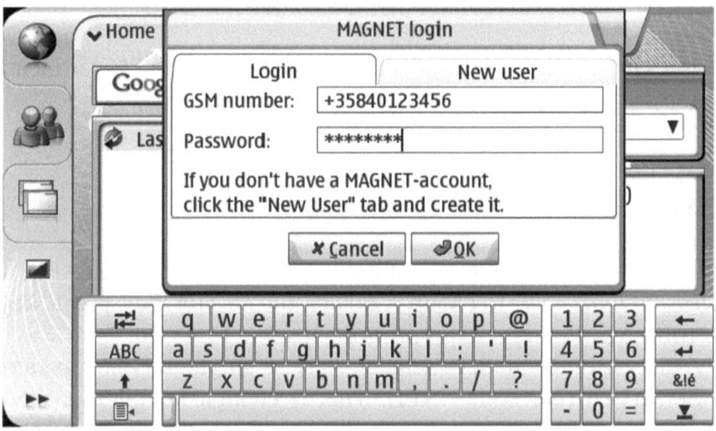

Fig. 7.19 Logging in to the PNDS client application

proceeding with fetching a PN certificate and creating PN federations, for example. The main login screen of the PNDS client application is shown in Fig. 7.19.

The above constitutes actually a so-called *single sign-on* system. The user signs on once to the PNDS client and can be authenticated via PN certificates ever since, without having to introduce any further user names and passwords. The PN certificate is written for the complete PN name which in this case is "+35840123456" Note that the PN name does not need to reveal the user's GSM number, but PN certificates can also be written for pseudonyms, such as "gym_owner". In this case other users do not know who is behind the pseudonym, but the PNDS service provider can still store this information for legislative purposes, for example.

PNDS Federations Using the PNDS

As mentioned already, the PNDS can be used to broaden the PN concept to include also the so-called *PN Federations* (PN-F) where two or more different persons set up a shared virtual packet network, in order to achieve a common goal. The PNDS issued certificates, discussed in the previous section, provide a very good starting point for supporting PN federations. Some additional data structures still need to be defined so that the participants of the federation can be authenticated and authorised. For this the so-called *PN federation profile* is introduced. PN-F profile is stored somewhere in the IP network, for example in the PNDS service provider's database. The PN-F profile contains for example the following information:

- Name of the federation (and a corresponding PN-F certificate)
- Owner of the federation
- Deputies (i.e. additional federation administrators)
- Invitees (i.e. who are allowed to join the federation)
- Who have joined the federation
- Passphrase
- Federation's private key (corresponding to the federation's PN certificate)
- Federation's group key

7.2.8.1 SW Architecture and Implementation Details

The idea is that via the PNDS Application Programming Interface (API), users can contact a *trusted third party* to obtain X.509 certificates for their public keys and to validate certificates of others [23]. This is depicted in Fig. 7.20.

Initially the business model includes only a single service operator which provides the service (i.e. PNDS API). This is enough for piloting the concept. Nevertheless the ultimate goal is that multiple service operators could engage in offering the service together.

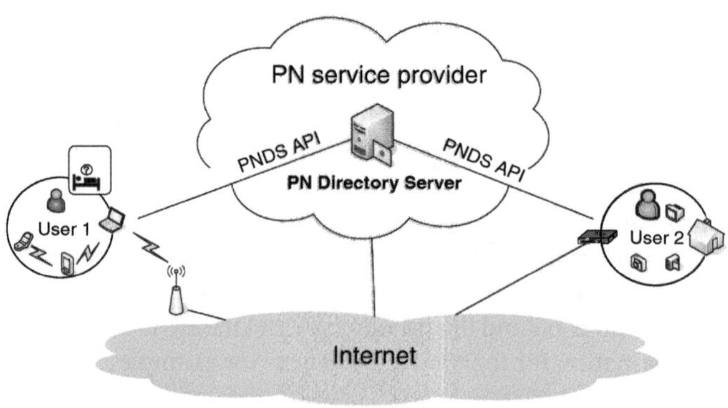

Fig. 7.20 High level PNDS view

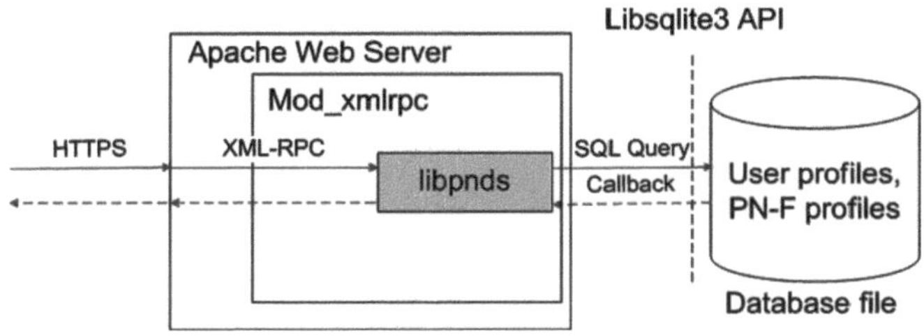

Fig. 7.21 PN directory server

Each certificate is written for a particular *PN name* that is unique. The default PN name is the GSM number, since it is personal and globally unique identifier that is readily available (for majority of people, at least). Also pseudonyms will be supported in order to enable anonymity. The PNDS service provider will always be able to track the GSM number behind each pseudonym.

Figure 7.21 illustrates the internals of the implemented PN Directory Server. It is based on the well-known Apache2 web server. Extra modules for SSL and XML-RPC support are used and a shared library called libpnds has been added to implement the PNDS API. In the background there is a database file which holds the user profiles and PN-F profiles.

7.2.9 Federation Manager

7.2.9.1 SW Architecture and Implementation Details

The Federation Manager (FM) is the central component to create, control and maintain federations with one or several PNs. Each PN has one FM that can work in ad hoc mode (meaning that federations with other PNs are initiated through discovery of neighbour nodes) or in infrastructure mode (meaning that the PNDS directory component (see Section 7.2.8) is first contacted both by PNs wishing to invite other PNs and by PN wishing to join inviting PNs).

The main design aspects considered in the realization of the FM software are:

- The simultaneous handling of several federations, leading to the management of several finite state machines (see Fig. 7.22).
- The FM shall be able to play either the role of a creator or that of a participant in a certain federation.
- Simple modelling of federation states. Only three macroscopic states are required, however in the waiting state more PN may join the federation. The further transition into the state "in-use" is triggered by a policy engine decision.

We illustrate the state transitions for an ad hoc scenario for both roles of the FM: creator and participant in Figs. 7.23 and 7.24.

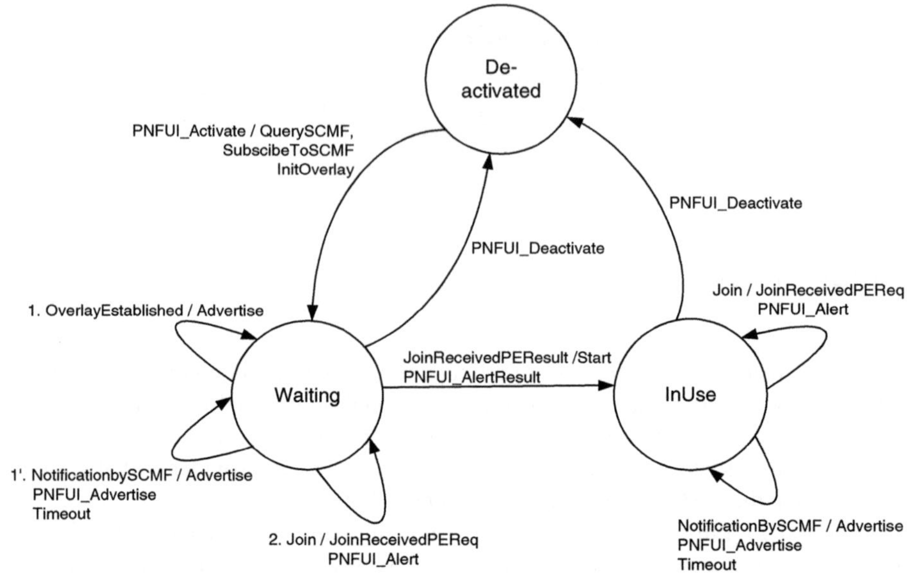

Fig. 7.22 Architecture of the Federation Manager

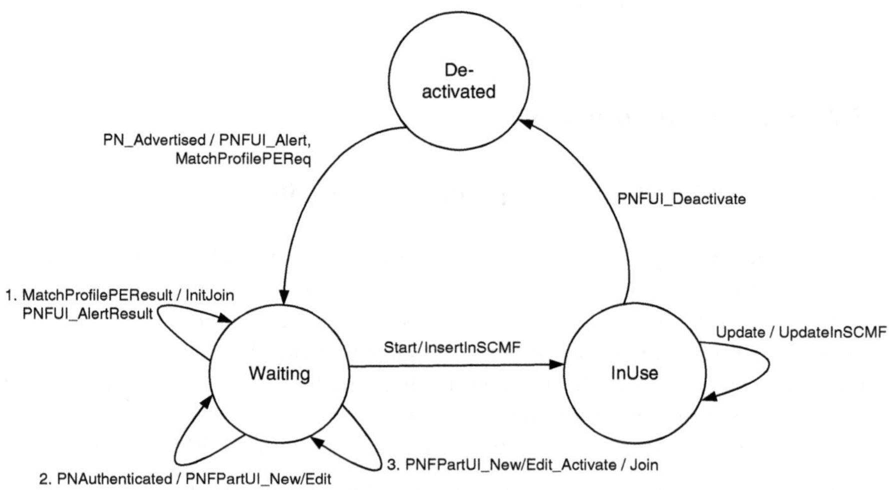

Fig. 7.23 Creator FM state diagram in ad-hoc scenario

The FM communicates with many other components of the MAGNET Beyond system. Here is an overview without going into the application protocol details:

- The policy engine: the FM sends policy requests and receives the policy decisions
- The gateway node: mediates the PN-F formation protocol to another PN

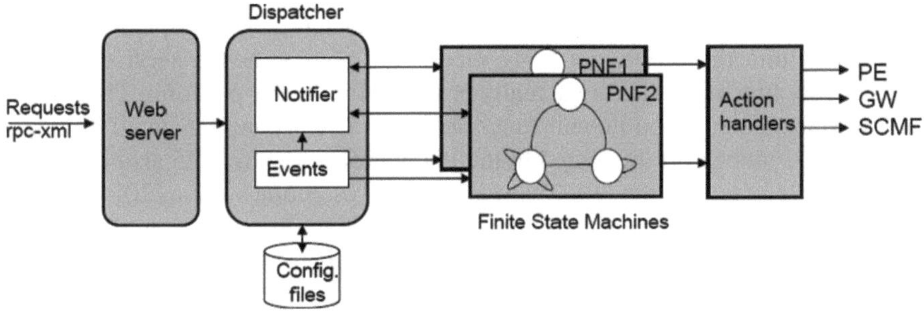

Fig. 7.24 Participant FM state diagram in ad-hoc scenario

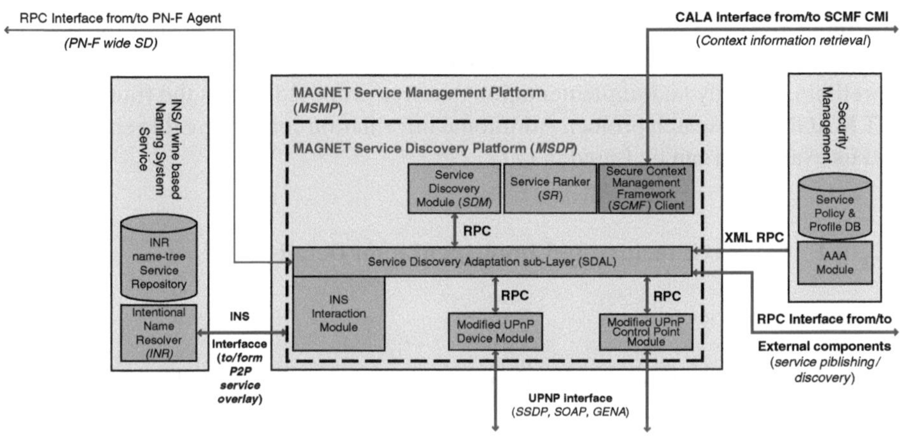

Fig. 7.25 Implemented MSMP framework for pilot system

- MSMP: informs the other PNs about the PNF-Agent, the basis for the service overlay
- PN Manager: receives user commands such as PNF creation, and inform the user with alerts
- SCMF: subscribes to context change events and to query profile and policy information
- Remote FMs: once other PNs are authenticated and a secure channel is established, the FMs communicate directly

7.2.10 MAGNET Service Management Platform (MSMP)

Figure 7.25 depicts the MSMP Service Management Node (SMN) architecture and modules implemented for the PN platform.

This figure points out that only the MAGNET Service Discovery Platform (MSDP) (including the context-aware service discovery modules), the Naming

System Service and the Security Manager blocks have been implemented and integrated within the SMNs, i.e. only secure and context aware service publishing/discovery functionalities are actually provided for the PN platform. The Service Session Management Module and functionalities, actually implemented on a specific test bed, are not yet integrated in the PN platform. Figure 7.25 also points out that we choose to rely on the INS/Twine P2P name resolution system [20] for implementing the SMN Naming System Service and the PN-wide service publishing and discovery, i.e. the distributed service repository and the SMN P2P service overlay. This choice was already made for implementing part of the PN Agent functionalities (see Section 7.2.5).

As already introduced in Chapter 3, the PN-F wide service discovery is handled through a P2P overlay network of PN-F Agent. The PN-F Agent is an overlay peer that implements all the SMN peer functionalities but that is dedicated for PN-Fs. The INS/Twine based MSMP implementation for the PN platform mainly relies on a preliminary study and implementation that was proposed within the framework of IST MAGNET research project. Additional information concerning this architecture and its evaluation can be found in [21].

7.2.10.1 SW Architecture and Implementation Details

Naming System Service

As aforementioned in the introduction part of Section 7.2.10, we choose to rely on the INS/Twine P2P name resolution system for implementing the MSMP distributed service repository, i.e. the SMN Naming System Service. This Naming System Service, already introduced in Fig. 7.25, is therefore mainly composed of the following sub-modules:

- An INS/Twine name tree for storing the PN Cluster service descriptions. This also means that the description format/language used for all the PN services/applications is of the type intentional name.
- An INR peer and an INR overlay network for handling the SMN P2P overlay and the PN wide service publishing and discovery operations. The PN SMN INRs self-configure into an application-level overlay network to exchange service descriptions.

Service Discovery Adaptation Sub-layer (SDAL)

The SDAL module introduced in Fig. 7.25 has also to implement an ins communication module for interacting with the Naming System Service module and can be viewed as an extended ins application. A generic RPC interface is also implemented within the SDAL for providing a generic interface between the SMN (i.e. MSMP) and any PN component that wants to interact with this SMN for service description – publishing, removal and discovery – purposes.

Context-Aware SD Modules

The Service Discovery Module (SDM), Service Ranker (SR) and SCMF client depicted in Fig. 7.25 are modules which ensure that services discovered are ordered according to the relevance to the user using context information. The SR is responsible for doing the service evaluation, i.e. it maps different context information related to the user and service being evaluated into a score value expressing the relevance of the service to the user. The higher score, the more relevant it is. At the same time the SDM is responsible for carrying out the activities required for the involved components to work together in order to achieve the context aware service discovery.

Security Management

The secure service discovery and provisioning operations are handled by the Security Management, as depicted in Fig. 7.25. This Security Management has been implemented by reusing the SCMF CASM modules. It provides service client authorization based on PN ids or PNDS certificates and service access control based on PN and PN-F security profiles and policies. The Security Management is not presented here since it is in fact a CASM module (dedicated in this case for a MSMP SMN) already detailed in Chapter 5.

Service Interworking Functionalities

Concerning the service interworking functionalities, the only legacy services that are actually handled within MSMP are of the type Universal Plug and Play (UPnP). This choice is mainly driven by the prominence of UPnP-enabled components and services (access points, printers, webcams, radio sets, PDAs, mobile phones...). Extra Legacy SD Modules can always be designed and activated within the SMN later. Two modified UPnP entities have thus been implemented within the MSMP SMN as depicted in Fig. 7.25: a Modified UPnP Device and a Modified UPnP Control Point (CP).

The Modified UPnP Device allows the PN Cluster SMN to be visible, i.e. discoverable and usable, within a UPnP framework. This is done through standard Simple Service Discovery Protocol (SSDP) messages (Discover and Notify messages) as depicted in Fig. 7.26. The modified Device also extends standard UPnP frameworks by offering the PN and PN-F wide service discovery functionality via MSMP, thus allowing UPnP-enable components to discover PN services registered within the MSMP-SMN. For that purpose:

- A transcoder module has been designed and implemented for converting UPnP XML descriptions into MSMP intentional name descriptions and vice versa
- A service discovery function has been implemented within the SMN Modified UPnP Device. This function can be called by any legacy UPnP CP through Simple Object Access Protocol (SOAP) control messages as depicted in Fig. 7.26.

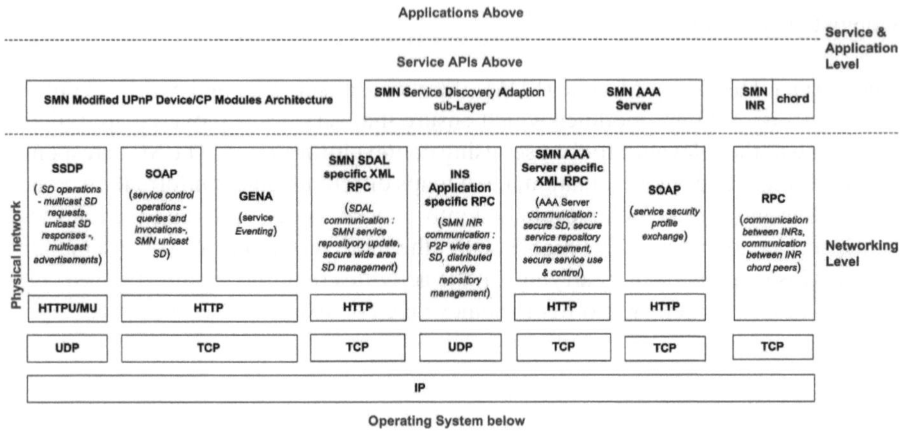

Fig. 7.26 Protocol stack of the PN platform MSMP

The modified UPnP CP, i.e. the UPnP service discovery client, mainly enables:

- The SMN to discover any available UPnP-enabled Devices/services in order to register their description within its repository. This is done through Simple Service Discovery Protocol (SSDP) messages (Discover message) as depicted in Fig. 7.26.
- Any UPnP Devices/services to advertise their descriptions within MSMP. This is done through SSDP Notification messages (Notify messages) as depicted in Fig. 7.26.

Thus any UPnP services can be discovered through MSMP.

Interactions Between SMN Internal Modules

All the interactions between the SMN internal modules and the SMN SDAL are handled though XML RPC and dedicated function calls. This is depicted both in Figs. 7.25 and 7.26.

Figure 7.26 summarizes the protocol stack that was carried out for the SMN service Gateway, the SMN peer and the SMN P2P overlay network.

Figure 7.26 points out that only one RPC interface is implemented for all the service discovery and publishing operations in the SDAL module. This means that this interface will be used by MSMP internal modules, namely the Modified UPnP CP Module, the Modified UPnP Device Module and the SDM, as well as by remote PN components. The communication interface between the SMN SDAL (an ins application) and the Naming System Service (i.e. the SMN INR peer), as well as the communication among the SMN peers within the P2P service overlay, is implemented using INS/Twine internal communication interfaces and ins announcement and query mechanisms. Figure 7.26 also points out that the SMN INR works on top of a Distributed Hash Table (DHT) process (namely Chord) to address scalability and to reduce latency. Binding updates mechanisms are also provided for maintaining up to date the service descriptions stored within the SMN INR name tree.

MSMP Service Descriptions

Within the implemented MSMP framework, the service descriptions are stored in the SMN INR name tree (i.e. the service repository of the INS-based SMN) as intentional name description strings. These description strings are similar to XML description strings, apart from the description syntax and the tree structure that are specific to INS/Twine. A fixed part of the description, with normalized attributes, has been specified in order to harmonize the context-aware service discovery process. On top of these normalized attributes, extra attributes related to.g. service profile, input/output modalities or communication interfaces can be added. The description fixed part is given below for a 'Presentation Service' application:

"[hasIdentifier=PS12334][resourceType=MAGNET_Pilot] [applicationType =Presentation_Service] [PNId=Ernoe] [PNFId=Ernoe_Fd1] [ClusrerId=Car] [nodeIPAddress=10.59.0.52]"

The *hasIdentifier* attribute is the service unique identifier (UUID). The *resourceType* attribute is used for storing the type of service environment provided by the service/application. The *applicationType* attribute is used for storing the kind of service offered. The *PNId* and *ClusterId* attributes provide minimum information on service/application location within a PN and are mainly used for restricting the service discoveries to a given PN or Cluster. The *PNFId* attribute is a PN-F scoped attribute added to the service description by the PN-F Agent when the service is registered at the federation level. Finally, the *nodeIPAddress* attribute is the IP address of the PN entity that holds the service/application. In the case of a PN-F service description registration, the PN-F Agent will automatically update the *nodeIPAddress* attribute value to the corresponding one in the PN-F addressing space.

The SMN also implements the same binding mechanisms as the ones previously introduced for the PN Agent in Section 7.2.5 in order to take into account PN Cluster services/applications mobility and for maintaining its service repository up to date.

Service Discovery Functionalities

The MSMP service discovery functionality implemented for the PN platform is name-based and carries out the standard INS/Twine query/response mechanisms. In that way, a service discovery query can be performed on any combination of value of the intentional name description attributes already depicted in the above paragraph, including wildcards. For example, the string *"[PNId=Ernoe] [applicationType=*]"* as attribute of an SMN service discovery query corresponds to the search of all the available services/applications within Ernoe's PN. The service/application intentional name descriptions matching the search attributes are

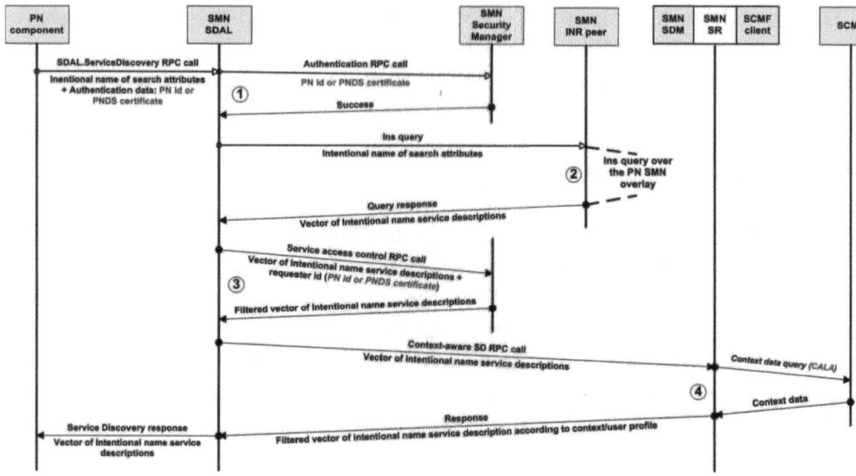

Fig. 7.27 Message flow of a service discovery performed via SMN SDAL

returned as service discovery results, by the SMN SDAL, in a string vector format. Figure 7.27 summarizes the message flow corresponding to a service discovery request triggered by a PN component, and performed via the SMN SDAL.

Figure 7.27 shows that the MSMP-SMN service/application discovery process can be broken down into four steps. Step 1 corresponds to the authentication of the requester (user, service client, component…). Step 2 corresponds to the service discovery operations within the service overlay. Step 3 corresponds to the service access control procedure (discovered service descriptions are filtered according to the requester access right and access policies). Finally, step 4 corresponds to the context-aware service discovery procedure (discovered service descriptions are filtered according to context information and requester preferences).

Obviously, the service discovery operation will be cancelled if the authentication performed by the Security Manager during step 1 fails. If the query attributes are PN-F scoped, the PN-F id is included in the search attributes and the SMN SDAL forwards the service discovery request to the corresponding PN-F Agent (its SDAL). This means that the search is performed within the PN-F service overlay. Otherwise, an INS query is directly forwarded to the SMN INR peer, as shown in Fig. 7.27, and the search is performed within the PN SMN overlay. This step was not depicted in Fig. 7.27 for drawing simplification purposes.

7.2.11 Secure Context Management Framework

As described in Chapter 3, context information can be used to adapt applications and services, but also the PN itself to the current situation of the user. In a PN, the context management has to work on the personal nodes that are currently available.

Nodes may go down, and connections may become unavailable. For the context management system this means the following:

- It cannot rely on a dedicated context infrastructure.
- It has to dynamically adapt to changes in the availability of nodes.

In the extreme case, the SCMF has to work on a single node, providing access to context information available from local context sources as well as to user profile information available from the local storage.

Therefore, a Context Agent is running on every node in a Personal Network. It provides applications and services with access to all context and user profile information that is currently available. The SCMF in a Personal Network is composed of the interacting Context Agents on all nodes. In the following, we give an overview of the internal structure of a Context Agent and show an example how a CALA request and response look like in the chosen XML representation.

7.2.11.1 Software Architecture and Implementation Details

Figure 7.28 shows the internal structure of a Context Agent. Requests for context information from local Context Aware Components (CACo) formulated in the Context Access LAnguage (CALA), which was described in Section 3.6.3, are accepted by the Context Management Interface (CMI) [I_a] and translated into an internal representation. The Context Management Interface exposes an XML-RPC interface for CALA encoded in XML. An example of such a request is shown at the end of this section.

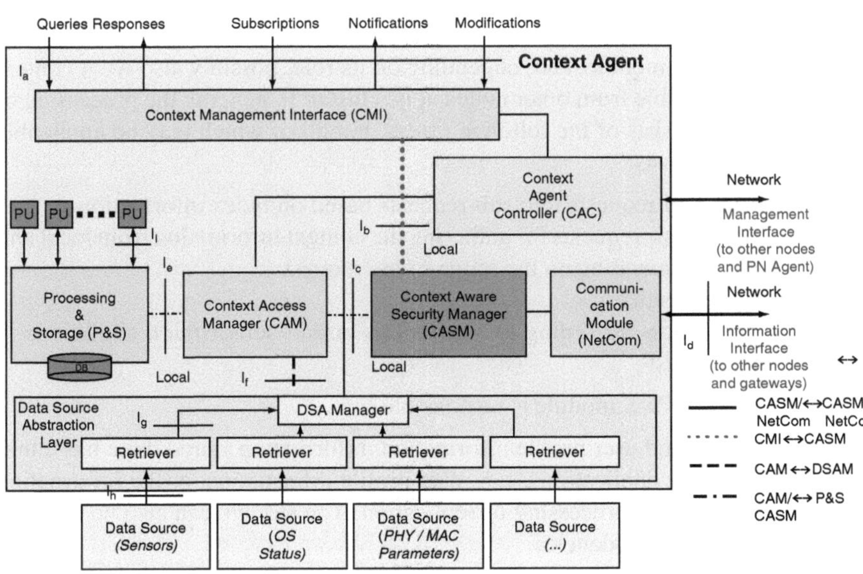

Fig. 7.28 High-level architecture of a context Agent

Thereafter the requests are passed to the Context-aware Security Manager (CASM) [I_b] which may enforce privacy policies to the Context Access Manager (CAM) [I_c]. The CAM has index information about locally available context information, either from Data Sources through the Data Source Abstraction Layer (DSAL) or the Processing & Storage module (P&S). If context information matching the request is available locally, the relevant parts of the request are forwarded to the DSAL [I_f] or the P&S [I_e]. In case the Context Agent acts as a Context Management Node (CMN), as explained in Section 3.6.1, the CAM also keeps index information about the information available on other nodes of the cluster. Depending on the configuration of the node in the SCMF and the scope of the request, the request is forwarded to other nodes through CASM via the Communication Module (NetCom) [I_d]. The CASM at the receiving side may again enforce privacy policies. Finally the CAM integrates and filters the returned results and returns them to the requesting CACo, through CASM for additional privacy control and CMI to be translated into a CALA response message.

After giving this high-level overview, we will now briefly describe the internal components of the Context Agent.

DSAL – The purpose of the DSAL is to provide a uniform interface to context information from all data sources to the Context Access Manager. For each data source there is a retriever that implements the proprietary data source interface [I_h]. The retriever also converts the data from its proprietary representation into the common SCMF representation that follows the ontology-based context model (see [26]) and, on request, provides it to the DSA Manager (DSAM) [I_g]. The DSAM manages the different retrievers and provides a single interface for synchronous and asynchronous requests to the CAM [I_f], thereby hiding details and distribution aspects of the DSAL.

CAM – The CAM is responsible for processing CALA requests. To efficiently do this, it keeps index structures about what context information is available from the DSAL, the P&S module, and, depending on its role, possibly also what context information is available from other nodes in its cluster. In general, the processing of CALA requests consists of the following steps, not all of which may be applicable for all types of requests:

1. Decomposition of requests into sub-requests based on index information
2. Forwarding the sub-requests for gathering the context information from local and distant sources depending on the scope of the request
3. Integrating the results
4. Filtering the results according to restrictions or/and subscription conditions in the original request

P&S Module – The P&S module is responsible for:

- Storing context and user profile information. Information stored there has either been put there by applications, e.g., user profile information, or the information is the result of some processing or replication. The storage can also be used for storing histories or tendencies.

- Processing context information, e.g., to derive higher-level context information like the current situation of the user or certain part of the meta data such as confidence. This is done via Processing Units (PU) connected via the I_I Interface.

CMI – The CMI handles the interaction between any CACo and the SCMF for accessing or modifying context information. These interactions are performed using CALA over a communication interface named I_a. This interface supports both synchronous and asynchronous access to information. Hence, the purpose of this module is to convert from the XML-based CALA representation used by CACos to the Context Agent-internal representation and also manage the call-back information in the case of asynchronous requests. The interface I_a is implemented using XML-RPC.

CASM – The purpose of *CASM* is to enforce the user's privacy policies with respect to accessing context and user profile information. This is of course especially relevant in the case of interactions in a PN Federation or with external components, which will be discussed below. The location of the CASM is selected so that all information going in and out needs to pass through this module. This ensures that all information and request messages are authenticated, authorized (and potentially accounted), and that information going out of the CAM is ensured to fulfil the privacy requirements of the owner of the device. CASM is described in more detail in Chapter 5.

CAC: The Context Agent Controller (CAC) is responsible for configuring and dynamically reconfiguring the Context Agent due to changes in the availability and roles of other Context Agents.

NetCom: The module which ensures serialization and deserialization of the internal representation of context information, requests, subscriptions and notifications.

7.2.11.2 CALA Application Interface

Figure 7.29 shows the XML encoding of a CALA request for the weight of a certain scale. The query has "CLUSTER" as a network scope, so only if the scale is connected to a node in the same cluster a result will be returned.

Figure 7.30 shows a possible result to the above query, i.e., the *MagnetEntity* of type *Scale* with identifier *Scale1234* and the attribute *showsWeight* that has *72.5* as a value.

A conceptual description of CALA requests and parameters can be found in Section 3.6.3.

7.2.11.3 Configuration of Retrievers and Processing Units

When developing retrievers and processing units, the Context Agent must be aware of the implementation, the offered attributes, the entity type etc. in order to establish the necessary index at the CAM. In order to do this the Context Agent relies on

```
<query>
     <selector>
             <hasIdentifier>http://www.ist-
magnet.org/Scale1234</hasIdentifier>
       <entityType>http://www.ist-
magnet.org/2007/12/17/IntegratedSCMFOntology.owl#Scale</entityType>
       <attributeName>http://www.ist-
magnet.org/2006/08/02/MagnetContextOntology.owl/showsWeight</attribu
teName>
      </selector>
      <restriction>
      <scope>
        <domain>CLUSTER</domain>
        </scope>
      <options/>
</query>
```

Fig. 7.29 Example of ID-based CALA query

```
<magnetEntities>
        <magnetEntity>
               <hasIdentifier>http://www.ist-
magnet.org/Scale1234</hasIdentifier>
               <type>http://www.ist-
magnet.org/2007/12/17/IntegratedSCMFOntology.owl#Scale</type>
               <attribute>
                        <name>http://www.ist-
magnet.org/2007/12/17/IntegratedSCMFOntology.owl#showsWeight</name>
                        <type>xsd:double</type>
                        <value><Double>72.5</Double></value>
                        <metadata><name>timestamp</name>
                                       <type>MetaData</type>
                                       <value><String>1221574202940
                                              </String></value>
                        </metadata>
               </attribute>
       </magnetEntity>
  </magnetEntities>
```

Fig. 7.30 Example of CALA result

configuration files which instruct respectively the DSAM and P& S module to register and correctly initiate retrievers and processing units. An example of a retriever configuration is shown below, and is taken from a scale retriever:

```
<dsam>
  <retriever>
  <ac name=''ScaleDiscovery''
       implClass=''ScaleDiscoveryAndAccessComponent''/>
    <mapper id=''toIdentifier'' implClass=''ToIdentifierMapper''/>
```

```
    <mapper id=''toWeight'' implClass=''ToWeightMapper''/>
    <mapper id=''toFriendlyName'' implClass=
      ''ToFriendlyNameMapper''/>
    <mapper id=''toStatus'' implClass=''ToStatusMapper''/>
    <map attributeName=''hasIdentifier'' entityType=''Scale''
       mapper=''toIdentifier ''source='
         'ScaleDiscoveryAndAccessSource''
       type=''EntityIdentifier'' />
    <map attributeName=''showsWeight'' entityType=''Scale''
mapper=''toWeight''  source=''ScaleDiscoveryAndAccessSource''
type=''double'' />
      <map attributeName=''hasFriendlyName'' entityType=''Scale''
       mapper=''toFriendlyName''
       source=''ScaleDiscoveryAndAccessSource'' type=''String'' />
  <map attributeName=''showsStatus'' entityType=''Scale''
mapper=''toStatus''  source=''ScaleDiscoveryAndAccessSource''
type=''String''/>
  </retriever>
</dsam>
```

The configuration instructs the Context Agent how to construct the retriever from an access component, what attributes it has and which mappers exists. The access component is the implementation of the access from the source and implements the native protocol, while the mappers are sub-components which map the native description into the SCMF format or magnet entities with attributes.

Similar approach and type of configuration exists for the P&S and its processing units, albeit processing units also includes context dependencies as the additional processing may require access to other context information.

In this way, a retriever or processing unit developer can focus on the core part, and leave much of the remaining work to the SCMF via these configurations.

7.2.12 MAGNET Air-Interfaces Driver

The generic architecture for a device capable of supporting MAGNET air interfaces has been developed with functional partitioning between the host and NIC (Network Interface Card). The NIC implements the MAGNET air interface prototype which consists of MAC and PHY layer with an appropriate interface to the host platform. USB is selected as the default physical interface to host. The network layer and applications are envisaged to be implemented on the host platform (Nokia 770 PDA or laptop). Figure 7.31 depicts the high level software architecture for interfacing the air interface prototypes. This is a generic architecture applicable to both HDR and LDR.

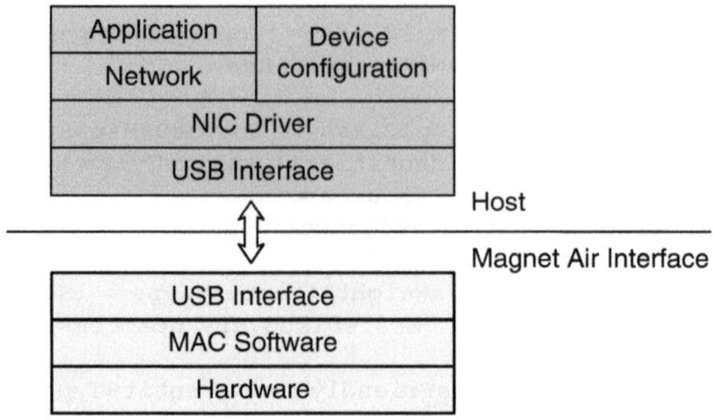

Fig. 7.31 Generic architecture of LDR and HDR bridging

7.2.12.1 MAGNET USB Driver for Logics and Usage

The MAGET USB driver supports both MAGNET Beyond LDR and HDR boards. The driver provides basic frame transfer service for the control application, which is responsible for the piconet formation and policies. In addition, the driver maps the HDR board as a network device for the Linux TCP/IP protocol stack.

To perform the legacy communication with the TCP/IP stack or with the control application, the driver sends and receives frames to and from the NIC over a USB interface. This is achieved by transmitting data using the provided facilities of the Linux kernel thanks to communication with USB devices. In this sense when plugging any MAGNET Air interface board it is automatically recognized by the driver provided the board has been previously and accordingly assigned with the vendor and product identifiers. This kind of configuration enables the plug and play paradigm for the MAGNET Air interface.

7.2.12.2 SW Architecture and Implementation Details

Each member of the piconet has the same basic control application, which manages the piconet formation. One of the control applications takes the role of the piconet master and others are just members. The control applications manage the local MAGNET LDR/HDR board via a character device (/dev/magnet/ctl1) as shown in Fig. 7.32. The driver automatically sets up a network device (mgn0) for the HDR boards. After suitable IP configuration, the hosts can use IP protocols in communication with each other over the piconet.

For the LDR board, the driver (Fig. 7.33) provides character device for the control application. Each write or read transfers a single frame from application to the LDR board or vice versa. In this case it is not required to create a network device, as the LDR board is going to be used attached to sensors transferring very low amount of data. Nevertheless, if required the driver (Fig. 7.34) can be configured so that the

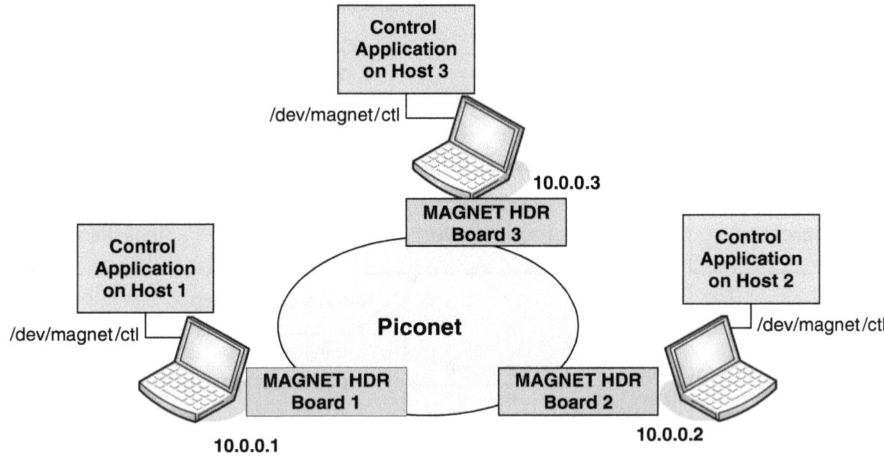

Fig. 7.32 HDR piconet model

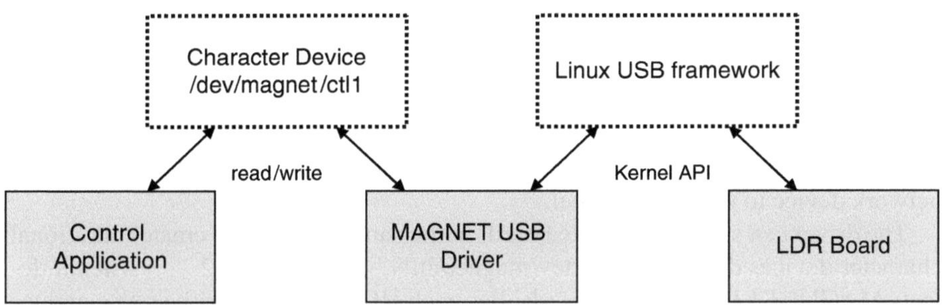

Fig. 7.33 Model interfaces for LDR driver

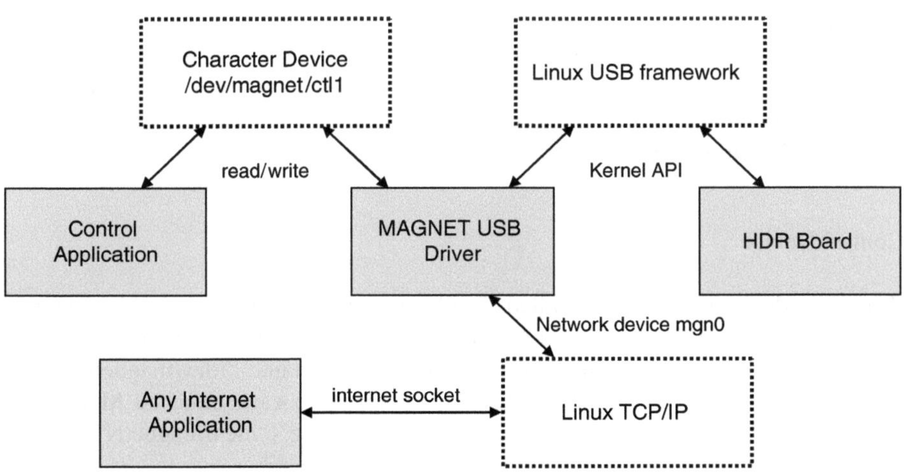

Fig. 7.34 Model interfaces for HDR driver

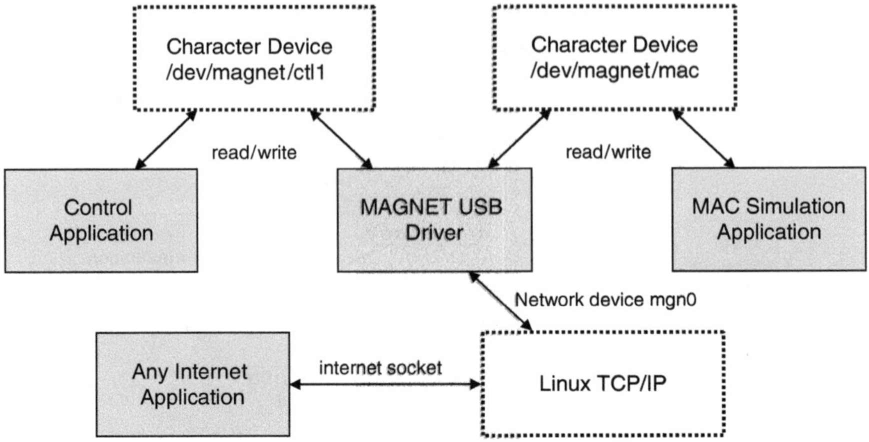

Fig. 7.35 Model interfaces for driver testing environment

LDR behaves in a similar way to the HDR board offering a network device, but with fewer communication capabilities.

For the HDR board, the driver provides the same character device for the control application. However, a subset of the frames (the data transfer) are used to implement IPv4 and IPv6 over the HDR. The driver introduces the HDR board as a network device to the Linux kernel.

The driver can support multiple boards at the same time, and it creates additional character devices dynamically (/dev/magnet/ctlN, where $N = 1, 2, \ldots$) when it detects MAGNET LDR or HDR boards. For each HDR board, the driver also creates a new network device (mgnM, where $N = 0, 1, \ldots$).

For debugging and simulation purposes, the driver also creates one additional character device/dev/magnet/mac. If any application opens this device, the driver (Fig. 7.35) behaves as if it had detected a new HDR USB board and creates a new control and network device. The driver relays the frames written to this character device to the newly created MAC device and vice versa. This allows creation of "simulated" LDR or HDR device.

Control Device

When the driver module is installed, it allocates a major number for the magnet character device and using the name "magnet". The driver uses minor number 0 for the simulation device, and if created, it gets the name "/dev/magnet0". The driver assigns non-zero minor numbers dynamically to each attached MAGNET HDR or LDR board, and creates one character device ("/dev/magnetN", where $N = 1, 2, \ldots$") for each attached board.

By using the UDEV configuration file more user friendly device names are generated from the default ones in the from "/dev/magnet/*".

Network Device

The driver creates and registers a network device "mgnN" (N = 0, 1, 2...) for each detected MAGNET HDR board, which responds to the MAC address query.

The current version of the driver presents the beginning of the MAGNET data frames as "MAC Header" towards the Linux TCP/IP stack. This does cause some inconveniences:

- The linux kernel code is not really prepared to handle "unknown" MAC headers properly. Thus, only the IPv4 side works. Fixing the Linux kernel for this would be trivial and then it would work natively for both IPv4 and IPv6.
- The "MAGNET MAC Header" is not recognized by the packet tracing tools (like Wireshark [24] or TCPDUMP [25]).

However, using all the features available from the MAGNET Air interface has the advantage that the MAC Header contains the full 64 bits MAC addresses of the devices. Another alternate solution would be that the driver presents a standard Ethernet MAC Header to the stack. Nevertheless, Ethernet MAC only has 48 bits of device address, and even that includes some constraints and not all bits are freely usable. In order to solve this issue, a proprietary algorithm to map MAGNET 64 bit device addresses into unique 48 bits Ethernet addresses has been used as this way compatibility with the already existing communication stack is provided.

Configuration Options

The features of the driver can be slightly adjusted at installation time with module parameters:

- max_frame = N

Sets the maximum frame length to N. This implicitly defines the MTU, which is visible for the network stack on the network device. MTU is the frame length (N) minus the link layer header overhead for the data frames. The default value of max_frame is 2048.

- Network = yes/no

Controls whether any network device (mgnN) gets created. If "No", then the driver does not create any network device for any detected HDR boards. The default value is "yes".

- fake_ether = yes/no

A "hack mode", which is not intended for real use. When "yes", the driver lies about the MAC header format, claiming it is "Ethernet" with 48 bits addresses without really providing an Ethernet MAC header. Interestingly, this change alone causes the Linux TCP/IP stack to fully support IPv4 and IPv6 over the MAGNET interface, but naturally all of the packet dumping utilities will be greatly confused by this "hack". The default value is "no". This option might be later changed to provide true Ethernet MAC header simulation.

7.3 Testbed Description

Testing the functionality of even a single PN needs several clusters with a reasonable number of devices in each one of them. Therefore, a minimum set of hardware devices was defined as requisite to set up and test a PN and PN-F platform. Indeed, a fully operational system has now been built through a process of conformance verification, integration, and interoperability testing of the different baseline components described before.

Different partners are hosting different parts of the PN/PN-F platform in their premises, as shown in Fig. 7.36, with all individual parts interconnected via the Internet to form a distributed testbed, as prerequisite to validate how the developed PN/PN-F solutions and pilot applications function over real-life network conditions.

In order to promote and ease the usage of the system among the partners and developers of pilot services in particular, a set of system installation and usage guidelines have been developed. The distributed testbed is set up on four different laboratories across Europe and has been the cornerstone for the integration process. The testbed is composed of both laptops and PDAs in order to showcase the feasibility of the system to be run on real user equipment. All the integrated components are implemented to run over the Linux Operating System. For the high capable devices like laptops, the Ubuntu distribution was selected while for the PDA-like devices,

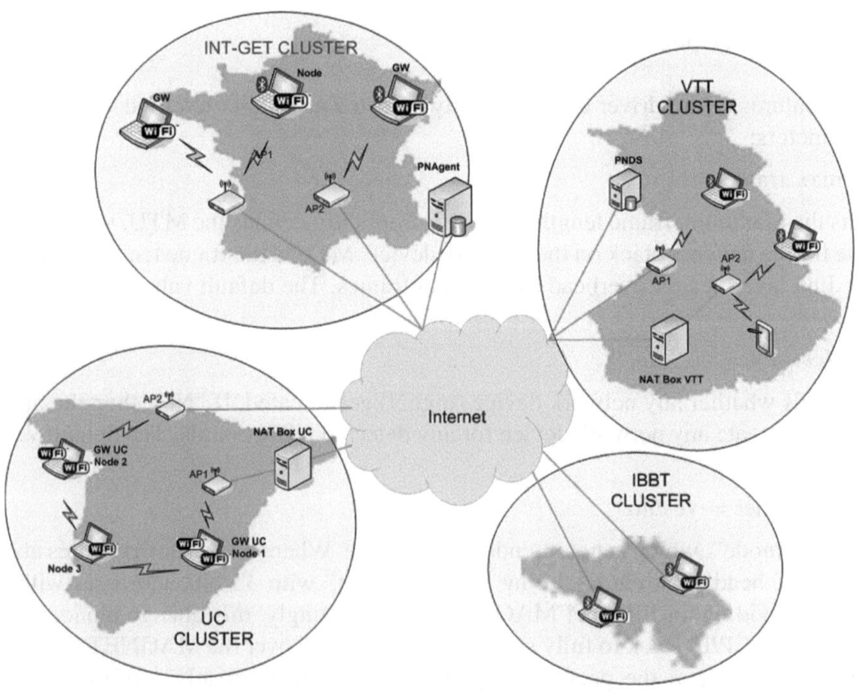

Fig. 7.36 Physical location of the remote testbed

the project decided to use Nokia Internet Tablets. Accordingly, easy to install SW packages have been created for the two selected kinds of MAGNET nodes, while the respective installation guides are planned to set all software from scratch. Thereby, the PN/PN-F Platform towards pilot services is now a reality, and the necessary means have been set to promote its use among application developers and potential end users.

7.3.1 Testbed Objectives

The PN/PN-F platform is currently being used for testing the developed PN and PN-F functionality, as described in the previous sections. Figure 7.37 depicts the test environment used for testing different network functionality (e.g. mobility, throughput, etc.).

The availability of this dedicated "always on" testbed was the only viable way to guarantee that all participating partners can assess the real usability of the pilot applications and the performance of the underlying platform. Since the objective of the platform is not only to prove the feasibility of a PN system but to support the pilot services atop of it, it is possible to assess the usability of the PN concept from a user-centric viewpoint. Indeed, the same platform will evolve via further performance testing to serve also as the platform to support the pilot services.

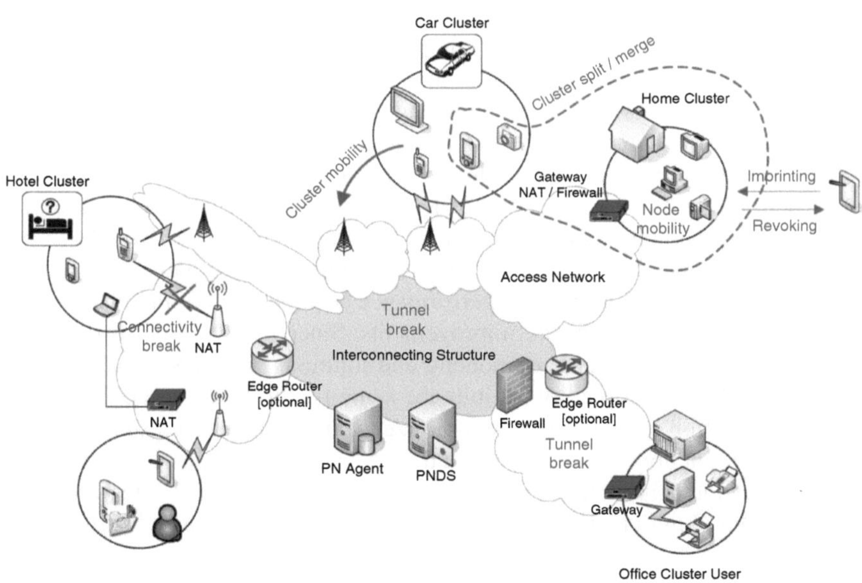

Fig. 7.37 Different supported test cases

7.3.2 Test Cases

For the actual component interoperability testing to be executed with overall PN system it is necessary to prepare and perform an interoperability testing plan. It includes end-to-end performance, robustness and reliability testing. In addition, it must be assured that the integrated components fulfil requirements of the system specification as well as support all the selected Pilot Services. A complete set of test cases were set up and carried out to guarantee the appropriateness of the solutions implemented and integrated.

For the realization of the PN-F connectivity and networking, the same concept of a network overlay was used. As such, in order to realize secure PN-F communication, all PN nodes of the PN-F members become part of a PN-F overlay. This means that essentially the same components, solutions and protocols are being used for both the PN and PN-F overlays. Therefore, similar test scenarios could be defined for evaluating the behaviour of the PN-F solutions at connectivity and network level.

7.3.3 Performance Evaluation

The functionality and performance of the PN/PN-F pilot system implementation has been analysed extensively and the most important results will be summarized here. For this analysis, as already stated in the previous section, a detailed test plan has been written that describes the desired functionalities and foreseen performance measurements for the different components in the PN/PN-F architecture (see Table 7.3). First, through a strict definition of scenarios, this test plan served as the basis for verifying whether the implemented platform indeed realizes the foreseen functionalities. Later, after the validation of these functionalities, the test plan has been used to offer a detailed quantitative evaluation of the effectiveness of the proposed solutions for a wide range of test-bed setups and parameter values. In addition, where possible, comparisons with existing technologies have been provided or performance has been assessed through theoretical models. The obtained results offer a good insight into the performance achieved by the PN platform and offers valuable lessons for future improvements. Since the PN-F network overlay solution makes use of the same components and solutions as the PN network overlay solution, the PN-F functionality exhibits the same performance as the PN components. In this section we will briefly summarize the most important results, highlight the most important novel contributions and where possible, compare with the state-of-the-art. It should be noted that the successful realization and integration of a working PN/PN-F pilot platform itself, is already a major step beyond current state-of-the-art.

First the interoperability and performance of the PN node initialization was tested with all basic operations that users must perform to obtain a fully functional node. These tests included creation and storage of the PN profile, detecting the neighboring nodes, and personalization of communication between personal devices

Table 7.3 MAGNET system prototype test scenarios

Test case	Components involved	Scope	Functionalities tested
Node personalization	Trust establishment module	Cluster	Imprinting of new personal node and transitive generation of long-term trust relationship with former personal devices
Secure intra-cluster communication	Neighbour discovery module and UCL	Cluster	Secure link layer establishment procedure and packet protection so that privacy, origin authentication and integrity are assured
Cluster formation	Neighbour discovery module and PN/PN-F routing module	Cluster	Establishment of links to neighbouring personal nodes and routing information exchange so that all other nodes in the cluster have a route to the new personal node
PN formation	PN/PN-F routing module, Dynamic tunnel establishment module and PN Agent framework	PN	PN Gateway node successfully registers within PN Agent. Establishment of tunnels between remote clusters and inter-cluster routing protocol is able to send messages over tunnel(s) so that a communication routes to all other nodes within the PN are created
PN dynamics	PN/PN-F routing module, Dynamic tunnel establishment module and PN Agent framework	PN	Re-configuration of the overlay on presence of events caused by cluster mobility (cluster splitting and merge, etc.)
Network specifics	Dynamic tunnel establishment module and PN Agent framework	PN	Maintenance of the network overlay under special circumstances of the access network such as NAT traversal or firewalls
Service discovery and usage	MSMP	Cluster/PN	Service/Application is registered within the service platform so that it is discoverable through its cluster's SMN. All the descriptions of PN services/applications that match specified service attributes are returned from the Cluster SMN in an SD response

(continued)

Table 7.3 (continued)

Test case	Components involved	Scope	Functionalities tested
PN-F establishment (ad-hoc case)	Federation Manager, CPFP, PN/PN-F routing module, MSMP, SCMF	Cluster	PN-F Member discovers published federation and checks in it. Upon acceptance from the PN-F Creator the federation overlay is automatically configured on member's nodes (nodes from all PN-F members become part of the same federation cluster). If no trust relationship exists in advance common Certificate Authority certificates are leveraged to set this relationship. Federation parameters are exchanged through corresponding profiles. Service and context is enabled if such resources are shared in the federation
PN-F establishment (infrastructure case)	PNDS, Federation Manager, CPFP, Neighbour discovery, UCL, PN/PN-F routing module and Dynamic tunnel establishment	PN-F	PN-F Member fetches federation information from PNDS and starts federation establishment through the Internet. Upon acceptance from the PN-F Creator the federation overlay is automatically configured on member's nodes (Members' clusters are interconnected through tunnels that are dynamically established). If no trust relationship exists in advance common Certificate Authority certificates are leveraged to set this relationship. Federation parameters are exchanged through corresponding profiles. Service and context is enabled if such resources are shared in the federation
Secure context management	SCMF	Cluster /PN/PN-F	Access to context information (both synchronously and asynchronously) at all levels of the SCMF architecture

(imprinting), and these were tested by measuring how long the initialization and starting up phases took. As can be concluded from the results, there were many factors that affected the performance. In the PN node initialization, the user interaction is in important role, and affects the performance directly. Therefore, these tests were more concerning interoperability and usability of the system.

When starting up a PN node, the PN Manager seamlessly initializes all needed baseline components. This enhances the user experience even if the user interface is not fully designed for the end users, but collaborates also in the PN node testing work. Baseline components automatically set up the PN networking environment, and the user is not burdened by technical complexity to realize this ubiquitous connectivity. This is an important achievement, since no major technological skills can be expected from future PN users, requiring an as easy as possible to use solution.

The first step on PN formation is the discovery of neighbouring personal nodes for the cluster formation. Based on the pair-wise keys exchanged during the imprinting procedure, nodes in the same radio domain not only discover each other but also are able to authenticate those nodes belonging to the PN. This way, a secure link is established between each pair of neighbouring personal nodes. The main advance beyond the state of the art is the homogenization of the session keys exchange method independently of the underlying wireless access technology by means of leveraging the potential of the Universal Convergence Layer [18]. At PN-F level, a similar procedure is followed. In this case, the Primary Master Key (PMK) used for deriving subsequent link layer session keys is not shared on a node-to-node basis as it was the case on PNs (i.e. each pair of personal nodes share a different PMK) but on a PN-to-PN basis (i.e. all nodes in the two PNs use the same PMK). Anyways, since the session keys are derived independently between each pair of nodes, the security of the communications at link level within the federation is assured. Both analytical and thorough experimental analyses have been carried out in order to assess the functionality and performance of the secure link establishment procedure defined. The system has been tested under different circumstances in terms of wireless access technology used and traffic conditions. As a general conclusion from the results shown, it can be stated that the complete secure link establishment process does not represent a task that would affect the overall system performance. Additionally, the measurement campaign carried out proves the functionalities of the implemented components.

Although the analyses have been carried out at PN level, similar conclusions can be extracted at PN-F level since the secure link establishment procedure used to create the links on ad-hoc federations (i.e. between any two nodes belonging to two PNs that are part of a federation) are the same as the ones used at PN level. This is one important feature of the system that can reuse existing functionalities to extend its scope, thus assuring its scalability.

After the secure link establishment process, personal nodes are ready to start communicating securely at cluster level. From a performance point of view, this is the link-level system feature that requires a more careful study. When traffic is exchanged between personal nodes at cluster level, additional mechanisms are enforced at the UCL [18]. These mechanisms, basically cryptographic processing, protects the packet so that privacy, origin authentication and integrity, among other issues, are assured. At cluster level, multihop scenarios can be given so that the end-to-end security is assured by securing each of the links of the communication. By definition, all the nodes in a cluster are personal, so the packet is protected by the security of each of the links that forms the end-to-end route. The counterpart is that the packet has to be encrypted and decrypted on every link of the route with the additional overhead that this implies.

The analytical and experimental evaluation that has been performed backs the approach taken for securing the communications on multihop wireless clusters of personal devices proving at the same time that the implementation done fulfils the requirements imposed. By comparing the approach taken with nowadays most spread solution for securing IP communications, namely IPSec, we have proven that it not only presents comparable performance on large clusters but it outperforms IPSec when typical small and dynamic clusters are considered.

Nevertheless, there are other important features that cannot be measured so directly and that imply a key security improvement for the ad hoc networking scenarios. Most of the attacks performed against ad hoc networks [19] would be prevented if the IP addresses of the nodes within the network would be hidden to those other nodes not allowed to be part of the network. This way, malicious nodes would not be able to inject traffic or spurious control information in the network so that not only the communication integrity is assured but also the network stability is protected. While the UCL solution provides this thanks to its hop-by-hop encryption behaviour, when IPSec end-to-end security is used, the IP header must be visible for each of the intermediate nodes so that they can route the datagrams without having to decrypt the actual payload. Hence, although IPSec has shown better performance for certain multihop configurations, the characteristic that allows this advantage (i.e. end-to-end secure association) represents a security disadvantage that leads to some security vulnerabilities when ad-hoc networks are considered. These vulnerabilities should be addressed using other security mechanisms that would increase the overhead resulting on worse performance than the one exhibited by the UCL implementation.

Finally, the solution adopted by the UCL does not require the nodes to have any other IP address than the private personal network addresses while when using IPSec in tunnel mode, the nodes must have been served by the access network or network manager with additional IP addresses. Of course, the IPSec transport mode might be used but in this case, the exposed IP address (as described in the previous paragraph) would not be the one from the access network but the private personal network one with all the security threats that this would lead to. This not only has security implications but from a network configuration point of view implies better performance.

At PN-F level, when two nodes from two different PNs that are part of a federation want to communicate directly on an ad-hoc manner, the UCL protects the communication using the same techniques described in this section. This is a key feature of the system when scalability is to be assured. Taking this into account, the analyses done for the communication between two personal nodes within the same cluster are also valid for the PN-F case. In this case, the cluster is formed by all the nodes belonging to PNs that are part of the federation.

For the PN/PN-F routing, an integrated hierarchical multi-mode routing protocol has been developed based on ad hoc networking techniques. It provides an innovative approach to exploit the flexibility of ad hoc routing technologies, but at the same time to optimize it for the needs of PN/PN-F communication. The routing framework is hierarchical, in the sense that a separation is made between intra-cluster and inter-cluster routing. It is also a multi-mode framework, meaning that it offers both reactive and proactive routing at the intra-cluster and inter-cluster level, allowing the use of different routing strategies. A similar approach has not been observed in literature before and goes further than traditional research in ad hoc routing protocols. Further the extension of ad hoc techniques over the infrastructure (i.e. over the tunnels between the different clusters) can also be considered as an innovation.

Within a cluster a modified version of a proactive and reactive ad hoc routing protocol has been implemented. The route and cluster setup time has been evaluated and care has been taken to also analyze this when background traffic is present, identifying the need for priority scheduling. Also the rerouting behavior upon link breaks has been analyzed. Finally, measurements also show that the software that is responsible for the routing, only has a minimal impact on CPU load and is highly performant compared to standard routing available in the OS.

For the inter-cluster routing solution, a new approach has been taken adapted to the specific topology of the PN/PN-F. Instead of routing based on next-hop information, inter-cluster routing is based on tunnel identifiers, which are generated by the dynamic tunneling solution. This tunneling solution also offers the possibility to reactively or proactively establish the overlay. In combination with the different possible routing strategies, this results in a highly flexible solution. The impact of the different routing and overlay establishment types has been evaluated.

Next, the message overhead of the routing framework has been analyzed theoretically. This theoretical analysis clearly shows that the design of a tailored inter-cluster routing solution outperforms classical next-hop based inter-cluster ad hoc routing. As such the proprietary inter-cluster routing protocol is an optimal design choice. Also, reactive and proactive inter-cluster routing (with proactive intra-cluster routing) have been compared and the resulting formulas help defining which strategy is most optimal as a function of the changes in cluster composition, the number of clusters, the number of gateways in every cluster and the frequency of inter-cluster traffic. For the combination of reactive intra-cluster and inter-cluster routing, the impact of the scope (i.e. propagation of the first route request in the cluster only or in the whole PN/PN-F) on the route establishment and route maintenance overhead has been analyzed. Again, this analysis helps defining on a theoretical basis which approach is most optimal in which PN/PN-F context, where the context consists of the PN/PN-F topology and the traffic patterns. Finally, a similar analysis has been made for the completely proactive versus the completely reactive solution. Using the theoretical overhead evaluation of our routing framework and taking into account the dynamics, topology and traffic characteristics of the PN/PN-F under consideration into account, guidelines about which routing strategy to select taking into account the message overhead can be derived. In addition, these guidelines can be obtained very easily and quickly compared to an approach based on simulations, making it a valuable tool to estimate routing performance.

Concerning the tunneling, a comparison has been made between IPSec, IPSec over UDP (used when behind NAT) and IPinIP and the impact of NAT boxes on achievable throughput has been analyzed. Finally, the behavior of the routing protocols has been evaluated in case of PN/PN-F dynamics such as cluster splits, cluster merges and changes in available gateways. From these measurements, some guidelines for possible improvements have been derived.

7.4 Conclusions

This chapter has described the main aspects of the implementation of a full-blown system fulfilling the key requirements imposed by the Personal Networking concept. It has presented the different components that compose the system and portrayed how they support the system functionalities. Additionally, the deployment of a pan-European Personal Networking testbed has been depicted. The implementation of a PN and PN Federations system has driven part of the MAGNET project research agenda that has its target on making Personal Networks happen. Indeed, some aspects of the initial specification have been revisited since particular issues have only shown up during the implementation and deployment phase. The deployment of the distributed testbed has been helpful not only because it has eased the system integration but also because it has set the basis for a larger scope platform that can be used to perform usability tests with real users.

The implemented system and the testbed have undergone a thorough system performance evaluation both from a network and user centric point of view. Extensive testing has been carried out in order to measure the response of the system under specific scenarios. Pilot services have been implemented atop of the system and usability tests based on these applications have been carried out.

References

1. I.G. Niemegeers, S. Heemstra de Groot, From personal area networks to personal networks: A user oriented approach. J. Wireless Pers. Commun. **22**, 175–186 (2002)
2. I. Niemegeers, S. Heemstra de Groot, Personal networks: Ad hoc distributed personal environments, Med-HocNet, IFIP Conference on Ad-Hoc Networks, Sept 2002
3. E. Gustafsson, A. Jonsson, Always best connected. IEEE Wireless Commun. **10**(1) (2003)
4. L. Munoz, L. Sanchez, J. Lanza, M. Alutoin. S. Lehtonen. D. Zeghlache, M. Girod Genet, W. Louati, J. Hoebeke, I. Moerman, G. Holderbeke, M. Ghader, M. Jacobsson, A proposal for self-organizing networks. Wireless World Research Forum Meeting 15 (SIG 3), White Paper, Paris, France, 8–9 Dec 2006
5. J. Hoebeke, G. Holderbeke, I. Moerman, M. Jacobsson, V. Prasad, N. Wangi, I. Niemegeers, S. Heemstra De Groot, Personal network federations. 15th IST Mobile and Wireless Communications Summit, Myconos, June 2006
6. J. Hoebeke, G. Holderbeke, I. Moerman, W. Louati, W. Louati, M. Girod Genet, D. Zeghlache, L. Sanchez, J. Lanza, M. Alutoin, K. Ahola, S. Lehtonen, J.J. Pallares, Personal networks: from concept to a demonstrator. 15th IST Mobile and Wireless Communications Summit, Myconos, June 2006

7. L. Sanchez, J. Lanza, R.L. Olsen, M. Bauer, M. Girod-Genet, A Generic Context Management Framework for Personal Networking Environments. Pernets Workshop 2006, Mobiquitous, July 2006
8. M. Alutoin, K. Ahola, S. Lehtonen, J. Paananen, Personal Network Directory Service. Telektronikk J. **1** (Mar 2007)
9. R.L. Olsen, M. Bauer, L. Sanchez, J. Lanza, Self organisation of context agents in personal networks and federations. 10th International Symposium on Wireless Personal Multimedia Communications, India, Dec 2007
10. E. Kohler, R. Morris, B. Chen, J. Jannotti, M.F. Kaashoek, The Click modular router. ACM Trans. Comp. Systems **18**(3), 263–297 (Aug 2000)
11. S. Murthy, J.J. Garcia-Luna-Aceves, An efficient routing protocol for wireless networks. Mobile Networks Appl. **1**(2), 183–197 (Oct 1996)
12. C.E. Perkins, E.M. Royer, Ad-hoc on-demand distance vector, in *Proceedings 2nd IEEE Workshop on Mobile Computing Systems and Applications*, New Orleans, LA, Feb 1999, pp. 90–100
13. NIST, Recommendation for the Triple Data Encryption Algorithm (TDEA) Block Cipher (PDF), Special Publication 800–67
14. IETF IP Security Protocol (IPSec), http://www.ietf.org/html.charters/OLD/ipsec-charter.html (RFCs 2401–2412)
15. L. Phifer, Slipping IPSec Past NAT, http://www.isp-planet.com/technology/2001/ipsec_nat.html
16. P. Srisuresh, M. Holdrege. IP network address translator (NAT) terminology and considerations, Aug 1999. RFC 2663
17. J. Postel, Internet Control Message Protocol, RFC 792, IETF, Sept 1981
18. J. Lanza, L. Sanchez, L. Muñoz, Experimental comparison of two solutions for securing heterogeneous ad-hoc network communications, in *Proceedings of Wireless Personal Multimedia Communications*, Lapland 2008
19. P. Papadimitratos, Z.J. Haas, Secure message transmission in mobile ad hoc networks. Elsevier Ad Hoc Networks J., Elsevier **1**(1), 193–209 (2003)
20. M. Balazinska, H. Balakrishnan, D. Karger, INS/Twine: A scalable peer-to-peer architecture for intentional resource discovery, in *Proceedings of the First International Conference on Pervasive Computing*, pages 195–210, Zurich, Switzerland, Aug 2002 (Springer-Verlag)
21. W. Louati, M. Girod-Genet, D. Zeghlache, Implementation of UPnP and INS/Twine interworking for scalable wide-area service discovery, in *Proceeding of WPMC2005*, Aalborg, Denmark, Sept 2005
22. Certicom Research, Standards for efficient Cryptography. SEC 2 Recommended Elliptic Curve Domain Parameters, 2000
23. D. Cooper, S. Santesson, S. Farrell, S. Boeyen, R. Housley, W. Polk, Internet X.509 Public Key Infrastructure Certificate and Certificate Revocation List (CRL) Profile, RFC 5280, IETF, May 2008
24. Wireshark Network Protocol Analyzer, http://www.wireshark.org/
25. Tcpdump, http://www.tcpdump.org/
26. IST-027396, Deliverable D2.3.1: Specification of PN networking and security components, M. Jacobsson et al., Dec 2006

Chapter 8
Standardisation and Exploitation

Liljana Gavilovska

8.1 Introduction

Standards represent an established norm or requirement for technical specifications, criteria, methods, processes, or practices. In electronic communication, a standard can be defined as an agreement between some players in a certain field where the topic of agreement is the technical specifications of various aspects of the communication technologies. Often a distinction is made between so called de jure and de facto standards. This distinction relates to the emergence of the standard – basically whether they emerge as a result of the specifications of a national or international authority or as the outcome of market introduction of a specification leading to the establishment of that specification as a particular standard.

Standards represent a vital tool for technologies in general and wireless networks in particular. Many viable techniques for wireless transmission are lying in a dark drawer, useless, because of the lack of standards. Moreover, not every standard is a success. There is a social/economic/technical condition that has to be met for a standard to become a success. We can say that a social need, an economical support and a technical solution together build a successful standard. A standard is a passport to a faster technology road towards exploitation and market visibility.

MAGNET/MAGNET Beyond is a leading project in personal area networking, one of the cornerstones of the future wireless technologies known as 4G. Wireless personal and body area networks are expected to play an increasing role in applications such as health, personal safety, secure wireless data exchanges or home entertainment.

The MAGNET Beyond partners have been actively involved in relevant standardisation activities. Major MAGNET Beyond ideas and achievements have been presented to different standardisation bodies and fora in order to push towards the inclusion of technical outcomes from this project in draft standards and better exploitation of its results. The MAGNET Beyond experts realised visibility and

L. Gavilovska (✉)
Aalborg University/University of Skopje, Niels J. Vej 12, Aalborg 9220, Denmark
e-mail: liljana@feit.ukim.edu.mk

R. Prasad (ed.), *My Personal Adaptive Global NET (MAGNET)*, Signals and Communication Technology, DOI 10.1007/978-90-481-3437-3_8,

direct participation in the following relevant standardisation bodies and fora: IEEE, ETSI, IETF, Ecma, OMA, 3GPP, ITU, ISO and WWRF. They have also participated in the standardisation activities of the IST Cluster 'Broadband Air Interfaces','Systems and Architectures Beyond 3G,' and 'Mesh and Sensor Networks' (see www.cordis.lu/ist/ka4).

The achievements of the MAGNET Beyond effort (patents, demo-platform, pilots, test bed, etc.) have been offered to the worldwide community and have and will inspire further project cooperations, education, implementation and other exploitation activities.

The aim of MAGNET/MAGNET Beyond is to have a significant influence in specifying the PN/PN-F (Personal Network/Personal Network Federation) environment with MAGNET Beyond ideas and results, and validate it for industrial commercialisation, which involves a clear understanding on how the PN technology can be aligned with the business and the potential PN market.

8.2 Standardisation Activities and Impact

Wireless connectivity has already enabled computer users to profit from a new convenient mobile lifestyle. Consumers are now demanding the same simplicity throughout their homes, connecting personal computers (PCs), personal digital recorders, MP3 recorders and players, digital camcorders and digital cameras, high-definition TVs, set-up boxes, game systems, personal digital assistants and mobile phones to each other in versatile domestic wireless personal area (WPAN) and body area (WBAN) networks. However, current wireless local area network (WLAN) and WPAN technologies cannot yet meet the needs of tomorrow's connectivity for the host of emerging consumer electronic devices that offer full mobility while requiring low power, quality of service (QoS) and security. So, as computing, communications and consumer applications converge to provide domestic consumers with extensive new services in an intelligent ambient environment, there is an urgent need to develop short-range user-centred wireless networks.

MAGNET Beyond represents an innovative concept that addresses the challenge to deliver the next generation of ubiquitous and converged network and service infrastructures for communication, computing and media. It provides a new type of infrastructure that can overcome the scalability, flexibility, dependability and security bottlenecks of current technologies and permits the emergence of dynamic, pervasive and robust new communication services. The MAGNET Beyond concept introduces the Personal Network (PN) and the Personal Network Federation (PN-F) as enablers of this new infrastructure. The PAN (Personal Area Network) as a basic component of the PN relies on suitable air interfaces to ensure the communication process.

Even though wireless communication has exploded in the last decade, wireless standards are dominated by a few protocol types. For example, most cellular networks use fixed-capacity channels, whereas data networking standards (e.g. 802.11,

802.15) are often contention-based so they can exploit statistical multiplexing of traffic. The use of simple, traffic-specific protocols has helped the rapid growth of mobile networks, but it stifles innovation and has lead to inefficient spectrum use. Today, in addition to satellite communication, only three wireless technologies have made an impact: WLANs, WPANs, and wireless wide area networks (WWANs). There are various ongoing standardisation activities supported mainly by the European Telecommunication Standardisation Union (ETSI) and the Institute of Electrical and Electronic Engineers (IEEE). Currently, the standardised WPAN technologies are BLUETOOTH, HIPERPAN and IEEE 802.15. These technologies are used for short distance ($\sim$10 m) with low data rates for different QoS requirements. It is envisaged that the WPANs will exist in all mobile terminals in the near future. The WPAN standards, IEEE 802.15.3 and 3a are developed and work is ongoing for paving the way towards broadband WPANs with envisioned data rates up to about 1 Gbps. IEEE 802.15.4 is focusing on very low data rate solutions, at only a few or a few 100 kbps, which is the first step towards body area networks.

Ultra Wideband (UWB) schemes have been considered for IEEE802.15.3 and IEEE 802.15.4. The working group IEEE 802.15.3a proposed Direct-Sequence UWB (DS-UWB) for low and medium data rates and MultiBand orthogonal frequency-division multiplexing (MB-OFDM) for high data rates. The latter is based on a transmission over 14 overlapping OFDM channels each having a bandwidth of 528 MHz for 128 subcarrier signals. Currently, UWB is under consideration for Body Area Networks (BAN) in IEEE 802.15.6.

Broadband wireless access is the third wireless revolution, after cell phones and WiFi. The broadcast nature of wireless transmission offers ubiquity and immediate access for both fixed and mobile users, clearly a vital element of next generation quadruple play (i.e., voice, video, data, and mobility) services. Unlike wired access (copper, coax, fiber), a large portion of the deployment costs is incurred only when a subscriber signs up for service. An increasing number of municipal governments around the world are financing the deployment of multihop wireless networks with the overall aim of providing ubiquitous Internet access and enhanced public services.

MAGNET proposed air interfaces for high data rate (HDR) and low data rate (LDR) applications (see Chapters 4 and 6). The HDR applications are enabled by a multi-carrier spread spectrum (MC-SS) air-interface solution. The only other available solution with similar capabilities at the moment is WiMedia, a radio platform standard for high-speed UWB wireless connectivity. For LDR applications, a low-power, low-complexity frequency modulation based UWB (FM-UWB) air-interface solution was proposed compatible to standards such as BLUETOOTH, ZigBee, and WiBree. The medium access control (MAC) of these two technologies is based on the IEEE 802.15.3 and IEEE 802.15.4 standards. The FM-UWB approach was adopted after being studied and compared with other solutions like ZigBee and Bluetooth. Accordingly, the MC-SS scheme was compared to the orthogonal frequency-division multiplexing (OFDM) based UWB PHY scheme in a WiMedia system. Results are reported in details and show that the developed air interfaces fulfill the requirements for next generation technologies (see Chapter 4).

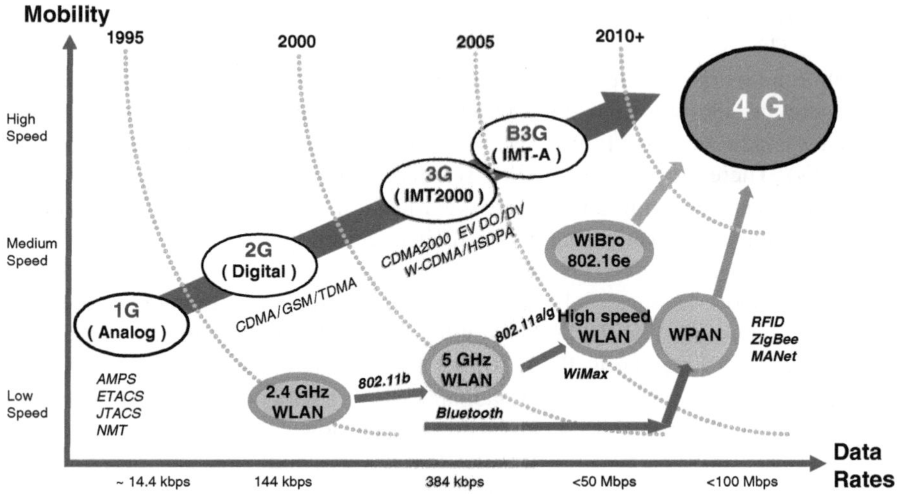

Fig. 8.1 Standardisation activities towards 4G communication systems

Security, availability, and reliability are three key requirements for the successful deployment of the MAGNET Beyond concept, especially in anticipated future applications. With a multitude of wireless standards in use, it is very important to ensure the dependability of the connections established through PNs and PN-Fs. Scalability also plays an important role. MAGNET Beyond proposes novel solutions for physical encryption applicable to the PN-F security architecture. The solutions include an efficient hybrid protocol that secures the federation. Furthermore, MAGNET Beyond presents a physical layer encryption mechanism designed for both LDR and HDR.

Figure 8.1 shows an overview of the current standardisation activities towards 4G and the position of WPAN standards in terms of mobility and data rates. The Personal Networks (PNs) introduced in MAGNET Beyond belong to the WPAN line and rely on the LDR and HDR capabilities.

8.2.1 Standardisation Bodies and Related Contributions

MAGNET Beyond has been acting dynamically to influence standards [1–4]. Several standardization bodies and fora were identified as relevant to the MAGNET Beyond activities. Significant contributions were submitted to the areas of eHealth (ETSI) and body area networks (BAN), including optimised radio interfaces, and specifications such as the PN/PN-F approach. Advancements in user-centric aspects and service creation were also identified through the standardisation activities.

The remainder of this section summarizes the activities and contributions to different standardisation bodies and fora.

8.2.1.1 IEEE Activities

The IEEE is a leading developer of standards that underpin many of today's technologies (http://standards.ieee.org/). The standards are developed in an environment that builds consensus in an open process based on input from all interested parties. With nearly 1,300 standards either completed or under development, IEEE is a central source of standardisation in both traditional and emerging fields, particularly telecommunications, information technology and power generation. IEEE is especially relevant as a forum for (de jure) standardisation of unlicensed wireless transmission (e.g., WiFi, WiMAX and BAN) and has been targeted by MAGNET-activities in this area.

MAGNET Beyond has participated on several 802.11, 802.15 and 802.16 meetings. Its most significant contributions include: (a) influencing the creation of the IEEE 802.15.6 BAN group [5], (b) promotion of the physical layer LDR, (c) the accompanying encryption system, (d) identification of key BAN user requirements and constraints, (e) providing insight into the usage of medical frequency bands and the UWB frequency band, and (f) initiating cooperation with TG30 for body affects.

The MAGNET Beyond's concepts were represented through several occasions together with ongoing IEEE 802.15 activities [6]. To discuss the role of standards for health BAN applications and regulatory consideration, several participants from MAGNET Beyond participated in the International Symposium on Medical Information and Communication Technology 2007 (ISMICT'07), Oulu, Finland. MAGNET Beyond partners have participated on IEEE 802.15.6 Regulation Subcommittee within the activity of the TG15.6 for body area networks (BAN) and numerous meetings and discussions held in 2008 with groups concerning the IEEE 802.15.6 standardisation. This key standard with respect to MAGNET Beyond targets low-power devices operating on, in, or around human body (but not limited to humans) and supports variety of applications including medical, consumer electronics/personal entertainment and others. Two members of MAGNET Beyond have become voting members in 802.15. A liaison has been established between IEEE 802.15.6 and ETSI EP eHealth [7].

The collaboration with these IEEE groups led towards the preparation of a Letter of Intent (LoI) and a planned submission of the standard. The IEEE 802.15.6 standard is expected to be published sometime in 2010. These collaborations are expected to extend beyond the end of the project.

8.2.1.2 ETSI Activities

The European Telecommunications Standards Institute (ETSI) [8] is an independent, non-profit, standardisation organisation of the telecommunication industry (equipment makers and network operators) in Europe, with worldwide projection. Significant ETSI standardisation bodies include TISPAN (for fixed networks and Internet convergence). ETSI inspired the creation of, and is a partner in the 3GPP (third Generation Partnership Project) [9]. ETSI has recently started the eHealth activity.

MAGNET Beyond has influenced and participated in all meeting of the new ETSI group, EP (ETSI Project) eHEALTH [7] that has been given high level attention and support within the ETSI secretariat. The main contributions are focused on the eHEALTH technical standardisation requirements, in particular regarding network architecture and radio interfaces. The ongoing work is in close co-operation with the IEEE 802.15.4 [10].

Magnet Beyond has been represented to the meetings of the ESO and the WCG regarding the implementation of EU mandate M403 on eHEALTH standardization and the meetings regarding the establishment and management of the ETSI project team STF 355. The Special Task Force (STF) was established to address additional work items (e.g., use case scenarios, applications and short range wireless).

The major activities within this standardization body were towards the adopting of the PN and PN-F concepts, new Radio Interfaces (RIs), use cases, and ongoing work on UWB.

8.2.1.3 Ecma Activities

Ecma (European Computer Manufacturers Association) [11] is a European standard body of interest to MAGNET, because of its low threshold and low overhead in standards development. The Ecma output typically is fed into ETSI and/or ISO/IEC (International Organization for Standardization/International Electrotechnical Commission).

The major activities were towards the possibilities of UWB standardisation and definition of the roadmap for personal networks (PNs) with particular focus on basic PN and PN-Federation.

8.2.1.4 IETF Activities

The last standardisation body that MAGNET Beyond is involved in is the Internet Engineering Task Force (IETF). It is an open standard organisation, with no formal membership or membership requirements. IETF is divided into several working areas (e.g., Application Area, General Area, Internet Area, Operation and Management Area, Routing Rae, Security Area, Transport Area).

The IETF Security Area is of special interest to MAGNET Beyond. The project has submitted a Request For Comments (RFC) for the Certified PAN Formation Protocol (CPFP) (draft-abiri-cpfp-01 – Certified Pan Formation Protocol).

8.2.1.5 WWRF Activities

The Wireless World Research Forum (WWRF) [12] is a global organization, founded in August 2001. It now has over 140 members from five continents, representing all sectors of the mobile communications industry and the research community. The objective of the forum is to formulate visions on strategic future research directions in the wireless field, among industry and academia, and to generate, identify, and promote research areas and technical trends for mobile and wireless system technologies.

MAGNET Beyond partners has chosen WWRF for presentation and validation of methodologies for scenario construction, user requirement development and business modelling. The work by the MAGNET partners has primarily focused on two areas: user centred scenarios and service creation.

The development of the scenarios was initiated in 2006. A draft outline of themes, global coverage, contents and focus was provided in the Working Group 1 on Human Perspectives. This was followed by a number of phone interviews with members of other Working Groups and Special Interest Groups, questionnaires on different trends and drivers; and finally a new draft was developed. This first draft was again followed by a number of phone interviews and more feedback resulting in a new draft – scenarios version 2.0. Up until this point the reference scenarios only focused on the narrative user stories. During the summer 2007, work was initiated on developing a machine-to-machine scenario to represent the large part of devices and user "invisible" networks and processors, inherent in the WWRF vision.

The results from MAGNET and MAGNET Beyond have clearly impacted the methodology for and the development of scenarios and visions within WWRF. Furthermore, the work has influenced other aspects, such as service creation, user profiles and business models. Finally, the creation of two new focus areas for WG1 in 2008, User requirements in developing/developed countries and Social Networks, is inspired by MAGNET Beyond.

8.2.1.6 Concluding Remarks

In addition to the aforementioned activities, the partners of MAGNET Beyond have been acting dynamically and have tried to influence an even broader standardisation community by cooperating with standardisation bodies such as: 3GPP (where there is an important interest in personal network architecture), ITU [13] (where the first personal cluster proposal have been tabled), and ISO (where currently input from national member organizations and affiliated standards bodies is awaited). MAGNET Beyond tutorials were presented to 3GPP and OMA (Open Mobile Aliance) [14], and have generated a considerable interest in PN/PN-F paradigms.

8.2.2 Impact for Further Developments

The MAGNET Beyond concept is paving a way to the real world through standardization activities and numerous presentations in front of different committees and groups. Participation in different standardization bodies and fora contributes to the project visibility and advertises its ideas. The participation in the IEEE 802.15.6 will create an opportunity to champion all MAGNET solutions within the group, and possibly attains as a result an IEEE standard with MAGNET specifications. The activity in the IEEE 802.15.6 BAN is focused on the development of a PHY-MAC standard. This is a long process – the schedule for development of the IEEE 802.15.6 standard extends beyond the end of the project. MAGNET Beyond, however, provided the initial platform for pursuit of this standard.

The major impact of the activities related to the ETSI EP eHealth includes work on: architecture and modelling (allowing different concepts such as PN/PN-F, VPON, HHA (Cruise) to be brought in); mapping of eHealth communication services on Telecom; RI (allowing UWB-FM to be brought in, but open for other RIs as well); inclusion of PN and PN-F architecture and services as reference for person centric network architecture for the future, in particular for eHealth applications. Through ETSI's input in the work under EU mandate the PN and PN-F architecture and protocols will be included in the work programme for Phase II, established under Phase I of the Mandate.

The impact in Ecma activities can be found in the early contributions mainly towards the formulation of the standardisation requirements and roadmap including the MAGNET Beyond ideas, PN related standardisation developments (particular focus on basic PN or PN-Federation), and reviewing the status of UWB in Ecma and in IEEE.

The resent establishment of the TC32 Editing Group on personal networking standardisation requirements, as a first step towards the establishment of a TC32 TG on the subject, is a first significant step towards the formalisation in the standards environment of the PN concepts into architecture, protocols and protocol stacks. A possible next step would be the establishment of a joint ETSI – Ecma work item under Phase II of Mandate M403.

Through IETF activities MAGNET Beyond is expected to standardize the security protocol developed within the project (CPFP).

The intended impact of the WWRF activities is to raise awareness of user centred service scenarios and service creation in MAGNET Beyond and, more broadly, of the activities and results of the project. It has promoted the MAGNET Beyond ideas and enabled the realization of the PN concepts and the creation of business built around PN services using MAGNET Beyond solutions.

MAGNET Beyond has achieved a significant impact in standardisation, and the project has received much credibility and recognition for its significant and effective standardisation efforts. These efforts are regarded by the EC as an important dissemination and exploitation activity, as a return on investment on research, and as an important step towards industrialisation and commercialisation.

Standardisation, however, is a difficult and sizeable task, and these activities represent a real and substantial effort to make a tangible measure of the commitment of MAGNET Beyond to "build the business". Standardisation is an important first step.

The partners that have carried the activity on behalf of MAGNET Beyond have been the drivers in the IEEE BAN, ETSI eHealth, and Ecma TC32, and continue their activity with a realistic opportunity and goal of achieving standards in these three bodies in the near future.

Further, MAGNET Beyond has contributed to the standardisation related activities of three clusters under the umbrella of the FP6 IST programme.

8.3 Exploitation Activities

MAGNET/MAGNET Beyond is a complex project targeting different areas of personal networks. It does not aim to produce a final product; rather focuses on the development of mature concepts and prototypes. During the first part of the project lifetime all results were centred on the consolidation of information and the integration of relevant and applicable concepts, methodologies and architectures. Accordingly, the knowledge exploitation focus of that period was on the dissemination of the MAGNET Beyond concepts, specifications and preliminary results and on initiating discussion within relevant communities.

The focus in the second period moved to the design and implementation of proposed architectures, protocols and algorithms. All of the partners developed plans on the further use of MAGNET Beyond results and generated Intellectual Property. Several patents were proposed and approved. The last project period was used to finalize the proof of the concepts, produce the project prototype and integrate the major ideas into the pilot. Extensive measurements support the developed PN/PN-F system. Initial steps towards commercialisation have already been made. The project results are reflected in ongoing standardization efforts, especially in the area of BAN (IEEE 802.15.6), e-Health (ETSI), and security (IETF). This opens a new platform for exploitation of the achievements of MAGNET.

The exploitable achievements are the pilots and produced hardware prototypes, protocols and algorithms, architecture solutions and concepts integrated into future networks and products. During the project lifetime MAGNET Beyond produced important results which may be further exploited in the area of air interfaces, IC block design, security, optimised algorithms, future wireless networking and Future Internet services, service development and personalization. Exploitation of MAGNET Beyond results can also be attained through their influence on education, cooperation with other EC and non EC projects, patents, standards, dissemination, business models and commercialization. Since the project is developing the concepts and not the market products, the exploitation of promising MAGNET Beyond results is yet to follow (announced patent application, standardization, etc.).

The following text highlights the possibilities for exploitation of the MAGNET Beyond solutions and results.

8.3.1 Exploitable Results

MAGNET Beyond is a complex project that addresses the whole protocol stack and broad research areas. The MAGNET Beyond team has achieved significant results in several areas, some of them resulting in patents and standards contributions.

In the area of **user-centric design** project completed an integration of the user profile into the overall enabling framework of the PN services, and provided usability and user experience evaluation methods (see Chapter 2). A novel activity based application and service **GUI** design for mobile device was finished and the possibility of patent application in 2009 is investigated.

Significant efforts were made to define the **business models** and the marketable results and to attract industrial partners as a step towards commercialization (see Chapter 2).

In the **networking** domain (see Chapter 3) the project developed the PN/PN-F concept and the enabling technologies. These include the network level functionalities (personalization, secure PN-formation, secure inter-cluster communication (Chapter 5), addressing and naming, and support for mobility), and application and service level functionality (such as intra-PN and extra-PN service discovery, context awareness, intra-PN content and device management, etc.). The project offers solutions for improved UCL (Universal Convergence Layer) functionality, integrated cluster gateway (GW) selection, and allows service management for external platforms through the SCMF-IMS Presence Gateway.

Proofs of concept were completed through extensive performance evaluation of all implemented solutions on a demo pilot and on a real testbed (see Chapter 7).

PN/PN-F introduces a new networking paradigm. It might open new communication horizons, and the concepts have already been integrated into other projects and have started to influence standardisation efforts.

MAGNET Beyond has defined two new optimised radio interfaces: for low (LDR) and high data rates (HDR). The LDR-UWB is proposed to the new group IEEE 802.15.6 on body area network and is proceeding towards a standard. In the area of radio **interfaces** (see Chapter 4) the major achievements are: the alternating wireless activity (AMC) collaborative coexistence scheme adapted to the MAGNET MC-SS wireless channel. The IAWA (Improved AWA) was developed to enable the coexistence of the specific MAGNET air interfaces: MC-SS and FM-UWB that allows evading interference in all HDR and LDR nodes.

In addition, MAGNET Beyond investigated solutions for alternative approaches to estimation of loss probabilities. Detailed comparisons were conducted between IEEE 802.11n and the proposed MC-SS air interface in order to sketch its standardization potentials. The enhancements with interleaver were investigated and the channel time allocation (CTA) algorithm was further enhanced with QoS (developed and integrated into the simulation model). The developed simulator will help WPAN users to deploy their own services.

A model of the channel gain which models shadowing, fading and correlation effects of the MAGNET channel was created. A new structure for joint channel and network decoding based on the proposed scheme shows the significant improvement in the overall system spectrum efficiency. These significant results open a platform for future optimized and enhanced PN solutions. Most of the result are already presented or will be presented on prestigious IEEE conferences and submitted to IEEE Transactions. This contributes to exploitation of knowledge through dissemination of MAGNET Beyond results.

In the security area the objective was to finalize the specifications of the different security protocols and architectures (Chapter 5). Extension requirements mainly support the federation concept providing two cryptographic systems for the federation management and physical layer encryption (Federation algorithm in the case of infrastructure modes and physical layer encryption adapted to MAGNET Beyond

LDR and HDR). A security threat analysis is proposed and the concept of virtual identity (VID) was introduced in the system. In order to meet the requirements of the secure context management framework (SCMF) in the PN environment, a context aware security manager (CASM) is provided, which relies on extended security profiles and the corresponding functionalities.

In the protocol domain, there are additional exploitable results such as: Certificate Revocation List (CRL) distribution method that allows simple PN node revocation, which resulted in a patent application. The protocol for secure pairing and transitive pairing of personal networks nodes, termed CPFP, was submitted as IETF (draft-abiri-cpfp-00.txt) and is planned for a patent.

Results in the security area have provided feedback to the standardization efforts. Many concepts will live beyond MAGNET Beyond, especially virtual identity, CASM and physical layer encryption.

8.3.2 Prototype

Significant exploitable interest exists in the hardware prototypes and building blocks produced in MAGNET Beyond. In order to prove the concept and to demonstrate the performance levels of the chosen interfaces and built-in PHY/MAC functionalities, MAGNET Beyond built two prototypes: for LDR and HDR (see Chapter 6).

Two LDR approaches were developed and demonstrated: a low band (LB) prototype as a proof of concept operating at 4.1 GHz and a high band (HB) approach using forefront IC technology to show ultimate performance for future market devices. The integration of IC blocks was fully completed. All LDR LB ICs and also the LDR HB ICs were fully implemented (the latest one in addition to the originally planed activities), and extensive measurements were performed and reported. High band transmitter performance results show that the full RF HB Transceiver power consumption is 10.9 mW, which is within the requirements (spec. $<$12 mW).

The LDR digital board implementing MAC schemes can be used in sensor and low data rate networks, and can enable piconet formation amongst several modes using the LDR PHY developed in MAGNET Beyond. It could be part of future integrated chipsets for FM-UWB High Band operation, and has strong impact to the standardization in IEEE 802.15.6 (and prepared for patent submission.).

One HDR approach was developed and the PHY was successfully tested. The HDR MC-SS prototype provides an open platform for R&D and demonstration (e.g., it is already transferred to the ORACLE and WHERE projects).

The project partners have succeeded in providing custom prototype boards that will be used as building blocks in the vertical integration process (see Chapter 7). Successful connection of the prototype to the UCL was also achieved. The prototypes were demonstrated to the broader academic and industry community (see the following section).

The relevance of the scientific and technical results is also validated by numerous in international journal and conference publications.

8.3.3 Testbed and Demo Platform

The project established a testbed spread over four European countries (see Chapter 7) to perform tests and analysis in a real distributed environment. The testbed was used for concept examination and practical performance measurements over the PN/PN-F. Supporting software was released and maintained in a software repository to facilitate easy interaction with the PN/PN-F testbed (essential for validation purposes as well as for development of pilot services). The distributed PN/PN-F testbed is available for assessing MAGNET Beyond solutions.

The MAGNET Beyond capabilities were demonstrated on this demonstration platform. During the project lifetime two pilots were demonstrated, "Icebreaker" and "Lifestyle Companion". The last pilot, "Lifestyle Companion", demonstrates the utilization of the MAGNET air interfaces (LDR and HDR) as part of the PN-wide platform, and represents a fully implemented, vertically integrated, demonstrator.

The obtained results offer a good insight into the performance achieved by the PN platform and valuable lessons for future improvements. The MAGNET Beyond prototypes and pilots were demonstrated on several occasions organized to disseminate the MAGNET Beyond results in front of academia and industry, and, as such, contribute to knowledge exploitation.

8.3.4 Dissemination Activities

The exploitation of the MAGNET Beyond achievements and results is also performed through dissemination activities and events. In order to strengthen the dissemination and to allow for a broader support for the MAGNET Beyond ideas, the project partners intensively participated with publications reflecting the project results in respectable journals and conferences and organized and participated in several dedicated events.

MAGNET Beyond concepts were presented on several workshops (in 2008: ICT-Mobile Summit, 9 June 2008, Stockholm: MAGNET Beyond workshop: "*Personalization in Future Ubiquitous Communication*"; http://www.ict-mobilesummit.eu/2008; TSOA workshop, 25 June 2008, Madrid, Spain, *First international workshop on user-centric service creation and execution* – joint workshop with OPUCE, SPICE, LOMS, MAGNET, PLASTIC, SMS).

In order to explain the MAGNET Beyond system in more details, the consortium prepared a tutorial: *Personal Networks and Personal Network Federations*. The tutorial was presented on TSOA workshop, 26 June 2008, and on the meeting of 3GPP, TISPAN, and OMA delegates, on 18 August 2008.

There were two organized round tables (*User centric service execution and management* and *User centric service creation*) where MAGNET Beyond demonstrated the usability of the PN concept to the service providers and operators, which attracted significant interest.

The MAGNET Beyond prototype and pilot was demonstrated on several occasions organized to disseminate the MAGNET Beyond achievements (IST 2006 Helsinki (21–23 November 2006), WWRF #18 (14 June 2007 in Helsinki), ITS 2007 (18–20 June 2007 in Aalborg), IST Mobile Summit (1–5 July 2007 in Budapest), ICT Mobile Summit (10–12 June 2008 in Stockholm) and TSOA Workshop (25–26 June 2008 in Madrid)).

Presentations of MAGNET Beyond (Personal Networks at a glance: FP6 project MAGNET) were given during the Second Joint NICT- CTIF Workshop (26–27 June 2008 Aalborg) and during the Seventh Triangular Co-operation Programme Workshop in Saariselka, Finland (10 September 2008) in front of researchers and industrial partners from Sweden, Finland, Denmark and Japan (Personal Networking towards the IMT Advanced: Where do they fit?). The auditorium expressed significant interest in the possibilities of PNs.

MAGNET Beyond was chosen among 60 participants to be showcased at the i-techpartner Communication and Mobile Applications Forum in Stockholm (28–29 April 2008), and was also presented at several industry meetings organized during the project lifetime.

The European Commission also demonstrated significant interest in MAGNET Beyond, as evident by the three articles it published about the project. They can be found on the EC portal for reporting results:

http://cordis.europa.eu/ictresults/index.cfm?section = home&tpl = home.

8.3.5 Project Influence

The MAGNET Beyond partners made substantial efforts to disseminate the project's achievements and ideas and to initiate interest in the area of personal area network. The project results have already been exploited through several initiatives, and have started influencing education programs, other projects and business initiatives.

Several education programs were motivated and initiated by the project ideas. For example, this happened at the Department of Information Technology – Broadband Communication Networks (INTEC – IBCN) of the Ghent University, where INTEC – IBCN is responsible for Bachelor and Master courses on telecommunication networks. IMEC further aims at exploiting the project results through the training of highly qualified engineers in Ph.D. programmes. University of Delft opened a Chair position for Personal Networks (Chair: Sonia Heemstra de Groot). Aalborg University has included the PN/PN-F topics in regular programs.

The enhanced knowledge and competence obtained through the participation in the MAGNET Beyond project will be exploited and used in future projects and partnerships with other projects and institutions (both in the academic and in the industry world). For instance, IMEC and the IBCN research group at the Ghent University are exploiting PN and PN-F solutions developed in MAGNET and MAGNET Beyond into a more generalized VPAN (Virtual Private Ad Hoc Networking) concept. Several national projects have been established that further build on the

VPAN concept (IBBT GBO SPAMM (Solutions Platform for Advanced Mobile Mesh), for more information see https://projects.ibbt.be/spamm/; IBBT ISBO VIN (Virtual individual Network), see https://projects.ibbt.be/vin, IBBT GBO TranseCare (Transparent ICT platforms for eCare), see https://projects.ibbt.be/transecare). Recently, the VPAN concept was introduced in the European project ITEA Usenet (Ubiquitous M2M Service Networks). WMC has also intiated several Dutch national projects in the area of personal networking and PN-F.

AAU has submitted Health projects/ICT-PSP-Project with AAU, NTUA, Alcatel-Lucent (PN-F between dementia patients, home-care department, nurses, and relatives to the patient). The ideas developed in MAGNET Beyond are already incorporated in the project proposals for new FP7 project and several national proposals.

Recently, IBBT (Interdisciplinary Institute for Broadband Technology) organized an iBoot camp. This camp is a highly focused approach to turn ideas and business opportunities into viable and executable business plans. It provides the participants with the necessary knowledge of how they can establish a business model, how they can find and interpret market information, how they can develop an estimation of the market size or a financial model in order to set up a business plan that is ready to be presented to investors. The evaluation of the feasibility of the idea from a business point of view is ongoing and will give feedback and advice for potential next steps in order to further exploit the body of knowledge. Twente Institute for Wireless and Mobile Communications (WMC) has focused its business interest on implementing the PN and PN-F concepts. The interest of the industrial partners is expressed and it is expected that they will exploit the MAGNET Beyond results in the respective domain of their interest.

Many MAGNET Beyond concepts are exploited by project partners. For example, NEC is trying to establish a market for context-aware services, e.g., by enhancing its own range of products with context-aware features. The have started an internal project with those ideas. NXP is exploiting the developed IC blocks. TCS and CEA-LETI (Laboratory for Electronics & Information Technology) are preparing a patent related to the HDR system. In addition, start-up company initiatives begun before the end of the project.

8.4 Conclusions

In order to bring its concepts closer to reality, MAGNET Beyond engaged in a broad range of activities and attained numerous achievements. It started with ideas and finished with demo prototypes, pilots, testbed, and a mature PN-PN/F concept. The final outcome – a system mosaic – a result of advanced research and prototyping of ideas, but also of the visibility the MAGNET Beyond partners fostered throughout the project's lifetime, and the recognition achieved through standardisation.

MAGNET Beyond brought on the global arena the concept of PN and PN-F which will continue to live through many yet-to-be-defined services and applications. It has already influenced several projects, and is continuing to motivate academic activities and new initiatives, to define a new footprint on personalisation and to show a roadmap towards future wireless networks, where personal networks will play an important role in personalized and ubiquitous communications.

References

1. D6.1.2 Review of current activities of relevant standardisation bodies and fora related to the candidate concepts (Dec 2004), http://www.ist-magnet.org/public + deliverables/phase1wp6
2. D6.1.3 Standardisation contributions (Dec 2005), http://www.ist-magnet.org/public + deliverables/phase1wp6
3. D7.1.2 1st series of standardisation contribution (Sept 2008), http://www.ist-magnet.org/public + deliverables/BeyondWP7
4. D.7.1.4 2nd series of standardisation contribution (Sept 2008), http://www.ist-magnet.org/public + deliverables/BeyondWP7
5. IEEE 802.15.6 (BAN), http://www.ieee802.org/15/pub/TG6.html
6. IEEE 802.15 Working Group for WPAN, http://ieee802.org/15/index.html
7. ETSI eHealth, http://portal.etsi.org/Portal_Common/home.asp
8. European Telecommunications Standards Institute (ETSI), www.etsi.org
9. Third Generation Partnerchip Project, http://www.3gpp.org/
10. IEEE 802.15.4 WPAN Task Group 4, http://www.ieee802.org/15/pub/TG4.html
11. Ecma International, http://www.ecma-international.org/
12. World Wireless Research Forum, http://www.wireless-world-research.org/
13. International Telecommunication Union, http://www.itu.int/net/home/index.aspx
14. Open Mobile Aliance, http://www.openmobilealliance.org/

Chapter 9
Conclusions and Future Work

Ramjee Prasad

9.1 Introduction

This book has proposed and introduced concepts providing scalable and affordable wireless networking for rich, personalised and easy-to-use communication services. These concepts were developed within the European funded project MAGNET/MAGNET Beyond. The topics of Personal Networks and Federations of Personal Networks, supporting security, identity and trust solutions and the optimized air interfaces are in line with the vision of the wireless research and development society towards ambient intelligence. This latter requires a radical change, which demands the definition of new interfaces and a multitude of standards in key areas of future media- and context-aware, multi-domain mobile networks.

9.2 Summary of Research Achievements

The solutions proposed by the research carried out within the scope of the IST project MAGNET/MAGNET Beyond created a framework for the dependability requirements expected by next generation networks. The protection of users and infrastructures, and the support by the security framework of the roles to be played by the various actors would strengthen the adoption and extensive use of PN services. The focus of research on security and privacy for PN services is a clear indication of awareness of societal and economic challenges ahead of upcoming wireless technologies.

The most significant contributions of this book in to advance the state of the art on the area of personal and group communications via personal networks and the federation of PNs.

In particular, the innovations are discovered in the proposed extension of the PN concept to multiple PNs constituents establishing common trust for group and

R. Prasad (✉)
Aalborg University, Niels J. Vej 12, Aalborg 9220, Denmark
e-mail: prasad@es.aau.dk

R. Prasad (ed.), *My Personal Adaptive Global NET (MAGNET)*, Signals and Communication Technology, DOI 10.1007/978-90-481-3437-3_9,

collaborative communications using the MAGNET PAN and PN technologies that were made available through the project research activities. The design of PAN technology and PN platforms was followed by the integration of pilot services that lead to the identification and definition of the system components and services required to support the vision of "optimally connected anywhere, anytime". The book gives an overview of the achievements reached in the projects, highlighting novel concepts that can be summarized in the next:

- Development of PN architectures optimised from a user's perspective and establishing trusted group communications involving multiple PNs and users.
- Support of the objective of novel air interfaces through the work on adaptive, flexible and efficient air interfaces to support and foster PN services.
- Producing the required infrastructure support for PAN and PN services via an IPv6 open framework, where tunnels and dynamic VPN can be established for intra and inter PN connectivity.
- Providing a resource management plane for handling, selecting and optimally using the multiple available radio technologies. Provisioning of middleware and configuration management support in the networks to support PN services.
- Production of adaptive and flexible architectures to provide the best available connectivity for each operating PN environment and scenario.
- Providing a mechanism for spectrum sharing, via advanced air interfaces addressing the coexistence in unlicensed spectrum through co-operation and opportunistic but fair usage of spectrum. By relying also on cognitive science concepts to select the greenest air interface technologies depending on environment and ambient conditions and PAN context. This was achieved via the flexible and adaptive PAN-optimised air interfaces that seek operation with the minimum power requirements and reduction in interference in the immediate user space and in the surrounding environment.
- Design of cooperative distributed resource management entities and supporting inter domain mobility and roaming, and vertical handover -through interlayer co-operation.
- Providing the needed functions and features required from the inter-connecting networks to support the PN services. In particular, the guidelines and technology framework to achieve composite network management regarding PN services was described.
- Providing reconfiguration capabilities at all levels in the PN architecture. This included the proposed PAN/PN reconfigurability (dynamic PAN/PN level formation) aspects at different levels: devices, network, protocols, and services. Further, PN reconfigurability was proposed via infrastructure networks with IPv6-based end-to-end connectivity support via tunnels and programmable nodes.
- Providing a framework for secure PAN multiple connectivity to remote devices and networks on the basis of required QoS from active PN applications and services. The PN concept was extended to multiple communicating PNs (PN-F), while retaining the user centric paradigm with security provision and user and environment context awareness.

- A user-centric approach for threat analysis methodology and security evaluation, and a security manager safeguarding transparently the user's privacy and enforcing user security preferences across personal overlay networks. Moreover, innovative authentication protocols and key management techniques have been provided to meet the specific needs of personal networking.
- Implementation and test of three prototypes dedicated to WPAN applications.
- Implementation of a full-blown system fulfilling the key requirements imposed by the Personal Networking concept.

Since the proposed concepts and solutions explore the interworking of licensed and unlicensed band communications, there is a huge potential for novel applications, services and technologies that can enrich peoples' lives. The biggest impact on the user's life follows from enabling the user to have easy, affordable and seamless control of their remote devices over heterogeneous communications networks. These are empowered to communicate efficiently with their selected interaction groups, no matter where the individual members are, or what kind of access is available for them to use.

Another aspect of the concepts and solutions proposed in this book is the controlled cooperation of the personal networks. Allowing trusted parties to share resources by coupling personal networks will improve and speed up processes in the safety, health, public and commercial sectors. For instance, public servants, such as, the police, the fire brigade and the medical response team federating their organisational PNs in a crisis situation will lead to improved coordination of the situation and efficient use of resources.

We are already witnessing increasing amounts of data traffic. Enabling full personal content creation and sharing will induce a noticeable increase in data volume and generate the need for new communication infrastructures and catalyse the creation of new applications and services enabled by the developments brought forth.

The MAGNET and MAGNET Beyond concepts and solutions open new social and economic opportunities through enabling full seamless and global user access to the following:

- New classes of feature-rich applications such as ubiquitous service provisioning in a secure heterogeneous networking environment for nomadic users.
- New form of person-to-person and person-to-group communications and interaction, offering fully mobile, personalised and adaptive access to services.
- New types of device-to-device applications based on adaptive and self-organised networks in the home, office and vehicular environments.
- New classes of person-to-device applications offering multi-access connectivity to remote devices and networks.

With all these new types of devices and service provision modalities, there is an increasing need for creating trust in the users and the communities. Advances in trust and usability, to a larger extent in security and privacy, will support e-Inclusion and remove acceptance barriers. The promotion of information and communication technologies through the proposed here concepts directly and indirectly enhances the economy and the quality of life of the citizens.

PNs comply with the requirements defined for IMT-A (International Mobile Communication-Advanced) systems and are a good approach towards ubiquitous service provision and extended coverage for IMT-A users. Solutions for current PNs provide also capability to protect the privacy information of the user, while enabling more efficient service discovery. A number of challenges that require to be solved relate to the efficient cooperation among heterogeneous technologies and to ensuring convergence of technologies and devices, while maintaining secure and trusted communications.

9.3 Future Directions of Research for PN and PN-F

Security, availability, and reliability are three key requirements for the successful deployment of next generation systems. With a multitude of wireless standards in use, a straightforward way assumed until now, for dependable wireless connections, has been to have software-defined radio (SDR) implementations of some of the existing licensed/unlicensed technologies, and to create a module that selects the appropriate technology depending on the perceived context e. g. in terms of security or interference. A significant qualitative novelty can be further brought by introducing dynamic spectrum access, whereby the wireless channels are opportunistically defined within the spectrum portions that are assessed to be available for communication. Being capable to work over a large spectrum, the mechanisms of dynamic spectrum access significantly increase the chances for achieving dependable wireless connections.

Therefore, for seamless connectivity, the future system user will need not only bandwidth availability but also a multitude of parameters that satisfy the demands of the PN user by flexibly and dynamically trading off the bandwidth for dependability.

It is good to remember that personal mobile networks oppose strict layered protocol design because of their dynamic nature, limited resources, mobility of nodes, time varying links and topology, the increasing complexity and the need to support a diverse range of multimode terminals, radio interfaces and protocols. All those aspect bring to new approaches in the future of the telecommunications Indeed a current trend for improving the performance of mobile systems, is also being supported today and implemented within the framework of a layered architecture that comprises an access, connectivity and application layer across the user, management and control planes. Layers are interworked across the network and at network elements with the lower layers, providing services to the upper layers. The functions in the layers have become increasingly adaptive, with focus on adaptivity in the lower or higher layers.

Radio Frequency Identification techniques (RFID) and related identification technologies will be the cornerstone of the upcoming Internet of Things (IoT). These emerging technologies can benefit from the proposed here concepts for personalised communications and build upon the presented solutions. Smart components will be able to execute different set of actions, according to their surroundings and the tasks they are designed for. There will be no limit to the actions and operations these smart "things" will be able to perform: for instance, devices will be able to direct their transport, adapt to their respective environments, self-configure, self-maintain, self-repair, and eventually even play an active role in their own disposal.

To reach such a level of ambient intelligence, however, major technological innovations and developments will need to take place. Governance, standardisation and interoperability are absolute necessities on the path towards the vision of things able to communicate with each other. In this respect, new power efficient, security centred and fully global communication protocols and sustainable standards must be developed, allowing vast amount of information to be shared amongst things and people. The ability of the smart devices to withstand any kind of harsh environment and harvest energy from their surroundings becomes crucial.

Today it is already possible to say that the Personal Networks have been a cornerstone of a new vision, being still the future for next generation of the wireless communication. In the view of the editor the fourth generation (**4G**) of wireless communication can be defined by the following equation:

$$\mathbf{4G} = \mathbf{IMT\text{-}A} + \mathbf{Pers}$$

Where **IMT-A** is a new global, unified wireless architecture which visualizes a hierarchy of interconnected access systems and envisions new radio interfaces including operation on new bands (licensed and maybe unlicenesed); **Pers** stands for Personalisation, topic of research in MAGNET/MAGNET Beyond.

Index

MIX
Papier aus verantwortungsvollen Quellen
Paper from responsible sources
FSC® C105338

If you have any concerns about our products,
you can contact us on
ProductSafety@springernature.com

In case Publisher is established outside the EU,
the EU authorized representative is:
Springer Nature Customer Service Center GmbH
Europaplatz 3, 69115 Heidelberg, Germany

Printed by Libri Plureos GmbH
in Hamburg, Germany